Enantioselective Nickel-Catalysed Transformations

RSC Catalysis Series

Editor-in-Chief:
Professor Chris Hardacre, *Queen's University Belfast, Northern Ireland, UK*

Series Editors:
Professor Bert Klein Gebbink, *Utrecht University, The Netherlands*
Professor Jose Rodriguez, *Brookhaven National Laboratory, USA*

Titles in the Series:
1: Carbons and Carbon Supported Catalysts in Hydroprocessing
2: Chiral Sulfur Ligands: Asymmetric Catalysis
3: Recent Developments in Asymmetric Organocatalysis
4: Catalysis in the Refining of Fischer–Tropsch Syncrude
5: Organocatalytic Enantioselective Conjugate Addition Reactions: A Powerful Tool for the Stereocontrolled Synthesis of Complex Molecules
6: *N*-Heterocyclic Carbenes: From Laboratory Curiosities to Efficient Synthetic Tools
7: *P*-Stereogenic Ligands in Enantioselective Catalysis
8: Chemistry of the Morita–Baylis–Hillman Reaction
9: Proton-Coupled Electron Transfer: A Carrefour of Chemical Reactivity Traditions
10: Asymmetric Domino Reactions
11: C-H and C-X Bond Functionalization: Transition Metal Mediation
12: Metal Organic Frameworks as Heterogeneous Catalysts
13: Environmental Catalysis Over Gold-Based Materials
14: Computational Catalysis
15: Catalysis in Ionic Liquids: From Catalyst Synthesis to Application
16: Economic Synthesis of Heterocycles: Zinc, Iron, Copper, Cobalt, Manganese and Nickel Catalysts
17: Metal Nanoparticles for Catalysis: Advances and Applications
18: Heterogeneous Gold Catalysts and Catalysis
19: Conjugated Linoleic Acids and Conjugated Vegetable Oils
20: Enantioselective Multicatalysed Tandem Reactions
21: New Trends in Cross-Coupling: Theory and Applications
22: Atomically-Precise Methods for Synthesis of Solid Catalysts
23: Nanostructured Carbon Materials for Catalysis

24: Heterocycles from Double-Functionalized Arenes: Transition Metal Catalyzed Coupling Reactions
25: Asymmetric Functionalization of C–H Bonds
26: Enantioselective Nickel-Catalysed Transformations

How to obtain future titles on publication:
A standing order plan is available for this series. A standing order will bring delivery of each new volume immediately on publication.

For further information please contact:
Book Sales Department, Royal Society of Chemistry, Thomas Graham House, Science Park, Milton Road, Cambridge, CB4 0WF, UK
Telephone: +44 (0)1223 420066, Fax: +44 (0)1223 420247
Email: booksales@rsc.org
Visit our website at www.rsc.org/books

Enantioselective Nickel-Catalysed Transformations

Hélène Pellissier
Aix Marseille Université, France
Email: h.pellissier@univ-amu.fr

ROYAL SOCIETY OF CHEMISTRY

RSC Catalysis Series No. 26

Print ISBN: 978-1-78262-425-7
PDF eISBN: 978-1-78262-670-1
EPUB eISBN: 978-1-78262-763-0
ISSN: 1757-6725

A catalogue record for this book is available from the British Library

Published by The Royal Society of Chemistry,
Thomas Graham House, Science Park, Milton Road,
Cambridge CB4 0WF, UK

Registered Charity Number 207890

For further information see our web site at www.rsc.org

Printed in the United Kingdom by CPI Group (UK) Ltd, Croydon, CR0 4YY, UK

Preface

The catalysis of organic reactions by metals still constitutes one of the most useful and powerful tools in organic synthesis. Although asymmetric synthesis is sometimes viewed as a subdiscipline of organic chemistry, actually this topical field transcends any narrow classification and pervades essentially all chemistry. Of the methods available for preparing chiral compounds, catalytic asymmetric synthesis has attracted most attention. In particular, asymmetric transition-metal catalysis has emerged as a powerful tool to perform reactions in a highly enantioselective fashion over the past few decades. Efforts to develop new asymmetric transformations have focused preponderantly on the use of a few metals, such as titanium, copper, ruthenium, rhodium, palladium, iridium, and more recently gold. However, by the very fact of the lower costs of nickel catalysts in comparison with other transition metals, enantioselective nickel-catalysed transformations have received a continuous ever-growing attention during recent decades that has led to exciting and fruitful research. This interest might also be related to the fact that nickel complexes are of high abundance, exhibit a remarkably diverse chemical reactivity, and constitute one of the most useful Lewis acids in asymmetric catalysis. However, it must be noted that nickel has long been viewed as just a low-cost replacement catalyst for palladium for cross-coupling reactions as a group 10 metal like palladium. Actually, the use of nickel in organometallic chemistry precedes many other examples of transition metal catalysis.

Nickel was first isolated and classified as a chemical element in 1751 by Cronstedt. In 1898, Mond discovered tetracarbonylnickel [$Ni(CO)_4$], a highly toxic liquid at room temperature, which decomposes back to nickel and carbon monoxide on heating. This behaviour was exploited in Mond's process for purifying nickel. Later in 1912, Sabatier reported the first hydrogenation

RSC Catalysis Series No. 26
Enantioselective Nickel-Catalysed Transformations
By Hélène Pellissier

Published by the Royal Society of Chemistry, www.rsc.org

of ethylene using nickel, for which he was awarded the 1912 Nobel Prize in Chemistry. Starting from the 1970s, nickel found extensive use for cross-coupling reactions. Moreover, many nickel complexes have long been considered as privileged catalysts for reactions of alkenes and alkynes. Nickel readily donates d-electrons to π-acceptors, so alkene bonding is generally strong. In this context, reactions of alkenes and alkynes, such as allylations, reductive couplings, oligomerisations, and cycloisomerisations, have also been widely investigated.

Among important work, Wilke reported in 1988 seminal contributions to the structure and reactivity of nickel complexes, including the synthesis of $Ni(cod)_2$, and its investigation in alkene oligomerisation reactions. Ever since, the remarkable properties of nickel, such as facile oxidative addition, ready access to multiple oxidation states, and facile activation and transformation of molecules that are chemically less reactive, have allowed the development of a broad range of innovative transformations for which other metals are inefficient and which have been long considered exceptionally challenging. Indeed, since nickel is a relatively electropositive late transition metal, oxidative addition, which results in loss of electron density around nickel, tends to occur quite readily. This facile oxidative addition allows for example the use of cross-coupling electrophiles that would be considerably less reactive under palladium catalysis such as phenols. Another key advantage of nickel is its large variability of electronic states [Ni(0)/Ni(I)/Ni(II)/Ni(III)]. Like palladium, for which most reactions are based on Pd(0)/Pd(II) catalytic cycles, Ni(0)/Ni(II) catalytic cycles are widely spread, but the easy accessibility of Ni(I) and Ni(III) oxidation states allows different modes of reactivity and mechanisms to occur. As a result, many transformations are based on Ni(I)/Ni(III), Ni(0)/Ni(II)/Ni(I), or even cycles in which nickel remains in the Ni(I) state for the entire catalytic cycle.

Nickel has a small atomic radius, and Ni–ligand bond lengths are often relatively short, producing solid and dense complexes. Nickel(II) forms compounds with all common anions, *i.e.* sulfide, sulfate, carbonate, hydroxide, carboxylates, and halides. Common salts of nickel, such as the chloride, nitrate, and sulfate, dissolve in water to give green solutions containing the metal aqua complex $[Ni(H_2O)_6]^{2+}$. The four halides form nickel complexes featuring octahedral Ni centres. Tetracoordinate nickel(II) complexes exist both in tetrahedral and square planar geometries. Another important advantage of nickel is related to its cost, which is roughly 2000 times lower than palladium and 10 000 times lower than platinum on a mole-for-mole basis.

In the past decade, chemists have taken advantage of all of the properties of nickel to develop novel powerful transformations, as demonstrated in this book. Its goal is to provide a comprehensive overview of the major developments in enantioselective nickel-catalysed transformations reported since the beginning of 2004, since this area was previously reviewed in 2005 by Hayashi and Shintani in a book chapter dealing with asymmetric synthesis based on the use of organonickel chemistry. This present book demonstrates the impressive amount of enantioselective synthetic uses that have been found for novel and already known nickel chiral catalysts in the last 10 years, from basic organic transformations, such as cycloadditions, conjugate additions, cross-couplings, hydrovinylations, hydrocyanations, α-functionalisation/arylation reactions of carbonyl compounds, additions of organometallic reagents to aldehydes, aldol- and Mannich-type reactions, and hydrogenations, to completely novel methodologies including domino reactions, for example.

The book is divided into 10 main chapters, according to the different types of reactions catalysed by chiral nickel catalysts, such as enantioselective cycloaddition reactions for the first chapter, enantioselective conjugate additions for the second chapter, enantioselective cross-coupling reactions for the third chapter, enantioselective domino, multicomponent, and tandem reactions for the fourth chapter, enantioselective hydrovinylation, hydrophosphination, hydrocyanation, and hydroalkynylation reactions for the fifth chapter, enantioselective α-functionalisation and α-arylation/alkylation reactions of carbonyl compounds for the sixth chapter, enantioselective additions of organometallic reagents to aldehydes for the seventh chapter, enantioselective aldol-type and Mannich-type reactions for the eighth chapter, enantioselective hydrogenation reactions for the ninth chapter, and enantioselective miscellaneous reactions for the tenth chapter. A final eleventh chapter includes the general conclusions.

Hélène Pellissier
Aix Marseille Université, France

Abbreviations

acac Acetylacetone
Am Amyl
Ar Aryl
BArF Tetrakis[3,5-bis(trifluoromethyl)phenyl]borate
BBN 9-Borabicyclo[3.3.1]nonane
BINAP 2,2′-Bis(diphenylphosphino)-1,1′-binaphthyl
(*S*)-Binapine (3*S*,3′*S*,4*S*,4′*S*,11b*S*,-11′b*S*)-(+)-4,4′-Di-*tert*-butyl-4,4′,5,5′-tetrahydro-3,3′-bi-3*H*-dinaphtho[2,1-*c*:1′,2′-*e*]phosphepin
BINIM Binapthyldiimine
BINOL 1,1′-Bi-2-naphthol
BIPHEP 2,2′-Bis(diphenylphosphino)-1,1′-biphenyl
Bn Benzyl
Boc *tert*-Butoxycarbonyl
BOX Bisoxazoline
(*R*,*R*)-Me-BPE 1,2-Bis[(2*R*,5*R*)-2,5-dimethylphospholano]ethane
BQd Benzoylquinidine
Cbz Benzyloxycarbonyl
CHIRAPHOS 2,3-Bis(diphenylphosphino)butane
CMOF Chiral metal-organic framework
cod Cyclooctadiene
Cp Cyclopentadienyl
Cp* Pentamethylcyclopentadienyl
CPME Cyclopentyl methyl ether
Cy Cyclohexyl
Cys Cysteine
DABCO 1,4-Diazabicyclo[2.2.2]octane
DBFOX 4,6-Dibenzofurandiyl-2,2′-bis(4-phenyloxazoline)
DBMA Dimethylbenzoic acid
DBU 1,8-Diazabicyclo[5.4.0]undec-7-ene
DCOH 3,5-Dichloro-2-hydroxybenzaldehyde
de Diastereomeric excess

RSC Catalysis Series No. 26
Enantioselective Nickel-Catalysed Transformations
By Hélène Pellissier

Published by the Royal Society of Chemistry, www.rsc.org

DIBAL	Diisobutylaluminum hydride
Difluorphos	5,5′-Bis(diphenylphosphino)-2,2,2′,2′-tetrafluoro-4,4′-bi-1,3-benzodioxole
DIOP	2,3-*O*-Isopropylidene-2,3-dihydroxy-1,4-bis(diphenylphosphino)butane
DIPAMP	1,2-Bis[(2-methoxyphenyl)(phenylphosphino)]ethane
DIPEA	Diisopropylethylamine
DMA	*N*,*N*-Dimethylacetamide
DME	Dimethoxyethane
DMF	*N*,*N*-Dimethylformamide
DMI	1,3-Dimethylimidazolidin-2-one
DOSP	*N*-(*p*-Dodecylphenylsulfonyl)prolinate
DPEN	1,2-Diphenylethylenediamine
dppp	1,3-Bis(diphenylphosphino)propane
dr	Diastereomeric ratio
DUPHOS	1,2-Bis(phospholano) benzene
ee	Enantiomeric excess
FOXAP	Ferrocenyloxazolinylphosphine
Hex	Hexyl
HFIP	Hexafluoroisopropanol
HIPT	Hexaisopropylterphenyl
HMPA	Hexamethylphosphoramide
HOMO	Highest occupied molecular orbital
INDABOX	2,2′-Methylenebis-(3a,8a-dihydro-8*H*-indeno[1,2-*d*]oxazole)
Josiphos	1-[2-(Diphenylphosphino)ferrocenyl]ethyldicyclohexylphosphine
L	Ligand
LUMO	Lowest occupied molecular orbital
Mes	Mesyl (methanesulfonyl)
MOM	Methoxymethyl
MOP	2-(Diphenylphosphino)-1,1′-binaphthyl
MS	Molecular sieves
MTBE	Methyl *tert*-butyl ether
Naph	Naphthyl
NBS	*N*-Bromosuccinimide
NFSI	*N*-Fluorobenzenesulfonimide
NMM	*N*-Methylmorpholine
NOBIN	2-Amino-2-hydroxy-1,1′-binaphthalene
NORPHOS	2,3-Bis(diphenylphosphino)bicyclo[2.2.1] hept-5-ene
Ns	Nosyl (4-nitrobenzenesulfonyl)
Nu	Nucleophile
Oct	Octyl
PCC	Pyridinium chlorochromate
Pent	Pentyl
PG	Protecting group
PHOX	Phosphinooxazoline
Phth	Phthalimido
Pigiphos	Bis{1-[2-(diphenylphosphino)ferrocenyl]ethyl} cyclohexylphosphine
Pin	Pinacolato
Piv	Pivalate
PFP	Pentafluorophenol
PMB	*para*-Methoxybenzyl
ppfa	*N*,*N*-Dimethyl-1-[2-(diphenylphosphino) ferrocenyl]ethylamine
PyBidine	Bis(imidazolidine) pyridine
PYBOX	Pyridine-bisoxazoline
QUINAP	1-[2-(Diphenylphosphino)-1-naphthyl] isoquinoline

Quinaphos 8-(Diphenylphosphino)-1-(3,5-dioxa-4-phosphacyclohepta-[2,1-*a*:3,4-*a*′]dinaphthalen-4-yl)-1,2-dihydroquinoline
QN 8-Quinoline
rs Regioselectivity
r.t. Room temperature
Salen 1,2-Bis(salicylidenamino)ethane
Segphos (*R*)-(+)-5,5′-Bis(diphenylphosphino)-4,4′-bi-1,3-benzodioxole
TADDOL α,α,α′,α′-Tetraphenyl-2,2-dimethyl-1,3-dioxolane-4,5-dimethanol
TANGPHOS 1,1′-Di-*tert*-butyl-2,2′-diphospholane
TBAT Tetrabutylammonium difluorotriphenylsilicate
TBHP *tert*-Butyl hydroperoxide
TBS *tert*-Butyldimethylsilyl
TEA Triethylamine
Tf Triflyl (trifluoromethanesulfonyl)
TFA Trifluoroacetic acid
TFE Trifluoroethanol
THF Tetrahydrofuran
TIPS Triisopropylsilyl
TMEDA Tetramethylethylenediamine
TMP 2,2,6,6-Tetramethylpiperidine
TMS Trimethylsilyl
Tol Tolyl
Ts Tosyl (*p*-toluenesulfonyl)
TsDPEN *N*-(*p*-Toluenesulfonyl)-1,2-diphenylethylenediamine
Xyl Xylyl (3,5-dimethylphenyl)

Contents

RSC Catalysis Series No. 26
Enantioselective Nickel-Catalysed Transformations
By Hélène Pellissier

Published by the Royal Society of Chemistry, www.rsc.org

CHAPTER 1

Enantioselective Nickel-Catalysed Cycloaddition Reactions

1.1 Introduction

Reactions which form multiple bonds, rings, and stereocentres are particularly important tools for the efficient assembly of complex molecular structures.[1] Of the many families of reactions discovered over the past century, cycloaddition reactions hold a prominent place in the arsenal of synthetic methods currently available to organic chemists, and research activity in this field shows no signs of abatement.[2] Among the metals used to catalyse cycloadditions,[3] nickel has been found competent to promote enantioselectively the formation of carbo- and heterocycles of various ring sizes.

1.2 1,3-Dipolar Cycloadditions

Heterocyclic compounds, which represent almost two-thirds of all known organic compounds, include some of the most significant for human beings. It is not surprising, therefore, that this class of compound has received special attention by chemists to provide selective synthetic access to the enormous variety of structural features typical of this class. The 1,3-dipolar cycloaddition, also known as the Huisgen cycloaddition,[4] is a classic reaction in organic chemistry consisting of the reaction of a dipolarophile with a 1,3-dipolar compound that allows the production of various five-membered heterocycles. This reaction represents one of the most productive fields of modern synthetic organic chemistry. Most dipolarophiles are alkenes,

RSC Catalysis Series No. 26
Enantioselective Nickel-Catalysed Transformations
By Hélène Pellissier

Published by the Royal Society of Chemistry, www.rsc.org

alkynes, or molecules possessing related heteroatom functional groups such as carbonyls and nitriles. The 1,3-dipoles can be basically divided into two different types: (1) the allyl anion type such as nitrones, azomethine ylides, and nitro compounds, bearing a nitrogen atom in the middle of the dipole, and carbonyl ylides and carbonyl imines, bearing an oxygen atom in the middle of the dipole, and (2) the linear propargyl/allenyl anion type such as nitrile oxides, nitrilimines, nitrile ylides, diazoalkanes, and azides. Two π-electrons of the dipolarophile and four electrons of the dipolar compound participate in a concerted, pericyclic shift. The addition is stereoconservative (suprafacial), and the reaction is therefore a $[2_s + 4_s]$ cycloaddition (Scheme 1.1).

However, the dipole might be stabilised by the adjacent central heteroatom X (nitrogen, oxygen, or sulfur) through resonance, and a non-concerted reaction pathway might also occur. Consequently, in some cases, the original stereochemistry of the alkene is not necessarily conserved, as depicted in Scheme 1.2.

The transition state of the concerted 1,3-dipolar cycloaddition reaction is controlled by the frontier molecular orbitals of the substrates. Hence, the reaction of dipoles with dipolarophiles involves either a LUMO-dipole/HOMO-dipolarophile reaction or a HOMO-dipole/LUMO-dipolarophile interaction, depending on the nature of the dipole and the dipolarophile.

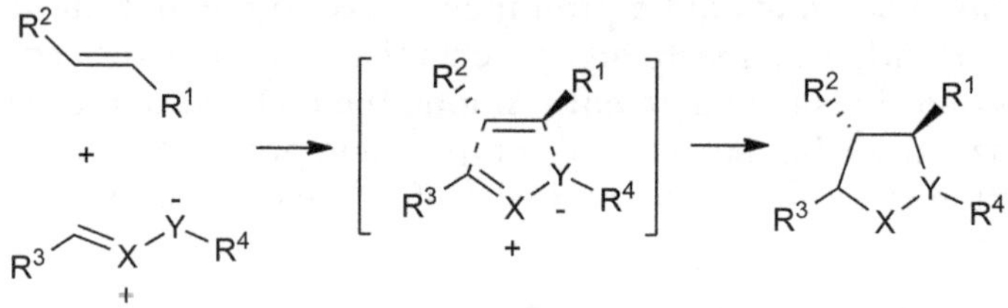

Scheme 1.1 General concerted 1,3-dipolar cycloaddition.

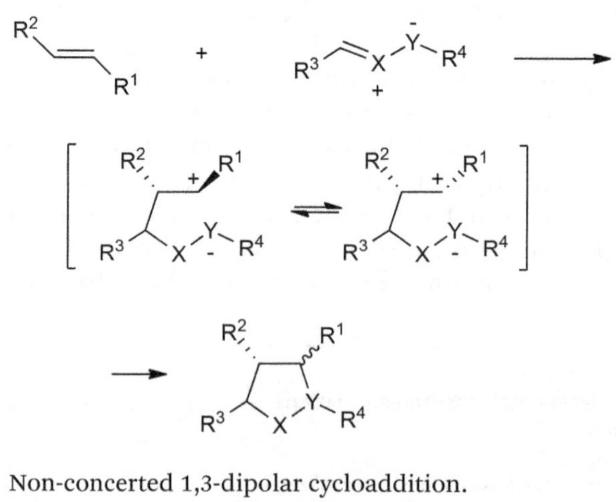

Scheme 1.2 Non-concerted 1,3-dipolar cycloaddition.

In some cases, when the frontier molecular orbital energies of the dipole and the dipolarophile are very similar, a combination of both modes of interactions can occur. These interactions can also be referred to as either *exo* or *endo*, where the *endo* transition state is stabilised by small secondary π-orbital interactions or *via* an *exo* transition state lacking such a stabilisation. However, steric effects can also be important factors for the *endo*/*exo* selectivity and override the secondary orbital interactions.[5] Depending on the substitution pattern in the reacting partners, the stereochemical outcome of the process gives rise to either *endo* or *exo* cycloadducts. Moreover, the presence of a metal, such as a Lewis acid, in 1,3-dipolar cycloaddition reactions can alter both the orbital coefficients of the reacting atoms and the energy of the frontier orbitals of both the 1,3-dipole and the dipolarophile, depending on the electronic properties of these reagents or the Lewis acid. In particular, the coordination of a Lewis acid to one of the two partners of the cycloaddition is of fundamental importance for asymmetric 1,3-dipolar cycloadditions, since the metal can catalyse the reaction.[6] Furthermore, the Lewis acid may also have influence on the selectivity of the cycloaddition reaction, since the regio-, diastereo-, and enantioselectivity can all be controlled by the presence of a metal–ligand complex. Thus, up to four stereocentres can be introduced in a stereoselective manner in only one single step. In recent years, asymmetric 1,3-dipolar cycloadditions have become one of the most powerful tools for the construction of enantiomerically pure five-membered heterocycles.[7] In particular, the asymmetric 1,3-dipolar cycloaddition reaction of nitrones with dipolarophiles, such as alkenes, has received considerable attention over the past 20 years.[7a,b,8]

Regio- and stereoselective nitrone cycloaddition, followed by reduction of the N–O bond to produce both an amino and a hydroxyl function, allows the synthesis of many products of potential interest. One of the reasons for the success of the synthetic applications of nitrones is that, contrary to the majority of the other 1,3-dipoles, most nitrones are stable compounds that do not require *in situ* formation. Another synthetic utility of this reaction is the variety of attractive nitrogenated compounds which are available from the thus-formed isoxazolidines. In particular, these products can be easily reduced under mild conditions to give the corresponding chiral 1,3-amino alcohols. The absolute majority of the 1,3-dipolar cycloaddition reactions are diastereoselective and involve chiral alkenes or nitrones. However, the catalytic enantioselective 1,3-dipolar cycloaddition reaction of nitrones has gone through rapid developments during the last 15 years.[9] In particular, metal-catalysed asymmetric 1,3-dipolar cycloadditions have only recently become an important research field.[7c,d,10] The efficiency of chiral catalysts relies not only on the capability of the enantiopure catalyst to help discriminate between the two π-faces of the dipolarophile, but also on its ability to control both the *exo*/*endo* selectivity and the regiochemistry as well as the yield. When coordinating to the dipole or the dipolarophile, the Lewis acid catalysts lower the energy difference between the LUMO–HOMO of the reacting species. The result is that the LUMO energy of one of the reacting species is lowered. This decreases the energy gap between the HOMO and the LUMO of the dipole and the dipolarophile, leading to increased reactivity. Effective catalysis by the use of a wide variety of chiral Lewis acid catalysts, including nickel complexes, has been reported for nitrone cycloaddition reactions using both electron-deficient and electron-rich alkene dipolarophiles. Early in 1997, Kanemasa *et al.* reported enantioselective 1,3-dipolar cycloadditions of nitrones with 3-crotonoyloxazolidin-2-one catalysed at room temperature by 10 mol% of Kanemasa's chiral ligand (*R*,*R*)-DBFOX-Ph,[11] which provided excellent yields (up to 100%) and high *endo* selectivities (up to >98% de), along with uniformly excellent enantioselectivities (up to >99% ee).[12] Later, comparable excellent results were described by Iwasa *et al.* by using chiral PYBOX ligands at 20 mol% of catalyst loading.[13] Inspired by their early work,[12] Kanemasa *et al.* reported in 2004 the enantioselective 1,3-dipolar cycloaddition of diphenyl nitrone with α-alkyl- and α-arylacroleins catalysed by a chiral nickel complex generated *in situ* from (*R*,*R*)-DBFOX-Ph and $Ni(ClO_4)_2{\cdot}6H_2O$.[14] The reaction afforded the corresponding chiral isoxazolidine-5-carbaldehydes, which were further submitted to reduction by treatment with $NaBH_4$ to give the corresponding alcohols in good to quantitative yields, moderate to excellent diastereoselectivities of up to >99% de, and good to excellent enantioselectivities of up to 98% ee, as shown in Scheme 1.3. The authors compared the reactivity of this chiral nickel catalyst to that of the corresponding zinc(II) complex and found that the latter provided even higher enantioselectivities of up to >99% ee.

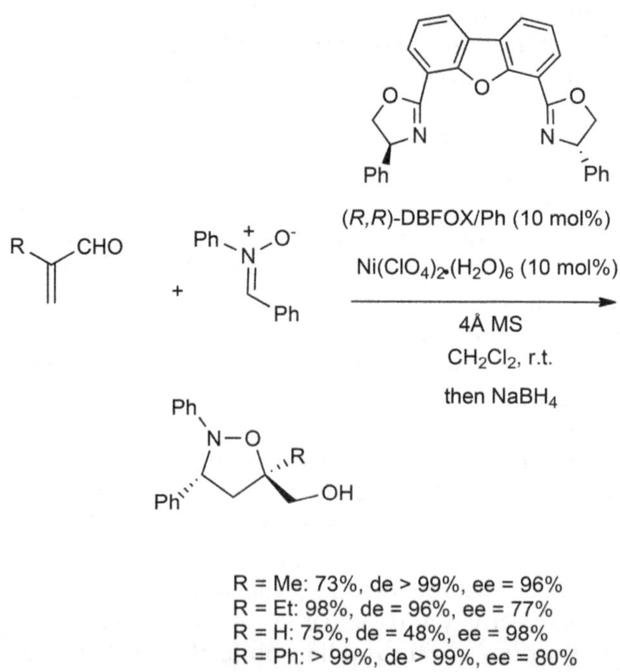

Scheme 1.3 Cycloaddition of diphenyl nitrone with α,β-unsaturated aldehydes.

In 2005, Desimoni and Faita investigated the enantioselective 1,3-dipolar cycloaddition of the same nitrone with an acryloyloxazolidinone catalysed by a combination of nickel perchlorate and chiral bisoxazoline ligands to give the corresponding chiral isoxazolidine-4-oxazolidin-2-one as a mixture of *endo* and *exo* diastereomers.[15] When the reaction was performed with 10 mol% of ligand **1**, it provided the *endo* diastereomer as the major product with a moderate enantioselectivity of 42% ee, whereas the minor *exo* diastereomer was obtained in a higher enantioselectivity of 85% ee (Scheme 1.4). On the other hand, using *trans*-diphenyl-substituted bisoxazoline **2** as chiral ligand reversed the diastereoselectivity of the reaction, since the *exo*-isoxazolidine-4-oxazolidin-2-one was obtained as the major product with good diastereoselectivity of 80% de and excellent enantioselectivity of 99% ee. In both cases the yields of the processes were quantitative. The shift towards *exo* selectivity observed when **2** was the chiral ligand was explained by the authors by considering the steric interactions between the phenyl groups, one on the C-5 position of the ligand and one on the nitrogen atom of the nitrone, in the *endo* transition state. The nitrone *exo* approach did not suffer from this unfavourable contribution, and the *exo* product was the preferred

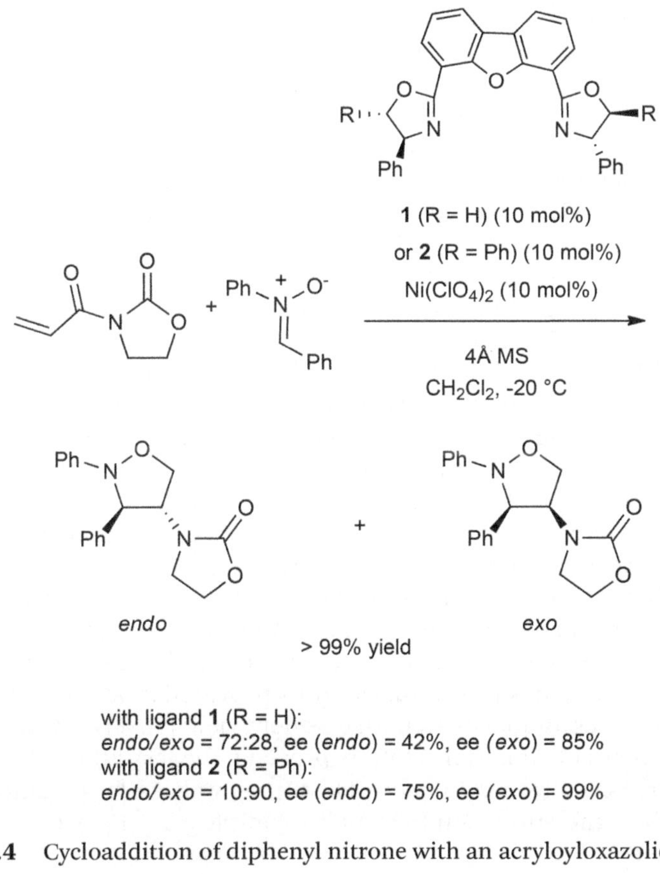

Scheme 1.4 Cycloaddition of diphenyl nitrone with an acryloyloxazolidinone.

stereoisomer. In this study the authors also investigated other metals, such as cobalt, zinc, and magnesium, in combination with the same two chiral ligands. As for the nickel catalyst derived from ligand **1**, *endo* selectivity (*endo*:*exo* of up to 70:30) was observed in the case of using the magnesium complex of ligand **1** along with enantioselectivity of up to 70% ee. On the other hand, the use of cobalt and zinc catalysts of the same chiral ligand provided good levels of *exo* enantioselectivity (*exo*:*endo* = 76:24 for cobalt, and *exo*:*endo* = 73:27 for zinc) with an enantioselectivity of up to 84% ee in both cases of metals. Concerning the involvement of ligand **2**, complexes of magnesium and cobalt favoured, like nickel, the formation of the *exo* cycloadducts in diastereoselectivities of 48 and 68% de and enantioselectivities of 94 and 92% ee, respectively (*vs.* 80% de and 99% ee with nickel), while the zinc complex of ligand **2** favoured the formation of the *endo* cycloadduct in 70% de and 90% ee.

Extremely high *exo* selectivity combined with high enantioselectivity was reached by Suga *et al.* in the enantioselective nickel-catalysed 1,3-dipolar cycloaddition of various nitrones with 3-(alk-2-enoyl)thiazolidine-2-thiones by using chiral binaphthyldiimine ligands.[16] Among a range of this type of ligand, the authors selected (*R*)-BINIM-DCOH as the optimal one, providing the corresponding *exo* cycloadducts in generally good yields, with enantioselectivities of up to 95% ee and with *exo*:*endo* ratios of up to >99:1, as shown in Scheme 1.5. This methodology offered remarkable *exo* selectivity associated with high enantioselectivity for a number of nitrones, in contrast to previously reported methodologies using other chiral Lewis acids. Furthermore, the *in situ* generated catalyst was used at a catalyst loading as low as 5 mol%.

Later, Feng *et al.* employed alkylidenemalonates as dipolarophiles in an enantioselective 1,3-dipolar cycloaddition with nitrones to give the corresponding chiral multisubstituted isoxazolidines.[17] The process was induced by a chiral nickel catalyst generated *in situ* from $Ni(ClO_4)_2{\cdot}6H_2O$ and chiral *N*,*N'*-dioxide **3** employed at low catalyst loading (5.5 mol%), which provided the cycloadducts in good yields (up to 99%) and with both high diastereo- and enantioselectivities of up to >98% de and 99% ee, respectively (Scheme 1.6). The scope of the reaction was broad, and it was insensitive to air or moisture. To explain the results, the authors proposed the transition state depicted in Scheme 1.6 in which the tetradentate ligand and the bidentate alkylidenemalonate coordinated with nickel(II) and formed an octahedral geometry. Then, the dipole could only attack at the *Si* face because *Re* face attack was unfavourable due to the steric hindrance between the *tert*-butyl group and the *C*-phenyl group of the nitrone. In this study the authors investigated the efficiency of other metals, such as magnesium and cobalt, before selecting nickel. Indeed, magnesium chiral catalysts provided moderate enantioselectivity (≤58% ee) combined with both excellent yield (98%) and *exo*:*endo* ratio (94:6), whereas cobalt chiral catalysts proved to be less efficient (22% yield), less diastereoselective (*exo*:*endo* = 58:42), as well as less enantioselective (81% ee) than the corresponding nickel complexes (91% yield, *exo*:*endo* = 93:7, 92% ee).

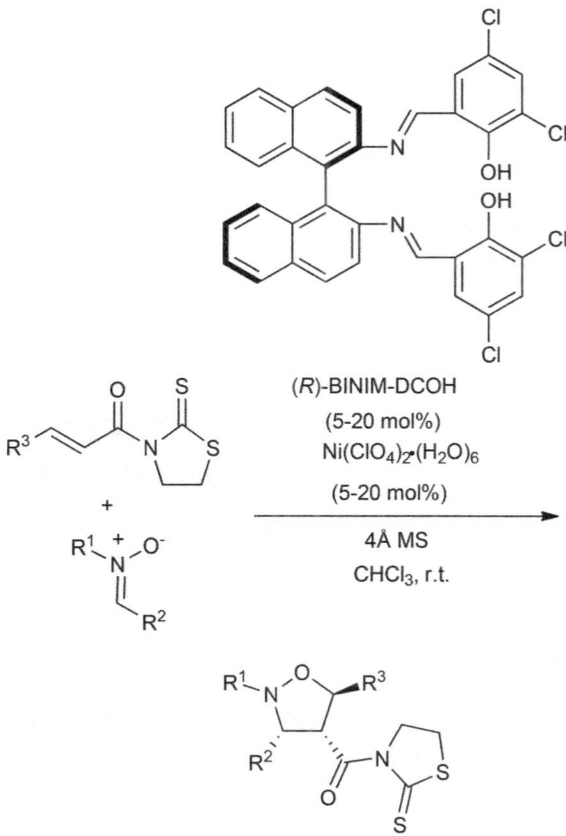

$R^1 = R^2 = Ph$, $R^3 = Me$: 70%, *exo/endo* = 99:1,
ee (*endo*) = 90%
R^1 = *p*-Tol, $R^2 = Ph$, $R^3 = Me$: 78%, *exo/endo* = 93:7
ee (*endo*) = 88%
R^1 = *p*-$MeOC_6H_4$, $R^2 = Ph$, $R^3 = Me$: 36%, *exo/endo* > 99:1,
ee (*endo*) > 90%
R^1 = *p*-ClC_6H_4, $R^2 = Ph$, $R^3 = Me$: 98%, *exo/endo* > 99:1,
ee (*endo*) = 95%
R^1 = *p*-BrC_6H_4, $R^2 = Ph$, $R^3 = Me$: 90%, *exo/endo* = 98:2,
ee (*endo*) = 92%
$R^1 = Ph$, R^2 = *p*-Tol, $R^3 = Me$: 78%, *exo/endo* > 99:1,
ee (*endo*) = 93%
$R^1 = Ph$, R^2 = *p*-ClC_6H_4, $R^3 = Me$: 85%, *exo/endo* > 99:1,
ee (*endo*) = 95%
$R^1 = Bn$, $R^2 = Ph$, $R^3 = Me$: 60%, *exo/endo* = 93:7,
ee (*endo*) = 90%
$R^1 = R^3 = Me$, $R^2 = Ph$: 68%, *exo/endo* = 86:14,
ee (*endo*) = 85%
$R^1 = Bn$, $R^2 = Et$, $R^3 = Me$: 42%, *exo/endo* = 90:10,
ee (*endo*) = 82%
$R^1 = R^2 = Ph$, $R^3 = Et$: 74%, *exo/endo* = 97:3,
ee (*endo*) = 91%
$R^1 = Ph$, R^2 = *p*-ClC_6H_4, $R^3 = Et$: 64%, *exo/endo* = 94:6,
ee (*endo*) = 92%
$R^1 = R^2 = Ph$, R^3 = *n*-Pr: 75%, *exo/endo* > 99:1, ee (*endo*) = 93%
R^1 = *p*-ClC_6H_4, $R^2 = Ph$, R^3 = *n*-Pr: 75%, *exo/endo* > 99:1,
ee (*endo*) = 93%
$R^1 = R^2 = R^3 = Ph$: 59%, *exo/endo* > 99:1, ee (*endo*) = 87%

Scheme 1.5 Cycloaddition of nitrones with 3-(alk-2-enoyl)thiazolidine-2-thiones.

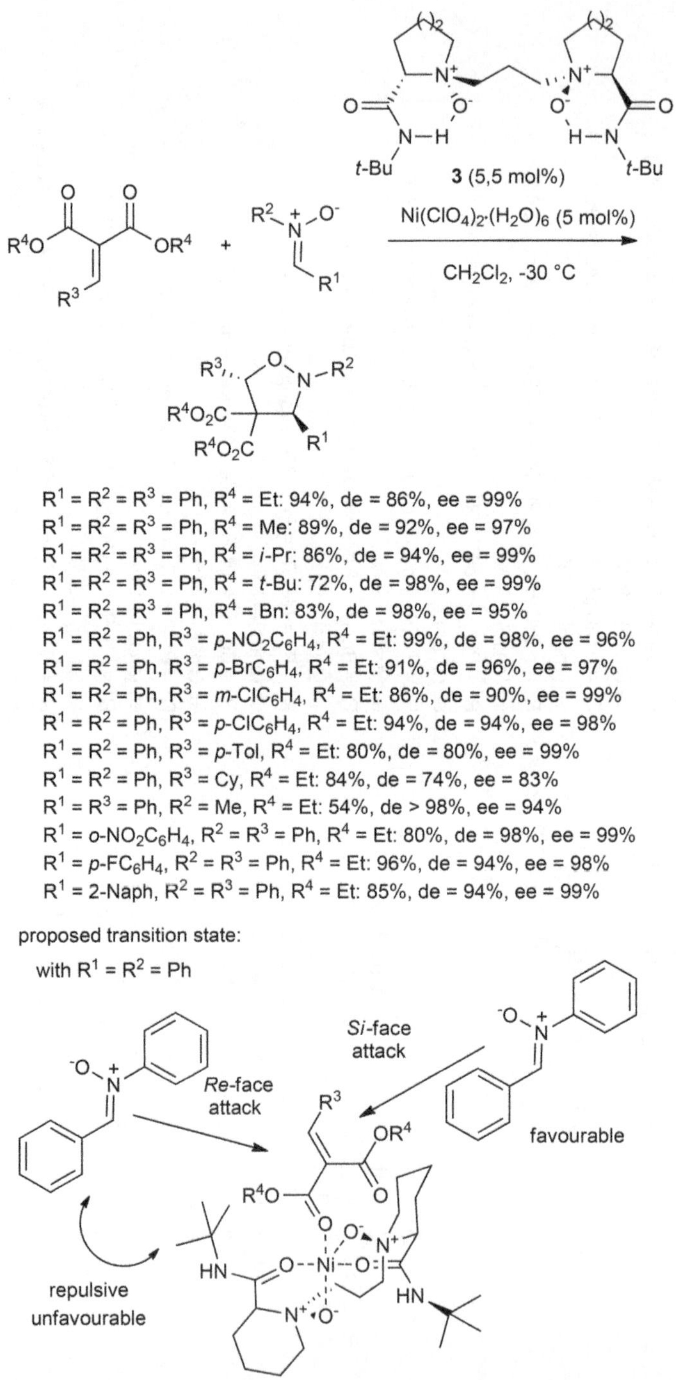

Scheme 1.6 Cycloaddition of nitrones with alkylidenemalonates.

Nitrile oxides are also commonly used as dipoles in 1,3-dipolar cycloadditions. They are reactive, relatively unstable, linear molecules, which may be generated from nitro compounds by treatment with aromatic isocyanates, or from aldoximes by halogenation followed by *in situ* dehydrohalogenation using a base. It is important to note that nitrile oxides are prone to dimerisation or polymerisation, especially upon heating. In order to avoid the dimerisation process, nitrile oxides are usually generated *in situ*. Only a few examples of cycloadditions between nitrile oxides and alkenes mediated by chiral Lewis acids have been reported. In 1996, Ukaji *et al.* described the first example occurring between nitrile oxides and allylic alcohols using chiral zinc catalysts.[18] However, highly enantioselective chiral Lewis acid-catalysed asymmetric cycloadditions between nitrile oxides and electron-deficient alkenes did not appear until 2004 with work published by Sibi *et al.*[19] This dealt with the cycloaddition of mesitylenenitrile oxide with 2-crotonoylpyrazolidinones in the presence of a chiral Lewis acid prepared from magnesium iodide and INDABOX as a chiral ligand. The corresponding pyrazolidinones were obtained in good yields (41–98%) and with high regioselectivity (up to 99 : 1 ratio) and enantioselectivity of up to 99% ee. Later, Suga *et al.* developed asymmetric 1,3-dipolar cycloadditions of 2-crotonoyl- and 2-acryloylpyrazolidinones with nitrile oxides generated *in situ* from the corresponding hydroximoyl chlorides in the presence of nickel complexes of chiral binaphthyldiimine (BINIM) derivatives as Lewis acid catalysts (Scheme 1.7).[20] The best results in terms of enantioselectivity (up to 95% ee) were reached by using (*R*)-BINIM-4-(3,5-xylyl)-2QN as chiral ligand in a catalyst loading of 10–30 mol%. The reactions of both 2-crotonoyl- and 2-acryloylpyrazolidinones with a range of alkyl as well as aromatic nitrile oxides afforded the corresponding highly functionalised chiral cycloadducts with good regioselectivity (**4**/**5** > 98 : 2 in all cases of substrates studied), along with good to high enantioselectivities for the major 4-R^2-substituted regioisomers **4** which were isolated in excellent yields (up to quantitative) most of the time.

Feng *et al.* have also shown that nitrile oxides undergo cycloaddition enantioselectively to other types of dipolarophiles, such as 3-arylideneoxindoles, to provide the corresponding chiral spiro[isoxazoline-3,3′-oxindoles] (Scheme 1.8).[21] Among a series of variously substituted *N*,*N*′-dioxide chiral ligands investigated, ligand **6**, with sterically hindered amide moieties derived from 2,6-diisopropylaniline, was selected as the most efficient to induce the reaction, which afforded the C-adduct **7** as the major regioisomer along with the O-adduct **8** as the minor one, often obtained as a trace amount. In spite of moderate yields, the products were obtained in excellent regio-, diastereo-, and enantioselectivities of up to 99 : 1 (**7**/**8**), >98% de, and >99% ee. In this study, nickel was selected as the most efficient metal compared with magnesium or zinc, which both provided lower enantioselectivities (≤5% ee) in spite of good yields (50–54%) and high diastereoselectivities of up to 94% de.

In addition, Sibi *et al.* have developed the enantioselective nickel-catalysed 1,3-dipolar cycloaddition of mesitylenenitrile oxide with an α,β-disubstituted acrylamide to give completely regioselectively the corresponding

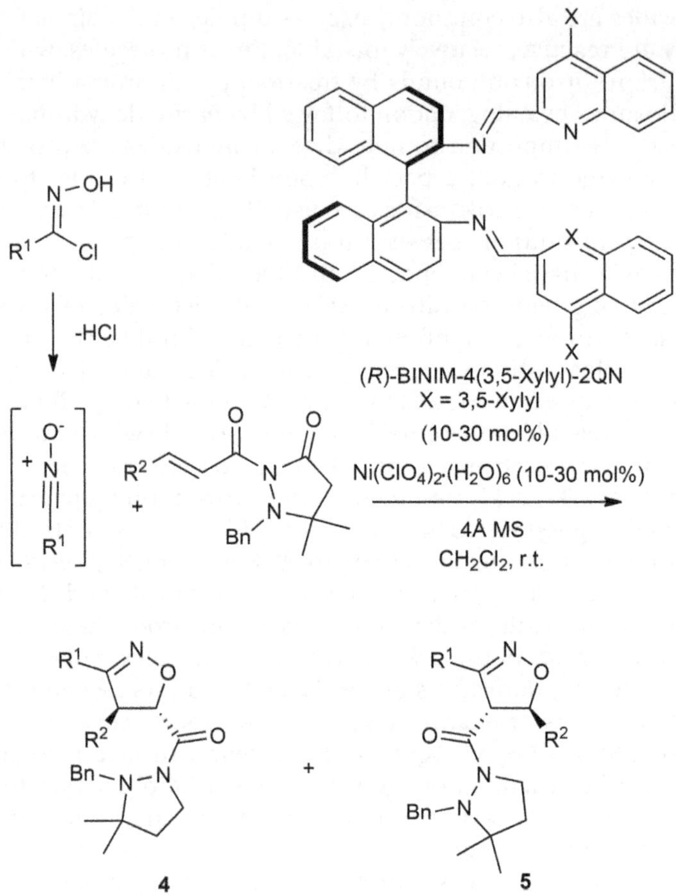

major regioisomer

minor regioisomer

4/**5** > 98:2

R^1 = Ph, R^2 = Me: 94%, ee = 95%
R^1 = *p*-MeOC_6H_4, R^2 = Me: 97%, ee = 92%
R^1 = *p*-ClC_6H_4, R^2 = Me: 93%, ee = 91%
R^1 = *o*-ClC_6H_4, R^2 = Me: > 99%, ee = 90%
R^1 = *i*-Bu, R^2 = Me: 99%, ee = 87%
R^1 = *t*-Bu, R^2 = Me: 94%, ee = 77%
R^1 = Ph, R^2 = H: 93%, ee = 93%
R^1 = *p*-MeOC_6H_4, R^2 = H: 67%, ee = 90%
R^1 = *p*-ClC_6H_4, R^2 = H: > 99%, ee = 84%
R^1 = *i*-Bu, R^2 = H: 91%, ee = 91%
R^1 = *n*-Bu, R^2 = H: 73%, ee = 87%
R^1 = *t*-Bu, R^2 = H: 92%, ee = 86%

Scheme 1.7 Cycloaddition of nitrile oxides with 2-crotonoyl- and 2-acryloylpyrazolidinones.

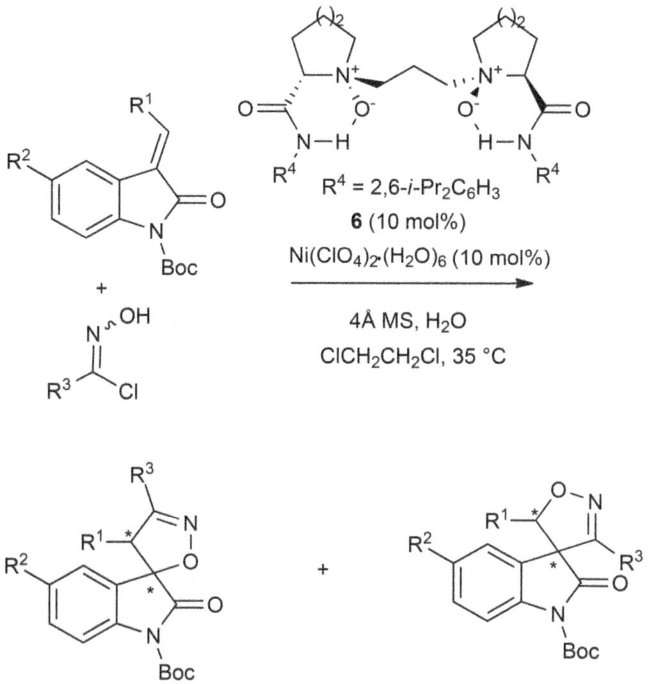

7 major regioisomer de > 98%

8 minor regioisomer de > 98%

$R^1 = R^3 = Ph$, $R^2 = H$: 43%, **7**/**8** = 97:3, ee (**7**) > 99%
R^1 = *m*-FC_6H_4, $R^2 = H$, $R^3 = Ph$: 30%, **7**/**8** = 98:2, ee (**7**) > 99%
R^1 = *p*-FC_6H_4, $R^2 = H$, $R^3 = Ph$: 45%, **7**/**8** = 98:2, ee (**7**) > 99%
R^1 = 2,6-$Cl_2C_6H_3$, $R^2 = H$, $R^3 = Ph$: 53%, **7**/**8** > 99:1, ee (**7**) = 99%
R^1 = *m*-BrC_6H_4, $R^2 = H$, $R^3 = Ph$: 65%, **7**/**8** = 68:32, ee (**7**) = 99%
R^1 = *p*-$CF_3C_6H_4$, $R^2 = H$, $R^3 = Ph$: 32%, **7**/**8** = 99:1, ee (**7**) > 99%
R^1 = 2-Naph, $R^2 = H$, $R^3 = Ph$: 40%, **7**/**8** = 71:29, ee (**7**) = 93%
$R^1 = R^3 = Ph$, $R^2 = F$: 35%, **7**/**8** = 98:2, ee (**7**) > 99%
$R^1 = R^3 = Ph$, R^2 = MeO: 38%, **7**/**8** = 94:6, ee (**7**) = 97%
$R^1 = Ph$, $R^2 = H$, R^3 = *p*-FC_6H_4: 60%, **7**/**8** = 96:4, ee (**7**) = 97%
$R^1 = Ph$, $R^2 = H$, R^3 = *m*-ClC_6H_4: 44%, **7**/**8** > 99:1, ee (**7**) = 87%
$R^1 = Ph$, $R^2 = H$, R^3 = *m*-BrC_6H_4: 46%, **7**/**8** = 95:5, ee (**7**) = 99%
$R^1 = Ph$, $R^2 = H$, R^3 = *p*-Tol: 40%, **7**/**8** = 96:4, ee (**7**) = 99%
$R^1 = Ph$, $R^2 = H$, R^3 = *p*-$MeOC_6H_4$: 43%, **7**/**8** = 90:10, ee (**7**) = 95%

Scheme 1.8 Cycloaddition of nitrile oxides with 3-arylideneoxindoles.

C-cycloadduct in which the carbon terminus of the dipole added to the β-carbon of the acceptor.[22] When the process was catalysed by a combination of nickel perchlorate with chiral bisoxazoline ligand **9**, the product was obtained in almost quantitative yield and with moderate enantioselectivity of 77% ee, as shown in Scheme 1.9. The authors showed that the enantioselectivity of

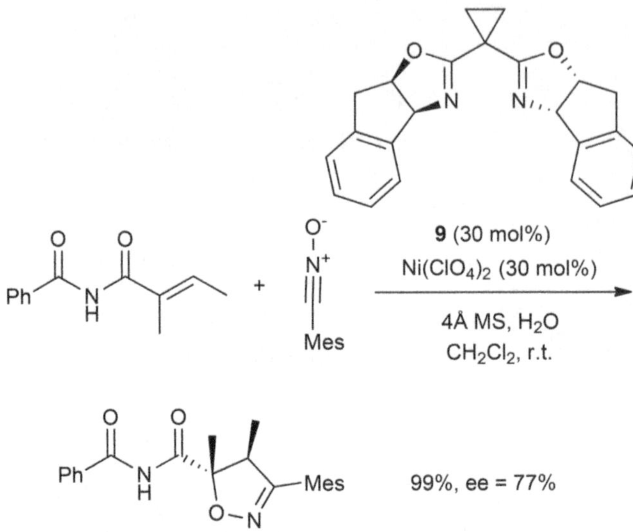

Scheme 1.9 Cycloaddition of mesitylenenitrile oxide with an α,β-disubstituted acrylamide.

the reaction could be slightly increased up to 82% ee, albeit with lower yield (51%), by using a magnesium Lewis acid generated *in situ* from $Mg(ClO_4)_2$ and INDABOX as a chiral ligand employed at 30 mol% of catalyst loading in dichloromethane at room temperature. It must be noted that this work was inspired by previously reported excellent results obtained by the same authors for the same reactions using a copper catalyst of ligand **9** under the same reaction conditions.[23] Indeed, using the catalyst generated *in situ* from this ligand and $Cu(OTf)_2$ at 30 mol% allowed a range of cycloadducts to be achieved in 50–89% yield, 62–98% de, and 86–98% ee.

Pyrrolidines are important structural units in organic chemistry and are frequently found in primary and secondary metabolites, as well as in other biomolecules and synthetic pharmaceuticals.[24] Therefore, significant effort has been devoted to the efficient asymmetric synthesis of functionalised and substituted pyrrolidines. In this context, azomethine ylides have become in recent years one of the most investigated classes of 1,3-dipoles and, based on their cycloaddition chemistry, various methods for the synthesis of pyrrolidine derivatives have been developed.[25] Azomethine ylides are planar 1,3-dipoles composed of one nitrogen and two terminal sp^2 carbon atoms. Their cycloadditions to alkenic dipolarophiles provide a direct and general method for the synthesis of pyrrolidine derivatives. Although there are examples of stable, isolable azomethine ylides, they are normally generated *in situ* and trapped by almost any multiple C–C or C–X (X = heteroatom) bond. A number of methods have been developed for their generation, including the ring opening of aziridines, the desilylation of various silylamino derivatives, the decarboxylation condensation of amino acids, the 1,2-prototropy/metallo-azomethine ylides of amino acid-derived imines, and the deprotonation of iminium salts. Advances in this area, over the last few decades, have made cycloaddition reactions of azomethine ylides a powerful synthetic tool, extensively used in the synthesis of natural products as well as other biologically interesting compounds.[26] However, it must be recognised that such reactions catalysed by chiral nickel catalysts are still in their infancy, especially those involving azomethine ylides as dipoles. In 2008, Shi *et al.* employed chiral binaphthyldiimine ligand **10** to induce the cycloaddition of azomethine ylides, generated *in situ* from the corresponding imino esters, with *N*-arylmaleimides to give the corresponding *endo* cycloadducts in moderate to good yields and with good to high enantioselectivities of up to 95% ee, as shown in Scheme 1.10.[27] Previous to these results, several groups had investigated these reactions by using other chiral metal catalysts. Among the best results are those reported by Carretero *et al.* in 2005, which were based on the use of chiral phosphine–copper complexes, providing up to 97% yield, >96% de, and >99% ee.[28] Comparable excellent results were also described by Kobayashi *et al.* in 2007 by using a calcium complex of a chiral bisoxazoline (up to 99% yield, >96% de, and 99% ee).[29] Excellent results were also reported by Najera *et al.* with a silver catalyst derived from BINAP (up to 90% yield, >96% de, and >99% ee).[30] On the other hand, the use of

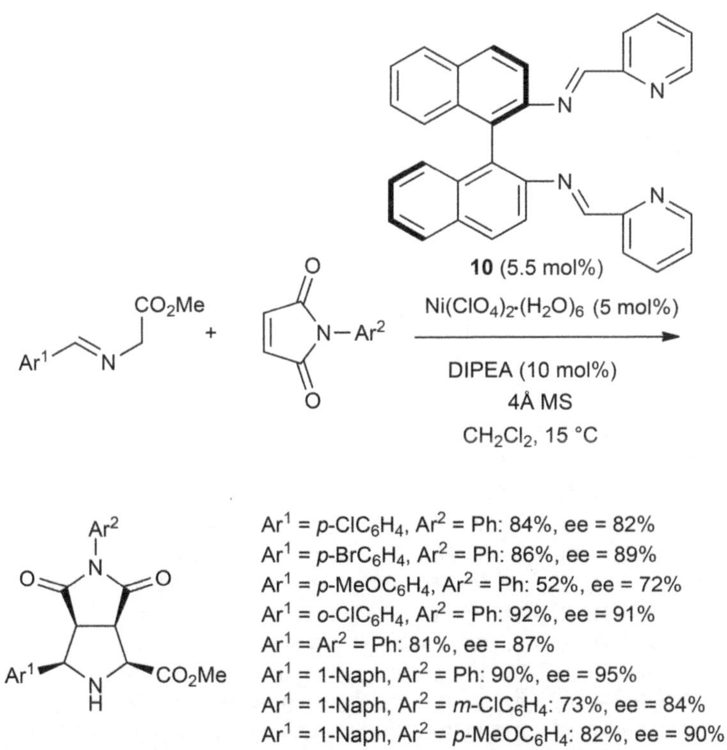

Scheme 1.10 Cycloaddition of azomethine ylides derived from imino esters with *N*-arylmaleimides.

zinc complexes of chiral bisoxazolines by Jørgensen *et al.* in 2002 provided slightly lower yields and enantioselectivities (up to 80% yield and 88% ee), along with excellent diastereoselectivities (up to >96% de).[31]

In the same context, azomethine ylides derived from imino esters were reacted by Arai and Awata with methyleneindolinones in the presence of an *in situ* generated chiral nickel catalyst derived from their original chiral imidazoline-aminophenol ligand **11**.[32] Remarkably, the process afforded the corresponding chiral spiro[pyrrolidine-3,3′-oxindoles] in excellent yields and *exo* selectivities, as well as enantioselectivities in almost all cases of substrates studied, as summarised in Scheme 1.11. Since the spiro-oxindole skeleton is

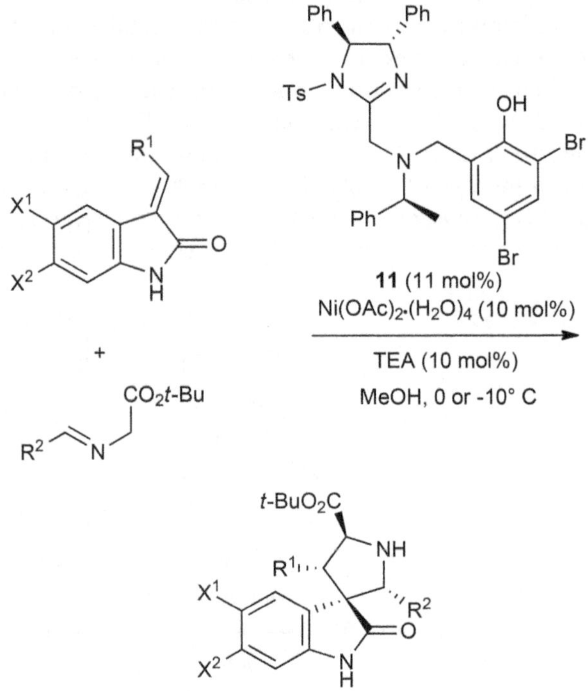

$X^1 = X^2 = H$, $R^1 = R^2 = Ph$: 98%, de > 98%, ee = 98%
$X^1 = X^2 = H$, $R^1 = p\text{-}BrC_6H_4$, $R^2 = Ph$: 92%, de > 98%, ee = 97%
$X^1 = X^2 = H$, $R^1 = o\text{-}BrC_6H_4$, $R^2 = Ph$: 90%, de > 98%, ee = 99%
$X^1 = Cl$, $X^2 = H$, $R^1 = R^2 = Ph$: 95%, de = 90%, ee = 94%
$X^1 = H$, $X^2 = Cl$, $R^1 = R^2 = Ph$: 96%, de = 94%, ee = 94%
$X^1 = X^2 = H$, $R^1 = Bz$, $R^2 = Ph$: > 99%, de = 70%, ee = 98%
$X^1 = X^2 = H$, $R^1 = Ph$, $R^2 = o\text{-}BrC_6H_4$: 84%, de = 94%, ee = 99%
$X^1 = X^2 = H$, $R^1 = Ph$, $R^2 = m\text{-}ClC_6H_4$: 93%, de > 98%, ee = 97%
$X^1 = H$, $X^2 = Cl$, $R^1 = R^2 = p\text{-}ClC_6H_4$: 91%, de = 96%, ee = 96%
$X^1 = X^2 = H$, $R^1 = CO_2Et$, $R^2 = CH_2t\text{-}Bu$: > 99%, de = 76%, ee = 92%
$X^1 = X^2 = H$, $R^1 = CO_2Et$, $R^2 = CHC(Me)_2$: > 99%, de = 58%, ee = 71%

Scheme 1.11 Cycloaddition of azomethine ylides derived from imino esters with methyleneindolinones.

found in many natural biologically active alkaloids, these products represent potent pharmaceutical candidates. These reactions were previously investigated by Waldmann *et al.* by using chiral copper catalysts derived from ferrocenyl P,N-ligands.[33] The formed pyrrolidines were obtained in 80–97% yields and with enantioselectivities of 85–96% ee.

Catalytic asymmetric cycloadditions of azomethine imines are still rarely studied, but among them are a few examples involving chiral nickel catalysts. In 2007, Suga *et al.* developed the first example providing high levels of asymmetric induction (up to 96% ee) along with high diastereoselectivity (up to >98% de) in asymmetric 1,3-dipolar cycloadditions between fused azomethine imines and 3-acryloyloxazolidin-2-one (Scheme 1.12).[34] These processes employed a chiral BINIM/Ni(II) complex as chiral catalyst and afforded the corresponding chiral highly functionalised pyrazolidines

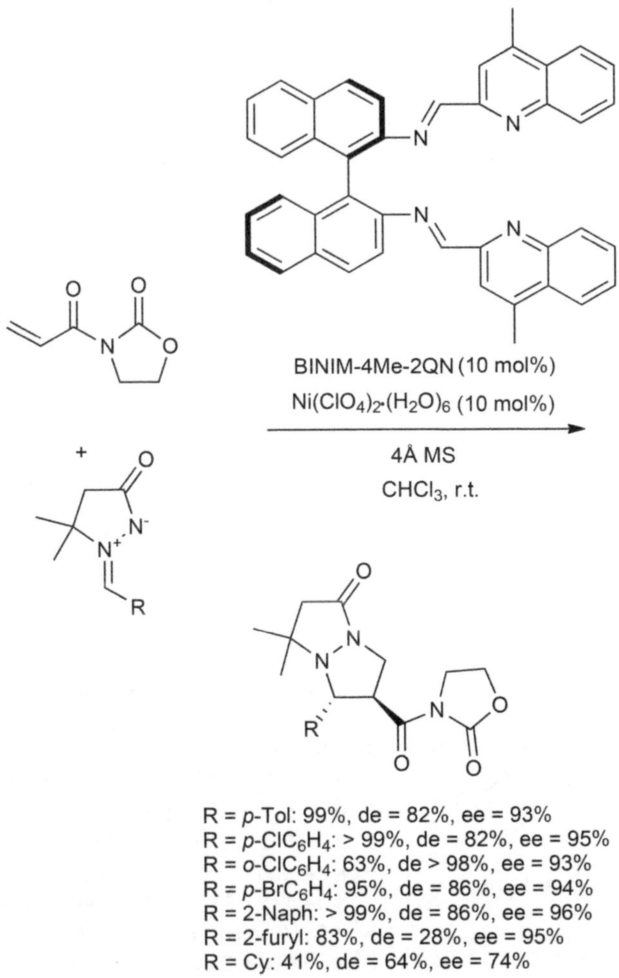

Scheme 1.12 Cycloaddition of azomethine imines with 3-acryloyloxazolidin-2-ones.

in good to quantitative yields, moderate to high diastereoselectivities, and generally high enantioselectivities, except in the case of the alkylimine (R = Cy). Among a range of BINIM ligands the authors selected BINIM-4Me-2QN as the optimal one when used in $CHCl_3$ as solvent. This novel methodology allowed a new route to medicinally important pyrazolidines to be achieved.

Later, Feng *et al.* successfully developed the first asymmetric 1,3-dipolar cycloaddition of azomethine imines with alkylidenemalonates by using an *in situ* generated nickel catalyst from chiral *N,N'*-dioxide **12** (Scheme 1.13).[35] Both aromatic- and aliphatic-substituted alkylidenemalonates were found suitable for the reaction, allowing the synthesis of a range of chiral *trans*-pyrazolone derivatives to be achieved in good to excellent yields (up to 99%) and with high to excellent enantioselectivities of up to 97% ee. It must be noted that in all cases of substrates studied the *trans* diastereomers were exclusively formed according to an *endo* approach. In this study the authors selected nickel as the most efficient source of metal among scandium, magnesium, and cobalt salts. Indeed, only traces of racemic cycloadducts were obtained by using a combination of $Sc(OTf)_3$ with a chiral *N,N'*-oxide, while the corresponding chiral complexes derived from $Mg(ClO_4)_2$ and

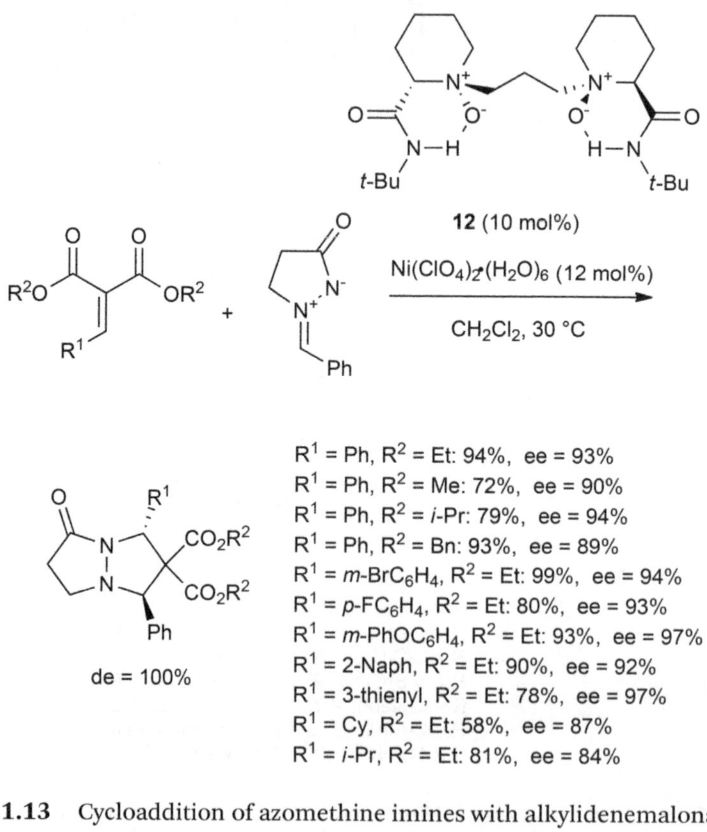

Scheme 1.13 Cycloaddition of azomethine imines with alkylidenemalonates.

$Co(ClO_4)_2{\cdot}6H_2O$ afforded the products in both low yields (32–37%) and enantioselectivities (13–16% ee) [*vs.* 80% yield and 93% ee with $Ni(ClO_4)_2{\cdot}6H_2O$].

On the other hand, 1,3-dipolar cycloaddition reactions between diazoalkanes and alkenes form relatively unstable 1-pyrazolines as the initial cycloadducts that either spontaneously release nitrogen to give the corresponding cyclopropanes, or undergo a 1,3-proton migration to give the thermodynamically more stable 2-pyrazoline derivatives. Diazo substrates, such as trimethylsilyldiazomethane and diazoacetates, undergo enantioselective diazoalkane cycloadditions in the presence of a chiral Lewis acid to give 2-pyrazolines with high levels of asymmetric induction. In the case of trimethylsilyldiazomethane, Kanemasa *et al.* reported the first successful examples of enantioselective cycloaddition reactions using 3-(alk-2-enoyl) oxazolidin-2-ones and (*R*,*R*)-DBFOX-Ph–transition metal aqua complexes as the chiral Lewis acids.[36] Comparable high yields of up to 89–93% combined with excellent enantioselectivities of up to 98–99% ee were reached by using chiral zinc and magnesium catalysts of the (*R*,*R*)-DBFOX-Ph ligand, while the corresponding nickel catalyst provided slightly lower results with 79% yield and 93% ee. For the diazoacetates, Maruoka reported on the highly enantioselective 1,3-dipolar cycloadditions using monodentate α-substituted acroleins and a chiral titanium BINOLate catalyst, providing up to 82% yield and 94% ee.[37] Furthermore, Sibi *et al.* reported the highly enantioselective synthesis of 2-pyrazolines *via* magnesium-catalysed cycloadditions of diazo esters with β-substituted, α-substituted, and α,β-disubstituted α,β-unsaturated pyrazolidinone imides using INDABOX as a chiral ligand.[38] The products were obtained in 52–91% yields and remarkably uniform enantioselectivities of 90–99% ee. In 2011, Suga *et al.* investigated the catalytic activity of chiral BINIM-derived nickel complexes for enantioselective 1,3-dipolar cycloadditions between ethyl diazoacetate and 3-acryloyloxazolidin-2-ones as well as 2-acryloylpyrazolidin-3-ones.[39] Among a range of BINIM ligands investigated, the authors found that (*R*)-BINIM-4Ph-2QN was the most efficient to provide the corresponding 2-pyrazolines **13** having a methane carbon substituted with an oxazolidinonyl group as major products, along with 2-pyrazolines **14** as minor products (Scheme 1.14). Actually, the reaction formed relatively unstable 1-pyrazolines **15** as the initial cycloadducts, which spontaneously underwent a 1,3-proton migration to give the thermodynamically more stable 2-pyrazoline derivatives **13** and **14**. The major cycloadducts **14** were achieved in high yields and with enantioselectivities of up to 97% ee, combined with moderate **13**/**14** ratios.

In 1987, Tsuji reported for the first time the racemic palladium-catalysed 1,3-dipolar cycloaddition of vinylcyclopropanes with aryl isocyanates to give the corresponding δ-lactams.[40] In 2008, Johnson *et al.* developed racemic palladium-catalysed cycloadditions of vinylcyclopropanes with aldehydes for the formation of tetrahydrofurans.[41] Inspired by these pioneering results, Kurahashi and Matsubara have recently developed nickel-catalysed 1,3-dipolar cycloaddition of vinylcyclopropanes with imines to give regioselectively the corresponding substituted pyrrolidine derivatives.[42] As shown in Scheme 1.15,

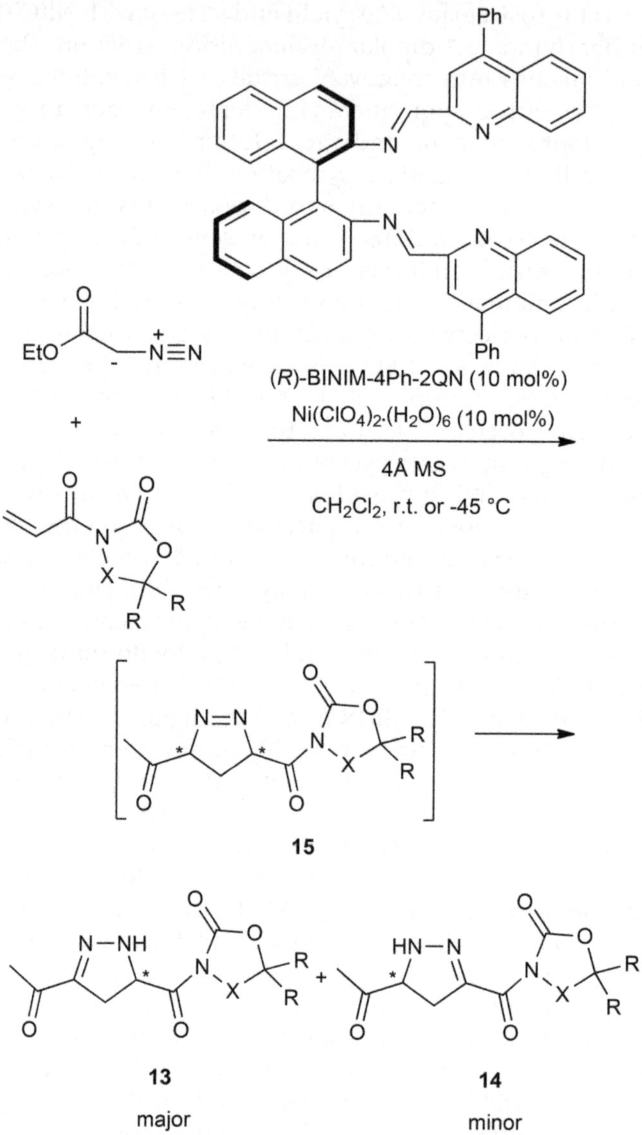

R = H, X = CH_2: 93%, **13/14** = 72:28, ee (**13**) = 90%
R = H, X = CMe_2: 91%, **13/14**= 85:15, ee (**13**) = 92%
R = Me, X = CH_2: 77%, **13/14** = 70:30, ee (**13**) = 92%
R = Me, X = NBn: 87%, **13/14** = 85:15, ee (**13**) = 97%
R = 1-Naph, X = NBn: 85%, **13/14** = 84:16, ee (**13**) = 95%

Scheme 1.14 Cycloaddition of ethyl diazoacetate with 3-acryloyloxazolidin-2-ones and 2-acryloylpyrazolidin-3-ones.

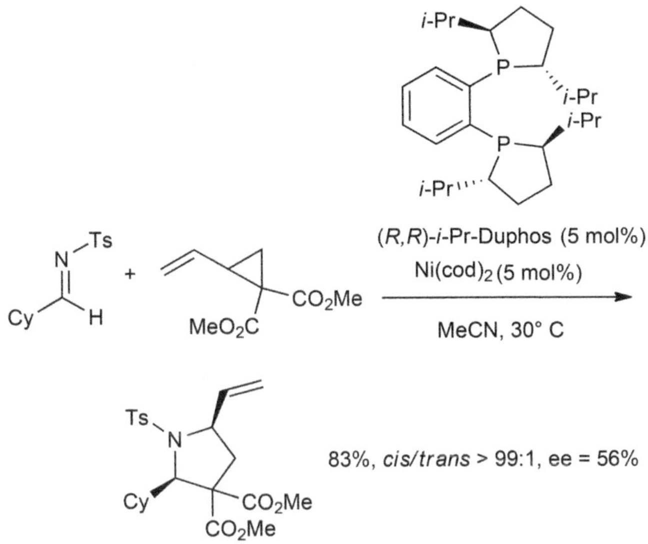

Scheme 1.15 Cycloaddition of an imine with a vinylcyclopropane.

performing the reaction in the presence of (*R*,*R*)-i-Pr-DUPHOS as a chiral ligand of $Ni(cod)_2$ allowed a chiral functionalised pyrrolidine to be produced as a single *cis* diastereomer in good yield (83%) and with moderate enantioselectivity of 56% ee.

Chiral furan derivatives, including tetrahydrofurans and 2,5-dihydrofurans, are useful building blocks and are frequently found as important structural units in many bioactive natural and non-natural products, and which exhibit a wide range of biological activities.[43] 1,3-Dipolar cycloaddition is one of the best ways to achieve furans. In 2011, Zhang *et al.* reported the first racemic version of cycloaddition between epoxides and alkynes which was catalysed by $Sc(OTf)_3$.[44] More recently, Feng *et al.* reported the first efficient asymmetric version of this process.[45] This nice reaction evolved through C–C bond cleavage of oxiranes using an *in situ* generated chiral nickel catalyst from $Ni(ClO_4)_2{\cdot}6H_2O$ and chiral *N*,*N'*-dioxide **16** (Scheme 1.16). A variety of chiral 2,5-dihydrofurans were obtained in good to excellent yields (up to 99%), along with uniformly very high enantioselectivities of up to 95% ee. In this study the authors compared the efficiency of nickel with other metals and lanthanides, including ytterbium and gadolinium, and found that nickel salts were much more enantioselective (77% ee *vs.* 13–33% ee with the corresponding Yb and Gd complexes). The substrate scope of the reaction was found to be broad since a range of substituted oxiranes could undergo the reaction. Indeed, aromatic epoxides with either electron-withdrawing or electron-donating substituents on the aromatic ring performed well, affording the corresponding products in both high yields and enantioselectivities (up to 93% ee). It seemed that epoxides with electron-donating groups at the *meta* positions exhibited higher activities (81–99%)

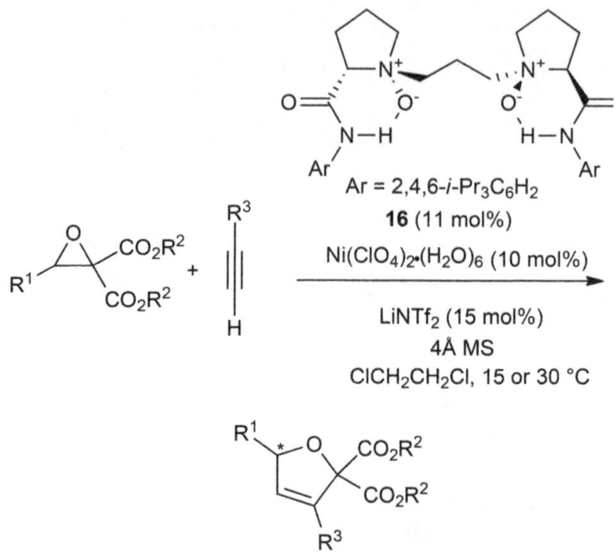

$R^1 = R^2$ = Ph, R^3 = *p*-MeOC$_6$H$_4$: 93%, ee = 92%
R^1 = *o*-FC$_6$H$_4$, R^2 = Ph, R^3 = *p*-MeOC$_6$H$_4$: 62%, ee = 92%
R^1 = *m*-MeOC$_6$H$_4$, R^2 = Ph, R^3 = *p*-MeOC$_6$H$_4$: 99%, ee = 92%
R^1 = *m*-PhOC$_6$H$_4$, R^2 = Ph, R^3 = *p*-MeOC$_6$H$_4$: 90%, ee = 93%
R^1 = *m*-FC$_6$H$_4$, R^2 = Ph, R^3 = *p*-MeOC$_6$H$_4$: 57%, ee = 92%
R^1 = *m*-ClC$_6$H$_4$, R^2 = Ph, R^3 = *p*-MeOC$_6$H$_4$: 67%, ee = 91%
R^1 = *m*-BrC$_6$H$_4$, R^2 = Ph, R^3 = *p*-MeOC$_6$H$_4$: 81%, ee = 89%
R^1 = *p*-ClC$_6$H$_4$, R^2 = Ph, R^3 = *p*-MeOC$_6$H$_4$: 95%, ee = 93%
R^1 = 1-Naph, R^2 = Ph, R^3 = *p*-MeOC$_6$H$_4$: 95%, ee = 91%
R^1 = 2-Naph, R^2 = Ph, R^3 = *p*-MeOC$_6$H$_4$: 94%, ee = 93%
R^1 = 3-furyl, R^2 = Ph, R^3 = *p*-MeOC$_6$H$_4$: 81%, ee = 95%
R^1 = 3-thienyl, R^2 = Ph, R^3 = *p*-MeOC$_6$H$_4$: 90%, ee = 95%
$R^1 = R^2$ = Ph, R^3 = *p*-EtOC$_6$H$_4$: 93%, ee = 91%
$R^1 = R^2$ = Ph, R^3 = 2-thienyl: 90%, ee = 92%
$R^1 = R^2$ = Ph, R^3 = 3-thienyl: 85%, ee = 95%

Scheme 1.16 Cycloaddition of alkynes with epoxides.

than electron-withdrawing groups (57–67%). Moreover, fused-ring and heteroaromatic substrates were also tolerated in the system, affording the desired cycloadducts with high yields (81–95%) and excellent enantioselectivities (91–95% ee). In addition, a variety of alkynes were demonstrated to give comparable results.

1.3 Diels–Alder Cycloadditions

Since its discovery in 1928 by Diels and Alder,[46] the Diels–Alder reaction has become one of the cornerstone reactions in organic chemistry for the construction of six-membered rings.[47] Few reactions can compete with the

[4 + 2] cycloaddition with respect to the degree of structural complexity that can be achieved in a single step.[48] The high regio- and stereoselectivity typically displayed by this cycloaddition, the ease of its execution, and the feature that, during its course, up to four new stereocentres may be created simultaneously have resulted in innumerable applications of this transformation in the construction of highly complex targets. Indeed, this reaction has undergone intensive development, becoming of fundamental importance for synthetic, physical, and theoretical chemists. Today, this powerful reaction is one of the most examined and well appreciated reactions, having an enormous spectrum of applications in chemistry. In particular, the Diels–Alder reaction has been found to be an excellent tool to build up chiral cyclic systems.[49] The reason for the interest in obtaining optically active compounds using the Diels–Alder methodology is that these reactions are normally easy to perform and proceed generally in a highly regio- and diastereoselective manner. Furthermore, the Diels–Alder reaction can give up to four new chiral centres. Catalytic enantioselective Diels–Alder reactions can be achieved by various Lewis acid transition metals.[50] Among them, chiral nickel(II) complexes have been used as efficient catalysts based on various nitrogen-containing chelating ligands. Pioneering remarkable results in this area were reported by Kanemasa *et al.* in 1997, who introduced DBFOX-Ph as a novel tridentate ligand providing enantioselectivities of up to >99% ee along with diastereoselectivities of up to 94% de in the Diels–Alder reactions of cyclopentadiene with 3-alkenoyloxazolidin-2-ones (Scheme 1.17).[11,51] In this early work the authors selected nickel from among a range of other metals, including magnesium, manganese,

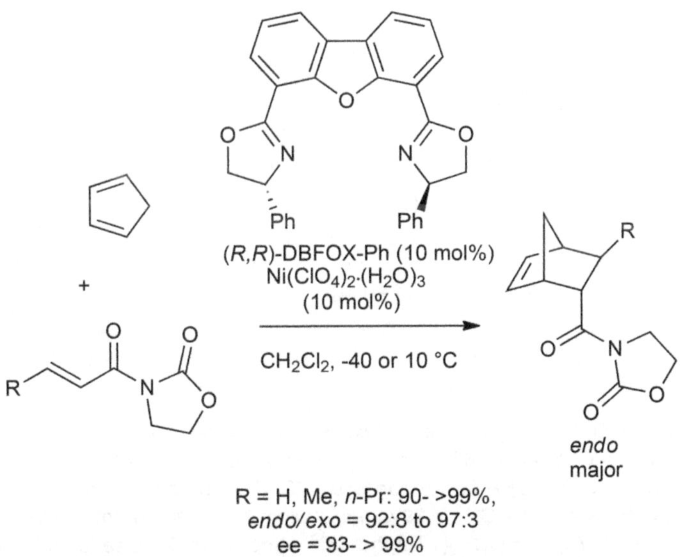

Scheme 1.17 Early Diels–Alder cycloaddition of cyclopentadiene with 3-alkenoyloxazolidin-2-ones reported by Kanemasa *et al.* in 1997.

iron, cobalt, copper, and zinc. While iron, cobalt, copper, zinc, and magnesium chiral complexes gave slightly lower enantioselectivities than the corresponding nickel catalysts, manganese complexes of DBFOX-Ph provided low enantioselectivities (≤25% ee).

In the following years, a range of other chiral amine ligands have been successfully applied in combination with nickel to catalyse enantioselective Diels–Alder reactions. For example, Suga *et al.* have applied binaphthyl-based chiral diimine ligands, such as (*R*)-BINIM-2QN, to induce the enantioselective nickel-catalysed Diels–Alder cycloaddition of cyclopentadiene with 3-alkenoyloxazolidin-2-ones.[52] The corresponding cycloadducts were achieved in good to quantitative yields, good to complete *endo* selectivity of up to >98% de, and high enantioselectivities of up to 96% ee, as shown in Scheme 1.18. In this study the authors compared the reactivity and selectivity of the chiral nickel catalyst with those of the corresponding cobalt, zinc, copper, and magnesium complexes of the same ligand under the same reaction conditions. The analogous cobalt catalyst generated *in situ* from (*R*)-BINIM-2QN and $Co(ClO_4)_2{\cdot}6H_2O$ provided the corresponding cycloadducts in excellent yields (up to 98%), and slightly lower diastereo- and enantioselectivities of up to 90% de (*vs.* >98% de with nickel) and 90% ee (*vs.* 94% ee with nickel), respectively. On the other hand, the zinc catalyst generated *in situ* from (*R*)-BINIM-2QN and $Zn(ClO_4)_2{\cdot}6H_2O$ gave comparable

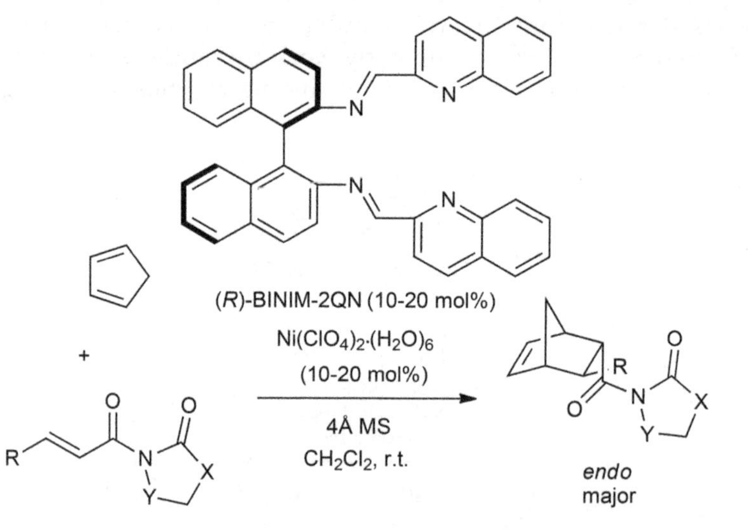

R = H, X = O, Y = CH_2: 94%, *endo/exo* > 99:1, ee (*endo*) = 94%
R = Me, X = O, Y = CH_2: 85%, *endo/exo* = 80:20, ee (*endo*) = 92%
R = Ph, X = O, Y = CH_2: 96%, *endo/exo* = 73:27, ee (*endo*) = 90%
R = CO_2Et, X = O, Y = CH_2: > 99%, *endo/exo* = 78:22, ee (*endo*) = 96%
R = H, X = CMe_2, Y = NCH_2(1-Naph): 73%, *endo/exo* = 92:8, ee (*endo*) = 94%

Scheme 1.18 Diels–Alder cycloaddition of cyclopentadiene with 3-alkenoyloxazolodin-2-ones and a 2-acryloylpyrazolidin-3-one.

yields and diastereoselectivities (up to 97% yield and 90% de), albeit with lower enantioselectivities of up to 84% ee, while the analogous copper and magnesium complexes afforded the cycloadducts in low enantioselectivities of 33% and 5% ee, respectively. The substrate scope of the nickel-catalysed reaction was extended to the reaction of a 2-acryloylpyrazolidin-3-one with cyclopentadiene, which afforded the corresponding *endo* cycloadduct in 73% yield, 84% de, and 94% ee (Scheme 1.18).

Recently, Feng *et al.* developed a useful, mild, and simple asymmetric Diels–Alder cycloaddition between 3-vinylindoles and methyleneindolinones for the construction of chiral spiro[carbazole-oxindoles].[53] The process, employing 10 mol% of a catalyst generated *in situ* from $Ni(OTf)_2$ and chiral *N,N'*-dioxide **17**, afforded the corresponding cycloadducts in uniformly excellent yields and enantioselectivities of up to 97% and 98% ee, respectively (Scheme 1.19). A wide variety of substrates were readily tolerated, generating exclusively the *exo*-spiro[carbazole-oxindole] derivatives in >98% de under mild reaction conditions. In this study the authors showed that poor enantioselectivities (11–20% ee) along with lower yields (17–71%) were obtained by using the corresponding chiral complexes derived from $Sc(OTf)_3$, $Yb(OTf)_3$, or $Cu(OTf)_2$.

The aza-Diels–Alder reaction is among the most powerful and convergent strategies for the stereoselective construction of piperidine derivatives. Although in recent years very important progress has been achieved in the catalytic asymmetric aza-Diels–Alder reaction of dienes with imines, the complementary alternative involving the asymmetric cycloaddition between azadienes and alkenes has been hardly studied. Actually, Carretero *et al.* only recently described the first enantioselective aza-Diels–Alder reaction of 1-azadienes catalysed by a chiral Lewis acid.[54] As shown in Scheme 1.20, this work consisted of the nickel-catalysed highly enantioselective aza-Diels–Alder reaction of *N*-sulfonyl-1-azadienes with vinyl ethers under mild reaction conditions. The success of this process relied on the use of DBFOX-Ph as the chiral ligand and the choice of the *N*-(quinoline-8-sulfonyl) group at the imine nitrogen. This ligand was selected as the most efficient one among various chiral ligands, including BINAP, BOX, and PYBOX chiral ligands. The inverse-electron-demand Diels–Alder reaction of a range of *N*-sulfonyl-1-azadienes with vinyl ethers provided the corresponding highly functionalised piperidines in moderate to good yields (up to 75%), excellent *endo* selectivities of up to 96% de, and moderate to high enantioselectivities of up to 92% ee. The study of the substrate scope showed that aryl substituents of varied electronic and steric nature at the β-position (R^2) were well tolerated, although electron-rich groups led to a slight decrease in enantioselectivity (77–80% ee). Even the substrate with a *tert*-butyl group at R^2 proved to be suitable (84% ee). In contrast, substitution compatibility at the iminic carbon (R^1) proved to be more limited. While *p*-substituted aryl groups were compatible, a dramatic drop in the enantioselectivity was observed with the more sterically demanding 2-naphthyl group (6% ee).

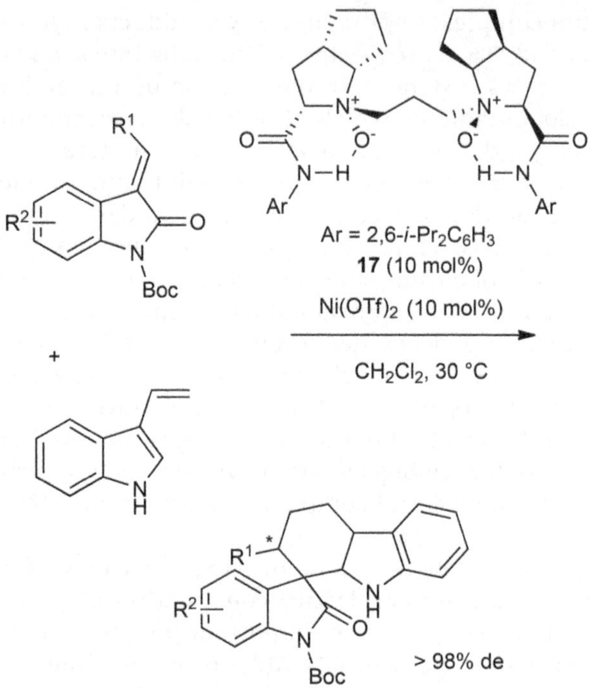

R^1 = CO_2Et, R^2 = H: 96%, ee = 96%
R^1 = CO_2Me, R^2 = H: 96%, ee = 96%
R^1 = CO_2*i*-Pr, R^2 = H: 97%, ee = 97%
R^1 = CO_2Ph, R^2 = H: 95%, ee = 96%
R^1 = CO_2*t*-Bu, R^2 = H: 97%, ee = 98%
R^1 = CO_2*t*-Bu, R^2 = 5-Me: 97%, ee = 98%
R^1 = CO_2*t*-Bu, R^2 = 5-MeO: 94%, ee = 98%
R^1 = CO_2*t*-Bu, R^2 = 5-F: 95%, ee = 97%
R^1 = CO_2*t*-Bu, R^2 = 5-Cl: 95%, ee = 98%
R^1 = CO_2*t*-Bu, R^2 = 5-Br: 93%, ee = 96%
R^1 = CO_2*t*-Bu, R^2 = 6-Br: 94%, ee = 96%
R^1 = CN, R^2 = H: 95%, ee = 93%
R^1 = *p*-Tol, R^2 = H: 83%, ee = 97%
R^1 = *p*-FC_6H_4, R^2 = H: 84%, ee = 94%
R^1 = *p*-ClC_6H_4, R^2 = H: 85%, ee = 93%
R^1 = 1-Naph, R^2 = H: 86%, ee = 92%
R^1 = *i*-Pr, R^2 = H: 90%, ee = 92%
R^1 = Cy, R^2 = H: 80%, ee = 90%

Scheme 1.19 Diels–Alder cycloaddition of 3-vinylindoles with methyleneindolinones.

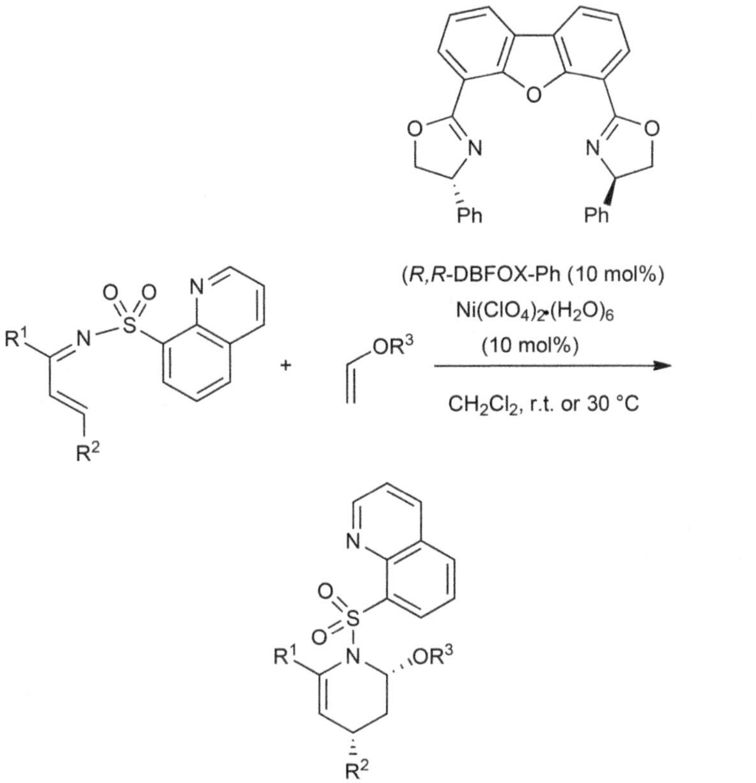

$R^1 = R^2$ = Ph, R^3 = *n*-Pr: 66%, *endo/exo* = 98:2, ee = 91%
$R^1 = R^2$ = Ph, R^3 = Et: 73%, *endo/exo* = 97:3, ee = 88%
$R^1 = R^2$ = Ph, R^3 = Cy: 70%, *endo/exo* = 98:2, ee = 88%
$R^1 = R^2$ = Ph, R^3 = *t*-Bu: 35%, *endo/exo* = 80:20, ee = 68%
R^1 = Ph, R^2 = *p*-FC_6H_4, R^3 = *i*-Pr: 75%, *endo/exo* = 98:2, ee = 92%
R^1 = Ph, R^2 = 2-Naph, R^3 = *i*-Pr: 69%, *endo/exo* = 97:3, ee = 90%
R^1 = Ph, R^2 = *t*-Bu, R^3 = *i*-Pr: 61%, *endo/exo* = 98:2, ee = 84%
R^1 = *p*-ClC_6H_4, R^2 = Ph, R^3 = *i*-Pr: 73%, *endo/exo* = 97:3, ee = 90%
R^1 = *p*-$CF_3C_6H_4$, R^2 = Ph, R^3 = *i*-Pr: 69%, *endo/exo* = 97:3, ee = 91%
R^1 = 2-Naph, R^2 = Ph, R^3 = *i*-Pr: 67%, *endo/exo* = 98:2, ee = 6%
R^1 = *p*-ClC_6H_4, R^2 = CH=CHPh, R^3 = *i*-Pr: 63%, *endo/exo* = 98:2, ee = 92%
R^1 = *p*-CNC_6H_4, R^2 = CH=CHPh, R^3 = *i*-Pr: 70%, *endo/exo* = 98:2, ee = 92%
R^1 = CH=CHPh, R^2 = Ph, R^3 = *i*-Pr: 68%, *endo/exo* = 90:10, ee = 20%
$R^1 = R^2$ = Ph, R^3 = *n*-Pr: 66%, *endo/exo* = 98:2, ee = 91%
R^1 = Ph, R^2 = *p*-$MeOC_6H_4$, R^3 = *i*-Pr: 65%, *endo/exo* = 98:2, ee = 80%
R^1 = Ph, R^2 = 2-furyl, R^3 = *i*-Pr: 52%, *endo/exo* = 97:3, ee = 77%

Scheme 1.20 Inverse-electron-demand Diels–Alder cycloaddition of *N*-sulfonyl-1-aza-1,3-dienes with vinyl ethers.

1.4 Other Cycloadditions

In the last decade, [2 + 2 + 2] cycloaddition reactions have evolved into a versatile member of the synthetic chemists' toolbox for the preparation of functionalised arenes. In particular, transition metal-catalysed [2 + 2 + 2] cycloadditions of unsaturated motifs, such as alkynes and alkenes, constitute the most atom-economical and facile protocol for the construction of a six-membered ring system.[55] Among them, enantioselective [2 + 2 + 2] cycloaddition is a fascinating protocol for the construction of chiral cyclic skeletons.[56] As an example, Stara and Stary have developed the enantioselective intramolecular nickel-catalysed [2 + 2 + 2] cycloaddition of aromatic triynes to provide the corresponding helicene derivatives.[57] A collection of mono- and bidentate phosphines, phosphites, phosphinites, and phosphorus amides possessing stereogenic units, such as a chiral centre, axis, or plane, have been tested as chiral ligands in these reactions, showing that axially the chiral binaphthyl-derived monodentate MOP-type phosphine ligand **18** was the optimal one (Scheme 1.21). Using a combination of 40 mol% of this ligand with 20 mol% of Ni(cod)$_2$ allowed the synthesis of chiral tetrahydro[6] helicene to be achieved at room temperature in 53% yield and with moderate enantioselectivity of 64% ee.

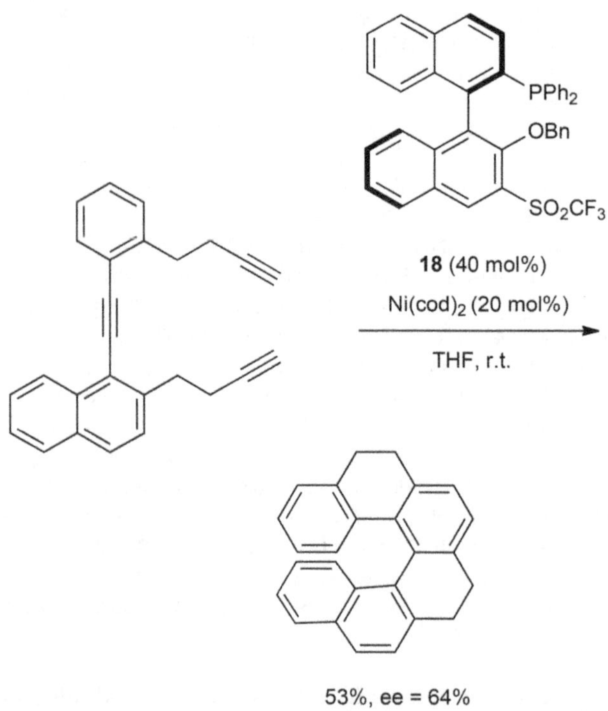

Scheme 1.21 [2 + 2 + 2] Intramolecular cycloaddition of an aromatic triyne.

More recently, remarkable levels of regio- and enantioselectivities of up to >95 : 5 and 99% ee, respectively, were reported by Murakami *et al.* in a novel intermolecular [2 + 2 + 2] cycloaddition reaction of two molecules of isocyanates with allenes.[58] This unprecedented pseudo-three-component reaction was catalysed by a combination of $Ni(cod)_2$ with unsymmetrical phosphino–oxazoline chiral ligand (*S*,*S*)-i-Pr-FOXAP (Scheme 1.22). The latter has been selected among a range of various chiral ligands, such as the C_2-symmetric bisphosphine ligands (*S*,*S*)-CHIRAPHOS, (*S*,*S*)-NORPHOS, and (*S*)-BINAP, which gave lower regioselectivities. This process provided an efficient access to chiral dihydropyrimidine-2,4-diones in moderate to good yields. Various combinations of monosubstituted allenes and isocyanates were investigated, demonstrating that allenes possessing a primary alkyl group readily reacted with high regio- and enantioselectivities, whereas the reaction of cyclohexylallene was sluggish to give the corresponding product in only 26% yield. Functional groups, such as benzyloxy, siloxy, and alkenyl, were tolerated, providing excellent enantioselectivities of up to 99% ee. Generally, higher regioselectivity was observed with electron-rich rather than electron-deficient aryl isocyanates. On the other hand, other alkyl isocyanates, including hexyl isocyanate, cyclohexyl isocyanate, and *tert*-butyl isocyanate, all failed to undergo the reaction. A plausible mechanism for the production of the dihydropyrimidine-2,4-dione from the corresponding allene and isocyanate is depicted in Scheme 1.22. Initially, the intermolecular oxidative cyclisation of a heteropair of the allene and isocyanate occurs on nickel(0) to give the five-membered-ring azanickelacyclic intermediate **19**. Subsequent insertion of another molecule of isocyanate into the nickel–nitrogen bond expands **19** to the seven-membered-ring azanickelacycle **20**, which is in equilibrium with zwitterionic π-allylnickel species **21**. Finally, an intramolecular recombination occurs at the more substituted carbon of the allyl moiety to afford the final formal cycloadduct along with nickel(0).

Tetrahydro-1,2-oxazine derivatives occur frequently in biologically active compounds, and also constitute valuable synthetic intermediates. Among the methods developed for the preparation of such compounds, the formal [3 + 3] cycloaddition of donor–acceptor cyclopropanes with nitrones catalysed by $Yb(OTf)_3$, pioneered by Kerr *et al.*, is a particularly elegant racemic approach.[59] A highly efficient asymmetric version of this reaction was developed by Sibi *et al.* in 2005.[60] In this pioneering work, nitrones reacted with cyclopropane-1,1-dicarboxylates in the presence of a chiral nickel catalyst of the DBFOX ligand to afford the corresponding chiral tetrahydro-1,2-oxazine derivatives in high yields (up to 90%) and with high enantioselectivity of up to 96% ee. When 2-substituted cyclopropane-1,1-dicarboxylates were used as substrates, the reactions proceeded smoothly to give the products with still high enantioselectivities albeit with low diastereoselectivities (the *cis*/*trans* ratios ranged from 1:1.4 to 1:0.8). Inspired by these results, Tang *et al.* reported two years later a highly enantioselective [3 + 3] cycloaddition of nitrones with

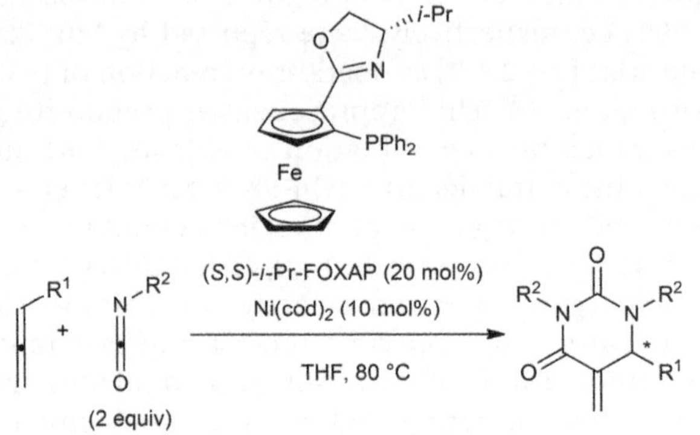

R^1 = *n*-Hex, R^2 = *p*-Tol: 67%, ee = 96%, regioselectivity > 95:5
R^1 = CH_2Cy, R^2 = *p*-Tol: 67%, ee = 94%, regioselectivity > 95:5
R^1 = Cy, R^2 = *p*-Tol: 26%, ee = 97%, regioselectivity > 83:17
R^1 = CH_2Bn, R^2 = *p*-Tol: 61%, ee = 97%, regioselectivity > 95:5
R^1 = $(CH_2)_4OBn$, R^2 = *p*-Tol: 61%, ee = 97%, regioselectivity > 95:5
R^1 = $(CH_2)_4OTBS$, R^2 = *p*-Tol: 60%, ee = 94%, regioselectivity > 95:5
R^1 = $(CH_2)_2CH{=}CH_2$, R^2 = *p*-Tol: 64%, ee = 99%, regioselectivity > 95:5
R^1 = *n*-Oct, R^2 = *p*-$Me_2NC_6H_4$: 57%, ee = 98%, regioselectivity > 95:5
R^1 = *n*-Oct, R^2 = *p*-$MeOC_6H_4$: 65%, ee = 99%, regioselectivity > 95:5
R^1 = *n*-Oct, R^2 = Ph: 70%, ee = 98%, regioselectivity > 95:5
R^1 = *n*-Oct, R^2 = *p*-$CO_2MeC_6H_4$: 76%, ee = 97%, regioselectivity = 94:6
R^1 = *n*-Oct, R^2 = *p*-$CF_3C_6H_4$: 79%, ee = 88%, regioselectivity = 86:14
R^1 = *n*-Oct, R^2 = *p*-AcC_6H_4: 55%, ee = 94%, regioselectivity > 95:5
R^1 = *n*-Oct, R^2 = 2-Naph: 82%, ee = 97%, regioselectivity > 95:5
R^1 = *n*-Oct, R^2 = Bn: 12%, ee = 94%, regioselectivity > 95:5
R^1 = *n*-Oct, R^2 = *m*-Tol: 65%, ee = 97%, regioselectivity > 95:5

proposed mechanism:

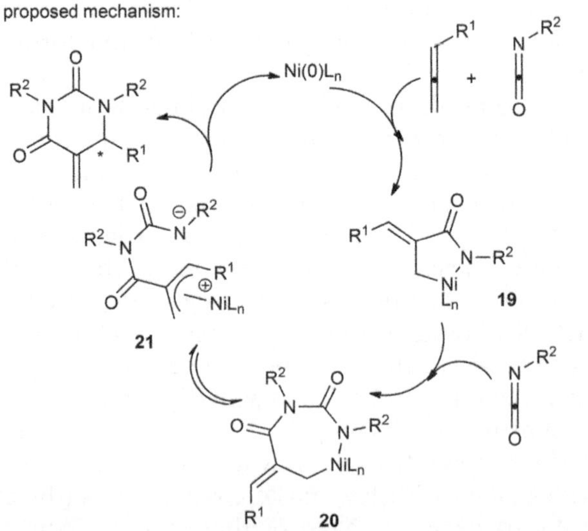

Scheme 1.22 Formal [2 + 2 + 2] cycloaddition of isocyanates and allenes.

2-substituted cyclopropane-1,1-dicarboxylates, providing a ready access to the corresponding chiral tetrahydro-1,2-oxazine derivatives with good diastereoselectivity (Scheme 1.23).[61] This process was induced by a chiral nickel catalyst generated *in situ* from $Ni(ClO_4)_2$ and the chiral trisoxazoline ligand **22** in DME at −30 °C. The diastereoselectivity of the reaction was improved greatly (up to 86% de) relative to that reported by Sibi in 2005. Studying the substrate scope of this reaction, the authors found that the ester groups of the cyclopropanes slightly influenced its enantioselectivity. Benzyl and ethyl diesters reacted with higher enantioselectivities with phenyl-substituted cyclopropanes than the corresponding methyl diester. All reactions of both electron-deficient and electron-rich α-aryl nitrones as well as α-heteroaryl nitrones proceeded with excellent enantio- and diastereoselectivities. The lowest enantioselectivity (80% ee)

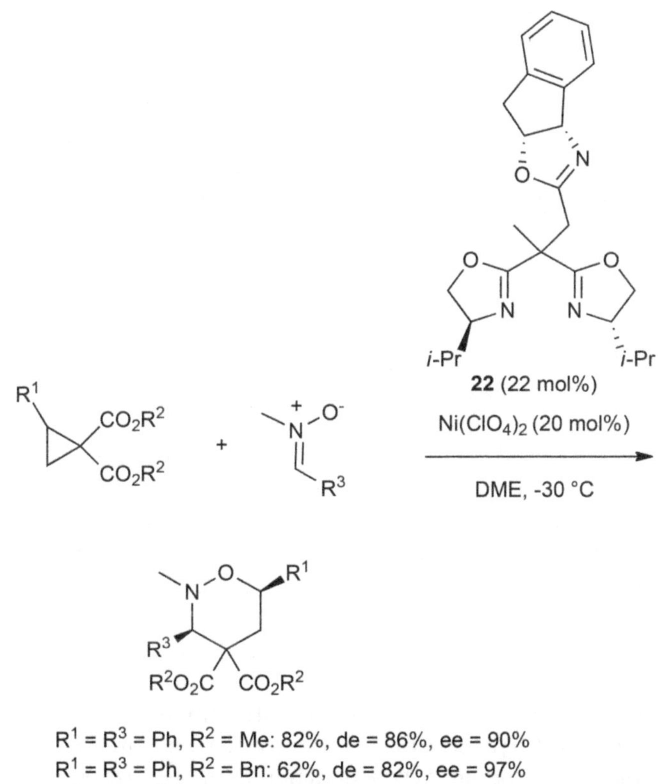

$R^1 = R^3 = Ph$, $R^2 = Me$: 82%, de = 86%, ee = 90%
$R^1 = R^3 = Ph$, $R^2 = Bn$: 62%, de = 82%, ee = 97%
$R^1 = Ph$, $R^2 = Et$, $R^3 = p\text{-}BrC_6H_4$: 85%, de = 84%, ee = 97%
$R^1 = Ph$, $R^2 = Et$, $R^3 = p\text{-}MeOC_6H_4$: 97%, de = 83%, ee = 97%
$R^1 = Ph$, $R^2 = Et$, $R^3 = p$-Tol: 80%, de = 84%, ee = 96%
R^1 = vinyl, $R^2 = Et$, $R^3 = Ph$: 88%, de = 72%, ee = 80%
R^1 = styryl, $R^2 = Et$, $R^3 = Ph$: 84%, de = 66%, ee = 80%
$R^1 = R^3 = Ph$, $R^2 = Et$: 74%, de = 84%, ee = 93%

Scheme 1.23 Formal [3 + 3] cycloaddition of nitrones with 2-substituted cyclopropane-1,1-dicarboxylates.

was observed for diethyl 2-vinyl- and 2-styrylcyclopropane-1,1-dicarboxylates as substrates.

1.5 Conclusions

This chapter concentrates on new developments achieved since the beginning of 2004 in asymmetric nickel-catalysed cycloaddition reactions, well demonstrating that this field constitutes an important tool for organic synthesis related to impressive progress made in the last decade in expanding this chemistry. Indeed, in the last 10 years a range of chiral nickel complexes, predominantly based on various nitrogen-containing chelating ligands, have been successfully applied as highly efficient catalysts in various enantioselective cycloadditions, including many 1,3-dipolar cycloadditions, various Diels–Alder cycloadditions, and other cycloadditions.

The asymmetric 1,3-dipolar cycloaddition reaction, which is undoubtedly one of the most important methods for the construction of chiral five-membered rings, is the reaction that has known the most developments in the last decade, with high levels of stereocontrol which is extremely important for constructing heterocyclic compounds from the viewpoint of the synthesis of biologically active compounds. The versatility of 1,3-dipoles and dipolarophiles, the regio- and stereoselectivity during the reaction, and the scope for further transformation of the cycloadducts to a variety of multifunctional molecules have elevated the asymmetric 1,3-dipolar cycloaddition reaction to an enviable methodology, not only for the construction of chiral functionalised normal-ring carbocycles, but also for the synthesis of complex natural products. Among important results recently reported in this field are 1,3-dipolar cycloadditions of diphenyl nitrone with α-alkyl- and α-arylacroleins catalysed by a chiral nickel complex generated *in situ* from (*R*,*R*)-DBFOX-Ph, which provided up to quantitative yield, >99% de, and 98% ee. The same nitrone also underwent cycloaddition to acryloyloxazolidinones, providing under comparable reaction conditions up to quantitative yield, 80% de, and very high enantioselectivities of up to 99% ee. Another good result concerned the use of (*R*)-BINIM-DCOH as a chiral ligand in the cycloaddition of nitrones with 3-(alk-2-enoyl)thiazolidine-2-thiones, providing up to 98% yield, >98% de, and 95% ee. Also in the context of [3 + 2] cycloadditions of nitrones, their reactions with alkylidenemalonates as dipolarophiles led to the corresponding chiral multisubstituted isoxazolidines in up to 98% yield, >98% de, and 99% ee when catalysed by chiral nickel complexes of *N*,*N*′-dioxide ligands.

In recent years, nitrile oxides have also undergone cycloadditions enantioselectively in the presence of *N*,*N*′-dioxide ligands, for example, to 3-arylideneoxindoles to provide regioselectively the corresponding chiral spiro[isoxazoline-3,3′-oxindoles]. Along with moderate to good yields, excellent diastereo- and enantioselectivities of up >98% de and >99% ee, respectively, were reached. In addition, excellent results (up to >99% yield, >98% de, and 99% ee) have also been described for 1,3-dipolar cycloadditions of azomethine ylides derived from imino esters with methyleneindolinones to give chiral spiro[pyrrolidine-3,3′-oxindoles] in the presence of an *in situ* generated nickel catalyst derived from chiral imidazoline-aminophenol ligands.

Importantly, the first example of highly enantioselective (96% ee) 1,3-dipolar cycloadditions of azomethine imines was only recently reported, employing 3-acryloyloxazolidin-2-ones as dipolarophiles and BINIM-4Me-2QN as a chiral nickel ligand. Moreover, the first asymmetric 1,3-dipolar cycloaddition of azomethine imines with alkylidenemalonates by using an *in situ* generated nickel catalyst derived from a chiral *N*,*N*′-dioxide ligand was successfully developed, providing up to 99% yield, 100% de, and 97% ee. Other dipoles such as diazoacetates have also undergone cycloaddition in the presence of chiral BINIM-derived nickel catalysts to give the corresponding 2-pyrazolines with high levels of asymmetric induction (up to 97% ee) by reaction with 3-acryloyloxazolidin-2-ones as well as 2-acryloylpyrazolidin-3-ones. In addition, the first efficient asymmetric 1,3-dipolar cycloaddition between alkynes and epoxides to afford chiral 2,3-dihydrofurans in up to 95% ee by using a chiral *N*,*N*′-dioxide ligand has been described.

In the last decade, several excellent results were also published in the area of enantioselective nickel-catalysed Diels–Alder cycloadditions. Among them, the reactions of cyclopentadiene with 3-alkenoyloxazolidin-2-ones induced by (*R*)-BINIM-2QN provided cycloadducts in up to >99% yield, >98% de, and 96% ee. Another excellent result was achieved by using a chiral *N*,*N*′-oxide-derived nickel catalyst in Diels–Alder cycloadditions of 3-vinylindoles with methyleneindolinones for the construction of chiral spiro[carbazole-oxindoles] in up to 97% yield, >98% de, and 98% ee. Moreover, the use of the chiral DBFOX-Ph ligand has allowed an inverse-electron-demand Diels–Alder reaction of a range of *N*-sulfonyl-1-azadienes with vinyl ethers to be achieved, providing highly functionalised piperidines in up to 75% yield, 96% de, and 92% ee.

In addition to 1,3-dipolar and Diels–Alder cycloaddition reactions, other types of cycloaddition have been successfully developed on the basis of chiral nickel catalysis. For example, remarkable levels of regio- and enantioselectivities of up to >95 : 5 and 99% ee, respectively, were reported in a novel intermolecular formal [2 + 2 + 2] cycloaddition of two molecules of isocyanates with allenes, giving an efficient access to chiral dihydropyrimidine-2,4-diones by using (*S*,*S*)-i-Pr-FOXAP as a chiral ligand. Another good result concerning a highly enantioselective [3 + 3] cycloaddition of nitrones with 2-substituted cyclopropane-1,1-dicarboxylates, providing ready access to chiral tetrahydro-1,2-oxazine derivatives, was described using a chiral trisoxazoline ligand, which allowed up to 97% yield, 86% de, and 97% ee. All these novel procedures have greatly improved the structural scope and synthetic utility of nickel-catalysed enantioselective cycloadditions, providing access to various functionalised important (poly)(hetero)cyclic compounds with high enantioselectivities. Further progress in this area would include the discovery of more reactive catalyst systems, allowing the use of lower catalyst loadings, and the cycloadditions of even more challenging substrates, such as non-activated alkenes or highly substituted dipolarophiles, as well as the development of applications in the synthesis of natural products and bioactive compounds.

References

1. S. H. Bertz, *J. Am. Chem. Soc.*, 1981, **103**, 3599–3601.
2. (a) *Advances in Cycloaddition*, ed. D. P. Curran, JAI Press, Greenwich, 1994, vol. I–III; (b) B. M. Trost, *Angew. Chem., Int. Ed. Engl.*, 1995, **34**, 259–281; (c) G.-J. Jiang, Y. Wang and Z.-X. Yu, *Science of Synthesis*, Thieme, Stuttgart, 2011, vol. 3, pp. 7–65; (d) *Methods and Applications of Cycloaddition Reactions in Organic Syntheses*, ed. N. Nishiwaki, Wiley, Hoboken, 2014.
3. M. Lautens, W. Klute and W. Tam, *Chem. Rev.*, 1996, **96**, 49–92.
4. R. Huisgen, *Angew. Chem., Int. Ed. Engl.*, 1963, **10**, 565–598.
5. A. Rastelli, R. Gandolfi and M. S. Amadè, *Adv. Quantum Chem.*, 1999, **36**, 151–167.
6. (a) G. Broggini, G. Molteni, A. Terraneo and G. Zecchi, *Heterocycles*, 2003, **59**, 823–858; (b) L. M. Stanley and M. P. Sibi, *Chem. Rev.*, 2008, **108**, 2887–2902; (c) M. Alvarez-Corral, M. Munoz-Dorado and I. Rodriguez-Garcia, *Chem. Rev.*, 2008, **108**, 3174–3198; (d) M. Naodovic and H. Yamamoto, *Chem. Rev.*, 2008, **108**, 3132–3148; (e) C. Najera, J. M. Sansano and M. Yus, *J. Braz. Chem. Soc.*, 2010, **21**, 377–412; (f) Y. Xing and N.-X. Wang, *Coord. Chem. Rev.*, 2012, **256**, 938–952.
7. For reviews on asymmetric 1,3-dipolar cycloadditions, see: (a) K. V. Gothelf and K. A. Jørgensen, *Chem. Rev.*, 1998, **98**, 863–909; (b) S. Karlsson and H.-E. Högberg, *Org. Prep. Proced. Int.*, 2001, **33**, 103–172; (c) S. Kanemasa, *Synlett*, 2002, **9**, 1371–1387; (d) K. V. Gothelf, *Synthesis*, 2002, 211–247; (e) *Synthetic Applications of 1,3-Dipolar Cycloaddition Chemistry Toward Heterocycles and Natural Products*, ed. A. Padwa and W. H. Pearson, Wiley, New York, 2003, vol. 59; (f) H. Pellissier, *Tetrahedron*, 2007, **63**, 3235–3285; (g) J. Adrio and J. C. Carretero, *Chem. Commun.*, 2011, **47**, 6784–6794; (h) C. Najera and J. M. Sansano, *J. Organomet. Chem.*, 2014, **771**, 78–92.
8. M. Frederickson, *Tetrahedron*, 1997, **53**, 403–425.
9. K. V. Gothelf and K. A. Jørgensen, *Chem. Commun.*, 2000, 1449–1458.
10. K. V. Gothelf and K. A. Jörgensen, *Acta Chem. Scand.*, 1996, **50**, 652–660.
11. S. Kanemasa, Y. Oderaotoshi, H. Yamamoto, J. Tanaka, E. Wada and D. P. Curran, *J. Org. Chem.*, 1997, **62**, 6454–6455.
12. S. Kanemasa, Y. Oderaotoshi, J. Tanaka and E. Wada, *J. Am. Chem. Soc.*, 1998, **120**, 12355–12356.
13. (a) S. Iwasa, S. Tsushima, T. Shimada and H. Nishiyama, *Tetrahedron Lett.*, 2001, **42**, 6715–6717; (b) S. Iwasa, S. Tsushima, T. Shimada and H. Nishiyama, *Tetrahedron*, 2002, **58**, 227–232.
14. M. Shirahase, S. Kanemasa and Y. Oderaotoshi, *Org. Lett.*, 2004, **6**, 675–678.

15. G. Desimoni, G. Faita, M. Mella and M. Boiocchi, *Eur. J. Org. Chem.*, 2005, 1020–1027.
16. H. Suga, T. Nakajima, K. Itoh and A. Kakehi, *Org. Lett.*, 2005, **7**, 1431–1434.
17. D. Chen, Z. Wang, J. Li, Z. Yang, L. Lin, X. Liu and X. Feng, *Chem.–Eur. J.*, 2011, **17**, 5226–5229.
18. M. Shimizu, Y. Ukaji and K. Inomata, *Chem. Lett.*, 1996, **25**, 455–456.
19. M. P. Sibi, K. Itoh and C. P. Jasperse, *J. Am. Chem. Soc.*, 2004, **126**, 5366–5367.
20. H. Suga, Y. Adachi, K. Fujimoto, Y. Furihata, T. Tsuchida, A. Kakehi and T. Baba, *J. Org. Chem.*, 2009, **74**, 1099–1113.
21. X. Lian, S. Guo, G. Wang, L. Lin, X. Liu and X. Feng, *J. Org. Chem.*, 2014, **79**, 7703–7710.
22. M. P. Sibi, Z. Ma, K. Itoh, N. Prabagaran and C. P. Jasperse, *Org. Lett.*, 2005, **7**, 2349–2352.
23. M. P. Sibi, Z. Ma and C. P. Jasperse, *J. Am. Chem. Soc.*, 2004, **126**, 718–719.
24. D. O'Hagan, *Nat. Prod. Rep.*, 2000, **17**, 435–446.
25. I. Coldham and R. Hufton, *Chem. Rev.*, 2005, **105**, 2765–2809.
26. C. Nàjera and J. M. Sansano, *Curr. Org. Chem.*, 2003, **7**, 1105–1150.
27. J.-W. Shi, M.-X. Zhao, Z.-Y. Lei and M. Shi, *J. Org. Chem.*, 2008, **73**, 305–308.
28. S. Cabrera, R. Gomez-Arrayas and J. C. Carretero, *J. Am. Chem. Soc.*, 2005, **127**, 16394–16395.
29. S. Saito, T. Tsubogo and S. Kobayashi, *J. Am. Chem. Soc.*, 2007, **129**, 5364–5365.
30. C. Najera, M. de Gracia-Retamosa and J. M. Sansano, *Org. Lett.*, 2007, **9**, 4025–4028.
31. A. S. Gothelf, K. V. Gothelf, R. G. Hazell and K. A. Jørgensen, *Angew. Chem., Int. Ed.*, 2002, **41**, 4236–4238.
32. A. Awata and T. Arai, *Chem.–Eur. J.*, 2012, **18**, 8278–8282.
33. A. P. Antonchick, C. Gerding-Reimers, M. Catarinella and H. Waldmann, *Nat. Chem.*, 2010, **2**, 735–740.
34. H. Suga, A. Funyu and A. Kakehi, *Org. Lett.*, 2007, **9**, 97–100.
35. J. Li, X. Lian, X. Liu, L. Lin and X. Feng, *Chem.–Eur. J.*, 2013, **19**, 5134–5140.
36. S. Kanemasa and T. Kanai, *J. Am. Chem. Soc.*, 2000, **122**, 10710–10711.
37. T. Kano, T. Hashimoto and K. Maruoka, *J. Am. Chem. Soc.*, 2006, **128**, 2174–2175.
38. M. P. Sibi, L. M. Stanley and T. Soeta, *Org. Lett.*, 2007, **9**, 1553–1556.
39. H. Suga, Y. Furihata, A. Sakamoto, K. Itoh, Y. Okumura, T. Tsuchida, A. Kakehi and T. Baba, *J. Org. Chem.*, 2011, **76**, 7377–7387.
40. K. Yamamoto, T. Ishida and J. Tsuji, *Chem. Lett.*, 1987, 1157–1158.
41. A. T. Parsons, M. J. Champbell and J. S. Johnson, *Org. Lett.*, 2008, **10**, 2541–2544.
42. R. Tombe, T. Kurahashi and S. Matsubara, *Org. Lett.*, 2013, **15**, 1791–1793.
43. C.-X. Cui, H. Li, X.-J. Yang, J. Yang and X.-Q. Li, *Org. Lett.*, 2013, **15**, 5944–5947.
44. R. Liu, M. Zhang and J. Zhang, *Chem. Commun.*, 2011, **47**, 12870–12872.

45. W. Chen, X. Fu, L. Lin, X. Yuan, W. Luo, J. Feng, X. Liu and X. Feng, *Chem. Commun.*, 2014, **50**, 11480–11483.
46. O. Diels and K. Alder, *Liebigs Ann. Chem.*, 1928, **460**, 98–122.
47. (a) *Cycloaddition Reactions in Organic Synthesis*, ed. W. Carruthers, Pergamon, Oxford, 1990; (b) *Comprehensive Organic Synthesis*, ed. B. M. Trost and I. Fleming, Pergamon Press, Oxford, 1991, vol. 5, W. Oppolzer, ch. 4.1, and W. R. Roush, ch. 4.4; (c) H. Waldmann, *Synthesis*, 1994, 535–551.
48. P. L. Fuchs, *Tetrahedron*, 2001, **57**, 6855–6875.
49. H. B. Kagan and O. Riant, *Chem. Rev.*, 1992, **92**, 1007–1019.
50. D. A. Evans and J. S. Johnson, in *Comprehensive Asymmetric Catalysis*, ed. E. N. Jacobsen, A. Pfaltz and H. Yamamoto, Springer-Verlag, New York, 1999, ch. 33.1.
51. S. Kanemasa, Y. Oderaotoshi, S.-i. Sakaguchi, H. Yamamoto, J. Tanaka, E. Wada and D. P. Curran, *J. Am. Chem. Soc.*, 1998, **120**, 3074–3088.
52. (a) H. Suga, A. Kakehi and M. Mitsuda, *Bull. Chem. Soc. Jpn.*, 2004, **77**, 561–568; (b) H. Suga, A. Kakehi and M. Mitsuda, *Chem. Lett.*, 2002, 900–901.
53. H. Zheng, P. He, Y. Liu, Y. Zhang, X. Liu, L. Lin and X. Feng, *Chem. Commun.*, 2014, **50**, 8794–8796.
54. (a) J. Esquivias, R. Gomez Arrayas and J. C. Carretero, *J. Am. Chem. Soc.*, 2007, **129**, 1480–1481; (b) J. Esquivias, I. Alonso, R. Gomez Arrayas and J. C. Carretero, *Synthesis*, 2009, **1**, 113–126.
55. (a) K. P. C. Vollhardt, *Angew. Chem., Int. Ed. Engl.*, 1984, **23**, 539–556; (b) N. E. Schore, in *Comprehensive Organic Synthesis*, ed. B. M. Trost and I. Fleming, Pergamon, Oxford, 1991, vol. 5, p. 1129; (c) R. Boese, A. P. V. Sickle and K. P. C. Vollhardt, *Synthesis*, 1994, 1374–1382; (d) D. B. Grotjahn, in *Comprehensive Organometallic Chemistry II*, ed. E. W. Abel, F. G. A. Stone, G. Wilkinson and L. Hegedus, Pergamon, Oxford, 1995, vol. 12, p. 741; (e) I. Ojima, M. Tzamarioudaki, Z. Li and R. J. Donovan, *Chem. Rev.*, 1996, **96**, 635–662; (f) C. Aubert, O. Buisine, M. Petit, F. Slowinski and M. Malacria, *Pure Appl. Chem.*, 1999, **71**, 1463–1470; (g) M. Malacria, C. Aubert and J.-L. Renaud, in *Science of Synthesis*, ed. M. Lautens and B. M. Trost, Thieme, New York, 2001, vol. 1, p. 439; (h) C. Aubert, O. Buisine and M. Malacria, *Chem. Rev.*, 2002, **102**, 813–834; (i) J. A. Varela and C. Saa, *Chem. Rev.*, 2003, **103**, 3787–3801; (j) Y. Yamamoto, *Curr. Org. Chem.*, 2005, **9**, 503–519; (k) S. Kotha, E. Brahmachary and K. Lahiri, *Eur. J. Org. Chem.*, 2005, 4741–4767; (l) V. Gandon, C. Aubert and M. Malacria, *Curr. Org. Chem.*, 2005, **9**, 1699–1712; (m) P. R. Chopade and J. Louie, *Adv. Synth. Catal.*, 2006, **348**, 2307–2327; (n) V. Gandon, C. Aubert and M. Malacria, *Chem. Commun.*, 2006, 2209–2217; (o) N. Agenet, O. Buisine, F. Slowinski, V. Gandon, C. Aubert and M. Malacria, in *Organic Reactions*, ed. L. E. Overman, Wiley, New York, 2007, vol. 68, p. 1; (p) N. Weding and M. Hapke, *Chem. Soc. Rev.*, 2011, **40**, 4525–4538; (q) C. Aubert, L. Fensterbank, P. Garcia, M. Malacria and A. Simmoneau, *Chem. Rev.*, 2011, **111**, 1954–1993; (r) G. Dominguez and J. Perez-Castells, *Chem. Soc. Rev.*, 2011, **40**, 3430–3444; (s) Y. Shibata and K. Tanaka, *Synthesis*, 2012, **44**, 323–350.

56. (a) T. Shibata and K. Tsuchikama, *Org. Biomol. Chem.*, 2008, **5**, 1317–1323; (b) K. Tanaka, *Chem.–Asian J.*, 2009, **4**, 508–518.
57. J. G. Stara, A. Andronova, A. Kollarovic, S. Vyskocil, S. Juge, G. C. Lloyd-Jones, P. J. Guiry and I. Stary, *Collect. Czech. Chem. Commun.*, 2011, **76**, 2005–2022.
58. T. Miura, M. Morimoto and M. Murakami, *J. Am. Chem. Soc.*, 2010, **132**, 15836–15838.
59. I. S. Young and M. A. Kerr, *Angew. Chem., Int. Ed.*, 2003, **42**, 3023–3026.
60. M. P. Sibi, Z. Ma and C. P. Jasperse, *J. Am. Chem. Soc.*, 2005, **127**, 5764–5765.
61. Y.-B. Kang, X.-L. Sun and Y. Tang, *Angew. Chem., Int. Ed.*, 2007, **46**, 3918–3921.

CHAPTER 2

Enantioselective Nickel(II)-Catalysed Conjugate Addition Reactions

2.1 Introduction

First reported by Komnenos in 1883,[1] the Michael reaction can be considered as one of the most powerful and reliable tools for the stereocontrolled formation of carbon–carbon bonds, as has been demonstrated by the huge number of examples in which it has been applied as a key strategic transformation in the total synthesis of optically active natural products, pharmaceuticals, and materials.[2] Although base catalysis is well known as a very efficient and high-yielding process in Michael reactions, the strongly basic conditions are often a limiting factor since they can lead to a number of side and subsequent reactions generating by-products. In order to circumvent these drawbacks, catalysis by transition metals, which work formally under neutral conditions, has attracted the attention of chemists in this area as a mild and efficient alternative to base catalysis.[2b,f,k,l] Actually, the catalysis of the Michael reaction by transition metals was first reported in 1972 by Saegusa *et al.*, who treated malonates and diketones with various α,β-unsaturated carbonyl compounds and derivatives using achiral copper complexes.[3] This reaction was quite general, both for nucleophiles and electrophiles. Another important historical precedent was the Michael addition of β-dicarbonyl compounds to a broad variety of electrophiles catalysed by nickel acetylacetonate [$Ni(acac)_2$], described by Nelson *et al.* in 1979–1980.[4] Since this pioneering contribution, nickel has become one of the preferred metals in catalytic Michael reactions.[5] The first enantioselective example of transition-metal catalysis in Michael reactions was reported by

RSC Catalysis Series No. 26
Enantioselective Nickel-Catalysed Transformations
By Hélène Pellissier

Published by the Royal Society of Chemistry, www.rsc.org

Brunner and Hammer in 1984, who employed $Co(acac)_2$ combined with a chiral C_2-symmetric diamine ligand which provided enantioselectivities of up to 66% ee.[6] Later in 1988, Soai *et al.* developed the first asymmetric conjugate additions performed under chiral nickel(II) catalysis.[7] The process, which involved diethylzinc addition to chalcones, provided the corresponding 1,4-products in low to moderate enantioselectivities (≤55% ee) by using an ephedrine derivative as the chiral ligand, as shown in Scheme 2.1 (first equation). Soon after, better enantioselectivities of up to 90% ee were achieved by the same authors using $Ni(acac)_2$ as the nickel source instead of $NiBr_2$, bipyridine as additive, and acetonitrile as solvent (Scheme 2.1, second equation).[8]

A few years later, good enantioselectivities were reported by several groups using other ephedrine-derived ligands[9] as well as other types of chiral ligands, including proline amides supported on zeolite[10] and various chiral amino alcohols.[11] Among other excellent early results are those reported in 1999 by Kanemasa *et al.*, dealing with the enantioselective conjugate addition of thiophenols to 3-crotonoyloxazolidin-2-ones catalysed by a nickel(II) catalyst derived from the chiral bisoxazoline (*R*,*R*)-DBFOX-Ph.[12] Under the optimised conditions, several aromatic thiols underwent the reaction to give the corresponding conjugate addition products in moderate to excellent yields and with enantioselectivities of up to 97% ee (Scheme 2.2, first equation). The same ligand could also be successfully applied to a nickel-catalysed enantioselective conjugate addition of *N*-methyl aromatic amines to 3-alkenoyloxazolidin-2-ones, providing

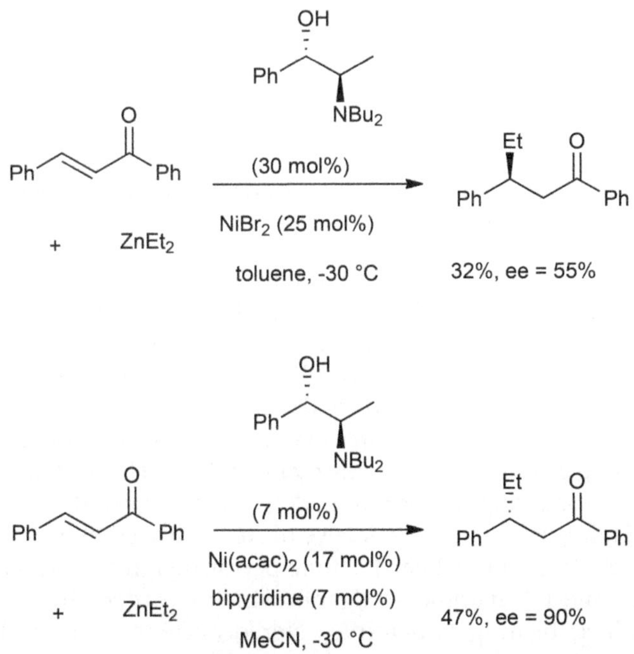

Scheme 2.1 First enantioselective nickel-catalysed conjugate additions reported by Soai *et al.* in 1988–1989.

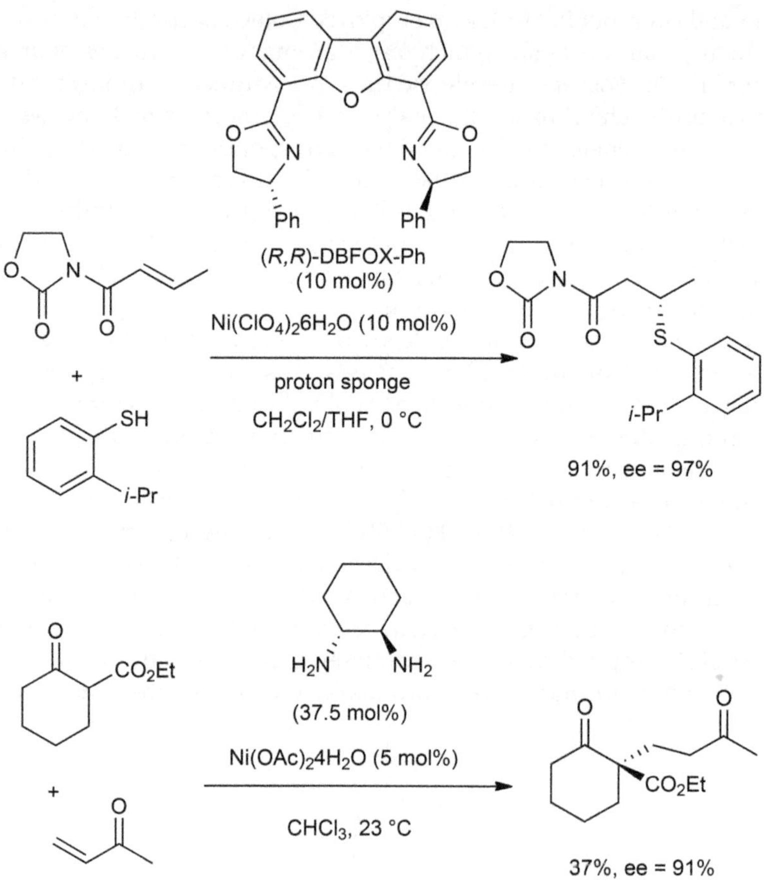

Scheme 2.2 Early asymmetric conjugate additions reported by Kanemasa in 1999 and Christoffers in 2000.

the corresponding chiral highly functionalised amines in good yields of up to 87% and enantioselectivities of up to 96% ee. This ligand was also used in 2001 by Jørgensen *et al.* to a nickel-catalysed enantioselective addition of *N*-methyl aromatic amines to 3-alkenoyloxazolidin-2-ones, giving the corresponding chiral amines in moderate to good yields (23–87%) and good enantioselectivities (48–96% ee).[13] The formed products could be easily converted into the corresponding β-amino acids by treatment with LiOH and H_2O_2 in aqueous THF. Another pioneering important work was that reported by Christoffers *et al.* in 2000, dealing with enantioselective nickel-catalysed Michael additions of cyclic β-keto esters to methyl vinyl ketone performed in the presence of chiral *trans*-cyclohexane-1,2-diamine as ligand, which afforded the corresponding Michael products bearing a quaternary stereocentre in enantioselectivities of up to 91% ee (Scheme 2.2, second equation).[14]

In the following years, many types of other nucleophile could be involved in asymmetric conjugate additions to a variety of electrophiles, including an increasing number of nitroalkenes, by using a range of chiral nickel catalysts. In particular, and as demonstrated in this chapter, the last decade has seen an important number of remarkable results achieved in this field, such as asymmetric nickel-catalysed conjugate additions of various 1,3-dicarbonyl compounds to nitroalkenes, including complex and functionalised ones such as 3-nitro-2*H*-chromenes, nitroenynes, and nitrodienynes, with a beautiful example employing a recyclable mesoporous catalyst. Other nucleophiles, such as γ-butyrolactams, α-keto esters, α-keto anilides, 3-substituted oxindoles, azaarylacetates, highly functionalised acetamides, and acetylazaarenes, *etc.*, also give excellent results in additions to nitroalkenes. Furthermore, organozinc reagents, β-keto esters, 2-silyloxyfurans, malononitriles, nitromethane, nitroacetates, and cyclic amines, among other nucleophiles, have been successfully added to various α,β-unsaturated carbonyl compounds and derivatives. Highly enantioselective intramolecular oxa-Michael additions to activated enones have also been described. Even more importantly, a range of powerful nickel-catalysed asymmetric domino reactions initiated by Michael additions, including multicomponent ones, have been successfully developed in the last 10 years. The goal of this chapter is to cover the advances in enantioselective nickel-catalysed conjugate additions reported in the last decade. This area was previously reviewed in 2005 by Hayashi and Shintani in a book chapter dealing with asymmetric synthesis based on the use of organonickel chemistry.[15] In addition, reviews focusing on metal-catalysed asymmetric conjugate additions have been published,[2b,f,k,l] along with more specialised ones[16] and more general ones.[17] This chapter is subdivided into four parts, dealing successively with enantioselective nickel-catalysed conjugate additions to nitroalkenes, enantioselective nickel-catalysed conjugate additions to α,β-unsaturated carbonyl compounds, enantioselective nickel-catalysed conjugate additions to other activated alkenes, and enantioselective nickel-catalysed domino and tandem processes initiated by a Michael reaction. The first part is subdivided into two sections, according to the different types of nucleophiles to be added to nitroalkenes, such as 1,3-dicarbonyl compounds and other nucleophiles. The second part is also divided into two sections, which successively concern additions to enones and additions to α,β-unsaturated amides.

2.2 Conjugate Additions to Nitroalkenes

2.2.1 1,3-Dicarbonyl Compounds as Nucleophiles

2.2.1.1 Additions to Simple Nitroalkenes

Among the most successful catalytic systems employed in enantioselective nickel-catalysed Michael additions of nucleophiles to nitroalkenes is that developed by Evans and Seidel in 2005, which allowed enantioselectivities of up to 95% ee to be achieved in the enantioselective conjugate additions of 1,3-dicarbonyl compounds to nitroalkenes (Scheme 2.3).[18] Products resulting

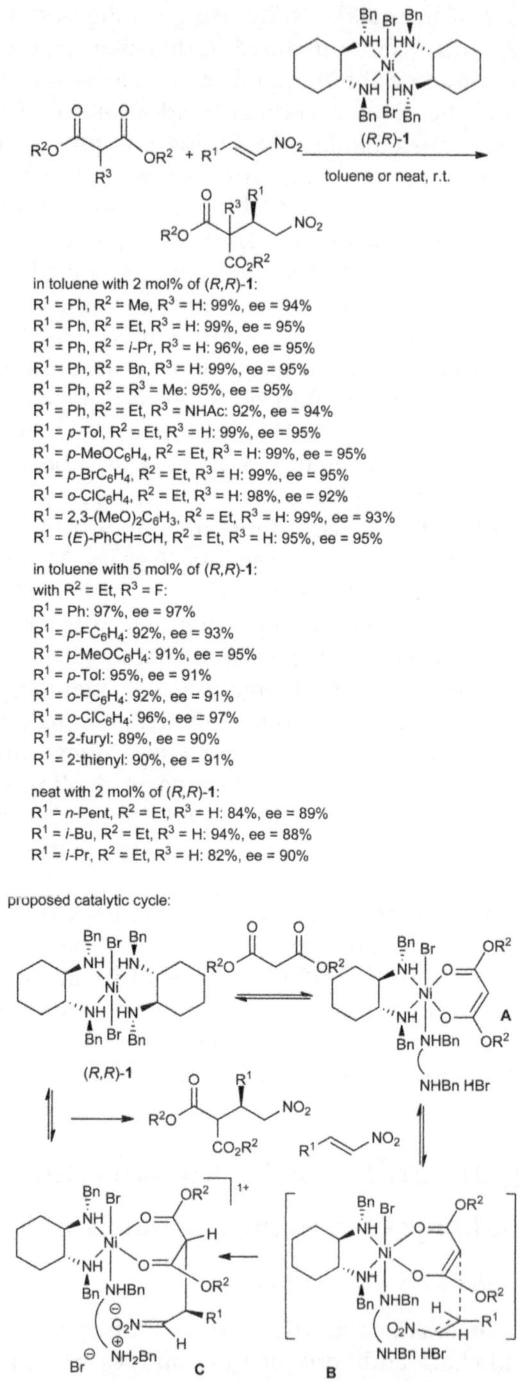

Scheme 2.3 Conjugate addition of malonates to nitroalkenes with a preformed diamine nickel catalyst.

from enantioselective additions of nucleophiles to nitroalkenes[2f,19] represent attractive targets, largely due to the attributes of the nitro group associated with its range of subsequent transformations.[20] To promote enantioselective conjugate additions of 1,3-dicarbonyl compounds to nitroalkenes, Evans employed a novel readily available chiral nickel(II) catalyst, (*R*,*R*)-**1**, consisting of a moderately Lewis acidic metal salt bound to two neutral chiral ligands. The catalyst design was based on the hypothetical catalytic cycle illustrated in Scheme 2.3. It should be basic enough to effect the substrate enolisation, alleviating the need for the addition of ancillary base. After substrate enolisation, one diamine ligand was partially released as its conjugate acid, resulting in the formation of enolate **A**. Nucleophilic attack of **A** on nitrostyrene could proceed through plausible transition structure **B**, where reinforcing steric and electrostatic effects could orient the nitrostyrene moiety. Intramolecular proton transfer from the pendant ammonium ion to the nitronate anion in **C** and subsequent product dissociation complete the catalytic cycle. Therefore, the two amine ligands in the catalyst system each play a distinct role: one serves as a chiral ligand to provide stereoinduction in the addition step, while the other functions as a base for substrate enolisation. Catalyst (*R*,*R*)-**1** was selected as the most efficient among a range of variously *N*-substituted cyclohexanediamine nickel complexes. For example, the authors found that complexes derived from unsubstituted cyclohexanediamine proved to be very poor catalysts (6% ee). When catalyst (*R*,*R*)-**1** was applied to promote the enantioselective conjugate additions of malonates to nitroalkenes, the reaction afforded the corresponding Michael products in remarkable yields and with enantioselectivities of up to 95% ee, as shown in Scheme 2.3. Uniformly high yields and enantioselectivities were obtained for a broad range of substituted and unsubstituted malonates. Variation of the aromatic residue on the nitroalkene was also well tolerated in reaction with diethyl malonate (92–95% ee). While the reactivities of alkyl-substituted nitroalkenes were diminished (82–84% yields) when the reactions were performed in toluene, reasonable reaction rates could be achieved upon performing the reactions under neat conditions, giving rise to products with yields of up to 94% and slightly lower enantioselectivities (88–90% ee). Introduction of a fluorine atom into biologically active compounds often leads to improvement of their biological characteristics due to unique properties of the fluorine atom. In this context, Kang and Kim have studied the conjugate addition of ethyl fluoromalonate (R^3 = F) to nitroalkenes catalysed by chiral nickel catalyst (*R*,*R*)-**1** employed at 5 mol% of catalyst loading in toluene at room temperature.[21] As shown in Scheme 2.3, a wide range of substituted aromatic and heteroaromatic nitroalkenes provided the corresponding chiral α-fluoro-γ-nitro carbonyl products in good to high yields (up to 97%) and with excellent enantioselectivities of 90–97% ee. This method provided an efficient route for the preparation of chiral α-fluoro-γ-nitro carboxylic acid derivatives.

More recently, Klimochkin *et al.* employed the same catalyst, (*R*,*R*)-**1**, at a lower catalyst loading of 1 mol% in the enantioselective conjugate addition of ethyl acetoacetate to nitroalkenes to give the corresponding chiral Michael products in only moderate yields but with generally high diastereoselectivities of 86–98% de, as shown in Scheme 2.4.[22] The utility of these products

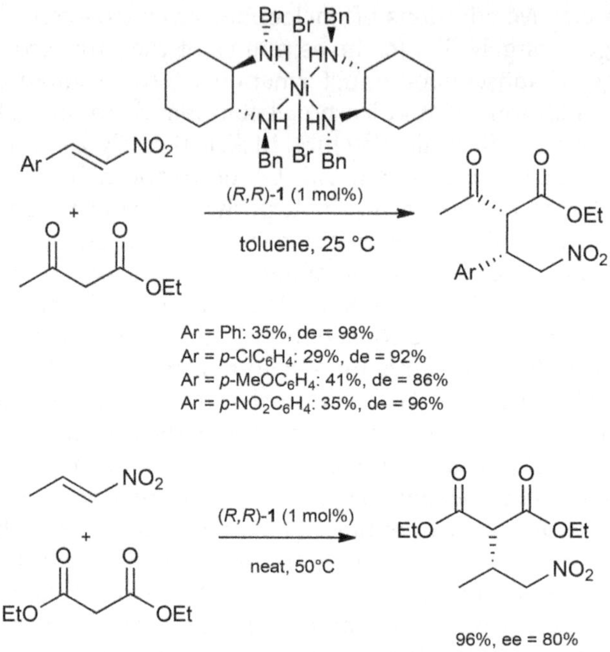

Scheme 2.4 Conjugate additions of ethyl acetoacetate to nitroalkenes and ethyl malonate to 1-nitropropene with a preformed diamine nickel catalyst.

was demonstrated by their conversion into chiral polysubstituted cyclohexanes through further domino Michael/intramolecular aldol reactions with cinnamaldehyde. In addition, the authors applied the same quantity of catalyst (*R*,*R*)-**1** to promote the addition of ethyl malonate to 1-nitropropene, which afforded at 50 °C the corresponding chiral nitro ester in excellent yield (96%) and with good enantioselectivity of 80% ee (Scheme 2.4). This product was further converted into (3*R*)-4-amino-3-methylbutanoic acid, which is a synthetic precursor of dipeptidyl peptidase-IV inhibitors that are promising for the treatment of type II diabetes mellitus.

Earlier in 2007, Evans *et al.* extended the scope of their early process (see Scheme 2.3) to the use of other nucleophiles such as β-keto esters and 1,3-diketones.[23] When β-keto esters were employed, excellent yields (94–99%) and high enantioselectivities (90–94% ee) at the position β to the nitro group were obtained, regardless of the nature of the two substituents on the β-keto esters. 1,3-Diketones were also applicable to this catalytic system, although the reactions proceeded slower and both yields and enantioselectivities were diminished (84–90% yields, 86–87% ee's). In order to improve these enantioselectivities, analogues of catalyst (*R*,*R*)-**1** were investigated in these reactions, and it was found that nickel(II)-bis[(*R*,*R*)-*N*,*N*′-(di-*p*-bromobenzyl) cyclohexane-1,2-diamine]Br_2 **2** was more effective in some cases with substrates such as 1,3-diketones. As shown in Scheme 2.5, the reactions of acetylacetone and heptane-3,5-dione with nitrostyrene afforded the corresponding products in good yields and slightly improved enantioselectivities of 90 and

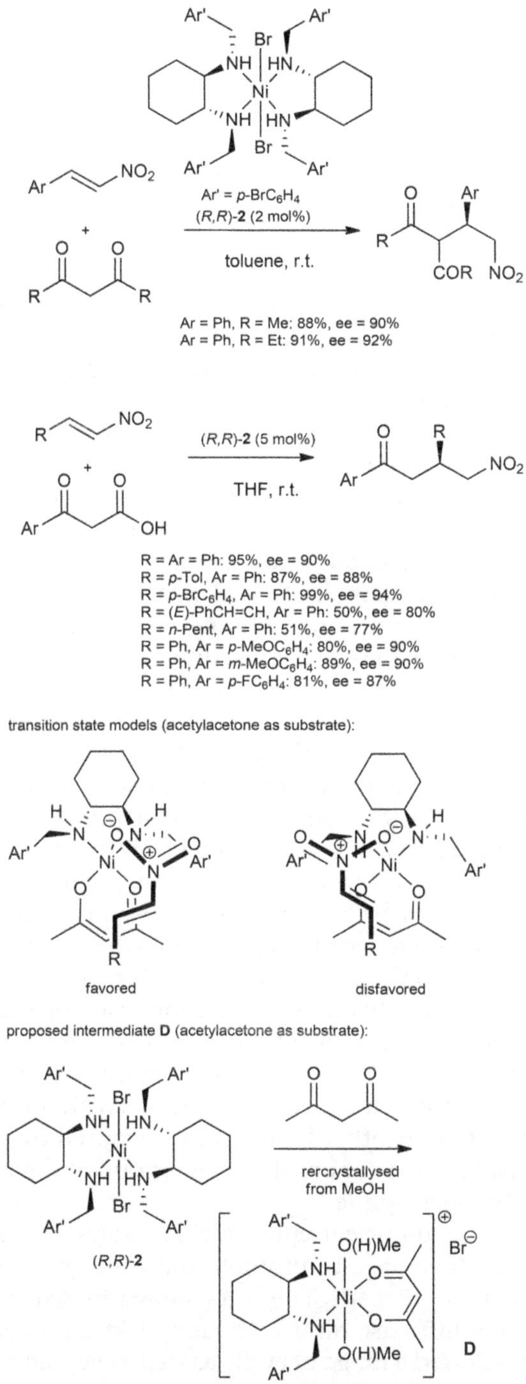

Scheme 2.5 Conjugate additions of 1,3-dicarbonyl compounds to nitroalkenes with a preformed diamine nickel catalyst.

92% ee, respectively, instead of 86 and 87% ee by using catalyst (*R*,*R*)-**1**. Moreover, the authors applied these conditions to the reaction of β-keto acids. While only rare examples of Lewis acid-catalysed enantioselective decarboxylative aldol reactions have been described, the authors found that the reactions of β-keto acids with various nitroalkenes occurred with decarboxylation under the reaction conditions, providing the corresponding chiral nitro ketones in moderate to excellent yields (50–99%) and with good to high enantioselectivities of 77–94% ee (Scheme 2.5). In this system, nitrostyrenes possessing an electron-deficient β-aryl substituent (R^1 = *p*-BrC_6H_4) yielded the best result (94% ee), while diminished yields and enantioselectivities were observed with *n*-pentyl and (*E*)-cinnamaldehyde-derived nitroalkenes (77 and 80% ee, respectively). Electron-donating groups on the aromatic ring of the β-keto acid were well tolerated (R^2 = *m*- and *p*-$MeOC_6H_4$) since an enantioselectivity of 90% ee was obtained in each case, whereas an electron-withdrawing substituent (R^2 = *p*-FC_6H_4) dramatically reduced the rate of the addition reaction, which required 138 hours instead of 21–80 hours for the other aromatic β-keto acids in reaction with cinnamaldehyde-derived nitroalkenes, giving both a good yield and enantioselectivity of 81 and 87% ee, respectively. In order to gain direct evidence for the existence of proposed intermediate **D**, the authors prepared single crystals of the combination of catalyst (*R*,*R*)-**2** with acetylacetone as substrate, which was subjected to X-ray analysis. The ORTEP diagram of this complex showed that the nickel exhibited octahedral geometry, with the chiral diamine ligand and enolate occupying the equatorial plane of the complex. Molecules of methanol used for the recrystallisation occupied each apical position and one bromide ion acted as a counterion. As the conditions in which the complex was generated mirror the reaction conditions, it seemed likely that only one diamine ligand was used during the reaction. It was presumed that the apical positions were occupied either by bromide ions or water, or one of each, under the reaction conditions. The crystal structure of the complex could be used to rationalise the sense of the stereoinduction observed in the addition reaction. The authors assumed that in the transition state the incipient nitronate anion was stabilised by interaction at the open apical position on the nickel. In the disfavoured transition state (pro *R* addition, Scheme 2.5), the nitro moiety of the electrophile faced steric interactions with the *N*-benzyl-derived group of the ligand. On the other hand, in the favoured transition state (pro *S* addition, Scheme 2.5), the *N*-benzyl-derived group of the ligand was orientated away from the electrophile.

On the other hand, a dinuclear chiral nickel catalyst, (*R*)-**3**, was developed by Mitsunuma and Matsunaga to promote the conjugate addition of α-substituted β-keto esters to nitroethylene.[24] As shown in Scheme 2.6 the use of 1–10 mol% of this Schiff base dinuclear catalyst in a mixture of EtOAc and toluene as solvent allowed a range of cyclic as well as acyclic β-keto esters to be added to nitroethylene, providing the corresponding chiral Michael products in moderate to almost quantitative yields (73–92%), combined with good to high enantioselectivities of up to 98% ee in the case of cyclic substrates, and

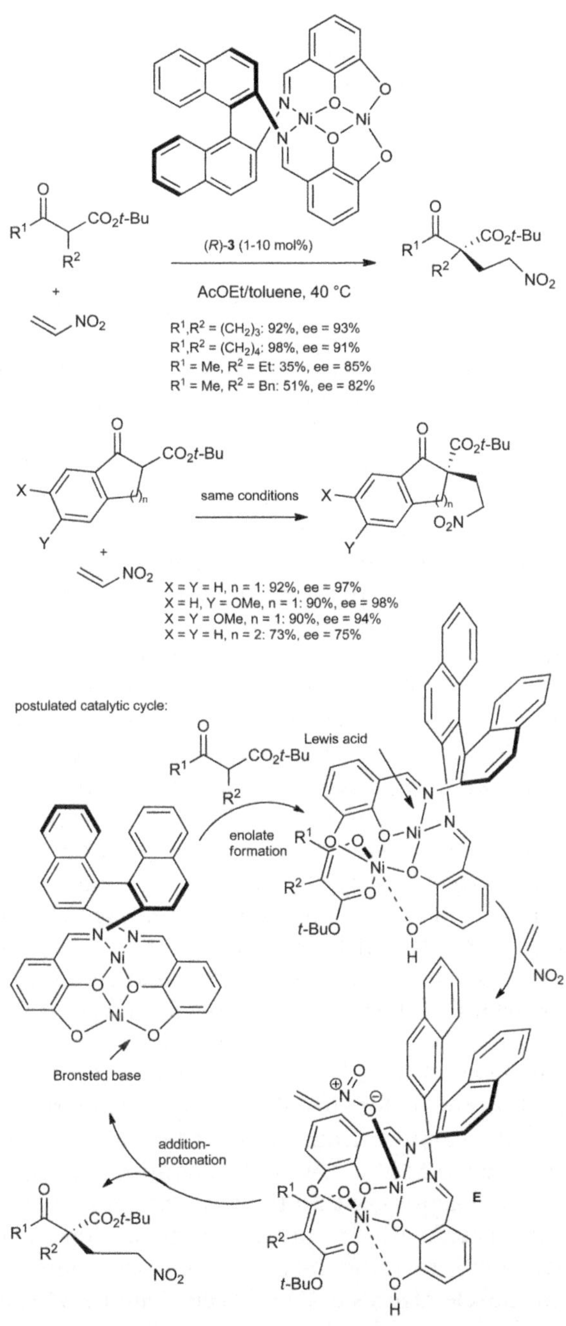

Scheme 2.6 Conjugate additions of α-substituted β-keto esters to nitroethylene with a preformed dinuclear nickel catalyst.

35–98% yields combined with enantioselectivities of 82–93% ee for the acyclic substrates. One advantage of the process was the formation of a quaternary carbon stereocentre adjacent to an ester group. The authors have proposed the mechanism depicted in Scheme 2.6 in which the two nickel centres functioned cooperatively. One of the Ni–O bonds in the outer O_2O_2 cavity was speculated to work as a Brønsted base to generate an Ni–enolate *in situ*. The other nickel in the inner N_2O_2 cavity functioned as a Lewis acid to control the position of nitroethylene, similar to conventional metal–Salen Lewis acid catalysis. The C–C bond formation *via* transition state **E**, followed by protonation, afforded the final product and regenerated the dinuclear catalyst.

In 2011, another preformed nickel catalyst **4** with a chiral diamine ligand incorporating a 3,3′-disubstituted 1,1′-bi(tetrahydroisoquinoline) scaffold was successfully employed by Czekelius *et al.* in the enantioselective addition of ethyl and methyl malonates to various substituted nitroalkenes, providing the corresponding chiral Michael products in remarkable general yields (92–99%) and enantioselectivities (91–99% ee) even at elevated temperature (80 °C), as shown in Scheme 2.7.[25] The choice of the appropriate base (*N*-methylmorpholine) as additive used at a catalytic amount proved to be essential for achieving the highest selectivity. Structural analysis of nickel complex **4** permitted insights into the putative catalyst structure and revealed both aromatic π-stacking interactions as well as close bromine–bromine interactions. The authors demonstrated that using a sterically more demanding complex (*p*-HIPT-C_6H_4 instead of *p*-BrC_6H_4) did not result in any improvement in the stereoselectivity of the reaction. *In situ* generated chiral nickel catalysts have also been applied to induce chirality in these reactions. As a recent example, Huang and Xia have designed new chiral diamine biisoindolines with a rigid backbone which were prepared through a rapid and reliable procedure based on a diaza-Cope rearrangement reaction with chiral 1,2-diamino-1,2-bis(2-hydroxyphenyl)ethane as starting material for the first time.[26] Among them, chiral biisoindoline **5** was found as the most efficient ligand for $NiBr_2$ to promote the nickel-catalysed conjugate addition of malonates to nitroalkenes, providing high yields and good enantioselectivities of up to 90% ee, as shown in Scheme 2.7. The reaction was performed in MTBE as solvent and in the presence of 1,2,2,6,6-pentamethylpiperidine as an additive. The reactivity and enantioselectivity were found to be dependent on the steric demands of the malonates. For example, when the alkyl ester group of the malonate was changed from methyl to *tert*-butyl, the required reaction time was increased to 24 hours, with the enantioselectivity increased from 78 to 87% ee in the reactions with nitrostyrene. The most reactive di-*tert*-butyl malonate was reacted with a number of electron-rich and electron-poor aromatic nitroalkenes, providing the corresponding products in excellent yields and with enantioselectivities of 79–90% ee (Scheme 2.7).

This type of reaction was also investigated by Liu and Li using a mesoporous organosilica chiral nickel catalyst.[27] This functionalised mesoporous organosilica, with chiral cyclohexanediamine-based nickel(II) complex **6** incorporated within the silica framework, was prepared through

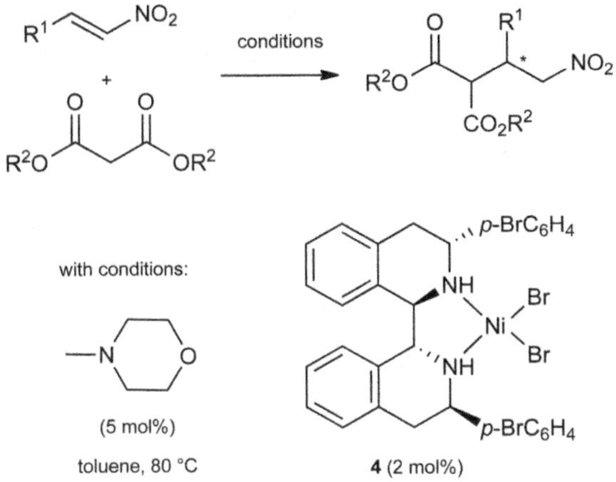

R^1 = Ph, R^2 = Et: 97%, ee = 97% (*R*)
R^1 = Ph, R^2 = Me: 97%, ee = 97% (*R*)
R^1 = *p*-$MeOC_6H_4$, R^2 = Et: 94%, ee = 95% (*R*)
R^1 = *p*-ClC_6H_4, R^2 = Et: 99%, ee = 97% (*R*)
R^1 = *o*-ClC_6H_4, R^2 = Et: 97%, ee = 99% (*R*)
R^1 = 2,4-$(OMe)_2C_6H_3$, R^2 = Et: 95%, ee = 91% (*R*)
R^1 = 2-furyl, R^2 = Et: 98%, ee = 97% (*R*)
R^1 = (*E*)-PhCH=CH, R^2 = Et: 98%, ee = 92% (*R*)
R^1 = 1-Naph, R^2 = Et: 92%, ee = 97% (*R*)

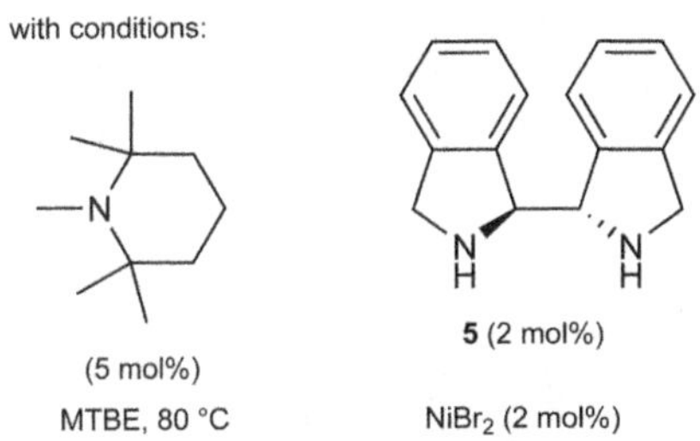

R^1 = Ph, R^2 = Et: 99%, ee = 80% (*S*)
R^1 = Ph, R^2 = Me: 97%, ee = 78% (*S*)
R^1 = Ph, R^2 = Bn: 99%, ee = 72% (*S*)
R^1 = Ph, R^2 = *t*-Bu: 96%, ee = 87% (*S*)
R^1 = *p*-$MeOC_6H_4$, R^2 = *t*-Bu: 93%, ee = 87% (*S*)
R^1 = *o*-$MeOC_6H_4$, R^2 = *t*-Bu: 98%, ee = 90% (*S*)
R^1 = 3,4-$(MeO)_2C_6H_3$, R^2 = *t*-Bu: 98%, ee = 87% (*S*)
R^1 = 1-Naph, R^2 = *t*-Bu: 98%, ee = 79% (*S*)
R^1 = (*Z*)-PhCH=CH, R^2 = *t*-Bu: 72%, ee = 85% (*S*)
R^1 = 2-furyl, R^2 = *t*-Bu: 99%, ee = 88% (*S*)

Scheme 2.7 Conjugate addition of malonates to nitroalkenes with a preformed 1,1′-bi(tetrahydroisoquinoline)-type diamine nickel catalyst, and an *in situ* generated nickel catalyst from a biisoindoline ligand.

co-condensation of a (1*R*,2*R*)-cyclohexanediamine-derived silane and a phenyl-bridged silane, followed by complexation of nickel(II) bromide in the presence of (1*R*,2*R*)-*N*,*N*′-dibenzylcyclohexanediamine. When applied to the heterogeneous catalysis of the enantioselective conjugate addition of alkyl malonates to aromatic nitroalkenes, it provided the corresponding chiral products in almost quantitative yields and with high enantioselectivities of up to 95% ee (Scheme 2.8). Excellent results were also achieved when using β-keto esters as donors, since the corresponding products were obtained in comparable yields and with even higher enantioselectivities of up to 97% ee. Importantly, this heterogeneous catalyst could be recovered easily and reused repeatedly for nine times without obviously affecting its enantioselectivity, thus demonstrating good potential for industrial applications.

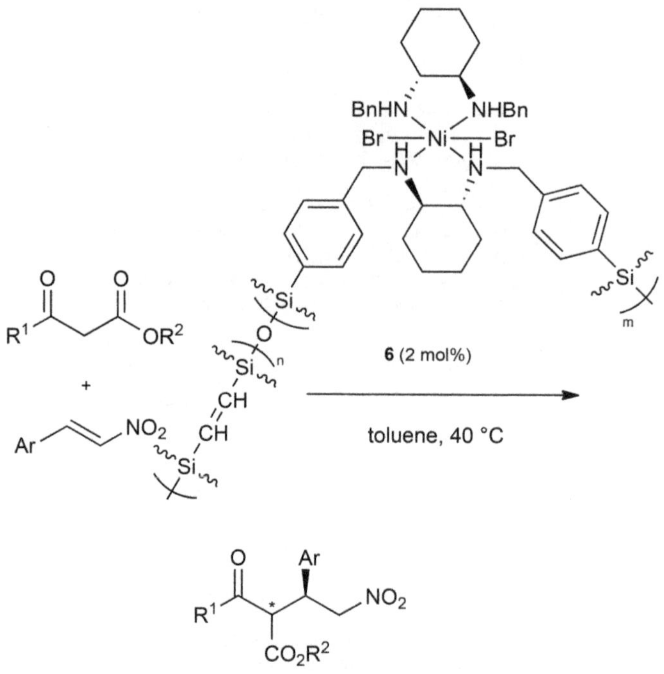

R^1 = OEt, R^2 = Et, Ar = Ph: > 99%, ee = 94%
R^1 = OMe, R^2 = Me, Ar = Ph: > 99%, ee = 95%
R^1 = OBn, R^2 = Bn, Ar = Ph: > 99%, ee = 94%
R^1 = OEt, R^2 = Et, Ar = *p*-$MeOC_6H_4$: 98%, ee = 95%
R^1 = OEt, R^2 = Et, Ar = *p*-ClC_6H_4: 99%, ee = 95%
R^1 = OEt, R^2 = Et, Ar = *p*-Tol: 99%, ee = 95%
R^1 = R^2 = Me, Ar = Ph: > 99%, dr = 1.2:1, ee = 95%
R^1 = Me, R^2 = Et, Ar = Ph: > 99%, dr = 1.3:1, ee = 94%
R^1 = Me, R^2 = *i*-Pr, Ar = Ph: > 99%, dr = 1.5:1, ee = 95%
R^1 = Me, R^2 = *t*-Bu, Ar = Ph: > 99%, dr = 1.7:1, ee = 95%
R^1 = *i*-Pr, R^2 = Et, Ar = Ph: 96%, dr = 1:1, ee = 97%

Scheme 2.8 Conjugate addition of malonates and β-keto esters to nitroalkenes with a mesoporous organosilica nickel catalyst.

2.2.1.2 *Additions to Other Nitroalkenes and Applications in Total Synthesis*

Despite the tremendous amount of work and effort devoted to the development of efficient and versatile Michael reactions, the structure of the electrophile has been often restricted to simple acyclic nitroalkenes, and the enantioselective Michael reaction using cyclic nitroalkenes as the acceptors has been relatively unexplored. In 2010, Chen and Li reported a rare example of enantioselective nickel-catalysed conjugate addition of malonates to readily available cyclic nitroalkenes such as 3-nitro-2*H*-chromenes.[28] This reaction was catalysed by 5 mol% of preformed catalyst (*R*,*R*)-**1** in toluene at 0 °C, affording the corresponding chiral highly functionalised products bearing two adjacent stereogenic centres as single *trans* diastereomers. As shown in Scheme 2.9 these interesting cyclic β-amino acid derivatives were achieved in high yields and with enantioselectivities of up to 95% ee. Other 1,3-dicarbonyl compounds, such as acetylacetone and dibenzoylmethane, were also tested under these conditions; unfortunately, only the racemic corresponding products were obtained in high yields. This novel methodology provided an efficient route for the preparation of chiral cyclic γ-aminobutyric acid derivatives, and the availability of these compounds could facilitate medicinal chemical studies in various fields.

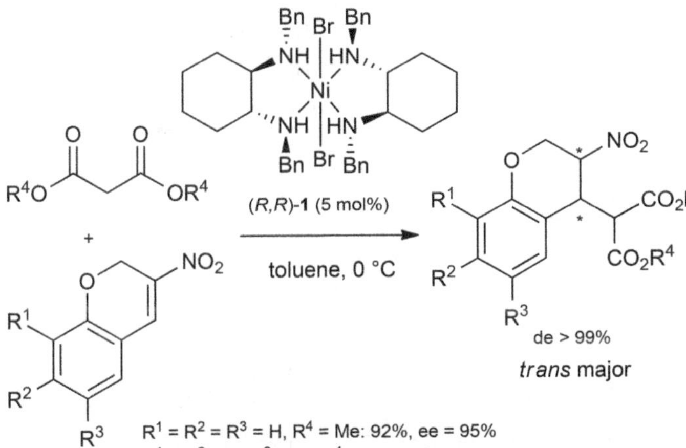

$R^1 = R^2 = R^3 = H$, $R^4 = Me$: 92%, ee = 95%
$R^1 = R^2 = H$, $R^3 = Br$, $R^4 = Me$: 91%, ee = 92%
$R^1 = R^2 = H$, $R^3 = Cl$, $R^4 = Me$: 96%, ee = 92%
$R^1 = R^4 = Me$, $R^2 = R^3 = H$: 88%, ee = 84%
$R^1 = R^2 = H$, $R^3 = R^4 = Me$: 91%, ee = 86%
$R^1 = R^2 = R^3 = H$, $R^4 = Bn$: 88%, ee = 91%
$R^1 = R^2 = H$, $R^3 = Br$, $R^4 = Bn$: 90%, ee = 90%
$R^1 = Me$, $R^2 = R^3 = H$, $R^4 = Bn$: 88%, ee = 92%
$R^1 = R^2 = H$, $R^3 = Me$, $R^4 = Bn$: 91%, ee = 92%

Scheme 2.9 Conjugate addition of malonates to 3-nitro-2*H*-chromenes with a preformed diamine nickel catalyst.

In 2012, Wu *et al.* reported a practical large-scale preparation of the antidepressant drug (*R*)-rolipram from isovanilline, the key step of which was an enantioselective nickel-catalysed conjugate addition of diethyl malonate to a functionalised nitroalkene.[29] Using only 1 mol% of catalyst (*R*,*R*)-**1**, the reaction afforded the corresponding enantiopure key Michael product in almost quantitative yield (Scheme 2.10). The latter was further converted into (*R*)-rolipram through three supplementary steps.

Chiral β-alkynyl acids constitute an important class of pharmaceutical compounds with diverse biological activities; however, their asymmetric synthesis remains a significant challenge. In this context, Peng and Shao have reported the first metal-catalysed asymmetric conjugate addition of malonates to nitroenynes promoted by a simple chiral nickel catalyst, which allowed a novel route to this type of chiral product.[30] As shown in Scheme 2.11 the 1,4-conjugate addition of malonates to nitroenynes catalysed by 2 mol% of preformed diamine nickel catalyst (*R*,*R*)-**2** afforded the corresponding functionalised Michael adducts in both excellent yields and with enantioselectivities of up to >99% ee. Notably, the regioselectivity of the reaction was perfect since no bis-Michael addition by-product was found. A survey of solvents showed that *m*-xylene offered the best enantioselectivities

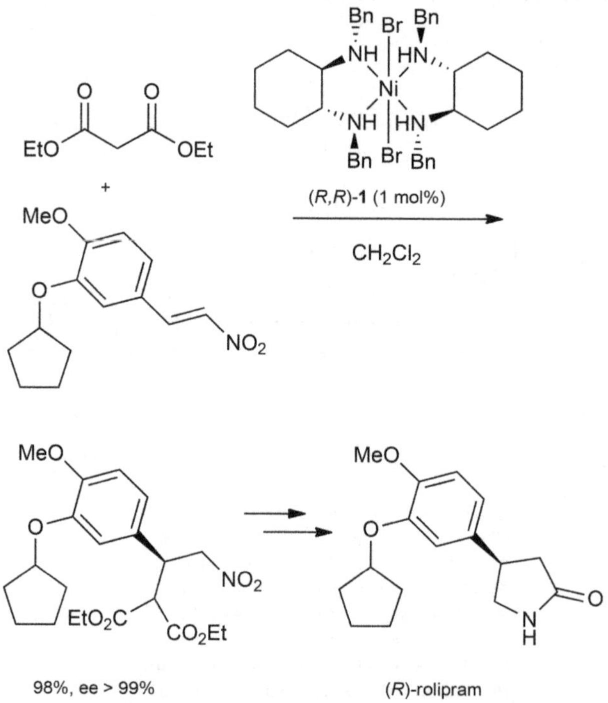

Scheme 2.10 Conjugate addition of ethyl malonate to a functionalised nitroalkene with a preformed diamine nickel catalyst for the synthesis of (*R*)-rolipram.

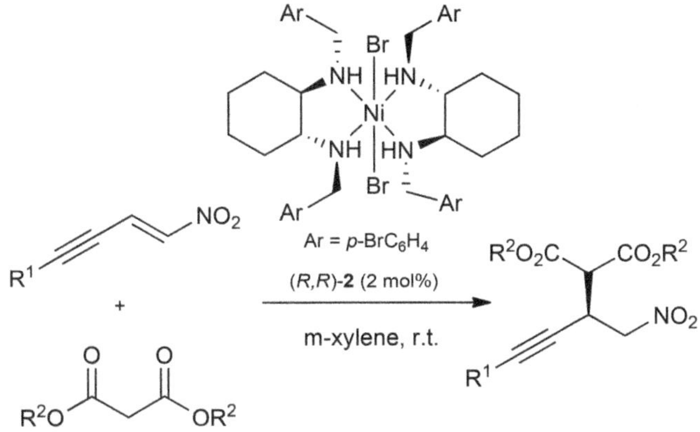

R^1 = Ph, R^2 = *t*-Bu: 81%, ee = 92%
R^1 = Ph, R^2 = Me: 98%, ee = 94%
R^1 = Ph, R^2 = Et: 96%, ee = 93%
R^1 = *p*-$MeOC_6H_4$, R^2 = Et: 95%, ee = 91%
R^1 = *p*-FC_6H_4, R^2 = Et: 95%, ee = 92%
R^1 = *m*-FC_6H_4, R^2 = Et: 88%, ee = 89%
R^1 = *n*-Bu, R^2 = Et: 92%, ee > 99%
R^1 = *n*-Pr, R^2 = Et: 90%, ee > 99%
R^1 = $BnCH_2$, R^2 = Et: 89%, ee = 95%

Scheme 2.11 1,4-Conjugate addition of malonates to nitroenynes with a preformed diamine nickel catalyst.

whereas DMF led interestingly to complete racemisation. The substrate scope was wide since comparable results were achieved with electron-neutral, electron-rich, and electron-deficient arylalkynyl substrates, as well as with alkyl-substituted alkynyl substrates. In addition, enantioselectivities of 90–92% ee at the β-position to the nitro group of the conjugate products combined with good yields (88–90%) were also obtained by using acetylacetone and ethyl 3-oxobutanoate as nucleophilic reagents in reaction with the phenyl-substituted nitroenyne. These products could be further converted by simple decarboxylation, performed in the presence of TsOH under reflux, into the corresponding chiral β-alkynyl acids. The latter were subsequently transformed into δ-keto γ-lactones and γ-alkylidene lactones, which represent two important types of scaffold frequently encountered as structural subunits in numerous biologically active products.

Very recently, the same authors reported the first catalytic asymmetric approach to octahydroindolones, which was based on the enantioselective nickel-catalysed conjugate addition of *tert*-butyl malonate to a nitrodienyne.[31] In the presence of a chiral catalyst generated *in situ* from $NiBr_2$ and chiral diamine 7, the 1,4-conjugate addition proceeded smoothly to provide the corresponding enantioenriched 1,3-enyne in 91% yield and with 93% ee, as shown in Scheme 2.12. It was worth noting that 1,6- or 1,8-addition was not observed. The formed functionalised product could be further decarboxylated and then cyclised into the corresponding cyclohexanone, which was

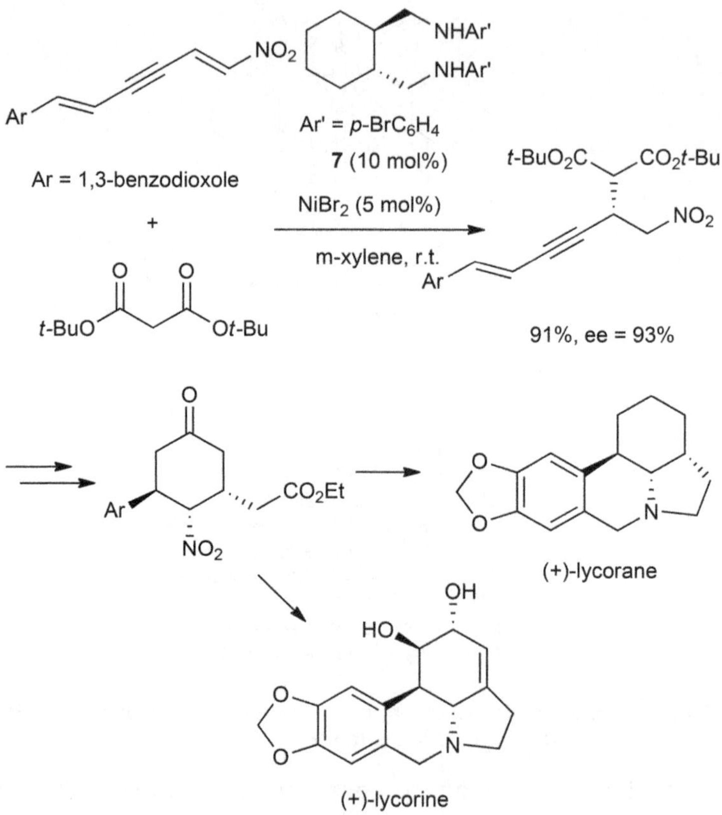

Scheme 2.12 1,4-Conjugate addition of *tert*-butyl malonate to a nitrodienyne with an *in situ* generated diamine nickel catalyst for the synthesis of (+)-lycorine and (+)-lycorane.

transformed into chiral octahydroindolones from which were synthesised the perhydroindole alkaloids (+)-lycorine and (+)-lycorane. It must be noted that the key enantioselective nickel-catalysed conjugate addition of malonates to nitrodienynes was found to be a general method for the synthesis of functionalised 1,3-enynes. Indeed, the reaction was compatible with a range of electron-rich and electron-deficient aryl-substituted nitrodienynes, as well as heteroaromatic and even alkyl-substituted nitrodienynes. Moreover, in addition to *tert*-butyl and ethyl malonates, a range of β-keto esters was suitable for this remarkable process, providing both high yields (88–94%) and enantioselectivities (90–94% ee) under the same conditions.[32]

2.2.2 Other Nucleophiles

In addition to 1,3-dicarbonyl compounds, a range of other nucleophiles has been successfully enantioselectively added to nitroalkenes. For example, Matsunaga and Shibasaki have used the vinylogous nucleophilicity of an α,β-unsaturated γ-butyrolactam to develop its enantioselective Michael

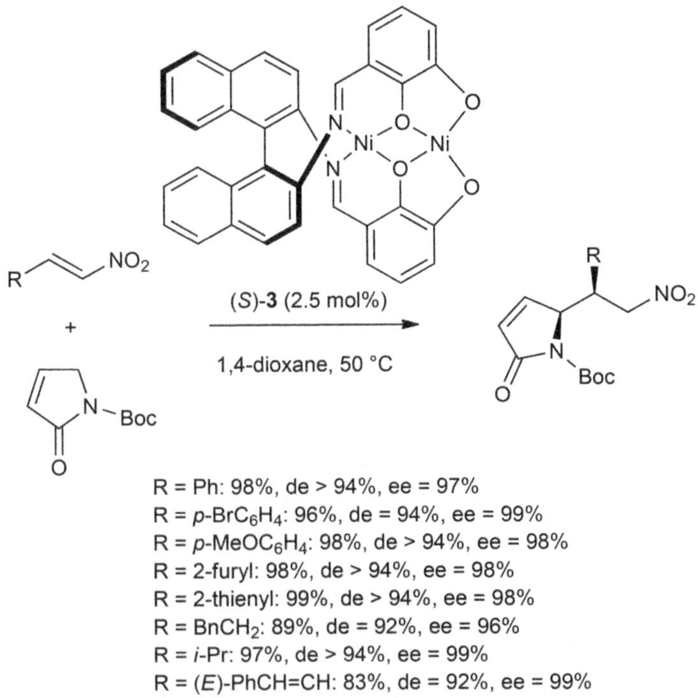

Scheme 2.13 Conjugate addition of a α,β-unsaturated γ-butyrolactam to nitroalkenes with a preformed dinuclear nickel catalyst.

addition to nitroalkenes under bimetallic Schiff base catalysis.[33] As shown in Scheme 2.13 the conjugate addition of this γ-butyrolactam to aryl-, heteroaryl-, and alkyl-substituted nitroalkenes proceeds smoothly in the presence of 2.5 mol% of dinuclear chiral catalyst (*S*)-**3** to give the corresponding γ-butyrolactams in excellent yields, with diastereoselectivities of up to >94% de and enantioselectivities of up to 99% ee. Moreover, a nitrodiene was also found compatible with the process since the corresponding 1,4-adduct was predominantly obtained in 83% yield, 92% de, and 99% ee, as shown in Scheme 2.13. The importance of this novel direct and remarkably efficient catalytic vinylogous Michael process is related in part to the fact that chiral butyrolactams are ubiquitous heterocyclic structural motifs found in many natural products and biologically active compounds.

In biosynthesis, pyruvic acid, a representative 1,2-dicarbonyl compound, is used as a key C_2 and C_3 donor unit. The use of related 1,2-dicarbonyl compounds, such as α-keto esters and α-keto anilides, as nucleophiles in catalytic asymmetric synthesis, however, is rather limited due to their high reactivity as electrophiles. Chemoselective activation of 1,2-dicarbonyl compounds as nucleophiles is required to avoid undesired self-condensation reactions of 1,2-dicarbonyl compounds. Applications of 1,2-dicarbonyl compounds as donors in asymmetric Michael reactions remained unsolved until a recent report by Sodeoka *et al.*[34] Indeed, these authors have described the first example of a diastereo- and enantioselective conjugate addition of α-keto esters to nitroalkenes with a broad generality, which employed a chiral cyclohexanediamine as ligand. This simple ligand **8** was selected as the most effective among other ligands investigated, such as BINAP, a chiral bisoxazoline, and an acyclic chiral diamine. As shown in Scheme 2.14, using 1 mol% of ligand **8** in combination with $Ni(OAc)_2$ allowed the conjugate additions of a broad range of α-keto esters to nitroalkenes to be achieved in good to high yields (up to 93%), remarkable *trans* diastereoselectivities of >94% de in almost all cases of substrates studied, along with high enantioselectivities of up to 94% ee. The process was performed in an environmentally friendly solvent such as isopropanol, and in the presence of a catalytic amount of triethylamine. Since the reaction conditions were mild, acid- or base-sensitive functional groups were well-tolerated, and even an unprotected hydroxyl group could be used without difficulty, providing the corresponding product in 70% yield, >94% de, and 93% ee.

On the other hand, Matsunaga and Shibasaki have observed an opposite *syn* diastereoselectivity in the enantioselective conjugate addition of α-keto anilides to nitroalkenes under dinuclear nickel catalysis.[35] Indeed, the use of 10 mol% of dinuclear chiral Schiff base nickel catalyst **9** in the presence of HFIP and 5 Å MS as additives in 1,4-dioxane as solvent allowed the corresponding Michael adducts to be achieved in moderate to good yields, with good *syn* diastereoselectivities of up to >90% de and combined with good to excellent enantioselectivities of up to 98% ee (Scheme 2.15). The substrate

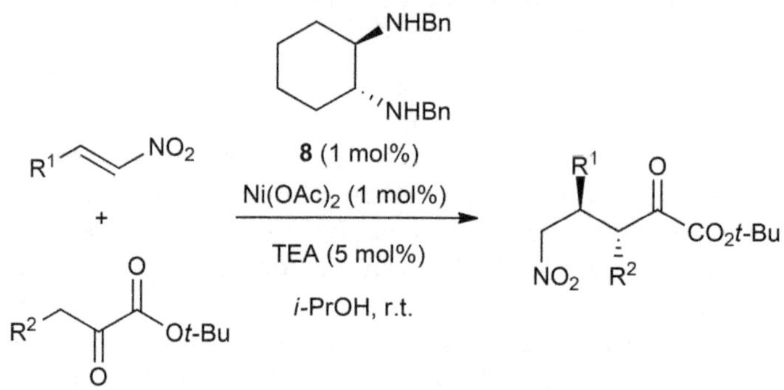

R^1 = Ph, R^2 = Me: 77%, de > 94%, ee = 94%
R^1 = Ph, R^2 = Cy: 60%, de > 94%, ee = 93%
R^1 = Ph, R^2 = $(CH_2)_2OH$: 84%, de > 94%, ee = 92%
R^1 = Ph, R^2 = $(CH_2)_2OBn$: 93%, de > 94%, ee = 93%
R^1 = $(CH_2)_2OTBS$, R^2 = Me: 70%, de > 94%, ee = 94%
R^1 = *p*-BrC_6H_4, R^2 = Bn: 80%, de > 94%, ee = 92%
R^1 = *p*-$MeOC_6H_4$, R^2 = Bn: 81%, de > 94%, ee = 90%
R^1 = $(CH_2)_2OH$, R^2 = Bn: 70%, de > 94%, ee = 93%
R^1 = CON(Me)OMe, R^2 = $(CH_2)_2OBn$: 89%, de > 94%, ee = 92%

Scheme 2.14 Conjugate addition of α-keto esters to nitroalkenes with an *in situ* generated diamine nickel catalyst.

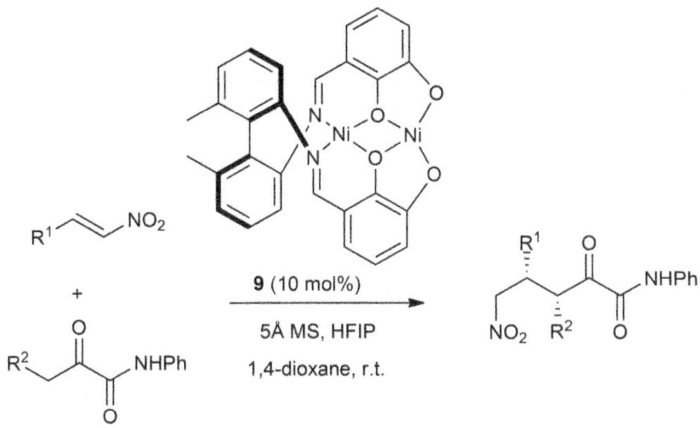

R^1 = Ph, R^2 = Me: 71%, de = 86%, ee = 90%
R^1 = *p*-$MeOC_6H_4$, R^2 = Me: 92%, de > 90%, ee = 98%
R^1 = *m*-$MeOC_6H_4$, R^2 = Me: 63%, de = 82%, ee = 85%
R^1 = 2-thienyl, R^2 = Me: 82%, de = 82%, ee = 72%
R^1 = (*E*)-PhCH=CH, R^2 = Me: 70%, de > 90%, ee = 90%
R^1 = $BnCH_2$, R^2 = Me: 60%, de = 82%, ee = 92%
R^1 = *p*-$MeOC_6H_4$, R^2 = Et: 74%, de > 90%, ee = 92%
R^1 = *p*-$MeOC_6H_4$, R^2 = Bn: 61%, de > 90%, ee = 83%

postulated catalytic cycle:

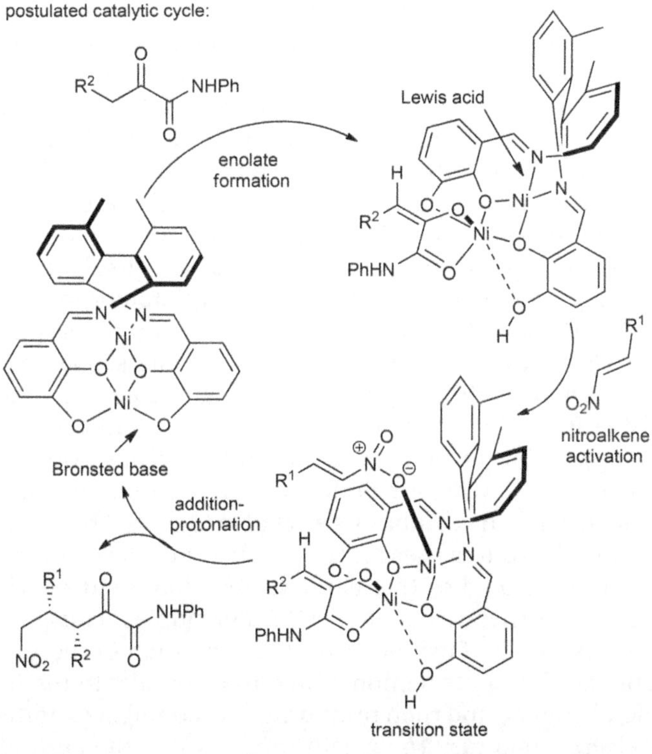

Scheme 2.15 Conjugate addition of α-keto anilides to nitroalkenes with a preformed dinuclear nickel catalyst.

scope showed that a broad range of nitroalkenes was tolerated. Nitrostyrenes with either an electron-withdrawing or electron-donating substituent on the aromatic ring gave products with good to excellent results, while a 2-thienyl-substituted nitroalkene provided a slightly decreased enantioselectivity (72% ee). A nitrodiene predominantly afforded the 1,4-adduct in high selectivities (>90% de and 90% ee) and a β-alkyl-substituted nitroalkene ($R^1 = BnCH_2$) was also applicable, giving 82% de and 92% ee. The authors assumed that the two nickel centres functioned cooperatively. The postulated reaction mechanism depicted in Scheme 2.15 shows that one of the Ni–O bonds in the outer O_2O_2 cavity was speculated to work as a Brønsted base to generate an Ni–enolate *in situ*. The other Ni in the inner N_2O_2 cavity functioned as a Lewis acid to control the position of the nitroalkene, similar to conventional metal–Salen Lewis acid catalysis. The C–C bond formation *via* the transition state, followed by protonation, afforded the *syn* adduct and regenerated the dinuclear catalyst.

3,3′-Disubstituted oxindoles with β-amino functionality constitute a key structural feature of several classes of pharmaceuticals and natural products and are extremely versatile building blocks that can undergo synthetically useful transformations. In view of the high nucleophilicity of the oxindole 3-position, the catalytic asymmetric conjugate addition of 3-substituted oxindoles to nitroalkenes represents one of the most powerful and straightforward approaches toward chiral 3,3′-disubstituted oxindoles with β-amino functionality. Among the rare examples describing these reactions is a nickel-catalysed reaction reported by Yuan *et al.* in 2011.[36] It involves a simple catalyst system assembled from chiral diamine ligand **10** and $Ni(OAc)_2$, which efficiently generated chiral nickel enolates derived from 3-substituted oxindoles bearing an N-1 carbonyl group such as an *N*-ethoxycarbonyl function (Scheme 2.16). The enolates smoothly underwent diastereo- and enantioselective conjugate addition to a wide range of nitroalkenes under mild reaction conditions, furnishing the corresponding chiral 3,3′-disubstituted oxindole products bearing two vicinal quaternary/tertiary stereocentres in good to high yields (74–95%), with diastereoselectivities of up to 98% de and along with enantioselectivities of 71–97% ee. The substrate scope was wide since the nitroalkenes with both electron-withdrawing and electron-donating aromatic substituents at the β-position gave excellent results. Additionally, the process was applicable to heteroaromatic nitroalkenes. Moreover, similar good results were achieved for oxindoles with different substituents at the C-3 and C-5 positions. The conditions were also tolerated by *N*-Boc-oxindoles and *N*-Cbz-oxindoles, but in these cases of substrates the enantioselectivities were not determined. On the other hand, no reaction occurred with aliphatic nitroalkenes, and neither with *N*-H- nor *N*-Bn-oxindoles. Thus, incorporating a carbonyl group on N-1 of oxindoles for the formation of the 1,3-dicarbonyl framework of oxindoles was demonstrated to be crucial to the formation of the chiral metal enolate and then promoting the conjugate addition reaction in a stereoselective manner. To explain their results, the authors proposed that in the transition state the nitronate anion was stabilised by interaction at the open apical position on the nickel. In the disfavoured transition state depicted in Scheme 2.16, the aromatic group of the nitroalkene faced steric

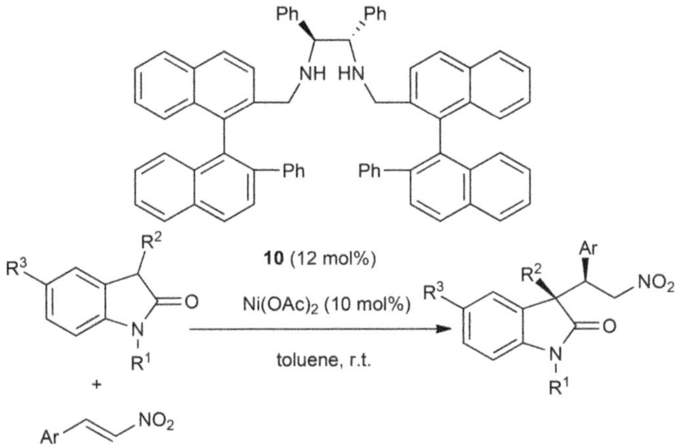

$R^1 = CO_2Et$, R^2 = Me, R^3 = H, Ar = Ph: 91%, de = 98%, ee = 95%
$R^1 = CO_2Et$, R^2 = Me, R^3 = H, Ar = *p*-BrC_6H_4: 94%, de = 98%, ee = 95%
$R^1 = CO_2Et$, R^2 = Me, R^3 = H, Ar = *p*-ClC_6H_4: 93%, de = 98%, ee = 93%
$R^1 = CO_2Et$, R^2 = Me, R^3 = H, Ar = *p*-FC_6H_4: 91%, de = 98%, ee = 95%
$R^1 = CO_2Et$, R^2 = Me, R^3 = H, Ar = *o*-BrC_6H_4: 91%, de = 80%, ee = 95%
$R^1 = CO_2Et$, R^2 = Me, R^3 = H, Ar = *o*-$MeOC_6H_4$: 74%, de = 50%, ee = 71%
$R^1 = CO_2Et$, R^2 = Me, R^3 = H, Ar = 2-furyl: 90%, de = 80%, ee = 83%
$R^1 = CO_2Et$, R^2 = Me, R^3 = H, Ar = 4-thienyl: 91%, de = 98%, ee = 97%
$R^1 = CO_2Et$, R^2 = Bn, R^3 = H, Ar = Ph: 79%, de = 80%, ee = 76%
$R^1 = CO_2Et$, $R^2 = R^3$ = Me, Ar = Ph: 95%, de = 98%, ee = 90%
R^1 = Boc, R^2 = Bn, R^3 = H, Ar = Ph: 74%, de = 20%, ee = not determined
R^1 = Cbz, R^2 = Me, R^3 = H, Ar = Ph: 83%, de = 98%, ee = not determined

transition states:

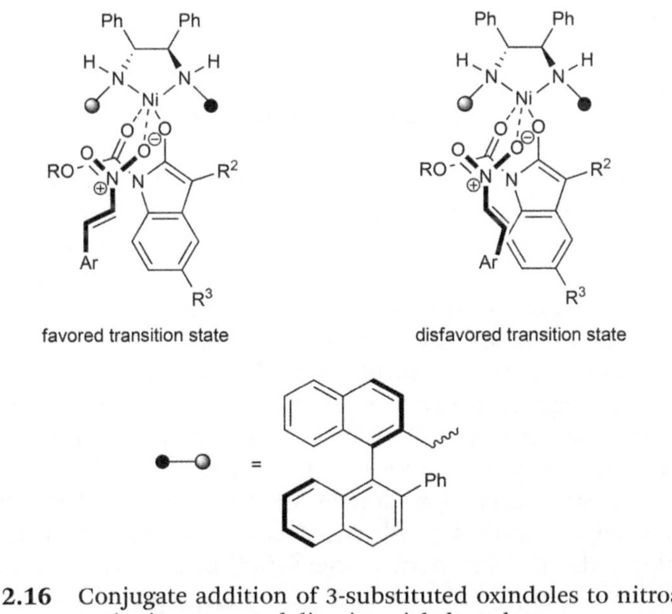

Scheme 2.16 Conjugate addition of 3-substituted oxindoles to nitroalkenes with an *in situ* generated diamine nickel catalyst.

interactions with the bulky binaphthalene moiety of the chiral ligand; thus, the activated metal enolate attacked the Michael acceptor from the *Re* face. On the other hand, in the favoured transition state (Scheme 2.16), the bulky binaphthalene moiety of the chiral ligand was oriented away from the aromatic group of the nitroalkene. Consequently, the activated metal enolate approached the Michael acceptor from its *Si* face, accounting for the absolute configuration of the final product. Additionally, the π–π stacking interaction between the phenyl ring of the 3-substituted oxindole and the aromatic group on the nitroalkene probably had an important effect on the reactivity of the addition reaction, since no reaction was observed when an aliphatic nitroalkene substrate was used.

In addition to these advances, Arai *et al.* have very recently developed enantioselective conjugate additions of 3′-indolyl-3-oxindoles to nitroethylene, enabling a rapid access to chiral indole-containing pyrrolidinoindolines.[37] As shown in Scheme 2.17 the reaction of a range of 3′-indolyl-3-oxindoles with nitroethylene was catalysed by a novel nickel catalyst, generated *in situ* from a chiral imidazoline-aminophenol ligand **11** and $Ni(OAc)_2{\cdot}4H_2O$, in the presence of HFIP as an additive in *o*-xylene at −20 °C, affording the corresponding chiral mixed 3,3′-biindoles in good yields of up to 95% and with enantioselectivities of up to 95% ee. The substrate scope was found to be wide since a range of 3′-indolyl-3-oxindoles containing various substituents at the 5- and 6-positions of the oxindole ring reacted smoothly to yield the corresponding products with good to excellent enantioselectivities. Only the 7-trifluoromethyl-substituted substrate afforded the desired product with a slightly reduced enantioselectivity (44% ee). In addition to *N*-Boc-protected substrates, *N*-methoxycarbonyl- and *N*-Cbz-protected substrates gave good results. Nevertheless, unprotected substrates (R^2 = H) and *N*-acetyl substrates (R^2 = Ac) resulted in no reaction. The authors have proposed a plausible catalytic cycle depicted in Scheme 2.17, beginning with the formation of the $[Ni(L^*)_2]$ complex from $Ni(OAc)_2$ and two equivalents of the chiral ligand (L*). This complex acts as a base to generate the enolate of the 3′-indolyl-3-oxindole. Although this type of dynamic ligand exchange seems unlikely, this process is supported by ESI-MS analysis. It suggests the formation of the $[Ni(L^*)_2]$ complex that allows a reversible process for the exchange of the ligand and the 3′-indolyl-3-oxindole. Then, conjugate addition to nitroethylene occurs in the reaction sphere produced by the Ni–L* complex to give the zwitterionic complex of [Ni(L*)(product)]. Ligand exchange between the anion of the product with 3′-indolyl-3-oxindole gives the final chiral mixed 3,3′-biindole, with regeneration of the enolate. The elimination of the ligand, generated in the exchange with the 3′-indolyl-3-oxindole, contributes to the stabilisation of the catalytically active Ni–L* species.

Azaarenes are structures that appear in numerous biologically active compounds such as natural products, pharmaceuticals, and agrochemicals. Therefore, the development of new methods for the incorporation of azaarenes into compounds or to functionalise preexisting azaarenes is of high value. In this context, Fallan and Lam have developed the first

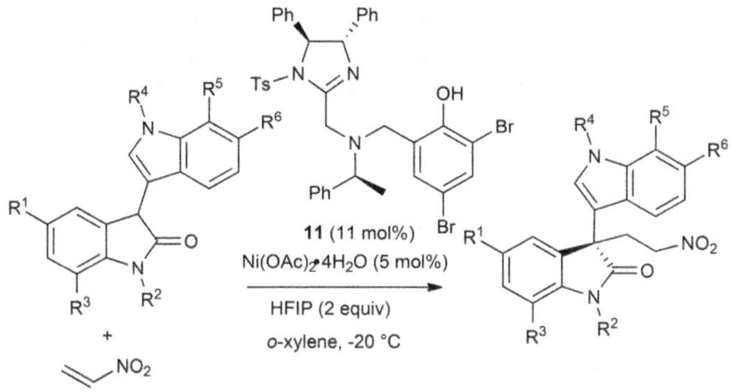

$R^1 = R^4 = Me$, $R^2 = Boc$, $R^3 = R^5 = R^6 = H$: 95%, ee = 90%
$R^1 = Br$, $R^2 = Boc$, $R^3 = R^5 = R^6 = H$, $R^4 = Me$: 95%, ee = 91%
$R^1 = F$, $R^2 = Boc$, $R^3 = R^5 = R^6 = H$, $R^4 = Me$: 76%, ee = 91%
$R^1 = R^4 = Me$, $R^2 = Boc$, $R^3 = R^5 = R^6 = H$: 82%, ee = 93%
$R^1 = OMe$, $R^2 = Boc$, $R^3 = R^5 = R^6 = H$, $R^4 = Me$: 94%, ee = 93%
$R^1 = R^5 = R^6 = H$, $R^2 = Boc$, $R^3 = CF_3$, $R^4 = Me$: 30%, ee = 44%
$R^1 = R^3 = R^5 = R^6 = H$, $R^2 = CO_2Me$, $R^4 = Me$: 85%, ee = 82%
$R^1 = R^3 = R^5 = R^6 = H$, $R^2 = Cbz$, $R^4 = Me$: 85%, ee = 82%
$R^1 = R^4 = R^6 = Me$, $R^2 = Boc$, $R^3 = R^5 = H$: 91%, ee = 92%
$R^1 = R^4 = R^5 = Me$, $R^2 = Boc$, $R^3 = R^6 = H$: 91%, ee = 95%
$R^1 = R^4 = Me$, $R^2 = Boc$, $R^3 = R^6 = H$, $R^5 = I$: 92%, ee = 92%
$R^1 = R^3 = R^4 = R^5 = R^6 = H$, $R^2 = Boc$: 85%, ee = 89%

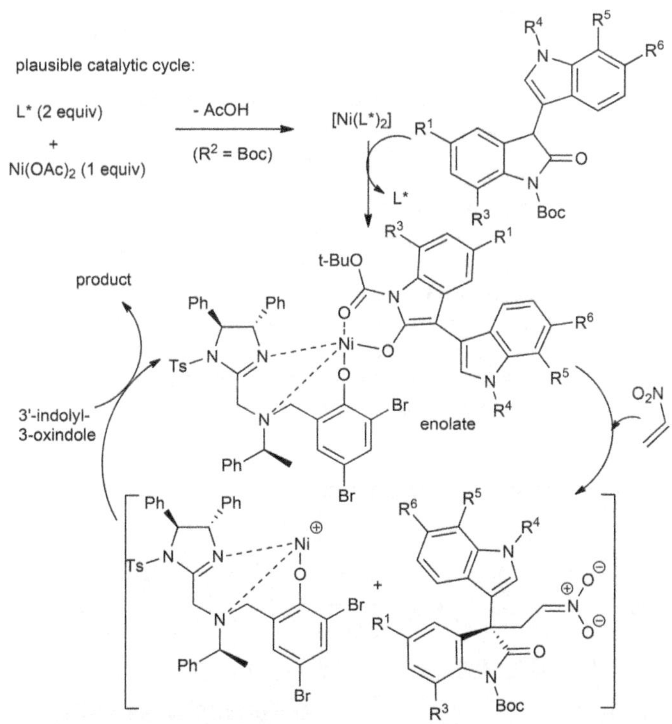

Scheme 2.17 Conjugate addition of 3′-indolyl-3-oxindoles to nitroethylene with an *in situ* generated imidazoline-aminophenol nickel catalyst.

nickel-catalysed enantioselective conjugate additions of azaarylacetates to nitroalkenes.[38] To promote these reactions, the authors employed preformed chiral nickel catalyst (*S*,*S*)-**1** in dioxane at room temperature in the presence of 3 Å MS, which led to the corresponding Michael products as inseparable mixtures of diastereomers due to epimerisation of the stereocentre α to their ester carbonyl group. The products were achieved in good yields and with good to high enantioselectivities of up to 99% ee for the major diastereomers (Scheme 2.18). The substrate scope of the reaction was extended to various β-(hetero)aryl-substituted nitroalkenes and to a range of azaarylacetates, including pyridine, chloropyrazine, dimethoxytriazine, isoquinoline, and quinazoline moieties. The subsequent decarboxylation of some Michael

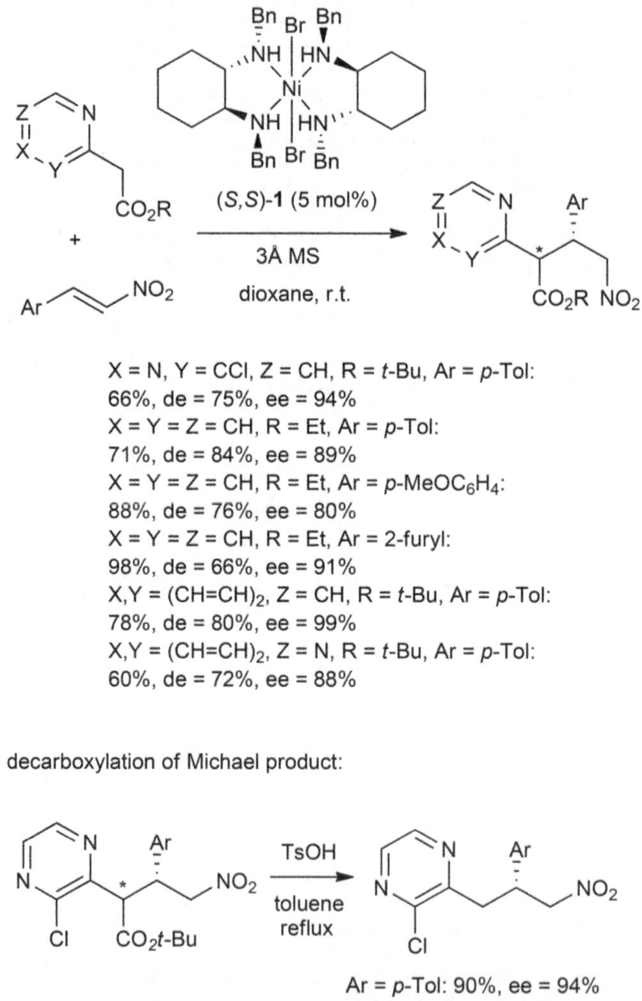

Scheme 2.18 Conjugate addition of azaarylacetates to nitroalkenes with a preformed diamine nickel catalyst.

products was readily achieved with no loss of enantiopurity upon heating with a catalytic amount of TsOH in toluene.

The authors demonstrated that this process was not limited to azaarylacetates by extending its scope to other nucleophiles such as acetamides. In the case of acetamides bearing a benzothiazole group, the corresponding Michael products were configurationally more stable and obtained in good yields, with good diastereoselectivities of up to >90% de and moderate to excellent enantioselectivities of up to >99% ee, as shown in Scheme 2.19. It was shown that when the nitroalkene carried a sterically more demanding *o*-tolyl substituent, the enantioselectivity of the reaction with benzothiazoles was diminished to 66% ee, while all other aryl substituents provided enantioselectivities of >86% ee. In addition, the reaction conditions could be applied to the reaction of other acetamides, such as azaaryl *N,N*-dimethylacetamides containing chloropyrazine, dimethoxytriazine, benzisoxazole, or 5-phenylisoxazole groups to give the corresponding products in generally good yields, with none to good diastereoselectivities and reasonable to high enantioselectivities of up to 83% ee, 93% ee, 88% ee, and 89% ee, respectively (Scheme 2.19). Being compatible with a wide range of azaarenes, including pyridines, pyrazines, triazines, isoquinolines, quinazolines, benzothiazoles, and benzisoxazoles, the novel methodologies depicted in Schemes 2.18 and 2.19 constitute a useful method for the preparation of enantioenriched azaarene-containing building blocks.

In 2013, Simpson and Lam employed 2-acetylazaarenes as nucleophiles in enantioselective nickel-catalysed conjugate additions to arylnitroalkenes.[39] After a survey of chiral metal complexes and reaction conditions, the authors found that the use of a chiral nickel catalyst generated *in situ* from $Ni(OAc)_2 \cdot 4H_2O$ and chiral bisoxazoline ligand **12** in isopropanol as solvent at room temperature allowed the reaction of 2-acetylpyridine with a phenylnitroalkene to be achieved in 86% yield and with enantioselectivity of 96% ee, as shown in Scheme 2.20. Further investigations revealed that, in addition to 2-acetylpyridine, 2-acetylazaarenes containing quinolone, pyrazine, thiazole, benzothiazole, or *N*-methylimidazole groups reacted smoothly with phenylnitroalkenes to provide the corresponding Michael products in good yields of up to 94% and with enantioselectivities of 94–99% ee (some relevant examples are shown in Scheme 2.20). Moreover, various (hetero)aryl-substituted nitroalkenes also provided uniformly high enantioselectivities (92–99% ee), but alkyl-substituted nitroalkenes did not react under the same reaction conditions. However, increasing the concentration of the reaction from 0.1 to 0.5 M with respect to the nitroalkene and using a 1 : 1 mixture of i-PrOH/CH_2Cl_2 as solvent was found to be beneficial, since reasonable yields of products derived from alkyl-substituted nitroalkenes could be produced under these conditions, albeit in moderate enantioselectivities (≤80% ee), as shown in Scheme 2.20.

In 2014, Hamashima and Kan reported the practical total syntheses of acromelic acids A and B, having potent neuro-excitatory activity, which were based on enantioselective nickel-catalysed asymmetric conjugate additions of α-keto esters to nitroalkenes.[40] Indeed, the key steps of the syntheses of these natural

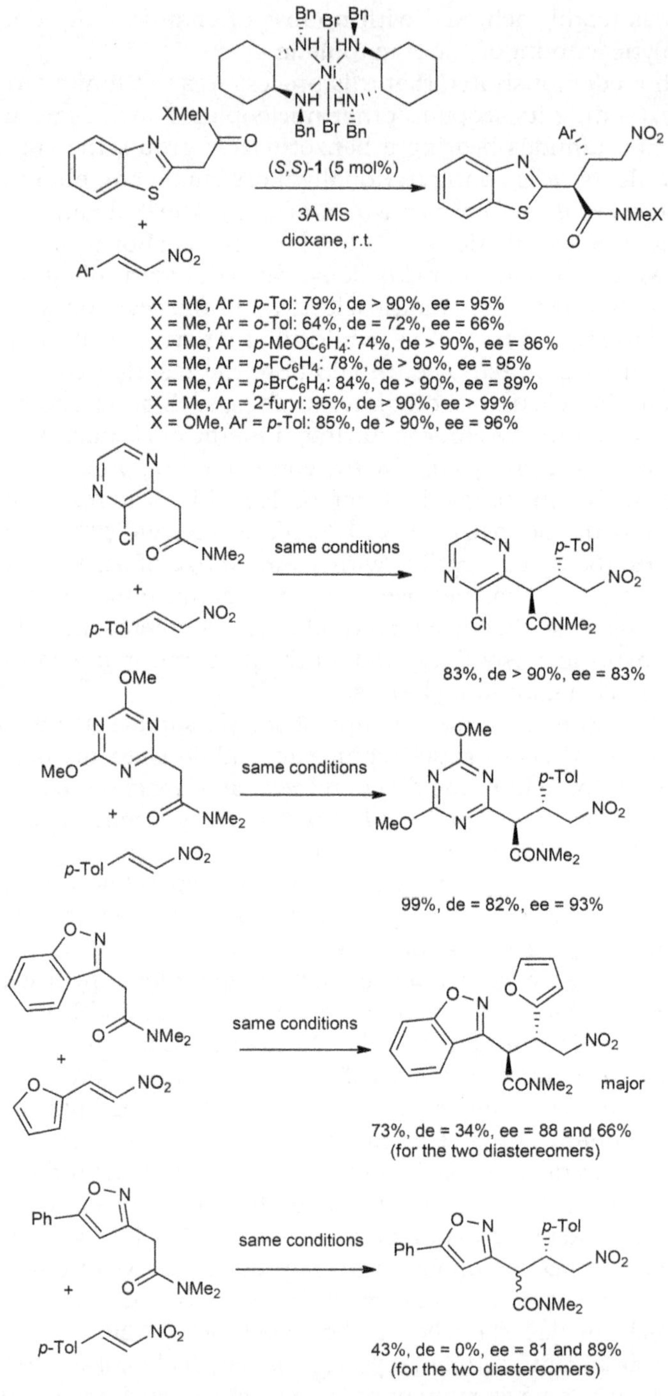

Scheme 2.19 Conjugate additions of acetamides to nitroalkenes with a preformed diamine nickel catalyst.

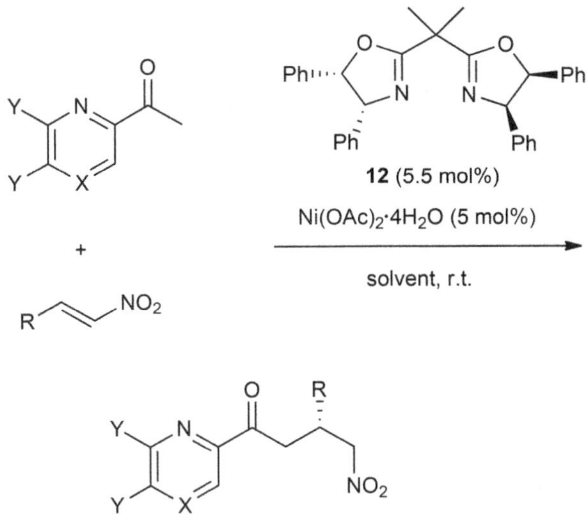

in *i*-PrOH:
X = CH, Y = H, R = Ph: 86%, ee = 96%
X = N, Y = H, R = Ph: 83%, ee = 99%
X = CH, Y,Y = $(CH{=}CH)_2$, R = Ph: 94%, ee = 94%
X = CH, Y = H, R = *p*-$CF_3C_6H_4$: 73%, ee = 92%
X = CH, Y,Y = $(CH{=}CH)_2$, R = *m*-$NO_2C_6H_4$: 62%, ee = 99%
X = CH, Y,Y = $(CH{=}CH)_2$, R = *p*-$MeOC_6H_4$: > 95%, ee = 95%
X = CH, Y = H, R = 2-thienyl: 51%, ee = 93%
X = CH, Y = H, R = Cy: 69%, ee = 65%
X = N, Y = H, R = *o*-Tol: 79%, ee = 98%
X = N, Y = H, R = *p*-FC_6H_4: 91%, ee = 97%
in *i*-PrOH/CH_2Cl_2 (1:1):
X = N, Y = H, R = Cy: 62%, ee = 67%
X = N, Y = H, R = CH_2Bn: 86%, ee = 69%
X = CH, Y,Y = $(CH{=}CH)_2$, R = Cy: 76%, ee = 78%
X = CH, Y,Y = $(CH{=}CH)_2$, R = CH_2Bn: 68%, ee = 80%

Scheme 2.20 Conjugate addition of 2-acetylazaarenes to nitroalkenes with an *in situ* generated bis(oxazoline) nickel catalyst.

products consisted of additions of α-keto esters **13** and **14** to nitroalkenes **15** and **16**, respectively, performed in the presence of 5 mol% of preformed diamine nickel catalyst **17** (Scheme 2.21). The corresponding Michael products **18** and **19** were obtained in 88% and quantitative yields, along with high diastereoselectivities of >92% and >90% de and high enantioselectivities of 95% and 91% ee, respectively. In the case of the synthesis of acromelic acid A starting from α-keto ester **13**, the reaction was performed in DME at −10 °C, whereas the reaction of α-keto ester **14** leading to acromelic acid B was carried out in isopropanol at −10 °C in the presence of triethylamine. Consecutive intramolecular reductive aminations of chiral products **18** and **19** allowed the construction of the pyrrolidine ring of acromelic acids A and B.

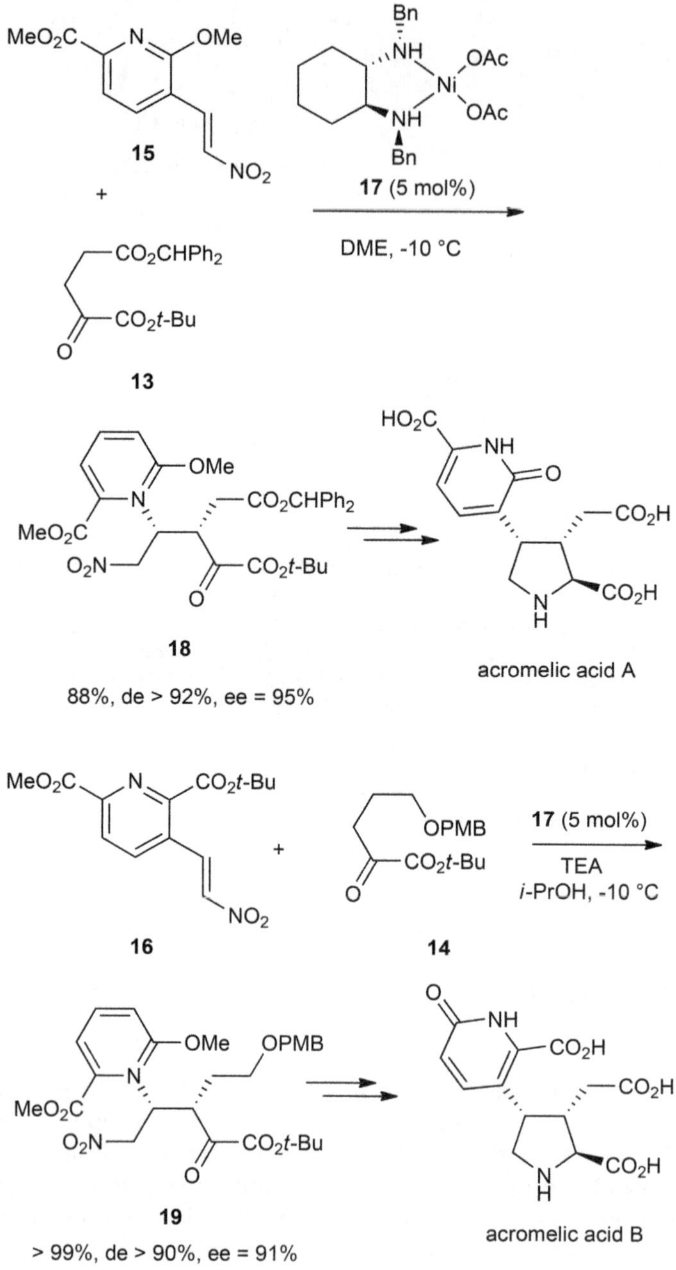

Scheme 2.21 Conjugate additions of α-keto esters to nitroalkenes with a preformed diamine nickel catalyst for the total synthesis of acromelic acids A and B.

2.3 Conjugate Additions to α,β-Unsaturated Carbonyl Compounds

2.3.1 Additions to Enones

Since the first enantioselective conjugate addition to enones catalysed by nickel reported in 1988 by Soai *et al.* using an ephedrine-derived ligand (see Scheme 2.1),[7] many types of chiral ligand have been successfully applied to induce nickel-catalysed conjugate additions of a wide variety of nucleophiles to α,β-unsaturated carbonyl compounds (see Introduction, above).[9–13] For example, Corey and Kwak reported in 2004 the first example of the catalytic asymmetric conjugate addition of TMS-acetylene to a cyclic α,β-unsaturated ketone, which was accomplished using chiral bisoxazoline nickel complex **20** as catalyst (Scheme 2.22).[41] This catalyst, possessing one acetylacetonate and one chiral bisoxazoline ligand, was prepared by reaction of this ligand with $Ni(acac)_2$. The reaction of cyclohex-2-enone with dimethylaluminum TMS-acetylide performed in the presence of 5 mol% of this preformed catalyst in *tert*-butyl alcohol at 0 °C provided the corresponding chiral Michael product in 86% yield and with a good enantioselectivity of 88% ee. The authors proposed the catalytic cycle depicted in Scheme 2.22, which involves the carbometalation of the enone by intermediate **21** as a key step.

Another useful C–C bond-forming reaction is the dialkylzinc addition to enones. In particular, the asymmetric metal-catalysed conjugate addition of organozinc reagents to enones is of great interest for the synthesis of optically active β-substituted carbonyl compounds, and has allowed the successful synthesis of biologically active products.[42] Copper and nickel chiral complexes have been the most widely investigated in these reactions. In the case of nickel, a wide variety of amino alcohols, pyridine, borneol, proline, and pyrrolidine derivatives, among other ligands, have been used after the pioneering studies described by Soai *et al.* in 1988 (see Scheme 2.1).[7] In particular, nitrogen-containing ligands have gained increasing importance in this area in the last decade.[43] For example, enantioselective catalytic addition of diethylzinc to chalcone was achieved in 2001–2002 by the Nayak and Kawanami groups, with enantioselectivity of up to 93% ee achieved by using combinations of nickel(II) acetylacetonate with chiral amino alcohols as ligands.[44] In 2007, Isleyen and Dogan investigated enantioselective diethylzinc conjugate addition to various enones by using chiral ferrocenyl-substituted aziridinylmethanols as ligands of preformed nickel catalysts.[45] The corresponding chiral products were achieved in moderate to good yields and enantioselectivities of up to 80% ee, as shown in Scheme 2.23. In spite of these moderate enantioselectivities, the process presented the advantage to use an easily prepared and recoverable ligand **22** which could be reused without losing its activity (Scheme 2.23). In 2008, Burguete and Luis developed novel α-amino amides derived from natural amino acids, which were further investigated as chiral nickel ligands in enantioselective diethylzinc additions to chalcones, providing the corresponding products in very good yields (93–99%) and with moderate enantioselectivities of up to 84% ee (Scheme 2.23).[46]

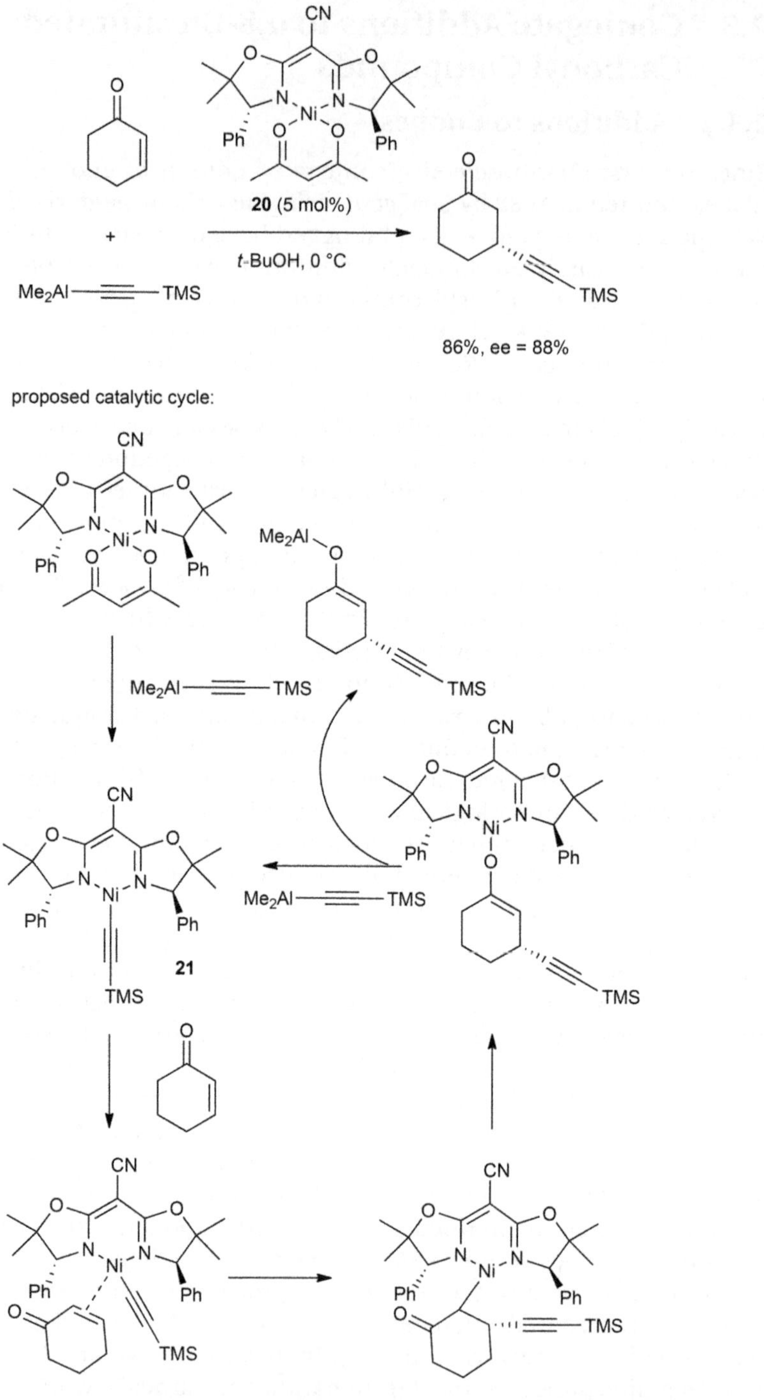

Scheme 2.22 Conjugate addition of dimethylaluminum TMS-acetylide to cyclohex-2-enone with a preformed bisoxazoline nickel catalyst.

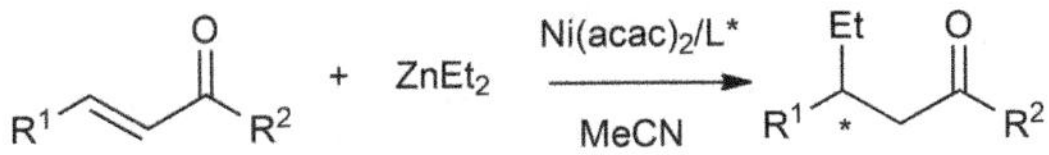

L* = 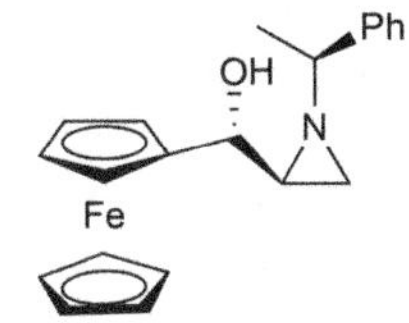

22 (15 or 25 mol%)
$Ni(acac)_2$ (0.5 or 1 mol%)
-35 °C

$R^1 = R^2$ = Ph: 85%, ee = 80%
R^1 = *p*-Tol, R^2 = Ph: 90%, ee = 75%
R^1 = *p*-$MeOC_6H_4$, R^2 = Ph: 75%, ee = 76%
R^1 = *m*-$MeOC_6H_4$, R^2 = Ph: 85%, ee = 80%
R^1 = *o*-ClC_6H_4, R^2 = Ph: 55%, ee = 50%
R^1 = 2-Naph, R^2 = Ph: 57%, ee = 78%
R^1 = Ph, R^2 = Me: 49%, ee = 22%
R^1 = Cy, R^2 = Ph: 45%, ee = 70%

L* = 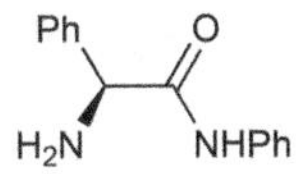

23 (10 mol%)
$Ni(acac)_2$ (5 mol%)
-20 °C

$R^1 = R^2$ = Ph: 97%, ee = 78%
R^1 = *p*-$MeOC_6H_4$, R^2 = Ph: 99%, ee = 84%
R^1 = *p*-ClC_6H_4, R^2 = Ph: 96%, ee = 70%
R^1 = Ph, R^2 = *p*-$MeOC_6H_4$: 95%, ee = 81%
R^1 = Ph, R^2 = *p*-ClC_6H_4: 93%, ee = 67%

L* = 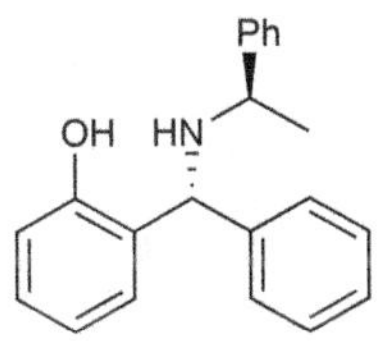

24 (15 mol%)
$Ni(acac)_2$ (1 mol%)
-30 °C

$R^1 = R^2$ = Ph: 92%, ee = 71%
R^1 = Ph, R^2 = *p*-ClC_6H_4: 41%, ee = 63%
R^1 = Ph, R^2 = *p*-$MeOC_6H_4$: 58%, ee = 59%
R^1 = *p*-ClC_6H_4, R^2 = Ph: 65%, ee = 64%

Scheme 2.23 Conjugate addition of diethylzinc to enones with an *in situ* generated ferrocenylaziridine nickel catalyst, an *in situ* generated amino amide nickel catalyst, and an *in situ* generated 1,3-amino alcohol nickel catalyst.

In this study, it was demonstrated that the side chain of the amino acid and the substituents in the amide nitrogen governed the enantioselectivity of the process. It resulted in the selection of phenylglycine-based ligand **23** bearing a phenyl group in the amide moiety as the most efficient ligand. In 2009, moderate enantioselectivities of up to 71% ee were described by Palmieri *et al.* in comparable reactions by using an *in situ* generated 1,3-amino alcohol nickel catalyst.[47] Several chiral 1-(aminoalkyl)naphthols and a 2-(aminoalkyl) phenol were tested as ligands in the diethylzinc conjugate addition to various chalcones, resulting in the selection of chiral 1,3-amino alcohol **24** as the most efficient ligand (Scheme 2.23).

In 2010, chiral tridentate ligand **25**, containing two stereogenic centres, one located on the sulfinyl sulfur atom and the other on the carbon atom in the aziridine moiety, was found by Lesniak and Kielbasinski to be a very efficient nickel ligand for the enantioselective conjugate addition of diethylzinc to chalcones as well as cyclohex-2-enone (Scheme 2.24).[48] The

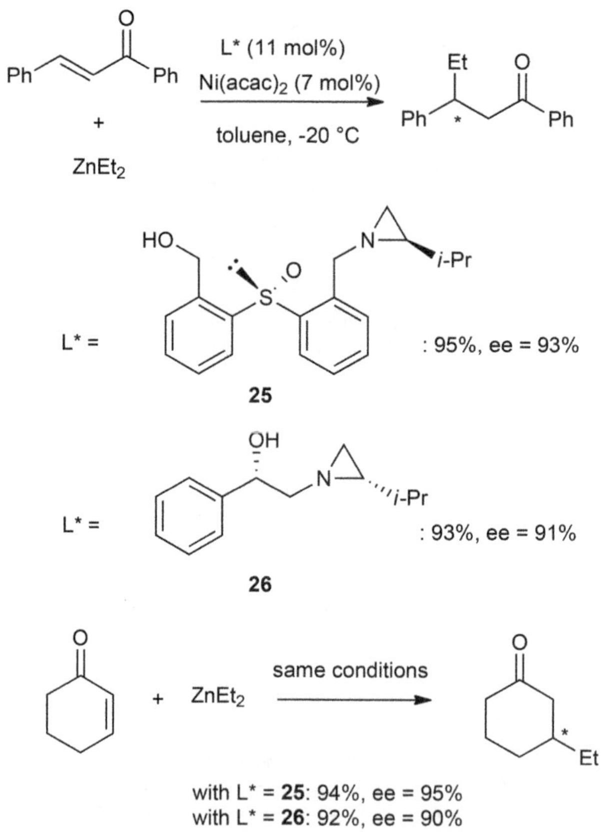

Scheme 2.24 Conjugate additions of diethylzinc to chalcones and cyclohex-2-enone with an *in situ* generated aziridine sulfoxide nickel catalyst and an *in situ* generated aziridine alcohol nickel catalyst.

reactions afforded the corresponding products in very high yields (95% and 94%, respectively) and with high enantioselectivities of 93 and 95% ee, respectively. Studying variously substituted ligands of this type, the authors found a decisive influence of the stereogenic centres located on the sulfinyl sulfur atom and in the aziridine moiety on the stereochemical course of the reactions. Each enantiomer of the products could be obtained by using the easily available diastereomeric ligands. Later, the same authors investigated chiral aziridine alcohol ligands derived from (*S*)-mandelic acid in the same reactions.[49] They found that chiral bidentate ligand **26** containing two stereogenic centres was the most effective in combination with $Ni(acac)_2$ to induce the reactions. High yields (92–93%) and enantioselectivities of 91 and 90% ee were respectively obtained for the diethylzinc conjugate additions to chalcone and cyclohex-2-enone, as shown in Scheme 2.24. As in the previous study, the authors found great influence of the stereogenic centre located at the aziridine moiety on the stereochemical outcome of the reactions.

Several groups have reported on nickel-catalysed asymmetric conjugate additions of organoboron reagents to α,β-unsaturated carbonyl compounds.[50] In 2008, Sieber and Morken successfully developed nickel-catalysed chemo- and enantioselective additions of allylboronic acid pinacol ester [allylB(Pin)] to nonsymmetric dialkylidene ketones.[51] With this aim, a collection of chiral phosphorus ligands was surveyed in the conjugate allylation of nonsymmetric dialkylidene ketones in the presence of $Ni(cod)_2$. Among TADDOL-derived phosphoramidites, phosphonites, and phosphites, TADDOL-derived phosphonite ligand **27** was selected as the most efficient to provide a unique combination of high enantioselection and high chemoselectivity (Scheme 2.25). The reaction exhibited high enantioselectivity of up to 96% ee, regardless of the nature of the arylidene group, and generally favoured chemoselective allylation of the arylidene site. Electron-deficient arenes were found efficient in terms of reaction rate; however, the chemoselectivity was lower for these substrates. On the other hand, electron-rich arenes reacted slower but with better chemoselectivities. Examination of various alkylidene groups revealed a correlation between the alkylidene size and both the stereo- and chemoselectivity. For example, when the pentyl group of the alkylidene was replaced with a methyl group, the chemoselectivity for arylidene allylation increased from 17 : 1 to >20 : 1 (with Ar = Ph); however, a corresponding decrease in the reaction enantioselectivity was noted. When the pentyl group was replaced with a larger cyclohexyl substituent, the opposite outcome was observed since the chemoselectivity was diminished and the enantioselectivity was enhanced (with Ar = Ph). The same outcome was observed as the size of the arylidene substituent was enhanced. Thus, in the series 2-furyl, phenyl, *o*-tolyl, and *o*-(trifluoromethyl)phenyl, the selectivity for the arylidene allylation increased from 5.1 : 1 to 32 : 1 (with R = *n*-Pent, Scheme 2.25). It is important to highlight that this work constituted the first catalytic enantioselective conjugate allylation reactions which could be applied to a range of activated aromatic enones.

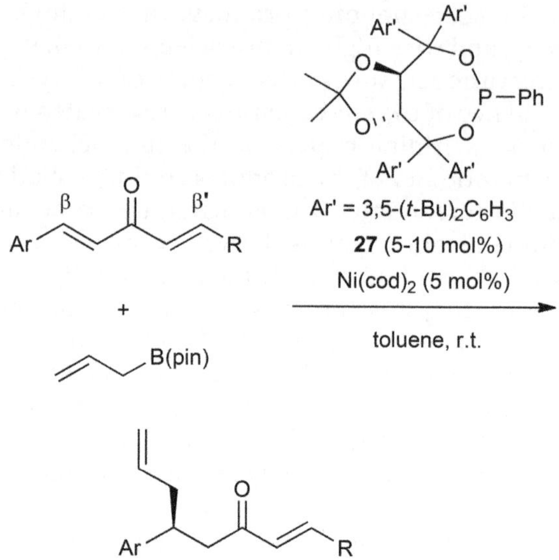

Ar = Ph, R = Cy: 76%, ee = 96%, β/β' = 7:1
Ar = Ph, R = Et: 82%, ee = 83%, β/β' > 20:1
Ar = Ph, R = Me: 60%, ee = 67%, β/β' > 20:1
Ar = Ph, R = *n*-Pent: 77%, ee = 93%, β/β' = 17:1
Ar = *m*-ClC_6H_4, R = *n*-Pent: 81%, ee = 93%, β/β' = 6.5:1
Ar = *p*-$MeOC_6H_4$, R = *n*-Pent: 81%, ee = 93%, β/β' = 20:1
Ar = *o*-$CF_3C_6H_4$, R = *n*-Pent: 45%, ee = 94%, β/β' = 32:1
Ar = *p*-$CF_3C_6H_4$, R = *n*-Pent: 78%, ee = 94%, β/β' = 7:1
Ar = *o*-Tol, R = *n*-Pent: 66%, ee = 93%, β/β' > 20:1
Ar = 2-furyl, R = *n*-Pent: 52%, ee = 93%, β/β' = 5.1:1
Ar = 2-Me-5-F-C_6H_3, R = *n*-Pent: 81%, ee = 92%, β/β' = 49:1

Scheme 2.25 Conjugate addition of allylboronic acid pinacol ester to nonsymmetric dialkylidene ketones with an *in situ* generated phosphonite nickel catalyst.

The conjugate addition of oxygen nucleophiles to electron-deficient alkenes has been a significant challenge in organic synthesis, owing to the low reactivity coupled with the reversibility of the reaction.[52] In particular, enantioselective intramolecular oxa-Michael addition, which provides a promising approach for the synthesis of pharmaceutically and biologically active chiral chromanone skeletons, has been rarely explored. For a long time, most reports have focused on catalysis through hydrogen bonding by employing organocatalysts. In 2008, Feng *et al.* reported an asymmetric intramolecular oxa-Michael addition of activated α,β-unsaturated ketones catalysed by a novel chiral *N*,*N*′-dioxide nickel(II) catalyst generated *in situ* from $Ni(acac)_2$ and the corresponding *N*,*N*′-dioxide ligand **28** (Scheme 2.26).[53] Screening various ligands of this type, the authors showed that a linker chain of three carbon atoms in the ligand was essential for the asymmetric addition. For example, ligands with a two-carbon atom linkage gave racemic products,

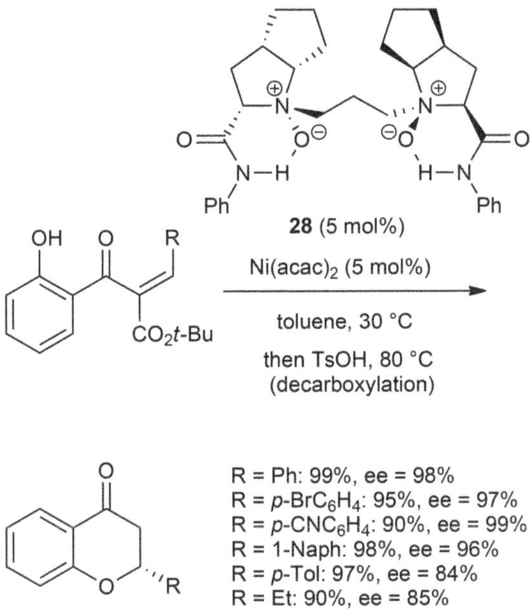

Scheme 2.26 Intramolecular oxa-Michael addition of activated enones with an *in situ* generated *N*,*N*′-dioxide nickel catalyst.

whereas ligands with a five-carbon atom linkage showed poor reactivity (only traces of products). Furthermore, it was demonstrated that the *N*-oxide group was essential for the reaction. Under the optimised reaction conditions using 5 mol% of *N*,*N*′-dioxide ligand **28** and 5 mol% of $Ni(acac)_2$ in toluene at 30 °C, the reaction of various aromatic alkene substrates led to the corresponding oxa-Michael products, which were subsequently decarboxylated by treatment with TsOH at 80 °C to give the final chromanones in both high yields and enantioselectivities of up to 99% ee, as shown in Scheme 2.26. Even an aliphatic enone (R = Et) was found to be suitable, affording the corresponding chromanone in 90% yield and with 85% ee. This novel process provided a promising approach for the synthesis of chiral flavanones with broad substrate scope and with tolerance to air and moisture.

With the aim of developing a novel route to amino acids, Seebach *et al.* reported the synthesis of a new type of substrate based on an achiral Ni(II) complex of a Schiff base of dehydroalanine.[54] An efficient catalytic method for asymmetric conjugate addition of CH acids to these novel Michael acceptors evolving through double induction was successfully developed using TADDOLs as chiral ligands and providing the corresponding chiral amino acids after hydrolysis of the intermediate nickel complexes. A series of NOBIN and TADDOL derivatives were tested as ligands, showing that TADDOL derivative **29** bearing 1-naphthyl groups invariably led to the best enantioselectivities of up to 80% ee, which were obtained in the presence of nucleophiles such as malonic ester derivatives, whereas nucleophiles such as thiophenol

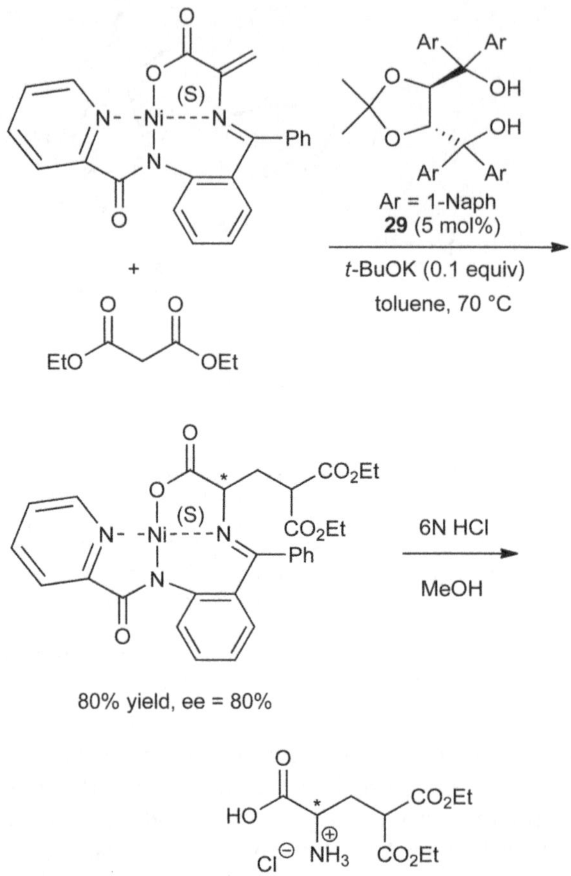

Scheme 2.27 Conjugate addition of diethyl malonate to a chiral Ni(II) complex of a Schiff base of dehydroalanine in the presence of a TADDOL ligand.

or amines reacted with the Schiff base of dehydroalanine without enantioselectivity (Scheme 2.27). The authors showed that racemic Michael products were achieved when the reactions were performed without TADDOLs. They have proposed that the function of the TADDOL could be to increase the malonic ester acidity by hydrogen bonding, and forming a chiral environment for recognition of the enantiotopic enolate of the Michael product in the proton transfer. It must be noted that even if the substrate of this reaction is not an enone, it was decided to situate this single work in this section for commodity.

2.3.2 Additions to α,β-Unsaturated Amides

In 2004, Suga *et al.* reported nickel-catalysed enantioselective Michael additions of 2-siloxyfurans to 3-alkenoyloxazolidin-2-ones using chiral binaphthyldiimine (BINIM) ligands.[55] Among a range of BINIM ligands,

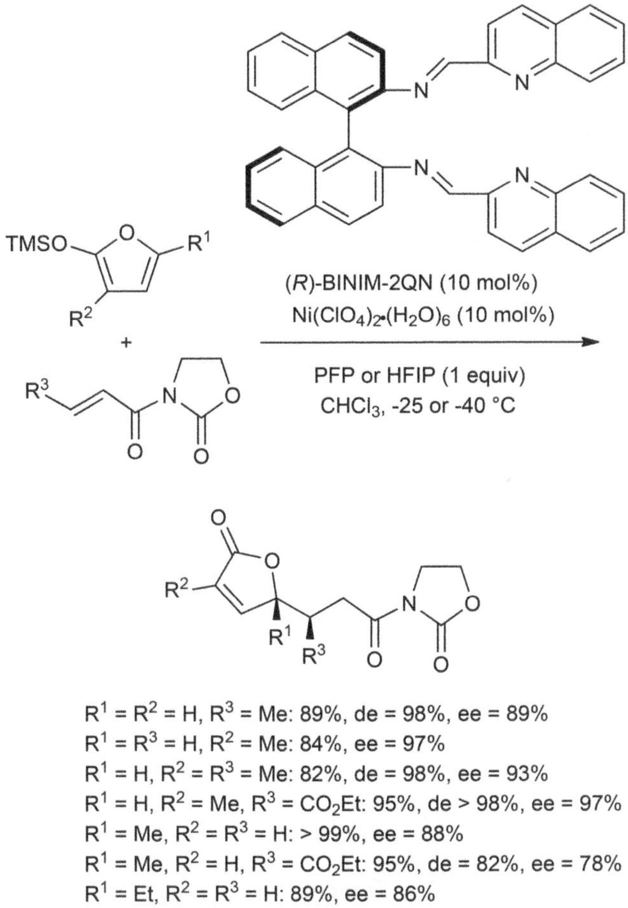

Scheme 2.28 Conjugate addition of 2-siloxyfurans to 3-alkenoyloxazolidin-2-ones with an *in situ* generated BINIM nickel catalyst.

(*R*)-BINIM-2QN was selected as the most efficient when used in the presence of $Ni(ClO_4)_2{\cdot}6H_2O$ in chloroform with HFIP or PFP as stoichiometric additive. As shown in Scheme 2.28 the reactions of various 2-siloxyfurans with variously substituted 3-alkenoyloxazolidin-2-ones afforded the corresponding Michael adducts with high *anti* selectivity of up to >98% de, combined with good to quantitative yields and good to high enantioselectivities of up to 97% ee. The best enantioselectivities (93–97% ee) were reached in the case of 3-methyl-2-(trimethylsiloxy)furan (R^1 = H, R^2 = Me) as substrate.

Inspired by their previous work dealing with enantioselective conjugate addition of *t*-butyl acetoacetate to crotonoylthiazolidinethione catalysed by a preformed nickel catalyst from *p*-Tol-BINAP ligand **30**,[56] and by those of Kanemasa on enantioselective nickel-catalysed additions of thiols and 1,3-diketones to unsaturated acylpyrazoles and oxazolidinones with a chiral amine ligand,[12,57] Evans *et al.* investigated the scope of enantioselective Michael additions of other β-keto esters to variously substituted unsaturated *N*-acylthiazolidinethiones, using catalyst **30** or related tetrafluoroborate catalyst **31**.[58] As shown in Scheme 2.29 the reactions promoted by 10 mol% of catalyst **30** or **31** afforded the corresponding Michael adducts **32** in both high yields and enantioselectivities of up to 97% ee. The tetrafluoroborate complex **31** was found slightly superior to catalyst **30** in terms of enantioselectivity. Branched as well as unbranched β-keto esters gave comparable results. The scope of the thione Michael acceptors showed that substitution at the δ-position

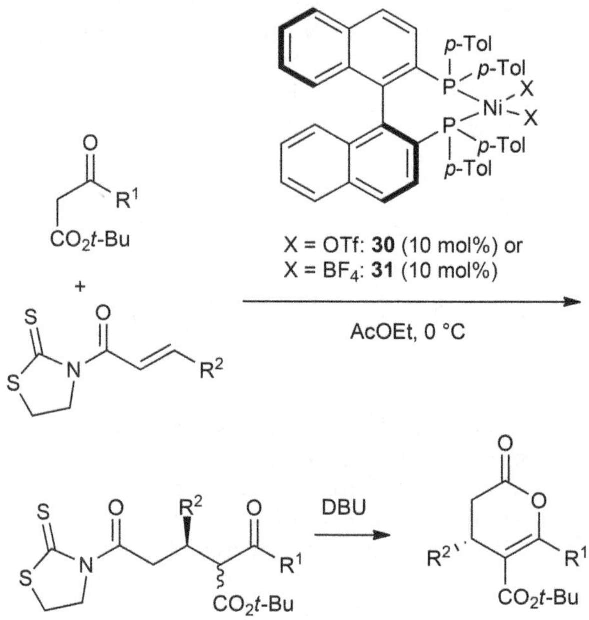

32 **33**

with catalyst **30**:
$R^1 = R^2$ = Me: 95% (**32**), ee (**33**) = 93%
R^1 = Et, R^2 = Me: 92% (**32**), ee (**33**) = 94%
R^1 = *n*-Pr, R^2 = Me: 86% (**32**), ee (**33**) = 93%
R^1 = *i*-Bu, R^2 = Me: 78% (**32**), ee (**33**) = 91%
R^1 = Cy, R^2 = Me: 97% (**32**), ee (**33**) = 90%
R^1 = Ph, R^2 = Me: 95% (**32**), ee (**33**) = 91%
with catalyst **31**:
R^1 = Me, R^2 = Et: 88% (**32**), ee (**33**) = 95%
R^1 = Me, R^2 = *n*-Pr: 91% (**32**), ee (**33**) = 95%
R^1 = Me, R^2 = *i*-Pr: 10% (**32**), ee (**33**) = 70%
R^1 = Me, R^2 = *i*-Bu 67% (**32**), ee (**33**) = 84%
R^1 = Me, R^2 = CO_2Et: 87% (**32**), ee (**33**) = 97%

Scheme 2.29 Conjugate addition of β-keto esters to unsaturated *N*-acylthiazolidinethiones with a preformed BINAP nickel catalyst.

resulted in a marked decrease in reactivity and enantioselectivity (67% yield and 84% ee for R^1 = Me, R^2 = i-Bu). Furthermore, for the γ-branched substrate (R^2 = i-Pr), the reaction with *t*-butyl acetoacetate was prohibitively slow (10% yield). Alternatively, the fumarate derivative (R^2 = CO_2Et) afforded by reaction with *t*-butyl acetoacetate the desired Michael product in high yield (87%) and with 97% ee. The Michael products were subsequently converted by treatment with DBU into the corresponding chiral dihydropyrones **33**, constituting excellent substrates for further enolate-based stereoselective transformations.

In 2003, Kanemasa *et al.* employed a chiral dibenzofuranbisoxazoline ligand, such as (*R*,*R*)-DBFOX-Ph, to generate *in situ* a nickel(II) aqua complex, which catalysed the Michael addition of malononitrile to α,β-unsaturated *N*-acyloxazolidinones with enantioselectivities of 55–93% ee.[59] Later in 2006, these authors demonstrated that enantioselective Michael additions of related tertiary nucleophilic precursors, such as substituted malononitriles, could be activated by a chiral nickel complex generated *in situ* from the (*R*,*R*)-DBFOX-Ph ligand and $Ni(ClO_4)_2{\cdot}6H_2O$.[60] The most efficient electrophilic substrates were found to be α,β-unsaturated amides of 3,5-dimethylpyrazoles, which afforded the corresponding Michael adducts in moderate to good yields and moderate to excellent enantioselectivities of up to 99% ee, as shown in Scheme 2.30. The authors have shown that the use of 3-crotonoyloxazolidin-2-one as acceptor provided the corresponding products in lower yields and enantioselectivities (5–84% ee). The reactions of more reactive α,β-unsaturated amides of 3,5-dimethylpyrazoles gave the best results when performed in 1 : 1 mixtures of an alcohol, such as i-PrOH or *t*-BuOH, and THF at room temperature in the presence of 10 mol% of catalyst loading. The lowest enantioselectivities (≤76% ee) were obtained for β,β-disubstituted electrophiles. The process presented the advantage to generate tertiary/quaternary and quaternary/quaternary carbon–carbon bond formations in one step.

Nitromethane is rather difficult to be activated through nickel catalysis due to its low enolisation ability. However, a nickel-catalysed highly enantioselective conjugate addition of nitromethane to α,β-unsaturated carbonyl compounds was achieved by Itoh and Kanemasa in 2002, using a chiral nickel catalyst generated *in situ* from the (*R*,*R*)-DBFOX-Ph ligand and $Ni(ClO_4)_2{\cdot}3H_2O$ in the presence of 2,2,6,6-tetramethylpiperidine (TMP) as an additive (Scheme 2.31).[61] A catalytic amount of this additive was crucial for nitromethane to undergo the Michael addition, which afforded the corresponding Michael adducts in moderate to high yields (39–93%) and with good to high enantioselectivities (83–97% ee).

More recently, the same authors reinvestigated these reactions by using a preformed bisoxazoline nickel catalyst **34** derived from the (*R*,*R*)-DBFOX-Ph ligand (Scheme 2.32).[62] The reaction also employed 10 mol% of catalyst loading and was performed in alcoholic media at room temperature in the presence of 4 Å MS, which was indispensable to achieve good yields. Indeed, molecular sieves worked effectively as a base, leading to the catalytic generation of nickel(II) enolate or nitronate nucleophiles through deprotonation of

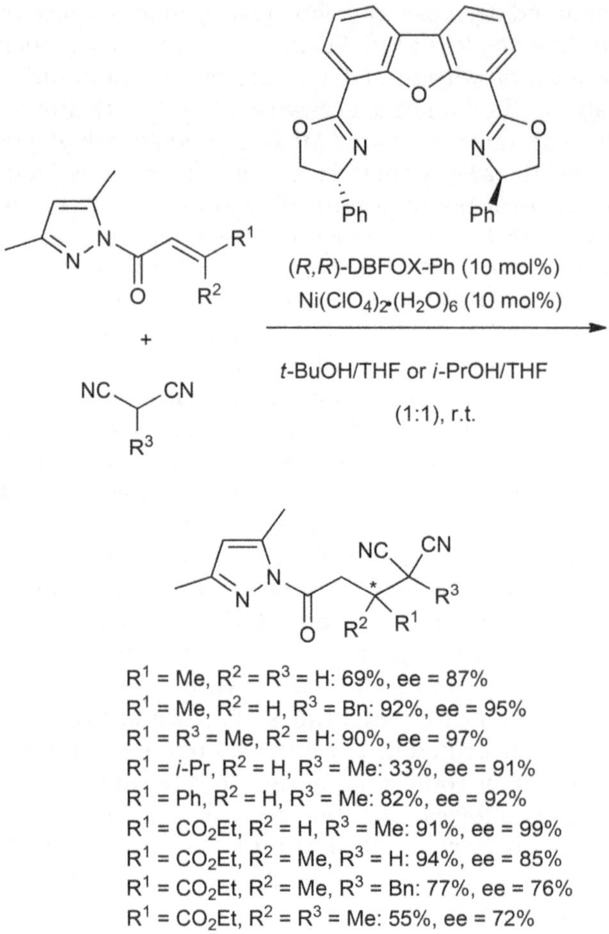

Scheme 2.30 Conjugate addition of substituted malononitriles to α,β-unsaturated amides of 3,5-dimethylpyrazoles with an *in situ* generated bisoxazoline nickel catalyst.

the α-hydrogen atom of the nucleophile precursors by treatment with a catalytic amount of chiral nickel ions. The resulting reactive intermediates could be successfully trapped with α,β-unsaturated amides of 3,5-dimethylpyrazoles to produce the corresponding chiral Michael products in moderate to good yields and with excellent general enantioselectivities of up to >99% ee, as shown in Scheme 2.32.

2.4 Conjugate Additions to Other Activated Alkenes

gem-Bisphosphonates have a high affinity for hydroxyapatite bone mineral surfaces and constitute an important class of biologically active compounds.[63] Highly enantioselective organocatalytic conjugate additions of

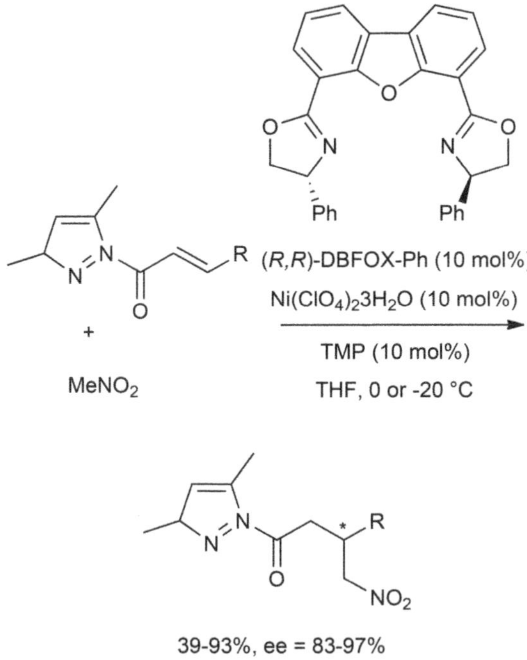

Scheme 2.31 Early conjugate addition of nitromethane to α,β-unsaturated amides of 3,5-dimethylpyrazole with an *in situ* generated bisoxazoline nickel catalyst reported by Kanemasa in 2002.

carbonyl compounds to ethylidenebisphosphonates have been reported, but the catalytic asymmetric synthesis of nitrogen-containing *gem*-bisphosphonates is in high demand since these products are often biologically much more active than *gem*-bisphosphonates without an amino group. In this context, Shibasaki *et al.* have developed enantioselective nickel-catalysed conjugate additions of nitroacetates **35** to ethylidenebisphosphonates **36**, using the preformed chiral dinuclear nickel catalyst (*R*)-**3** at 10 mol% of catalyst loading in toluene.[64] As shown in Scheme 2.33 the reaction tolerated methyl-substituted nitroacetate, as well as ethyl-, *n*-propyl-, and benzyl-substituted nitroacetates, which gave the corresponding products **37** in good yields and with enantioselectivities of 76–84% ee. It was noteworthy that the reaction also occurred using functionalised nitroacetate **35g** bearing a phthalimide moiety (81% yield, 93% ee). In addition to the ethyl group, benzyl and allyl groups were also applicable as protecting groups of bisphophonic acids, although the reactivity was somewhat decreased (65–69% yield), possibly due to steric hindrance. The authors assumed that a cooperative mechanism could be involved in which an Ni–aryl oxide moiety could function as a Brønsted base to deprotonate the α-proton of the nitroacetate to generate an Ni–enolate. The other Ni Lewis acidic metal centre could interact with the ethylidenebisphosphonate.

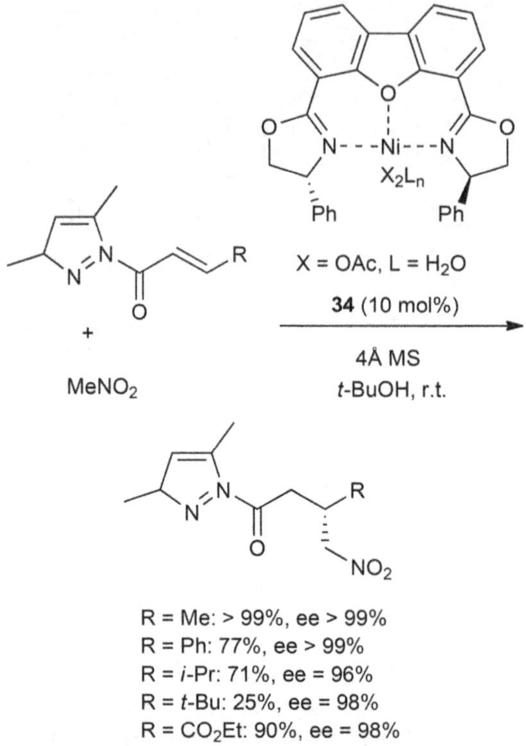

Scheme 2.32 Conjugate addition of nitromethane to α,β-unsaturated amides of 3,5-dimethylpyrazole with a preformed bisoxazoline nickel catalyst.

Hydroamination is an interesting and economical process that can be used to build secondary and tertiary amines constituting active components in drugs and agrochemicals, starting from the direct addition of amines to alkenes. This field is even more challenging if it is to afford chiral amines. Currently, metal catalysts that allow the formation of enantiomerically pure hydroamination products are still rare. In particular, only a few examples of nickel-catalysed enantioselective hydroamination reactions are known.[13,65] In a recent example, Fadini and Togni reported the enantioselective conjugate addition of amines to methacrylonitrile by using 5 mol% of preformed chiral ferrocenylphosphine nickel catalyst **38** (Scheme 2.34).[66] A range of aliphatic cyclic amines, such as morpholine, piperidine, thiomorpholine, and piperazine, afforded the corresponding products in comparable good yields and enantioselectivities of up to 96% ee. The use of other amines, such as pyrrolidine or benzylamine derivatives, provided the corresponding products in moderate to high yields (up to 99%), albeit with low enantioselectivity (≤24% ee). The authors also studied other α,β-unsaturated nitriles, such as acrylonitrile and crotononitrile, which led to almost racemic products. The substrate scope was also extended to electrophilic substrates other than nitriles, such as methyl acrylate and methyl or ethyl crotonates, but they also afforded racemic products.

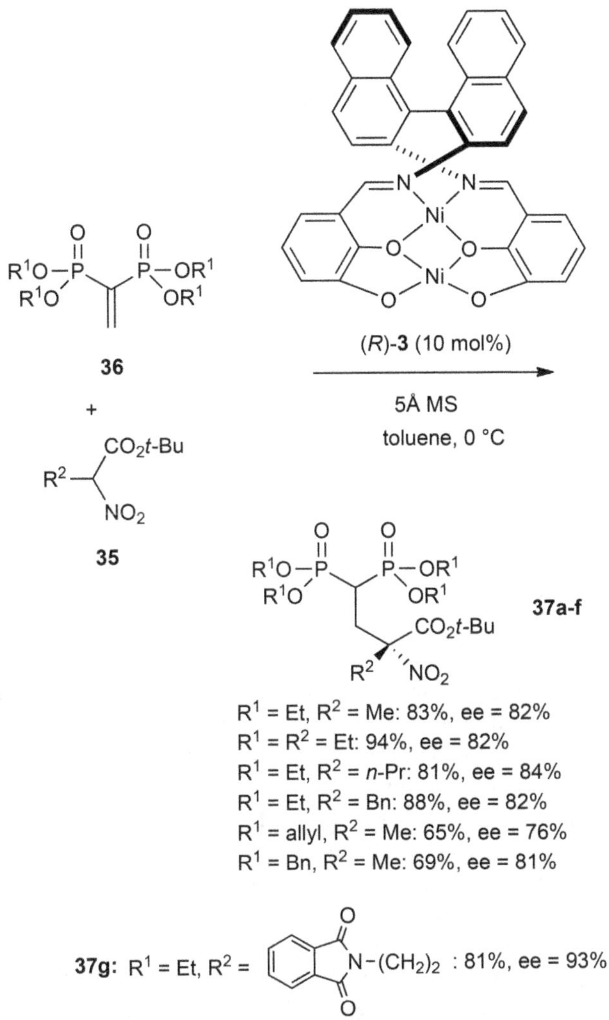

Scheme 2.33 Conjugate addition of nitroacetates to ethylidenebisphosphonates with a preformed dinuclear nickel catalyst.

2.5 Domino and Tandem Processes Initiated by a Michael Reaction

In spite of the explosive growth of enantioselective organocatalysed domino reactions in the last decade,[67] an increasing number of enantioselective metal-catalysed domino processes have been developed in the last few years.[68] A domino reaction has been defined by Tietze as a reaction which involves two or more bond-forming transformations, taking place under the same reaction conditions, without adding additional reagents and catalysts, and in which the subsequent reactions result as a consequence of the functionality

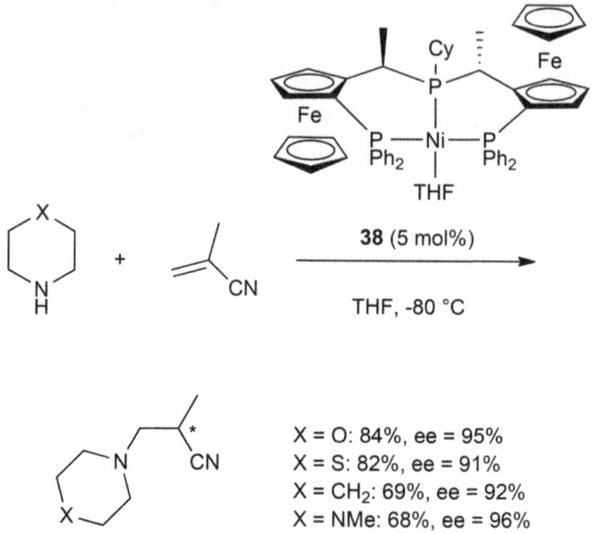

Scheme 2.34 Conjugate addition of cyclic amines to methacrylonitrile with a preformed ferrocenylphosphine nickel catalyst.

formed by bond formation or fragmentation in the previous step. It must be recognised that a relatively narrow distinction exists between domino and consecutive cascade or tandem reactions. From the point of view of an operator, the only difference between the two lies in the point along the sequence at which one or more catalysts or reagents had to be added to effect either the initiation of a sequence (that is, a domino reaction) or propagation to the next step (that is, a consecutive reaction or tandem reaction). It must be noted that the descriptors domino, cascade,[69] and tandem are often used indistinguishably from one another in the literature,[70] and a variety of opinions exist on how such reactions should be classified. According to Tietze, a domino reaction is strictly defined as a process in which two or more bond-forming transformations occur based on functionalities formed in the previous step and, moreover, no additional reagents, catalysts, or additives can be added to the reaction vessel, nor can reaction conditions be changed.[71] On the other hand, tandem sequences involve the isolation of intermediates, a change in reaction conditions, or the addition of reagents or coupling partners. The quality and importance of a domino reaction can be correlated to the number of bonds generated in such a process and the increase in molecular complexity. Domino reactions can be performed as single-, two-, and multicomponent transformations.[72] The use of one-, two-, and multicomponent domino reactions in organic synthesis is increasing constantly, since they allow the synthesis of a wide range of complex molecules, including natural products and biologically active compounds, in an economically favourable way by using processes that avoid the use of costly and time-consuming protection–deprotection processes, as well as purification procedures of intermediates.[73]

Since the first catalytic domino Michael–aldol reaction reported by Noyori *et al.* in 1996,[74] there have been numerous examples of domino and multicomponent domino reactions initiated by the Michael reaction. Among recent examples involving nickel catalysts, Kanemasa *et al.* have reported the enantioselective domino Michael/cyclisation reaction of dimedone to 1-(2-crotonoyl)-3,5-dimethylpyrazoles **39** catalysed by an *in situ* generated bisoxazoline chiral nickel catalyst of (*R*,*R*)-DBFOX-Ph (Scheme 2.35).[60,75] The reaction begins with the Michael addition of the nickel enolate of dimedone to 1-(2-crotonoyl)-3,5-dimethylpyrazole **39** to give nickel intermediate **40**, which undergoes cyclisation to provide intermediate **41**. The latter is then submitted to elimination of the pyrazole moiety, followed by dehydration, to finally afford the corresponding chiral enol lactone **42** in good to quantitative yields and with moderate to high enantioselectivities of up to 95% ee. The reaction employed one equivalent of acetic anhydride which trapped the pyrazole through *N*-acetylation, allowing yield and enantioselectivity to be improved. The role of the pyrazole chelating auxiliary of acceptor **39a** determined the reactivity of **39a** not only as electrophile but also as leaving group. Accordingly, the authors examined another acceptor **39b** (X = Br), which was expected to be more reactive than **39a** because of its less basic property as chelating auxiliary. Therefore, the reaction of **39b** with dimedone in isopropanol provided the corresponding enol lactone **42** in quantitative yield and 95% ee, whereas **42** was obtained in both lower yield and enantioselectivity (80% yield, 76% ee) starting from **39a**.

Earlier, these authors also performed this type of reaction with a closely related catalyst system, albeit in the presence of TMP as an additive and two equivalents of acetic anhydride in THF at room temperature, providing the corresponding enol lactones through the same domino Michael/cyclisation reaction (Scheme 2.36).[57] The chiral polyfunctionalised products were obtained in good to high yields (up to 99%) and with general high enantioselectivities of 89–99% ee. The use of acetic anhydride as additive was highly effective through the acetylation trapping of the liberated pyrazole, giving 1-acetyl-3,5-dimethylpyrazole. As shown in Scheme 2.36, a variety of 1-(alk-2-enoyl)-4-halo-3,5-dimethylpyrazoles having alkyl, aromatic, as well as heteroaromatic substituents were compatible with the reaction conditions, providing comparable results. The authors have proposed the structure depicted in Scheme 2.36 for the active catalyst.

These reaction conditions were also applied to the domino Michael/cyclisation reactions of other nucleophiles, such as 4-hydroxy-6-methyl-2-pyrone **43**, 4-hydroxycoumarin **44**, and 3-hydroxyperinaphthenone **45**, as shown in Scheme 2.37.[57] By reaction with variously substituted 1-(alk-2-enoyl)-4-bromo-3,5-dimethylpyrazoles, these compounds afforded the corresponding chiral domino products **46a–c**, **46d–h**, and **47**, respectively. These products arose from domino Michael/cyclisation reactions and were obtained in moderate to good yields and with good to high enantioselectivities of up to 98% ee.

In 2010, Arai *et al.* investigated a catalytic asymmetric *exo*′-selective virtual [3 + 2] cycloaddition of imino esters with *trans*-nitroalkenes.[76] Usually, when

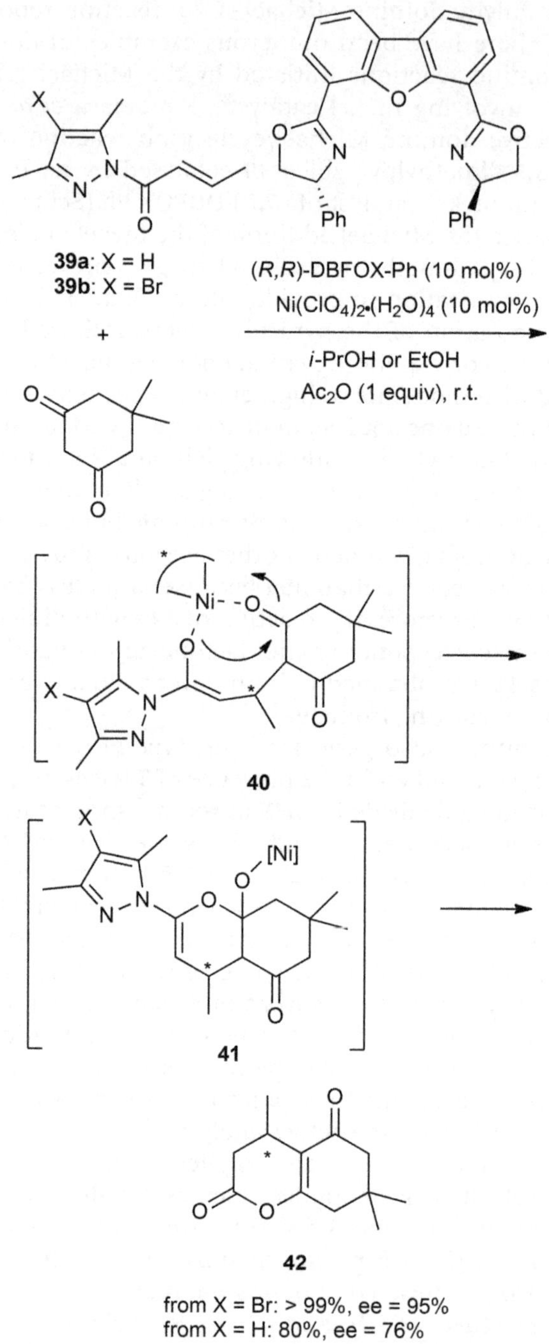

Scheme 2.35 Domino Michael/cyclisation reaction of dimedone with 1-(2-crotonoyl)-3,5-dimethylpyrazoles catalysed by an *in situ* generated (*R,R*)-DBFOX-Ph nickel catalyst.

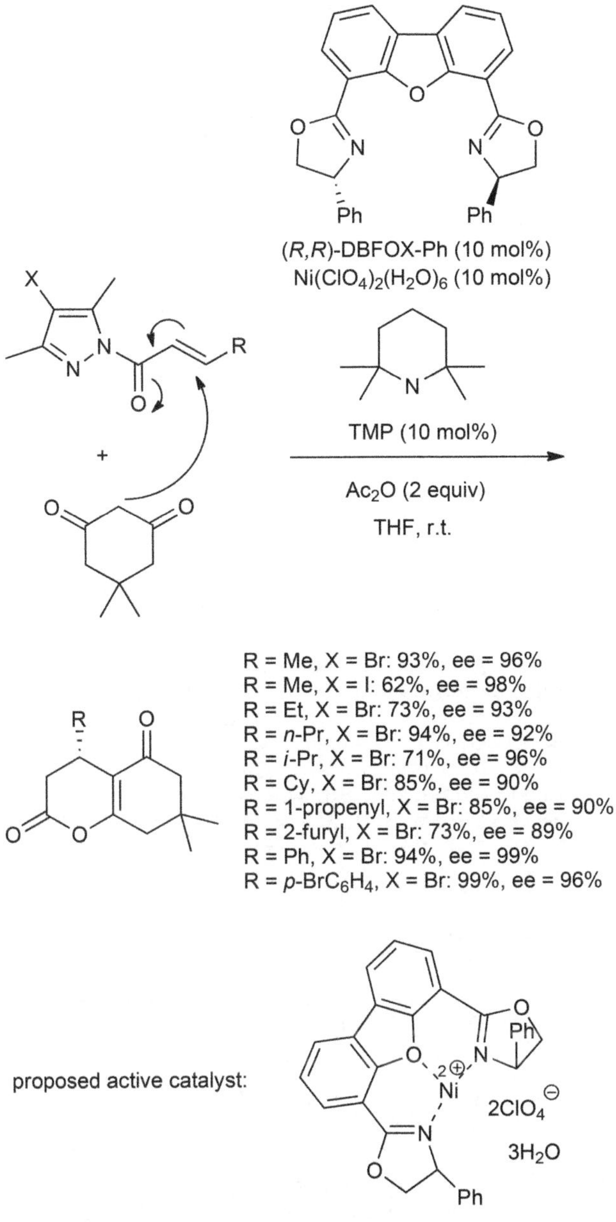

Scheme 2.36 Domino Michael/cyclisation reaction of dimedone with 1-(alk-2-enoyl)-4-halo-3,5-dimethylpyrazoles catalysed by an *in situ* generated (*R*,*R*)-DBFOX-Ph nickel catalyst.

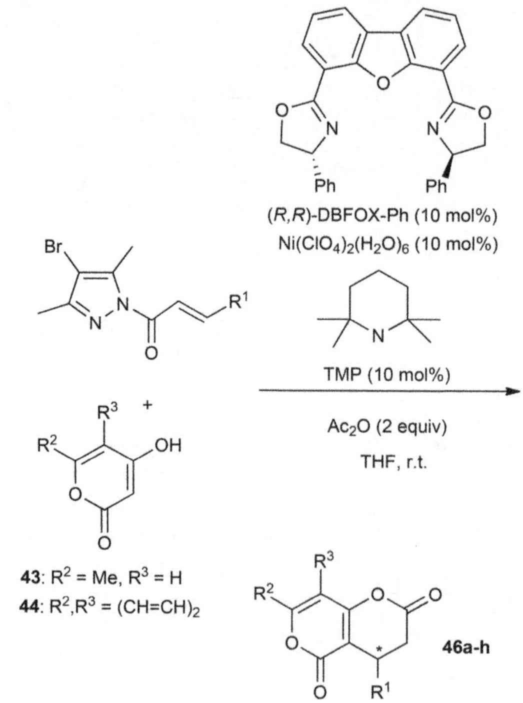

46a: $R^1 = R^2 = Me$, $R^3 = H$: 85%, ee = 98%
46b: R^1 = 5-Br-2-MeOC_6H_3, $R^2 = Me$, $R^3 = H$: 23%, ee = 84%
46c: R^1 = 3,5-$Br_2C_6H_3$, $R^2 = Me$, $R^3 = H$: 60%, ee = 96%
46d: $R^1 = Me$, $R^2,R^3 = (CH{=}CH)_2$: 66%, ee = 93%
46e: $R^1 = Ph$, $R^2,R^3 = (CH{=}CH)_2$: 64%, ee = 96%
46f: R^1 = *p*-BrC_6H_4, $R^2,R^3 = (CH{=}CH)_2$: 65%, ee = 95%
46g: R^1 = 5-Br-2-MeOC_6H_3, $R^2,R^3 = (CH{=}CH)_2$: 53%, ee = 84%
46h: R^1 = 3,5-$Br_2C_6H_3$, $R^2,R^3 = (CH{=}CH)_2$: 80%, ee = 97%

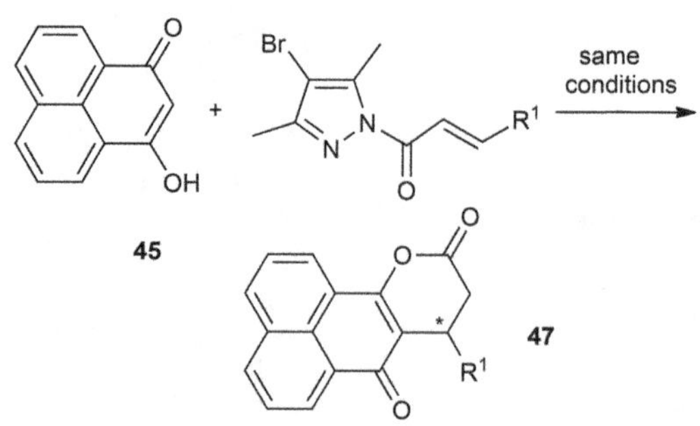

R^1 = Me: 81%, ee = 91%
R^1 = 5-Br-2-MeOC_6H_3: 31%, ee = 89%

Scheme 2.37 Domino Michael/cyclisation reactions of enols with 1-(alk-2-enoyl)-4-bromo-3,5-dimethylpyrazoles catalysed by an *in situ* generated (*R*,*R*)-DBFOX-Ph nickel catalyst.

a *trans*-nitroalkene is used in a [3 + 2] cycloaddition, the stereoconjunction between the 3- and 4-positions is fixed in a *trans* conformation, and four diastereomers are possible, classified as *endo*, *exo*, *endo'*, and *exo'* isomers. Screenings of the metal salts to study the *exo'* adduct ratio have found that nickel salts facilitated the selective production of the *exo'* products. Thus, these authors have performed the *exo'*-selective reaction of imino esters and *trans*-nitroalkenes by using a combination of $Ni(OAc)_2$ with the chiral imidazoline-aminophenol **11** as a catalyst system, which provided the corresponding pyrrolidines in good yields and with diastereoselectivities of up to 84% de combined with high enantioselectivities of up to 99% ee for the major isomers, as shown in Scheme 2.38. This novel methodology represents the first general success in the catalytic asymmetric *exo'*-selective reaction of imino esters and nitroalkenes. In order to explain the results, the authors have proposed that the products were generated from a domino Michael addition of the imino esters onto the *trans*-nitroalkenes, which was then followed by a Mannich reaction, as depicted in Scheme 2.38.

Coumarin derivatives are probably one of the most common skeletons found in natural products. Owing to their extensive array of biological activities and pharmacological properties, the synthesis of this class of compound has been a long-standing challenge in organic chemistry. Among the coumarin family members, the 4-hydroxycoumarin core, such as in the anticoagulant warfarin, is especially important. Among methodologies developed to achieve diversely structured chiral warfarins, the asymmetric domino Michael/cyclisation reaction of 4-hydroxycoumarin with α,β-unsaturated systems has become a very attractive methodology in recent years. Initially, Jørgensen *et al.* developed a bisoxazoline–copper(II) chiral complex to catalyse the reaction of a cyclic 1,3-dicarbonyl compound to a β,γ-unsaturated α-keto ester, which afforded a warfarin analogue.[77] Later, several groups reported the successful use of various organocatalysts in these reactions. In 2011, Lin and Feng applied an *N,N'*-dioxide chiral nickel catalyst to induce the enantioselective domino Michael/cyclisation reaction of cyclic 1,3-dicarbonyl compounds with β,γ-unsaturated α-keto esters.[78] The catalyst was generated *in situ* from $Ni(acac)_2$ and chiral *N,N'*-dioxide **48**, both employed at 5 mol% of catalyst loading in 1,2-dichloroethane in the presence of 4 Å MS at 0 °C. The reactions remarkably afforded the corresponding chiral warfarin analogues in nearly quantitative yields and with high enantioselectivities of up to 90% ee, as shown in Scheme 2.39. Interestingly, neither the steric hindrance nor the electronic nature of the aromatic ring (R^1) of the β,γ-unsaturated α-keto ester had any obvious effect on the enantioselectivity (87–90% ee). It was worth noting that the substrates with condensed-ring, heteroaromatic, and cinnamyl groups performed well, giving the corresponding products in excellent yields and with high enantioselectivities. Moreover, the α-keto ester with an ethyl substrate (R^2 = Et) also provided an excellent yield (98%) and a good enantioselectivity (85% ee), as well as the 4-hydroxycoumarin containing a 6-methyl group, which gave 98% yield combined with 87% ee. To explain these results, the authors proposed the transition state

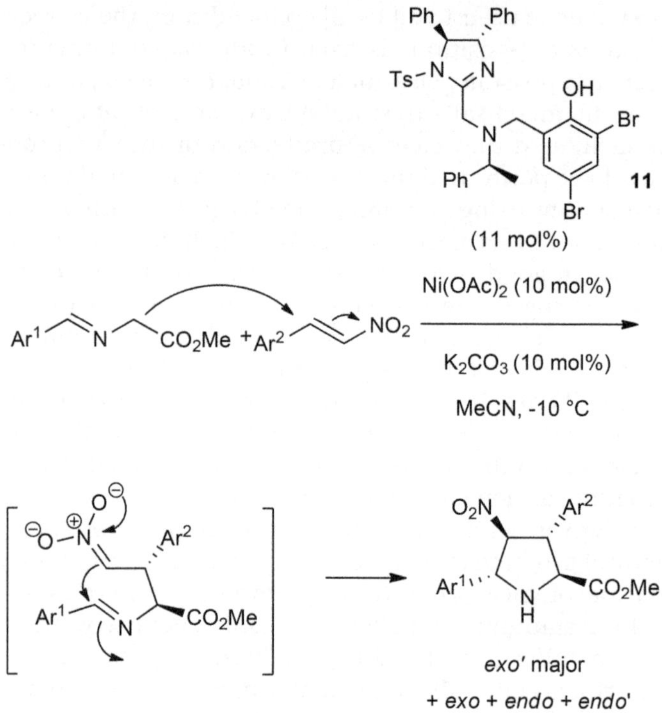

Ar^1	Ar^2	Yield (%)	*exo'/endo/exo/endo'*	ee (*exo'*) (%)
Ph	Ph	99	82:16:1:1	97
p-MeOC$_6$H$_4$	Ph	79	90:10:0:0	93
m-MeOC$_6$H$_4$	Ph	67	89:11:0:0	96
p-ClC$_6$H$_4$	Ph	84	87:9:0:4	97
2-Naph	Ph	70	92:7:0:1	99
Ph	*p*-MeOC$_6$H$_4$	93	86:13:0:1	96
Ph	*p*-BrC$_6$H$_4$	85	85:14:0:1	97
Ph	*m*-BrC$_6$H$_4$	94	80:17:0:3	96
Ph	*p*-NO$_2$C$_6$H$_4$	87	88:11:0:1	96

Scheme 2.38 Domino Michael/Mannich reaction of imino esters with nitroalkenes catalysed by an *in situ* generated imidazoline-aminophenol nickel catalyst.

depicted in Scheme 2.39. The authors speculated that *N*,*N'*-dioxide ligand **48** and the β,γ-unsaturated α-keto ester coordinated with Ni(acac)$_2$ to form a complex. Then, the 4-hydroxycoumarin could only attack the *Re* face of the double bond since the *Si* face of the double bond was hindered by the sterically bulky group. The corresponding domino product was afforded with the *S* configuration.

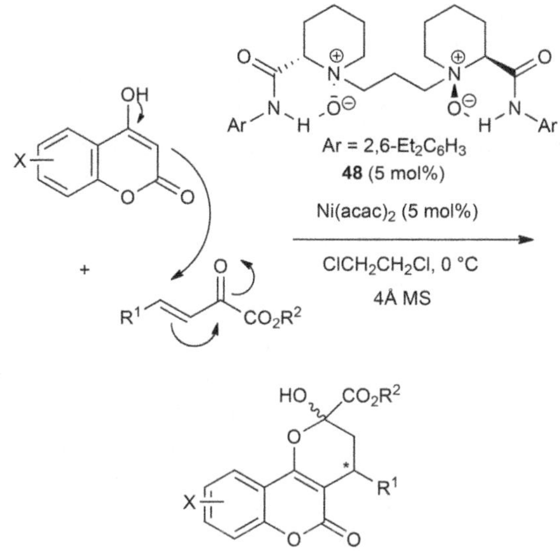

R^1 = Ph, R^2 = Me, X = H: 99%, ee = 89%
R^1 = *m*-Tol, R^2 = Me, X = H: 99%, ee = 89%
R^1 = *p*-Tol, R^2 = Me, X = H: 99%, ee = 89%
R^1 = *m*-$MeOC_6H_4$, R^2 = Me, X = H: 99%, ee = 87%
R^1 = *p*-$MeOC_6H_4$, R^2 = Me, X = H: 99%, ee = 88%
R^1 = *p*-PhC_6H_4, R^2 = Me, X = H: 98%, ee = 87%
R^1 = *m*-ClC_6H_4, R^2 = Me, X = H: 78%, ee = 89%
R^1 = *p*-ClC_6H_4, R^2 = Me, X = H: 99%, ee = 87%
R^1 = *m*-BrC_6H_4, R^2 = Me, X = H: 89%, ee = 87%
R^1 = *p*-BrC_6H_4, R^2 = Me, X = H: 99%, ee = 90%
R^1 = *p*-FC_6H_4, R^2 = Me, X = H: 99%, ee = 90%
R^1 = 2-Naph, R^2 = Me, X = H: 97%, ee = 89%
R^1 = 2-thienyl, R^2 = Me, X = H: 98%, ee = 87%
R^1 = (*E*)-PhCH=CH, R^2 = Me, X = H: 96%, ee = 89%
R^1 = Ph, R^2 = Et, X = H: 98%, ee = 85%
R^1 = Ph, R^2 = Me, X = 6-Me: 98%, ee = 87%

proposed transition state:

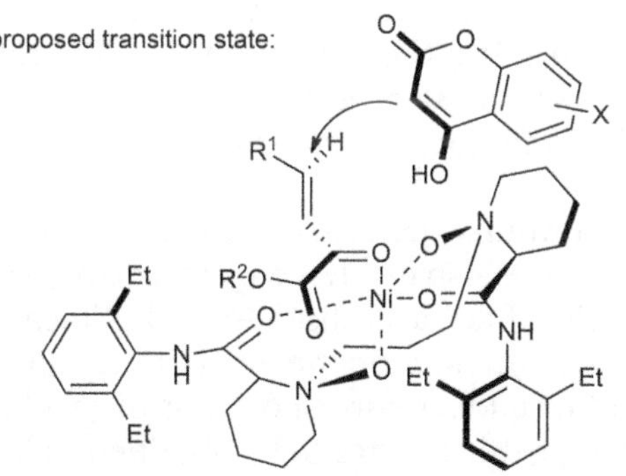

Scheme 2.39 Domino Michael/cyclisation reaction of cyclic 1,3-dicarbonyl compounds with β,γ-unsaturated α-keto esters catalysed by an *in situ* generated *N,N′*-dioxide nickel catalyst.

The Henry reaction has been often associated with the Michael reaction in a number of successful asymmetric domino sequences.[79] In this context, Wang *et al.* have developed enantioselective nickel-catalysed domino Michael/Henry reactions of 1,2-diones with nitroalkenes to afford chiral polyfunctionalised bicyclo[3.2.1]octane derivatives containing four stereogenic centres (Scheme 2.40).[80] These remarkable processes, which provided good to quantitative yields, good to high diastereoselectivities of up to 96% de, and good to excellent enantioselectivities of up to >99% ee, were induced for the first time by a combination of $Ni(OAc)_2$ with chiral cyclohexanediamine ligand **49**. For cyclic 1,2-diones, the best results were achieved with cyclohexane-1,2-dione, while cyclopentane-1,2-dione gave a lower yield, diastereo-, and enantioselectivity (76% yield, 82% de, and 51% ee with R^1 = Ph, R^2 = H). In the case of aromatic nitroalkenes in reaction with cyclohexane-1,2-dione, the aromatic ring of these substrates tolerated both electron-donating and electron-withdrawing functionalities at any position, although a 10% de was obtained when there was an *ortho*-bromine substituent at the aromatic ring. The additions of heteroaromatic nitroalkenes derived from furyl to cyclohexane-1,2-dione proceeded equally with excellent enantio- and diastereoselectivity. The more sterically hindered nitroalkene bearing a 2-naphthyl group was also a suitable substrate. Moreover, the domino product that arose from the reaction of α-bromo(phenyl)nitroalkene with cyclohexane-1,2-dione was obtained in 83% ee. Furthermore, several alkyl-substituted nitroalkenes provided the corresponding domino products in excellent enantioselectivities (97–98% ee) and moderate to good diastereoselectivities (78–96% de) by reaction with cyclohexane-1,2-dione. In addition to cyclic 1,2-diones such as cyclohexane-1,2-dione, several acyclic 1,2-diones also gave good results (90–97% ee). The authors have proposed the transition state depicted in Scheme 2.40 to explain the results. At the beginning of the transition state the nitronate anion is stabilised by interaction at the open apical position on the nickel ion. A Michael addition then occurs with the activated metal enolate structure of the 1,2-dione. In the favoured transition state, the substituted group of the nitroalkene is oriented away from the bulky aromatic group of the chiral ligand. The π–π stacking interaction between the enolate structure of the 1,2-dione and the double bond of the nitroalkene probably has a favourable effect on the stereoselectivity of the reaction. The resulting adduct subsequently undergoes an intramolecular Henry reaction to form the final product.

Later, asymmetric domino Michael/Henry reactions of cyclohexane-1,2-dione with nitroalkenes were also catalysed by an *in situ* generated nickel complex derived from chiral bisoxazolidine **50** and $Ni(acac)_2$ (Scheme 2.41).[81] Performing the reactions in isopropanol at room temperature with 5 mol% of catalyst loading allowed the corresponding bicyclo[3.2.1]octane derivatives bearing four stereogenic centres to be obtained in high yields (76–99%) with moderate to good diastereoselectivities of up to 80% de and excellent enantioselectivities (90–99% ee). The results showed that the nature of the substituents on the aryl group of the nitrostyrene influenced both yields and

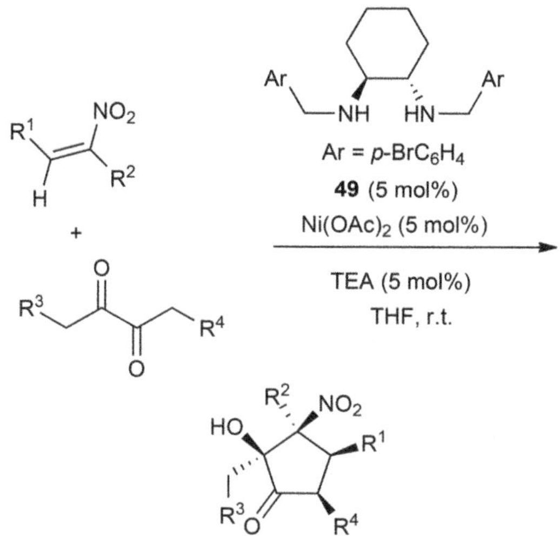

R^1 = Ph, R^2 = H, R^3,R^4 = $(CH_2)_2$: 98%, de = 82%, ee = 98%
R^1 = *p*-$MeOC_6H_4$, R^2 = H, R^3,R^4 = $(CH_2)_2$: 74%, de = 96%, ee = 98%
R^1 = 2-furyl, R^2 = H, R^3,R^4 = $(CH_2)_2$: 70%, de = 94%, ee > 99%
R^1 = *p*-FC_6H_4, R^2 = H, R^3,R^4 = $(CH_2)_2$: 98%, de = 96%, ee > 99%
R^1 = *m*-$MeOC_6H_4$, R^2 = H, R^3,R^4 = $(CH_2)_2$: 80%, de = 96%, ee = 98%
R^1 = 2-Naph, R^2 = H, R^3,R^4 = $(CH_2)_2$: > 99%, de = 78%, ee = 92%
R^1 = *p*-BrC_6H_4, R^2 = H, R^3,R^4 = $(CH_2)_2$: 77%, de = 96%, ee = 97%
R^1 = *p*-Tol, R^2 = H, R^3,R^4 = $(CH_2)_2$: 80%, de = 72%, ee > 99%
R^1 = Ph, R^2 = Br, R^3,R^4 = $(CH_2)_2$: > 99%, de = 90%, ee = 83%
R^1 = *o*-BrC_6H_4, R^2 = H, R^3,R^4 = $(CH_2)_2$: > 99%, de = 10%, ee = 90-91%
R^1 = 3,4-$Me_2C_6H_3$, R^2 = H, R^3,R^4 = $(CH_2)_2$: 91%, de = 75%, ee > 97%
R^1 = Ph, R^2 = H, R^3,R^4 = CH_2: 76%, de = 82%, ee = 51%
R^1 = Ph, R^2 = R^3 = H, R^4 = Et: 86%, de = 60%, ee = 90%
R^1 = Ph, R^2 = H, R^3 = R^4 = Me: 91%, de = 94%, ee = 97%
R^1 = Cy, R^2 = H, R^3,R^4 = $(CH_2)_2$: 51%, de = 78%, ee = 97%
R^1 = *n*-Bu, R^2 = H, R^3,R^4 = $(CH_2)_2$: 87%, de = 96%, ee = 98%

proposed transition state:

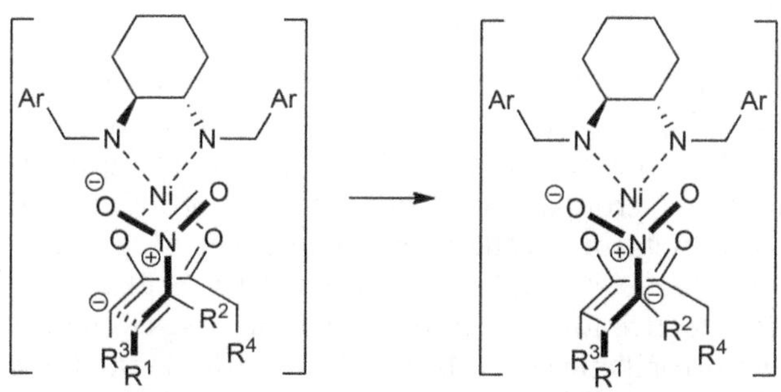

Scheme 2.40 Domino Michael/Henry reaction of 1,2-diones with nitroalkenes catalysed by an *in situ* generated diamine nickel catalyst.

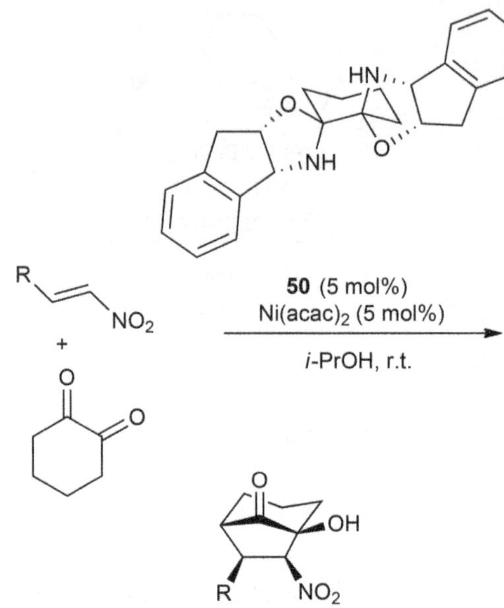

R = Ph: 99%, de = 75%, ee = 99%
R = *p*-BrC_6H_4: 88%, de = 60%, ee = 92%
R = *p*-ClC_6H_4: 80%, de = 75%, ee = 96%
R = *p*-FC_6H_4: 90%, de = 80%, ee = 99%
R = *o*-BrC_6H_4: 90%, de = 34%, ee = 98%
R = *p*-Tol: 98%, de = 75%, ee = 91%
R = *p*-$MeOC_6H_4$: 95%, de = 75%, ee = 90%
R = 2-furyl: 92%, de = 60%, ee = 91%
R = 1-Naph: 94%, de = 75%, ee = 91%
R = *n*-Bu: 76%, de = 75%, ee = 95%

Scheme 2.41 Domino Michael/Henry reaction of cyclohexane-1,2-dione with nitroalkenes catalysed by an *in situ* generated bisoxazolidine nickel catalyst.

stereoselectivities. For nitroalkenes with a 4-halo substitution, both diastereo- and enantioselectivities increased by changing Br to F. In the case of the 2-bromo-substituted nitrostyrene, excellent enantioselectivity was also observed (98% ee), but the diastereoselectivity was relatively low (34% de). The addition of nitroalkenes bearing electron-donating groups, heteroaromatic, and naphthyl groups also proceeded smoothly and afforded the corresponding products in excellent yields and with high diastereo- and enantioselectivities (92–98% yield, 60–75% de, 90–91% ee). Furthermore, the catalyst was also effective for the domino reaction of cyclohexane-1,2-dione with an aliphatic nitroalkene (R = *n*-Bu) under standard reaction conditions, providing the corresponding product in 76% yield, with a good diastereoselectivity of 75% de and a high enantioselectivity of 95% ee. Importantly, the authors demonstrated that it was possible to decrease the catalyst loading to 3 mol% without significantly affecting the stereoselectivity.

In 2014, Arai and Yamamoto described asymmetric nickel-catalysed domino Michael/Henry reactions between 2-sulfanylbenzaldehydes and nitroalkenes to give the corresponding chiral 2-aryl-3-nitrochroman-4-ols in most cases in almost quantitative yields and with good to high diastereo- and enantioselectivities of up to >98% de and 95% ee, respectively (R^2 = H, Scheme 2.42).[82] These reactions were promoted by an *in situ* generated catalyst from 10–11 mol% of a chiral imidazoline-aminophenol ligand **11** and $Ni(OAc)_2 \cdot 4H_2O$ in chloroform at −20 °C. The substrate scope of the process showed that various electron-withdrawing and electron-donating substituents were tolerated on the benzene ring of the nitroalkene, providing the products in enantioselectivities of 84–95% ee. In addition, (*E*)-2-(2-nitrovinyl)thiophene reacted with 2-sulfanylbenzaldehyde to give the corresponding product in 94% ee and 90% de. Two substituted 2-sulfanylbenzaldehydes (R^2 = Cl, *t*-Bu) also reacted smoothly with phenylnitroethylene in 80–87% ee. Unfortunately, the use of an aliphatic nitroalkene (R^1 = *n*-pent) resulted in a reduced yield (74%) of the desired product as well as low enantioselectivity of 4% ee. A proposed mechanism explaining the manner in which the chiral (2*S*,3*R*,4*R*)-thiochromanes are synthesised through domino Michael/Henry reactions is depicted in Scheme 2.42. The process begins with the formation of an L*–Ni–thiolate intermediate due to the high affinity of thiol for the nickel centre, which reacts with the nitrostyrene through a Michael addition to give the corresponding L*–Ni–nitronate intermediate. The latter leads to the formation of the six-membered ring of the thiochromane to adopt a strained half-boat-like conformation in which the eclipsed interaction between the carbonyl and nitro groups is increased if the carbonyl group remains in the equatorial position. The Henry reaction then proceeds from the transition-state complex (Scheme 2.42), in which the C4–O–Ni bond is in the pseudoaxial position, to give the final (2*S*,3*R*,4*R*)-thiochromane.

The cyclopropane ring is an important structural motif in a great number of natural products and biologically active agents.[83] In addition, cyclopropyl derivatives also constitute valuable synthetic building blocks in organic synthesis.[84] Consequently, significant effort has been made to develop efficient synthetic methods for chiral cyclopropanes.[85] Among them, asymmetric domino reactions (and tandem sequences) involving a Michael addition followed by an intramolecular alkylation have been developed by several groups for the synthesis of chiral functionalised cyclopropanes.[86] The reactions can involve domino Michael/intramolecular alkylation reactions of bromonitromethanes with α,β-unsaturated carbonyl derivatives, as well as additions of bromomalonates to nitroalkenes. So far, most of these reactions have been catalysed by organocatalysts. In 2012, Kim *et al.* reported the first example of an enantioselective tandem Michael/intramolecular alkylation sequence between bromomalonates and nitroalkenes promoted by a chiral nickel catalyst.[87] The process begins with the Michael addition of the bromomalonate to the nitroalkene in the presence of 5 mol% of chiral preformed diamine nickel catalyst **51** in dibromomethane at room temperature, which is followed by an intramolecular alkylation induced by addition of DBU to the reaction mixture to afford the final chiral cyclopropane (Scheme 2.43).

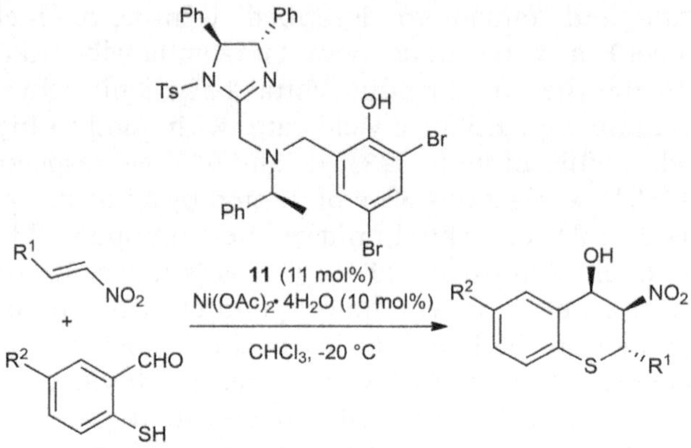

R^1 = Ph, R^2 = H: > 99%, de = 94%, ee = 95%
R^1 = *p*-ClC_6H_4, R^2 = H: > 99%, de > 98%, ee = 94%
R^1 = *m*-ClC_6H_4, R^2 = H: > 99%, de > 98%, ee = 95%
R^1 = *p*-BrC_6H_4, R^2 = H: > 99%, de > 98%, ee = 93%
R^1 = *m*-BrC_6H_4, R^2 = H: > 99%, de > 98%, ee = 94%
R^1 = *p*-FC_6H_4, R^2 = H: > 99%, de > 98%, ee = 90%
R^1 = *p*-$NO_2C_6H_4$, R^2 = H: 98%, de > 98%, ee = 91%
R^1 = *m*-$NO_2C_6H_4$, R^2 = H: > 99%, de = 98%, ee = 93%
R^1 = *p*-$MeOC_6H_4$, R^2 = H: 98%, de = 92%, ee = 84%
R^1 = *m*-$MeOC_6H_4$, R^2 = H: > 99%, de > 98%, ee = 94%
R^1 = *p*-Tol, R^2 = H: > 99%, de = 84%, ee = 84%
R^1 = *m*-Tol, R^2 = H: > 99%, de > 98%, ee = 94%
R^1 = 2-thienyl, R^2 = H: > 99%, de = 90%, ee = 94%
R^1 = Ph, R^2 = Cl: > 99%, de = 82%, ee = 80%
R^1 = Ph, R^2 = *t*-Bu: > 99%, de = 94%, ee = 87%
R^1 = *n*-Pent, R^2 = H: 74%, de = 56%, ee = 4%

proposed catalytic cycle:

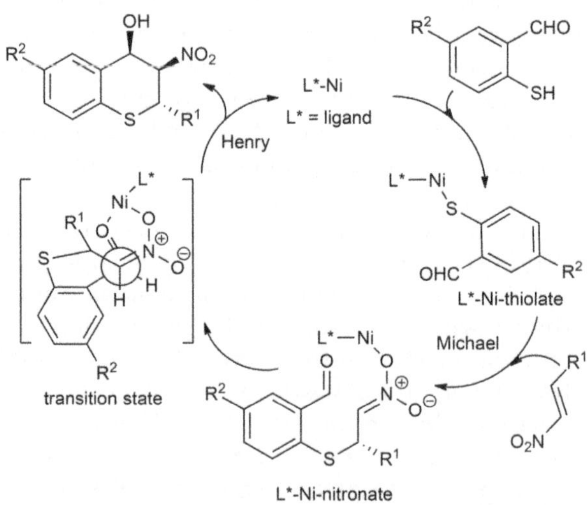

Scheme 2.42 Domino Michael/Henry reaction of sulfanylbenzaldehydes with nitroalkenes catalysed by an *in situ* generated imidazoline-aminophenol nickel catalyst.

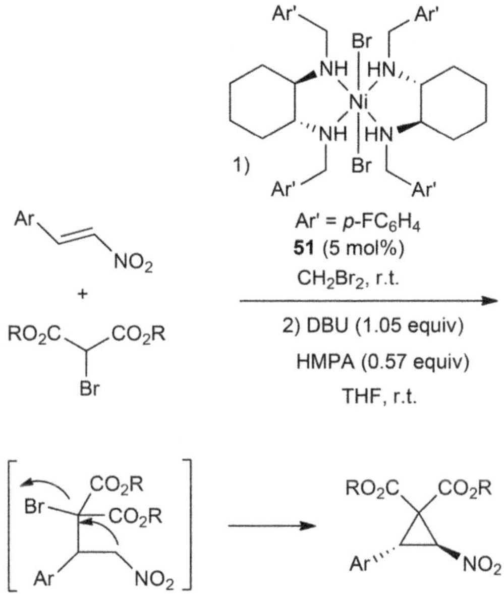

Ar = Ph, R = Me: 87%, ee = 94%
Ar = Ph, R = Et: 99%, ee = 97%
Ar = p-MeOC$_6$H$_4$, R = Me: 89%, ee = 96%
Ar = p-MeOC$_6$H$_4$, R = Et: 96%, ee = 99%
Ar = p-Tol, R = Me: 86%, ee = 97%
Ar = p-Tol, R = Et: 97%, ee = 99%
Ar = p-BrC$_6$H$_4$, R = Me: 85%, ee = 91%
Ar = p-ClC$_6$H$_4$, R = Et: 92%, ee = 98%
Ar = p-FC$_6$H$_4$, R = Et: 95%, ee = 98%
Ar = o-NO$_2$C$_6$H$_4$, R = Et: 93%, ee = 91%
Ar = o-BrC$_6$H$_4$, R = Et: 90%, ee = 93%
Ar = 2-thienyl, R = Et: 89%, ee = 99%
Ar = 2-furyl, R = Et: 91%, ee = 91%
Ar = 1-Naph, R = Et: 70%, ee = 85%

proposed transition state:

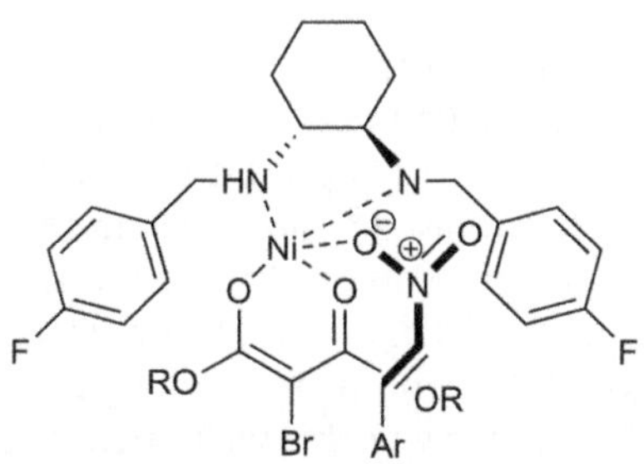

Scheme 2.43 Tandem Michael/intramolecular alkylation sequence between bromomalonate and nitroalkenes catalysed by a preformed diamine nickel catalyst.

A range of aromatic and heteroaromatic nitroalkenes reacted smoothly with ethyl as well as methyl bromomalonates to give the corresponding cyclopropanes in high yields and enantioselectivities (85–99% ee). To explain the stereoselectivity of the process, the authors have proposed that the bromomalonate was activated by the nickel catalyst through a bidentate fashion. Then, the bromomalonate anion attacked the *Si* face of the double bond of the nitroalkene, as shown in the transition state depicted in Scheme 2.43.

In the last few years, an explosive number of multiple-catalyst systems for various organic transformations have been developed.[88] In particular, the combination of organocatalysts and transition metal catalysts has evolved as a new strategy to carry out enantioselective transformations that could not be performed in a traditional way by simply employing one of the two catalysts. These transformations not only demonstrate the potential of this merged catalytic approach, but they also show that there are more options to render a reaction highly enantioselective than testing different chiral metal–ligand complexes, organocatalysts, or additives. By using appropriate combinations of an organocatalyst and an achiral or chiral transition metal catalyst, facile ways for reaction optimisation can be achieved by simply varying one of the two existing catalysts. The first example of combining a transition metal and an organocatalyst was reported by Ito *et al.* in 1986, dealing with a remarkable enantioselective domino aldol/cyclisation reaction of aldehydes with methyl isocyanoacetate catalysed by a combination of a gold complex and a chiral tertiary amine as organocatalyst, allowing diastereo- and enantioselectivities of up to >99% de and 97% ee, respectively, to be achieved in combination with yields of 83–100%.[89] Although the combination of transition metal catalysis with organocatalysis has allowed a range of novel and useful reactions to be achieved,[88d,90] the development of domino reactions induced by a combination of two types of catalysts still remains a challenge. While the organocatalysis is dominated by Lewis base catalysts, such as amines, carbenes, and tertiary phosphines, a metal catalyst usually has an empty coordination site to interact and activate a substrate. The challenge in combining an organocatalyst and a metal catalyst is in part to avoid the deactivation of the catalyst by Lewis acid/base interaction. Even in the absence of a catalyst poison, the presence of a Lewis base can erode the chiral environment of a chiral metal complex. Consequently, the success of tandem catalysis will need fine tuning of the hardness and softness of the metal catalyst and the organocatalyst to increase their compatibility. The combination of relay nickel catalysis with organocatalysis has been recently applied to develop highly efficient asymmetric multicomponent reactions.[91] As an example, McQuade and co-workers have developed an original one-pot multicomponent reaction catalysed by a microencapsulated amine catalyst **52** and chiral nickel complex (*R*,*R*)-**1** (Scheme 2.44).[92] Although the enantioselectivity of this process was not high (72% ee), the site isolation of two otherwise incompatible catalysts provided by microencapsulation brought new insight into the development of amine–Lewis acid tandem sequences. The encapsulation of the amine catalyst was the key for the success of the reaction for the following reasons: (1) the use of a soluble amine catalyst led to catalyst deactivation by complexation with the

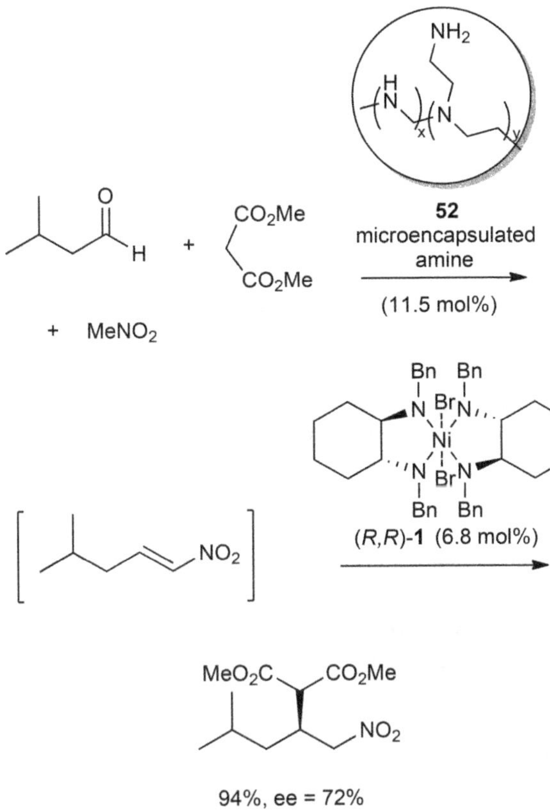

Scheme 2.44 Three-component domino Henry/Michael reaction catalysed by a combination of a chiral preformed nickel catalyst and a microencapsulated amine catalyst.

nickel catalyst; (2) a silica MCM-41 or polystyrene supported amine catalyst failed to catalyse the nitroalkene formation at room temperature, but the encapsulated poly(ethyleneimine) could; (3) the microencapsulated amine swollen in methanol retained its catalytic potency when in toluene, which allowed the one-pot reaction to be run in a mixture of two different solvents, and the microencapsulated amine and nickel catalyst could work under their respective ideal solvents of methanol and toluene. It must be noted that even if this nice and unwanted multicomponent domino reaction is not initiated but terminated by a Michael addition, it was decided to mention it in this paragraph dedicated to enantioselective nickel-catalysed domino reactions initiated by a Michael reaction owing to its rarity.

2.6 Conclusions

During the last 10 years, an important number of novel, highly efficient asymmetric conjugate additions of various nucleophiles to a wide variety of acceptor-activated alkenes has been developed on the basis of asymmetric nickel(II) catalysis by the very fact of the lower costs of nickel catalysts in comparison with other transition metals. These powerful processes can be considered as one of the most powerful and reliable tools for the stereocontrolled formation of carbon–carbon (and carbon–heteroatom) bonds, as has been demonstrated by the huge number of examples in which it has been applied as a key strategic transformation in total synthesis. Using optically active nickel(II) catalysts, these key reactions can be induced highly enantioselectively. In the last decade, a number of important developments have been achieved in this area, such as asymmetric nickel-catalysed conjugate additions of various 1,3-dicarbonyl compounds to nitroalkenes, including complex and functionalised ones, *e.g.* 3-nitro-2*H*-chromenes, nitroenynes, and nitrodienynes, with a beautiful example employing a recyclable mesoporous catalyst. Other nucleophiles, such as γ-butyrolactams, α-keto esters, α-keto anilides, 3-substituted oxindoles, azaarylacetates, highly functionalised acetamides, and acetylazaarenes, *etc.*, also give excellent results in additions to nitroalkenes. Furthermore, organozinc reagents, β-keto esters, 2-siloxyfurans, malononitriles, nitromethane, nitroacetates, and cyclic amines, among other nucleophiles, have been successfully added to various α,β-unsaturated carbonyl compounds and derivatives. Highly enantioselective intramolecular oxa-Michael additions to activated enones have also been described. Even more importantly, a range of powerful nickel-catalysed asymmetric domino reactions initiated by Michael additions, including multicomponent ones, have been successfully developed in the last 10 years. Undoubtedly, the future direction in this field is to continue expanding the scope of enantioselective nickel-catalysed conjugate additions through the employment of novel chiral nickel(II) catalysts, and to apply these powerful strategies, including fascinating domino processes based on Michael reactions, to the synthesis of biologically interesting molecules, including natural products.

References

1. T. Komnenos, *Justus Liebigs Ann. Chem.*, 1883, **218**, 145–169.
2. (a) P. Perlmutter, *Conjugate Addition Reactions in Organic Synthesis*, Pergamon Press, Elmsford, New York, 1992; (b) J. Christoffers, *Eur. J. Org. Chem.*, 1998, 1259–1266; (c) N. Krause, *Angew. Chem., Int. Ed.*, 1998, **37**, 283–285; (d) M. P. Sibi and S. Manyem, *Tetrahedron*, 2000, **56**, 8033–8061; (e) M. Kanai and M. Shibasaki, *Catalytic Asymmetric Synthesis*, 2nd edn, Wiley, New York, 2000, p. 569; (f) N. Krause and A. Hoffmann-Roder, *Synthesis*, 2001, 171–196; (g) O. M. Berner, L. Tedeschi and D. Enders, *Eur. J. Org. Chem.*, 2002, 1877–1894; (h) S. C. Jha and N. N. Joshi, *ARKIVOC*, 2002, 167–196; (i) J. Christoffers and A. Baro, *Angew. Chem., Int. Ed.*, 2003, **42**, 1688–1690; (j) T. Hayashi, *Bull. Chem. Soc. Jpn.*, 2004, **77**, 13–21; (k) J. Comelles, M. Moreno-Manas and A. Vallribera, *ARKIVOC*, 2005, **ix**, 207–238; (l) J. Christoffers, G. Koripelly, A. Rosiak and M. Rössle, *Synthesis*, 2007, **9**, 1279–1300; (m) A. Gutnov, *Eur. J. Org. Chem.*, 2008, **27**, 4547–4554.

3. T. Saegusa, Y. Ito, S. Tomitra and H. Kinoshita, *Bull. Chem. Soc. Jpn.*, 1972, **45**, 496–499.
4. (a) J. H. Nelson, P. N. Howells, G. L. Landen, G. C. DeLullo and R. A. Henry, *Fundam. Res. Homogeneous Catal.*, 1979, **3**, 921–939; (b) J. H. Nelson, P. N. Howells, G. C. DeLullo, G. L. Landen and R. A. Henry, *J. Org. Chem.*, 1980, **45**, 1246–1249.
5. (a) J. Clariana, N. Galvez, C. Marchi, M. Moreno-Manas, A. Vallribera and E. Molins, *Tetrahedron*, 1999, **55**, 7331–7344; (b) C. Marchi, E. Trepat, M. Moreno-Manas, A. Vallribera and E. Molins, *Tetrahedron*, 2002, **58**, 5699–5708; (c) V. Gallo, P. Mastrorilli, C. F. Nobile, G. Romanazzi and G. P. Suranna, *J. Chem. Soc., Dalton Trans.*, 2002, 4339–4342.
6. H. Brunner and B. Hammer, *Angew. Chem., Int. Ed. Engl.*, 1984, **32**, 312–313.
7. (a) K. Soai, S. Yokoyama, T. Hayasaka and K. Ebihara, *J. Org. Chem.*, 1988, **53**, 4148–4149; (b) K. Soai, T. Hayasaka, S. Ogajin and S. Yokoyama, *Chem. Lett.*, 1988, 1571–1572.
8. K. Soai, T. hayasaka, S. Ogajin and S. Yokoyama, *J. Chem. Soc., Chem. Commun.*, 1989, 516–517.
9. T. Fujisawa, S. Itoh and M. Shimizu, *Chem. Lett.*, 1994, 1777–1780.
10. A. Corma, M. Inglesias, M. V. Martin, J. Rubio and F. Sanchez, *Tetrahedron: Asymmetry*, 1992, **3**, 845–848.
11. (a) C. Bolm and M. Ewald, *Tetrahedron Lett.*, 1990, **31**, 5011–5012; (b) J. F. G. A. Jansen and B. L. Feringa, *Tetrahedron: Asymmetry*, 1992, **3**, 581–582; (c) A. H. M. De Vries, J. F. G. A. Jansen and B. L. Feringa, *Tetrahedron*, 1994, **50**, 4479–4491; (d) A. H. M. De Vries, R. Imbos and B. L. Feringa, *Tetrahedron: Asymmetry*, 1997, **8**, 1467–1473.
12. S. Kanemasa, Y. Oderaotoshi and E. Wada, *J. Am. Chem. Soc.*, 1999, **121**, 8675–8676.
13. W. Zhuang, R. G. Hazell and K. A. Jørgensen, *Chem. Commun.*, 2001, 1240–1241.
14. J. Christoffers, U. Rößler and T. Werner, *Eur. J. Org. Chem.*, 2000, 701–705.
15. R. Shintani and T. Hayashi, in *Modern Organonickel Chemistry*, ed. Y. Tamaru, Wiley-VCH, Weinheim, 2005, pp. 240–272.
16. For a review on metal-catalysed asymmetric conjugate addition reactions with formation of quaternary stereocentres, see: C. Hawner and A. Alexakis, *Chem. Commun.*, 2010, **46**, 7295–7306; for a review on rhodium and copper-catalysed asymmetric conjugate additions of alkenyl nucleophiles, see: D. Müller and A. Alexakis, *Chem. Commun.*, 2012, **48**, 12037–12049.
17. *Catalytic Asymmetric Conjugate Reactions*, ed. A. Cordova, Wiley-VCH, Weinheim, 2010.
18. D. A. Evans and D. Seidel, *J. Am. Chem. Soc.*, 2005, **127**, 9958–9959.
19. A. G. M. Barrett and G. G. Graboski, *Chem. Rev.*, 1986, **86**, 751–762.
20. (a) N. Ono, *The Nitro Group in Organic Synthesis*, Wiley-VCH, Weinheim, 2001; (b) D. Seebach, E. W. Colvin, F. Lehr and T. Weller, *Chimia*, 1979, **33**, 1–18.
21. S. H. Kang and D. Y. Kim, *Bull. Korean Chem. Soc.*, 2009, **30**, 1439–1440.

22. (a) A. N. Reznikov, E. A. Sidnin and Y. N. Klimochkin, *Russ. J. Org. Chem.*, 2013, **48**, 1600–1604; (b) A. N. Reznikov, E. V. Golovin and Y. N. Klimochkin, *Russ. J. Org. Chem.*, 2013, **49**, 663–668; (c) A. N. Reznikov and Y. N. Klimochkin, *Russ. J. Org. Chem.*, 2012, **48**, 1526–1529.
23. D. A. Evans, S. Mito and D. Seidel, *J. Am. Chem. Soc.*, 2007, **129**, 11583–11592.
24. H. Mitsunuma and S. Matsunaga, *Chem. Commun.*, 2011, **47**, 469–471.
25. K. Wilckens, M.-A. Duhs, D. Lentz and C. Czekelius, *Eur. J. Org. Chem.*, 2011, 5441–5446.
26. Q. Zhu, H. Huang, D. Shi, Z. Shen and C. Xia, *Org. Lett.*, 2009, **11**, 4536–4539.
27. (a) R. Jin, K. Liu, D. Xia, Q. Qian, G. Liu and H. Li, *Adv. Synth. Catal.*, 2012, **354**, 3265–3274; (b) K. Liu, R. Jin, T. Cheng, X. Xu, F. Gao, G. Liu and H. Li, *Chem.–Eur. J.*, 2012, **18**, 15546–15553.
28. W.-Y. Chen, L. Ouyang, R.-Y. Chen and X.-S. Li, *Tetrahedron Lett.*, 2010, **51**, 3972–3974.
29. L. Wen, F. Tang, C. Ge, X. Wang, Z. Han and J. Wu, *Synth. Commun.*, 2012, **42**, 3288–3295.
30. X. Li, X. Li, F. Peng and Z. Shao, *Adv. Synth. Catal.*, 2012, **354**, 2873–2885.
31. Z. Sun, M. Zhou, X. Li, X. Meng, F. Peng, H. Zhang and Z. Shao, *Chem.–Eur. J.*, 2014, **20**, 6112–6119.
32. X. Li, F. Peng, M. Zhou, M. Mo, R. Zhao and Z. Shao, *Chem. Commun.*, 2014, **50**, 1745–1747.
33. (a) N. E. Sheperd, H. Tanabe, Y. Xu, S. Matsunaga and M. Shibasaki, *J. Am. Chem. Soc.*, 2010, **132**, 3666–3667; (b) H. Tanabe, Y. Xu, B. Sun, S. Matsunaga and M. Shibasaki, *Heterocycles*, 2012, **86**, 611–622.
34. A. Nakamura, S. Lectard, D. Hashizume, Y. Hamashima and M. Sodeoka, *J. Am. Chem. Soc.*, 2010, **132**, 4036–4037.
35. Y. Xu, S. Matsunaga and M. Shibasaki, *Org. Lett.*, 2010, **12**, 3246–3249.
36. Y.-Y. Han, Z.-J. Wu, W.-B. Chen, X.-L. Du, X.-M. Zhang and W.-C. Yuan, *Org. Lett.*, 2011, **13**, 5064–5067.
37. A. Awata, M. Wasai, H. Masu, S. Kado and T. Arai, *Chem.–Eur. J.*, 2014, **20**, 2470–2477.
38. C. Fallan and H. W. Lam, *Chem.–Eur. J.*, 2012, **18**, 11214–11218.
39. A. J. Simpson and H. W. Lam, *Org. Lett.*, 2013, **15**, 2586–2589.
40. H. Ouchi, A. Asahina, T. Asakawa, M. Inai, Y. Hamashima and T. Kan, *Org. Lett.*, 2014, **16**, 1980–1983.
41. Y.-S. Kwak and E. J. Corey, *Org. Lett.*, 2004, **6**, 3385–3388.
42. A. Alexakis, J. E. Bäckvall, N. Krause, O. Pamies and M. Diéguez, *Chem. Rev.*, 2008, **108**, 2796–2823.
43. (a) A. Togni and M. Venanzi, *Angew. Chem., Int. Ed. Engl.*, 1994, **33**, 497–527; (b) F. Fache, E. Schulz, M. L. Tommasino and M. Lemaire, *Chem. Rev.*, 2000, **100**, 2159–2231.
44. (a) U. Shadakshari and S. K. Nayak, *Tetrahedron*, 2001, **57**, 8185–8190; (b) I. Wakimoto, Y. Tomioka and Y. Kawanami, *Tetrahedron*, 2002, **58**, 8095–8097.

45. A. Isleyen and O. Dogan, *Tetrahedron: Asymmetry*, 2007, **18**, 679–684.
46. J. Escorihuela, M. I. Burguete and S. V. Luis, *Tetrahedron Lett.*, 2008, **49**, 6885–6888.
47. C. Cimarelli, D. Fratoni and G. Palmieri, *Synth. Commun.*, 2009, **39**, 3184–3190.
48. M. Rachwalski, S. Lesniak and P. Kielbasinski, *Tetrahedron: Asymmetry*, 2010, **21**, 1890–1892.
49. M. Rachwalski, S. Jarzynski and S. Lesniak, *Tetrahedron: Asymmetry*, 2013, **24**, 1117–1119.
50. E. Shirakawa, Y. Yasuhara and T. Hayashi, *Chem. Lett.*, 2006, **35**, 768–769.
51. J. D. Sieber and J. P. Morken, *J. Am. Chem. Soc.*, 2008, **130**, 4978–4983.
52. C. F. Nising and S. Bräse, *Chem. Soc. Rev.*, 2008, **37**, 1218–1228.
53. L. Wang, X. Liu, Z. Dong, X. Fu and X. Feng, *Angew. Chem., Int. Ed.*, 2008, **47**, 8670–8673.
54. Y. N. Belokon, S. Harutyunyan, E. V. Vorontsov, A. S. Peregudov, V. N. Chrustalev, K. A. Kochetkov, D. Pripadchev, A. S. Sagyan, A. K. Beck and D. Seebach, *ARKIVOC*, 2004, **iii**, 132–150.
55. H. Suga, T. Kitamura, A. Kakehi and T. Baba, *Chem. Commun.*, 2004, 1414–1415.
56. D. A. Evans, S. J. Miller, T. Lectka and P. von Matt, *J. Am. Chem. Soc.*, 1999, **121**, 7559–7573.
57. K. Itoh, M. Hasegawa, J. Tanaka and S. Kanemasa, *Org. Lett.*, 2005, **7**, 979–981.
58. D. A. Evans, R. J. Thomson and F. Franco, *J. Am. Chem. Soc.*, 2005, **127**, 10816–10817.
59. K. Itoh, Y. Oderaotoshi and S. Kanemasa, *Tetrahedron: Asymmetry*, 2003, **14**, 635–639.
60. H. Yanagita, K. Kodama and S. Kanemasa, *Tetrahedron Lett.*, 2006, **47**, 9353–9357.
61. K. Itoh and S. Kanemasa, *J. Am. Chem. Soc.*, 2002, **124**, 13394–13395.
62. M. Hasegawa, F. Ono and S. Kanemasa, *Tetrahedron Lett.*, 2008, **49**, 5220–5223.
63. S. Zhang, G. Gangal and H. Uludag, *Chem. Soc. Rev.*, 2007, **36**, 507–531.
64. Y. Kato, Z. Chen, S. Matsunaga and M. Shibasaki, *Synlett*, 2009, **10**, 1635–1638.
65. J. Pawlas, Y. Nakao, M. Kawatsura and J. F. Hartwig, *J. Am. Chem. Soc.*, 2002, **124**, 3669–3679.
66. L. Fadini and A. Togni, *Tetrahedron: Asymmetry*, 2008, **19**, 2555–2562.
67. (a) P. I. Dalko and I. Moisan, *Angew. Chem., Int. Ed.*, 2001, **40**, 3726–3748; (b) A. Berkessel and H. Gröger, in *Asymmetric Organocatalysis–From Biomimetic Concepts to Powerful Methods for Asymmetric Synthesis*, Wiley-VCH, Weinheim, 2005; (c) M. S. Taylor and E. N. Jacobsen, *Angew. Chem., Int. Ed.*, 2006, **45**, 1520–1543; (d) *Enantioselective Organocatalysis*, ed. P. I. Dalko, Wiley-VCH, Weinheim, 2007; (e) H. Pellissier, *Tetrahedron*, 2007, **63**, 9267–9331; (f) *Chem. Rev.*, 2007, **107**, 5413–5883, Special Issue on Organocatalysis, B. List (ed.); (g) D. W. C. MacMillan, *Nature*, 2008,

455, 304–308; (h) X. Yu and W. Wang, *Chem.–Asian J.*, 2008, **3**, 516–532; (i) A. Dondoni and A. Massi, *Angew. Chem., Int. Ed.*, 2008, **47**, 4638–4660; (j) P. Melchiorre, M. Marigo, A. Carlone and G. Bartoli, *Angew. Chem., Int. Ed.*, 2008, **47**, 6138–6171; (k) C. Palomo, M. Oiarbide and R. Lopez, *Chem. Soc. Rev.*, 2009, **38**, 632–653; (l) S. Bertelsen and K. A. Jørgensen, *Chem. Soc. Rev.*, 2009, **38**, 2178–2189; (m) M. Bella and T. Gasperi, *Synthesis*, 2009, 1583–1586; (n) *Recent Developments in Asymmetric Organocatalysis*, ed. H. Pellissier, Royal Society of Chemistry, Cambridge, 2010; (o) *Enantioselective Organocatalyzed Reactions*, vol. I and II, ed. R. Mahrwald, Springer, Berlin, 2011; (p) *Comprehensive Enantioselective Organocatalysis; Catalysts, Reactions, and Applications*, ed. P. Dalko, Wiley-VCH, Weinheim, 2013.

68. H. Clavier and H. Pellissier, *Adv. Synth. Catal.*, 2012, **18**, 3347–3403.
69. K. C. Nicolaou and J. S. Chen, *Chem. Soc. Rev.*, 2009, **38**, 2993–3009.
70. K. C. Nicolaou, D. J. Edmonds and P. G. Bulger, *Angew. Chem., Int. Ed.*, 2006, **45**, 7134–7186.
71. (a) L. F. Tietze and U. Beifuss, *Angew. Chem., Int. Ed.*, 1993, **32**, 131–163; (b) L. F. Tietze, *Chem. Rev.*, 1996, **96**, 115–136; (c) L. F. Tietze, G. Brasche and K. Gericke, *Domino Reactions in Organic Synthesis*, Wiley-VCH, Weinheim, 2006; (d) *Domino Reactions-Concepts for Efficient Organic Synthesis*, ed. L. F. Tietze, Wiley-VCH, Weinheim, 2014.
72. (a) L. F. Tietze and A. Modi, *Med. Res. Rev.*, 2000, **20**, 304–322; (b) C. Hulme and V. Gore, *Curr. Med. Chem.*, 2003, **10**, 51–80; (c) *Multicomponent Reactions*, J. Zhu and H. Bienaymé, Wiley-VCH, Weinheim, 2005; (d) D. J. Ramon and M. Yus, *Angew. Chem., Int. Ed.*, 2005, **44**, 1602–1634; (e) B. B. Touré and D. G. Hall, *Chem. Rev.*, 2009, **109**, 4439–4486; (f) H. M. L. Davies and E. J. Sorensen, *Chem. Soc. Rev.*, 2009, **38**, 2981–2982, Special Issue on Rapid formation of molecular complexity in organic synthesis; (g) Synthesis of Heterocycles via Multicomponent Reactions, *Topics in Heterocyclic Chemistry*, ed. R. V. A. Orru and E. Ruijter, vol. I and II, Springer, Berlin, 2010.
73. (a) G. H. Posner, *Chem. Rev.*, 1986, **86**, 831–844; (b) T.-L. Ho, *Tandem Organic Reactions*, Wiley, New York, 1992; (c) H. Waldmann, *Nachr. Chem., Tech. Lab.*, 1992, **40**, 1133–1140; (d) K. Fukumoto, *Synth. Org. Chem. Jpn.*, 1994, **52**, 2–18; (e) R. A. Bunce, *Tetrahedron*, 1995, **51**, 13103–13159; (f) P. J. Parsons, C. S. Penkett and A. J. Shell, *Chem. Rev.*, 1996, **96**, 195–206; (g) P. C. F. Balaure and P. I. A. Filip, *Rev. Roum. Chim.*, 2002, **46**, 809–833; (h) E. Capdevila, J. Rayo, F. Carrion, I. Jové, J. I. Borrell and J. Teixido, *Afinidad*, 2003, **506**, 317–337; (i) L. F. Tietze and N. Rackelmann, *Pure Appl. Chem.*, 2004, **76**, 1967–1983; (j) P. I. Dalko and L. Moisan, *Angew. Chem., Int. Ed.*, 2004, **43**, 5138–5175; (k) H. Pellissier, *Tetrahedron*, 2006, **62**, 2143–2173; (l) H. Pellissier, *Tetrahedron*, 2006, **62**, 1619–1665; (m) D. Enders, C. Grondal and M. R. M. Hüttl, *Angew. Chem., Int. Ed.*, 2007, **46**, 1570–1581; (n) G. Guillena, D. J. Ramon and M. Yus, *Tetrahedron: Asymmetry*, 2007, **18**, 693–700; (o) D. M. D'Souza and T. J. J. Müller, *Chem. Soc. Rev.*, 2007, **36**, 1095–1108; (p) A. Padwa and S. K. Bur, *Tetrahedron*, 2007, **63**, 5341–5378; (q) M. Colombo and I. Peretto, *Drug Discovery Today*,

2008, **13**, 677–684; (r) A.-N. Alba, X. Companyo, M. Viciano and R. Rios, *Curr. Org. Chem.*, 2009, **13**, 1432–1474; (s) J. E. Biggs-Houck, A. Younai and J. T. Shaw, *Curr. Opin. Chem. Biol.*, 2010, **14**, 371–382; (t) C. Grondal, M. Jeanty and D. Enders, *Nat. Chem.*, 2010, **2**, 167–178; (u) M. Ruiz, P. Lopez-Alvarado, G. Giorgi and J. C. Menéndez, *Chem. Soc. Rev.*, 2011, **40**, 3445–3454; (v) L. Albrecht, H. Jiang and K. A. Jørgensen, *Angew. Chem., Int. Ed.*, 2011, **50**, 8492–8509; (w) H. Pellissier, *Adv. Synth. Catal.*, 2012, **354**, 237–294; (x) C. De Graaff, E. Ruijter and R. V. A. Orru, *Chem. Soc. Rev.*, 2012, **41**, 3969–4009.

74. M. Kitamura, T. Miki, K. Nakano and R. Noyori, *Tetrahedron Lett.*, 1996, **37**, 5141–5144.
75. F. Ono, M. Hasegawa, S. Kanemasa and J. Tanaka, *Tetrahedron Lett.*, 2008, **49**, 5105–5107.
76. T. Arai, N. Yokoyama, A. Mishiro and H. Sato, *Angew. Chem., Int. Ed.*, 2010, **49**, 7895–7898.
77. N. Halland, T. Velgaard and K. A. Jørgensen, *J. Org. Chem.*, 2003, **68**, 5067–5074.
78. Z. Dong, J. Feng, W. Cao, X. Liu, L. Lin and X. Feng, *Tetrahedron Lett.*, 2011, **52**, 3433–3436.
79. (a) J. Boruwa, N. Gogoi, P. P. Saikia and N. C. Barua, *Tetrahedron: Asymmetry*, 2006, **17**, 3315–3326; (b) C. Palomo, M. Oiarbide and A. Laso, *Eur. J. Org. Chem.*, 2007, 2561–2574; (c) E. Marquès-Lopez, P. Merino, T. Tejero and R. P. Herrera, *Eur. J. Org. Chem.*, 2009, 2401–2420; (d) Y. Alvarez-Casao, E. Marques-Lopez and R. P. Herrera, *Symmetry*, 2011, **3**, 220–245.
80. W. Li, X. Liu, Z. Mao, Q. Chen and R. Wang, *Org. Biomol. Chem.*, 2012, **10**, 4767–4773.
81. Q. Liang, J. He and B. Ni, *Tetrahedron: Asymmetry*, 2014, **25**, 1146–1149.
82. T. Arai and Y. Yamamoto, *Org. Lett.*, 2014, **16**, 1700–1703.
83. L. A. Wessjohann, W. Brandt and T. Thiemann, *Chem. Rev.*, 2003, **103**, 1625–1648.
84. H.-U. Reissig and R. Zimmer, *Chem. Rev.*, 2003, **103**, 1151–1196.
85. H. Pellissier, *Tetrahedron*, 2008, **64**, 7041–7095.
86. (a) H. Xie, L. Zu, H. Li, J. Wang and W. Wang, *J. Am. Chem. Soc.*, 2007, **129**, 10886–10894; (b) R. Rios, H. Sunden, J. Vesely, G.-L. Zhao, P. Dziedzic and A. Cordova, *Adv. Synth. Catal.*, 2007, **349**, 1028–1032; (c) H. M. Hansen, D. A. Longbottom and S. V. Ley, *Chem. Commun.*, 2006, 4838–4840.
87. H. J. Lee, S. M. Kim and D. Y. Kim, *Tetrahedron Lett.*, 2012, **53**, 3437–3439.
88. (a) J.-A. Ma and D. Cahard, *Angew. Chem., Int. Ed.*, 2004, **43**, 4566–1508; (b) M. Kanai, N. Kato, E. Ichikawa and M. Shibasaki, *Synlett*, 2005, 1491–1508; (c) D. H. Paull, C. J. Abraham, M. T. Scerba, E. Alden-Danforth and T. Leckta, *Acc. Chem. Res.*, 2008, **41**, 655–663; (d) Z. Shao and H. Zhang, *Chem. Soc. Rev.*, 2009, **38**, 2745–2755; (e) C. Zhong and X. Shi, *Eur. J. Org. Chem.*, 2010, 2999–3025; (f) M. Rueping, R. M. Koenigs and I. Atodiresei, *Chem.–Eur. J.*, 2010, **16**, 9350–9365; (g) J. Zhou, *Chem.–Asian J.*, 2010, **5**, 422–434; (h) L. M. Ambrosini and T. H. Lambert, *ChemCatChem*, 2010, **2**, 1373–1380; (i) S. Piovesana, D. M. Scarpino Schietroma and M. Bella,

Angew. Chem., Int. Ed., 2011, **50**, 6216–6232; (j) N. T. Patil, V. S. Shinde and B. Gajula, *Org. Biomol. Chem.*, 2012, **10**, 211–224; (k) A. E. Allen and D. W. C. MacMillan, *Chem. Sci.*, 2012, **3**, 633–658; (l) Z. Du and Z. Shao, *Chem. Soc. Rev.*, 2013, **42**, 1337–1378; (m) H. Pellissier, *Tetrahedron*, 2013, **69**, 7171–7210; (n) *Enantioselective Multicatalysed Tandem Reactions*, ed. H. Pellissier, Royal Society of Chemistry, Cambridge, 2014.

89. Y. Ito, M. Sawamura and T. Hayashi, *J. Am. Chem. Soc.*, 1986, **108**, 6405–6406.
90. M. Klussmann, *Angew. Chem., Int. Ed.*, 2009, **48**, 7124–7125.
91. (a) R. V. A. Orru and M. de Greef, *Synthesis*, 2003, 1471–1499; (b) E. Ruijter, R. Scheffelaar and R. V. A. Orru, *Angew. Chem., Int. Ed.*, 2011, **50**, 6234–6246.
92. (a) S. L. Poe, M. Kobaslija and D. T. McQuade, *J. Am. Chem. Soc.*, 2006, **128**, 15586–15587; (b) S. L. Poe, M. Kobaslija and D. T. McQuade, *J. Am. Chem. Soc.*, 2007, **129**, 9216–9221.

CHAPTER 3

Enantioselective Nickel-Catalysed Cross-Coupling Reactions

3.1 Introduction

The transition-metal-catalysed cross-coupling reactions represent a powerful approach for the construction of carbon–carbon bonds; consequently, these reactions have been widely studied during the last few decades.[1] Their development has reached a level of sophistication that allows for a wide range of coupling partners to be combined efficiently. The emergence of cross-coupling as a popular method in synthesis arises from both the diversity of organometallic reagents used in these reactions and the broad range of functional groups which can be incorporated into the reagents.[2] This paradigm for carbon–carbon bond construction has allowed chemists to assemble complex molecular frameworks of diversified interests encompassing the total synthesis of natural products, medicinal chemistry, and industrial process development, as well as chemical biology, materials, and nanotechnology. Among the transition metals employed in cross-coupling reactions, the majority of the investigations have focused on nickel- and palladium-catalysed cross-couplings of aryl and alkenyl halides with various organometallic reagents.[3] In addition to simple non-asymmetric cross-coupling reactions, enantioselective versions have also been the topics of interest in organic and organometallic chemistry.[4] Nickel is by far the most versatile metal for the cross-coupling reactions of alkyl halides. Since the ground-breaking studies of Negishi,[5] Kumada,[6] Kochi and Tamura,[7] Suzuki,[8] and Knochel,[9] the design of new catalyst systems has enabled the cross-coupling of a wide

RSC Catalysis Series No. 26
Enantioselective Nickel-Catalysed Transformations
By Hélène Pellissier

Published by the Royal Society of Chemistry, www.rsc.org

number of secondary alkyl halides and other electrophiles with a variety of organometallic reagents, including zinc, boron, silicon, magnesium, tin, zirconium, and indium compounds, among others. On the other hand, asymmetric nickel-catalysed reductive coupling reactions of alkynes or dienes with aldehydes have emerged as powerful synthetic tools for the selective preparation of chiral functionalised alkenes and derivatives with high control of regio- and enantioselectivities.

3.2 Negishi Cross-Coupling Reactions

The Negishi cross-coupling reaction was first reported for the synthesis of unsymmetrical biaryls, in good yields, in 1977.[5] Since then, this reaction has been applied as a versatile nickel- or palladium-catalyzed coupling of organometals (Al, Zn, or Zr) with various halide-containing moieties (aryl, vinyl, benzyl, or allyl).[10] Alkylzinc compounds are easily subjected to the cross-coupling process, extending the scope and potential use of the Negishi reaction beyond regular $C(sp^2)–C(sp^2)$ couplings. The major disadvantage of the Negishi reaction is the incompatibility of organozinc reagents with some common moieties in organic compounds, as well as being relatively sensitive towards oxygen and water. These drawbacks have led to the use of excess Zn organometals, which can be considered as a limitation to the applications of the Negishi reaction. The most widely used organometallic reagents in this reaction are R_2Zn or RZnX types. The first example of a nickel-catalysed reaction of secondary alkyl bromides or iodides with organozinc reagents was reported by Zhou and Fu in 2003.[11] Their study served as a starting point for the development of an impressive series of asymmetric Negishi coupling reactions of secondary alkyl halides. Later, Fischer and Fu reported an asymmetric version of the nickel-catalysed Negishi reaction of secondary electrophiles, using a catalyst generated *in situ* from $NiCl_2$·glyme and with (*R*)-i-Pr-PYBOX as the chiral ligand.[12] The reaction requires no special precautions and is performed in air by using 1,3-dimethylimidazolidin-2-one (DMI) as solvent. Unfunctionalised and functionalised organozinc reagents, including those bearing alkene, benzyl ether, acetal, imide, and nitrile groups, are compatible with the process, as shown in Scheme 3.1. The corresponding chiral products are obtained in good yields (51–90%) and with moderate to very high enantioselectivities of 77–96% ee. The reaction is selective for the α-bromoamide in the presence of an external, unactivated primary alkyl bromide and is stereoconvergent, as racemic substrates are converted preferentially into one major enantiomer.

Another asymmetric and stereoconvergent variant of the Negishi reaction was developed by the same authors, involving racemic secondary benzylic bromides.[13] By using the (*S*)-i-Pr-PYBOX ligand and $NiBr_2$ as the catalyst system in *N*,*N*-dimethylacetamide (DMA) as solvent, the coupling of 1-bromoindanes with alkylzinc reagents proceeded in moderate to excellent yields of up to 89%, and with excellent enantioselectivities of 91–98% ee, as shown in Scheme 3.2. Acyclic benzyl bromides were also coupled effectively, although with lower enantioselectivities (75% ee). This methodology was also suitable

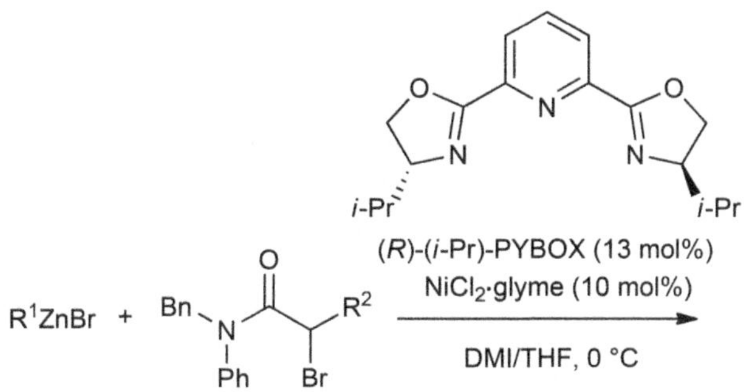

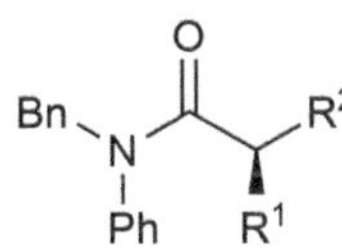

R^1 = *n*-Hex, R^2 = Et: 90%, ee = 96%
R^1 = *n*-Hex, R^2 = *n*-Bu: 85%, ee = 96%
R^1 = Me, R^2 = Et: 90%, ee = 91%
R^1 = Me, R^2 = *i*-Bu: 78%, ee = 87%
R^1 = $Bn(CH_2)_2$, R^2 = Et: 84%, ee = 96%
R^1 = $CyCH_2$, R^2 = Et: 58%, ee = 92%
R^1 = $(CH_2)_4OBn$, R^2 = Et: 77%, ee = 96%
R^1 = $(CH_2)_3CH(OCH_2CH_2O)$, R^2 = Me: 66%, ee = 77%
R^1 = $(CH_2)_4NPhth$, R^2 = Et: 51%, ee = 96%
R^1 = $(CH_2)_6CN$, R^2 = Et: 70%, ee = 93%

Scheme 3.1 Cross-coupling of secondary α-bromoamides with alkylzinc reagents.

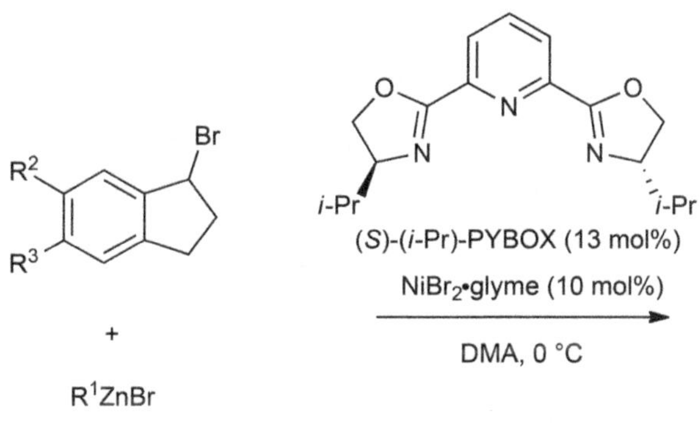

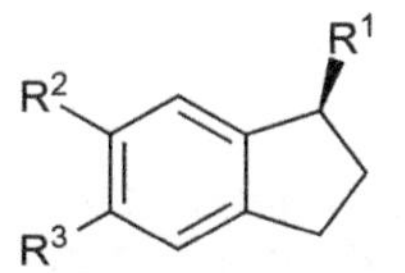

R^1 = *n*-Hex, R^2 = Et, R^3 = H: 89%, ee = 96%
R^1 = $(CH_2)_3CN$, R^2 = R^3 = H: 64%, ee = 91%
R^1 = $(CH_2)_6Cl$, R^2 = R^3 = H: 69%, ee = 94%
R^1 = $(CH_2)_3Ph$, R^2,R^3 = $(CH{=}CH)_2$: 76%, ee = 98%

Scheme 3.2 Cross-coupling of bromoindanes with alkylzinc reagents.

for secondary benzylic chlorides, such as 1-chloroindanes, with high enantioselectivities (91–93% ee) combined with moderate yields (56–61%), and was found to be insensitive to moisture and oxygen.

The same authors have described a third example of an asymmetric Negishi reaction involving racemic secondary allylic chlorides and a nickel catalyst generated *in situ* from $NiBr_2$·glyme and with (*S*)-Bn-PYBOX as the chiral ligand.[14] As shown in Scheme 3.3, the reaction was first tested with "symmetrical" allylic chlorides (R^1 = R^3), which should be transformed into the same product regardless of the reaction site. The yield and enantioselectivity dropped dramatically (95–69% ee) as the steric bulk of the R^3 substituent α to the chloro group increased (Me to i-Pr). Unsymmetrical allylic chlorides reacted at the carbon atom with the smallest substituent (R^1 or R^3) with a greater than 95 : 5 regioselectivity ratio and with enantioselectivities of 57–97% ee, regardless of the isomeric composition of the substrate. Conjugated allylic chlorides reacted preferentially at the γ-position.

With the aim of expanding the scope of nickel-catalysed cross-coupling reactions to other families of reaction partners, the same group developed highly efficient asymmetric cross-couplings of racemic secondary propargylic bromides with arylzinc reagents.[15] As shown in Scheme 3.4, a range of chiral alkynes were achieved by using a combination of $NiCl_2$·glyme and

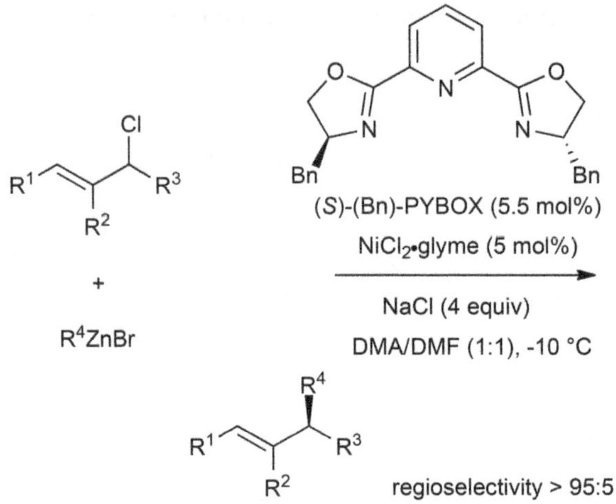

R^1 = R^3 = Me, R^2 = H, R^4 = *n*-Hex: 87%, ee = 95%
R^1 = R^3 = *i*-Pr, R^2 = H, R^4 = $(CH_2)_3$OTBS: 57%, ee = 69%
R^1 = R^2 = R^3 = Me, R^4 = *p*-MeO$C_6H_4(CH_2)_3$: 98%, ee = 54%
R^1 = *n*-Bu, R^2 = H, R^3 = Me, R^4 = $(CH_2)_2$Bn: 83%, ee = 97%
R^1 = CONEt_2, R^2 = H, R^3 = Me, R^4 = Et: 91%, ee = 57%
R^1 = CON(OMe)Me, R^2 = H, R^3 = Me, R^4 = $(CH_2)_3$OTBS: 93%, ee = 91%

Scheme 3.3 Cross-coupling of secondary allylic chlorides with alkylzinc reagents.

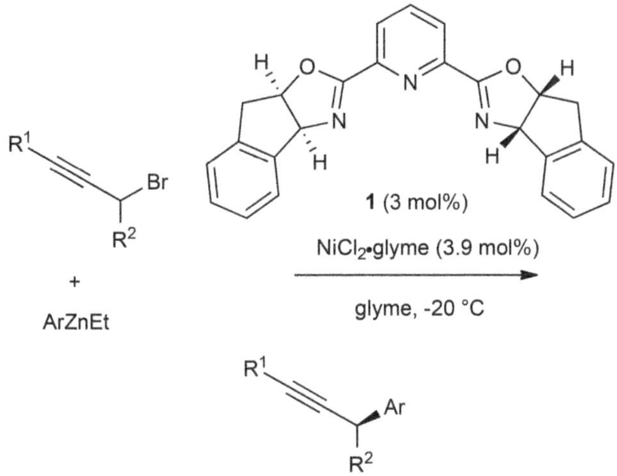

R^1 = TMS, R^2 = Me, Ar = *p*-Tol: 92%, ee = 93%
R^1 = TMS, R^2 = *n*-Bu, Ar = 2-Naph: 70%, ee = 93%
R^1 = TMS, R^2 = $(CH_2)_4CO_2Et$, Ar = *p*-(CH_2O*t*-Bu)C_6H_4: 76%, ee = 94%
R^1 = Et, R^2 = Me, Ar = *p*-(MeO)C_6H_4: 76%, ee = 91%
R^1 = *t*-Bu, R^2 = $(CH_2)_4Cl$, Ar = *p*-Tol: 81%, ee = 92%
R^1 = Ph, R^2 = Et, Ar = Ph: 83%, ee = 89%
R^1 = TMS, R^2 = $(CH_2)_5$-(Z)-CH=CHEt, Ar = *m*-(MeO)C_6H_4: 79%, ee = 94%

Scheme 3.4 Cross-coupling of secondary propargylic bromides with arylzinc reagents.

the commercially available chiral PYBOX ligand **1** employed at only 3 mol% of catalyst loading. The investigation of the substrate scope of the reaction showed that it was broad, since the presence of functional groups, such as ethers, esters, acetals, and alkenes on the propargylic bromides, was tolerated. Moreover, the reaction conditions could also be applied to the reactions of propargylic bromides that bear silyl protecting groups other than TMS. Thus, Me_2PhSi- and TIPS-substituted electrophiles also underwent enantioselective cross-coupling in good yields (81–88%) and with high enantioselectivities (88–93% ee).

Many interesting target molecules include ketones that bear an α-aryl substituent, making the development of methods for the synthesis of this structural motif an active area of investigation. In this context, Fu *et al.* have developed the first catalytic asymmetric cross-couplings of racemic secondary α-bromo ketones with arylzinc reagents to afford the corresponding chiral ketones bearing a potentially labile tertiary stereocentre (Scheme 3.5).[16] These stereoconvergent carbon–carbon bond-forming processes occurred under unusually mild conditions (−30 °C and no activators) in the presence of a nickel catalyst generated *in situ* from $NiCl_2$·glyme and the chiral PYBOX ligand **2**. The investigation of the substrate scope of the reaction showed that very good enantioselectivities of up to 96% ee and useful yields (76–93%)

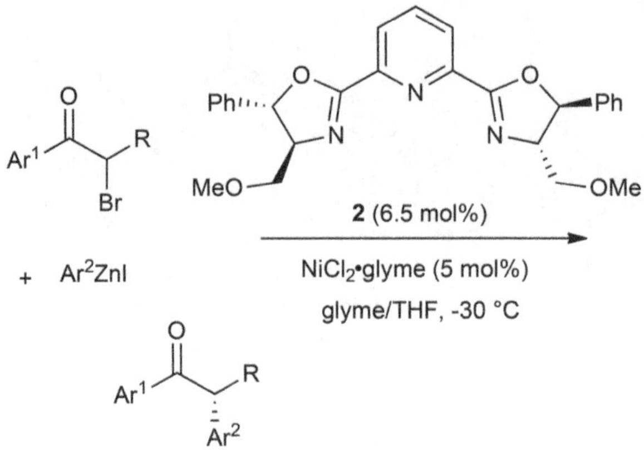

Ar^1 = Ar^2 = Ph, R = Me: 86%, ee = 96%
Ar^1 = Ph, Ar^2 = *m*-Tol, R = Me: 88%, ee = 94%
Ar^1 = Ph, Ar^2 = *p*-MeOC$_6$H$_4$, R = Me: 93%, ee = 96%
Ar^1 = Ar^2 = Ph, R = Et: 86%, ee = 94%
Ar^1 = Ar^2 = Ph, R = Bn: 76%, ee = 95%
Ar^1 = Ar^2 = Ph, R = (CH$_2$)$_2$Cl: 90%, ee = 92%
Ar^1 = *p*-MeOC$_6$H$_4$, Ar^2 = Ph, R = Me: 90%, ee = 96%
Ar^1 = *p*-CF$_3$C$_6$H$_4$, Ar^2 = Ph, R = Me: 76%, ee = 87%
Ar^1 = *o*-FC$_6$H$_4$, Ar^2 = Ph, R = Me: 80%, ee = 72%
Ar^1 = *o*-FC$_6$H$_4$, Ar^2 = Ph, R = Me: 72%, ee = 80%
Ar^1 = *o*-EtC$_6$H$_4$, Ar^2 = Ph, R = Me: 79%, ee = 75%
Ar^1 = 2-thienyl, Ar^2 = Ph, R = Me: 81%, ee = 96%
Ar^1 = Ar^2 = Ph, R = *i*-Pr: ee < 5%

Scheme 3.5 Cross-coupling of secondary α-bromo ketones with arylzinc reagents.

were observed with a variety of α-alkyl substituents, including those that were functionalised and β-branched; however, when R was large (i-Pr), little of the cross-coupling product was formed and in only 5% ee. If the aryl group of the ketone was bulky, the reaction proceeded with moderate enantioselectivity (79–80% ee). In contrast, good enantioselectivities (96% ee) were observed regardless of whether the group was electron-rich (96% ee) or electron-poor (87% ee). Moreover, a thiophene was compatible with the reaction, providing an enantioselectivity of 96% ee.

With the aim of adding a new dimension to their impressive enantioselective cross-coupling reactions of alkyl electrophiles, these authors investigated the possibility of using oxygen-based leaving groups instead of halides.[17] As shown in Scheme 3.6, they established that a diverse array of racemic propargylic carbonates were suitable coupling partners in nickel-catalysed Negishi reactions with arylzinc reagents in the presence of a nickel complex generated *in situ* from NiCl$_2$(PCy$_3$)$_2$ and the commercially available PYBOX ligand **1**. The carbonate of a phenyl group bearing a methoxy substituents at the 2-, 4-, and 6-positions was found optimal among a range of

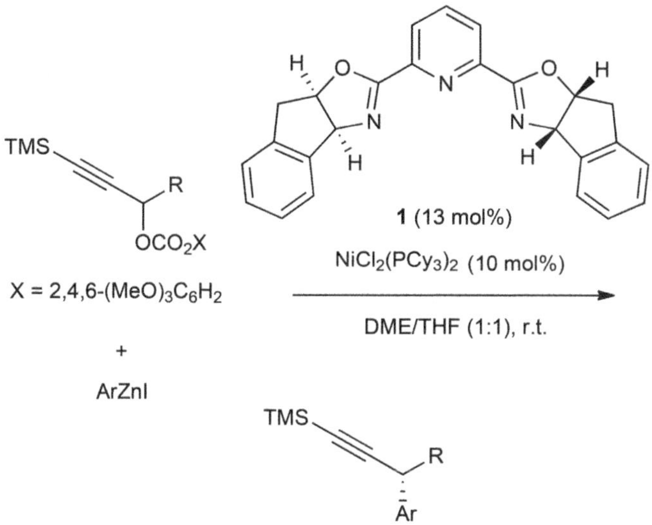

R = Me, Ar = *p*-Tol: 69%, ee = 93%
R = *n*-Bu, Ar = Ph: 81%, ee = 90%
R = *i*-Bu, Ar = *p*-Tol: 57%, ee = 93%
R = *n*-Bu, Ar = *o*-$MeOC_6H_4$: 66%, ee = 93%
R = *n*-Bu, Ar = *m*-$MeOC_6H_4$: 73%, ee = 92%
R = *n*-Bu, Ar = *p*-$MeOC_6H_4$: 76%, ee = 89%
R = $(CH_2)_4OPMB$, Ar = 3,5-$(CF_3)_2C_6H_3$: 85%, ee = 85%
R = $(CH_2)_4OTBS$, Ar = 7-MeO-2-Naph: 94%, ee = 91%
R = $(CH_2)_4OAc$, Ar = 4-Cl-5-F-C_6H_3: 81%, ee = 86%
R = $CH_2CH(Me)(CH_2)_2CH{=}C(Me)_2$, Ar = Ph: 79%, ee = 89%

Scheme 3.6 Cross-coupling of TMS-protected secondary propargylic carbonates with arylzinc reagents.

carbonates and xanthates. The method was compatible with a diverse set of functional groups, such as aryl methyl ethers (89–93% ee), acetals (84–92% ee), silyl esters (91% ee), esters (86% ee), aryl chlorides and fluorides (84% ee), alkenes (89% ee), and a Boc-protected nitrogen heterocycle (90% ee). Moreover, the process could be applied to other families of alkynes. Thus, regardless of whether the distal carbon of the propargylic carbonate bore a small or a large alkyl group (Me or *t*-Bu), or an aromatic substituent (Ph) instead of the TMS group, the stereoconvergent Negishi reaction proceeded with promising enantioselectivity (78–88% ee) and yield (62–95%).

Many bioactive compounds bear a cyano group attached to a stereogenic carbon. Furthermore, nitriles serve as versatile precursors to a diverse array of molecules, including heterocycles, aldehydes, ketones, amides, carboxylic acids, and amines.[18] Enantioenriched α-alkyl-α-aryl nitriles are particularly noteworthy targets, since they can be transformed into α-aryl carboxylic acids, such as α-arylpropionic acids which are widely used as nonsteroidal anti-inflammatory drugs. In 2012, Fu *et al.* reported a novel route to chiral α-aryl nitriles based on the highly enantioselective nickel-catalysed cross-coupling of α-bromo nitriles with arylzinc reagents.[19] This remarkable process was catalysed by chiral bisoxazoline ligand **3** because PYBOX ligands gave in this case low enantioselectivities (<40% ee). As shown in Scheme 3.7, a variety of useful enantioenriched secondary α-aryl nitriles bearing a wide range of functionalities were obtained in good to excellent yields (77–99%) and good to high enantioselectivities of 76–94% ee. This methodology was not limited to phenylation reactions since uniformly high enantioselectivities were achieved regardless of whether the aryl group was electron-rich or electron-poor. In addition, the same reaction conditions were applied to Negishi alkenylations using alkenylzinc reagents, with good yields and enantioselectivities of 80–92% ee, as shown in Scheme 3.7. This novel process opened a

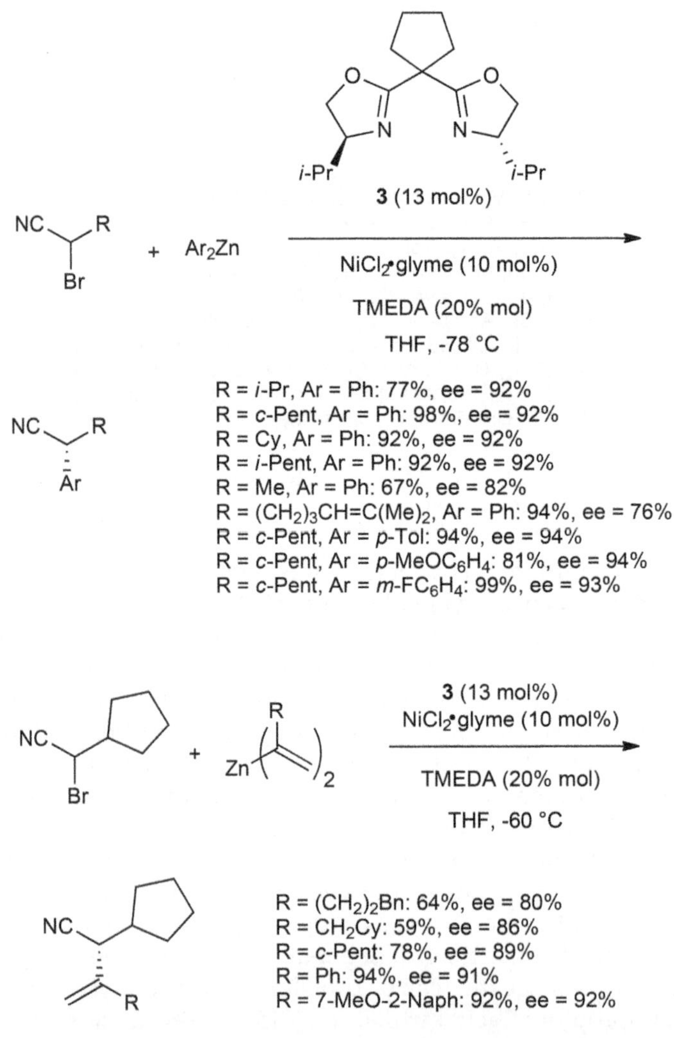

Scheme 3.7 Cross-couplings of α-bromo nitriles with aryl- and alkenylzinc reagents.

novel route to chiral allylic nitriles which are suitable substrates for stereoselective functionalisations. It is important to highlight that this excellent work represented the first method for the stereoconvergent cross-coupling of racemic α-halo nitriles, specifically, nickel-catalysed Negishi arylations and alkenylations.

Later, these authors developed other nickel-catalysed cross-coupling reactions involving secondary electrophiles with secondary nucleophiles.[20] Indeed, the reaction of racemic secondary benzylic bromides with secondary cycloalkylzinc reagents was achieved by using (*S*,*S*)-i-Pr-PYBOX as a ligand of $NiBr_2$·glyme in the presence of CsI as an additive, as shown in Scheme 3.8. The couplings proceeded in high enantioselectivities of up to 95% ee for

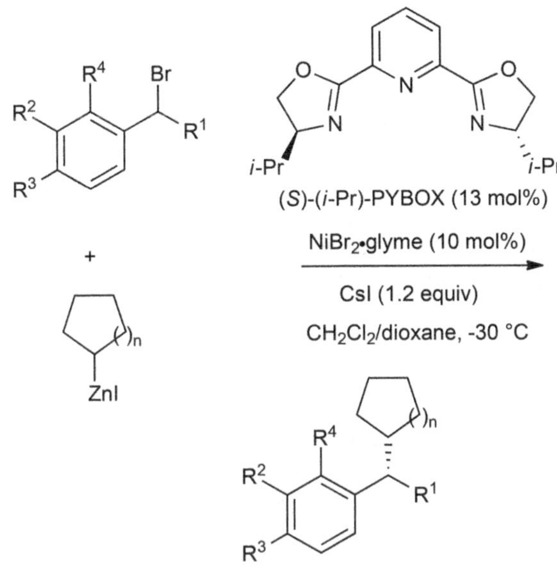

R^1 = Me, R^2 = R^3 = R^4 = H, n = 1: 91%, ee = 95%
R^1 = Et, R^2 = R^3 = R^4 = H, n = 1: 84%, ee = 93%
R^1 = *i*-Bu, R^2 = R^3 = R^4 = H, n = 1: 75%, ee = 93%
R^1 = Me, R^2 = R^3 = H, R^4 = F, n = 1: 48%, ee = 87%
R^1 = Me, R^2 = OMe, R^3 = R^4 = H, n = 1: 86%, ee = 95%
R^1 = Et, R^2 = OBn, R^3 = R^4 = H, n = 1: 65%, ee = 91%
R^1 = Me, R^2 = R^4 = H, R^3 = Ph, n = 1: 80%, ee = 95%
R^1 = Me, R^2 = R^4 = H, R^3 = Cl, n = 1: 74%, ee = 92%
R^1 = Me, R^2 = R^4 = H, R^3 = I, n = 1: 74%, ee = 91%
R^1 = Et, R^2 = R^4 = H, R^3 = Br, n = 1: 68%, ee = 92%
R^1 = Me, R^2 = CF_3, R^3 = R^4 = H, n = 1: 51%, ee = 91%
R^1 = Me, R^2 = OBn, R^3 = R^4 = H, n = 1: 65%, ee = 91%
R^1 = Et, R^2 = R^3 = R^4 = H, n = 3: 60%, ee = 87%
R^1 = Me, R^2 = OMe, R^3 = R^4 = H, n = 3: 52%, ee = 84%

Scheme 3.8 Cross-coupling of secondary aryl bromides with secondary alkylzinc reagents.

electrophiles in which the alkyl substituent (R^1) ranged in size from methyl to isobutyl. Furthermore, the aromatic ring could be *ortho*, *meta*, or *para* substituted, and it could bear an electron-withdrawing or an electron-donating group. In addition to the cyclopentylzinc reagent, a cycloheptylzinc reagent (n = 3) could be employed, providing lower but still good enantioselectivities (84–87% ee). Benzyl ethers, aryl ethers, aryl fluorides, aryl chlorides, and even aryl bromides and iodides were compatible with the cross-coupling conditions. Remarkably, the method was found to be insensitive to adventitious moisture, since the addition of 10 mol% of water had no effect on the enantioselectivity and yield.

Although enantioconvergent alkyl–alkyl couplings of racemic electrophiles have been widely developed, until 2013 there were no reports of the corresponding reactions of racemic nucleophiles.[21] In this context, Fu *et al.* developed Negishi cross-couplings of a racemic α-zincated *N*-Boc-pyrrolidine, generated *in situ* from commercially available *N*-Boc-pyrrolidine, with unactivated secondary iodides, thus providing a one-pot, catalytic asymmetric method for the synthesis of a range of chiral 2-alkylpyrrolidines in good yields (50–96%) and with good to high enantioselectivities of 58–94% ee, as shown in Scheme 3.9. In this case, the authors used a nickel catalyst generated *in situ* from $NiCl_2$·glyme and chiral diamine (*R*,*R*)-**4**. Cyclic as well as acyclic secondary alkyl iodides gave comparable high results, whereas a primary

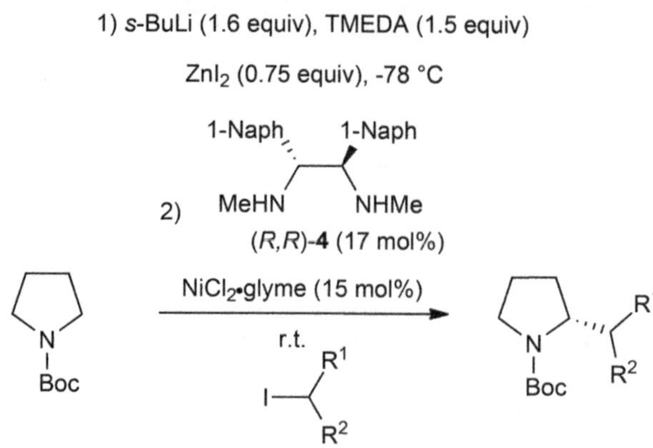

$R^1,R^2 = (CH_2)_5$: 80%, ee = 93%
$R^1,R^2 = (CH_2)_4$: 82%, ee = 91%
$R^1,R^2 = (CH_2)_6$: 50%, ee = 84%
$R^1,R^2 = (CH_2)_2O(CH_2)_2$: 96%, ee = 92%
$R^1,R^2 = (CH_2)_2N(Boc)(CH_2)_2$: 96%, ee = 94%
$R^1,R^2 = (CH_2)_2N(Cbz)(CH_2)_2$: 94%, ee = 91%
$R^1 = R^2$ = Me: 85%, ee = 90%
R^1 = H, R^2 = *n*-Pent: 85%, ee = 58%

Scheme 3.9 Cross-coupling of α-zincated *N*-Boc-pyrrolidine with unactivated secondary alkyl iodides.

alkyl iodide (R^1 = H, R^2 = *n*-Pent) provided a moderate enantioselectivity of 58% ee. Heterocyclic electrophiles were also coupled in high enantioselectivities (91–94% ee). This work represented the first enantioconvergent alkyl–alkyl cross-coupling of a racemic nucleophile, specifically, the asymmetric Negishi reaction of α-zincated *N*-Boc-pyrrolidine with unactivated secondary iodides, providing a one-pot route to an array of chiral 2-alkylpyrrolidines. Because the highest enantioselectivity was obtained for the incorporation of secondary alkyl substituents, this novel method complemented existing catalytic asymmetric approaches to the synthesis of these products, which are generally most effective for primary alkyl groups. It must be noted that pyrrolidines bearing an alkyl substituent at the 2-position constitute important subunits in bioactive natural and non-natural products, function as versatile intermediates in the synthesis of other useful classes of compounds, and serve as effective chiral organocatalysts and ligands in asymmetric catalysis.[22]

A tertiary stereogenic centre that bears two different aryl substituents is found in a variety of bioactive compounds, including medicines such as Zoloft and Dedrol. With the aim of developing a synthesis of this family of products on the basis on asymmetric Negishi cross-coupling reactions, Fu and co-workers have applied their previously reported methodologies involving alkylzinc reagents and benzylic halides as substrates, but neither proved effective when an arylzinc reagent was employed as the nucleophile.[23] On the other hand, they found that the cross-coupling reactions proceeded smoothly with readily available benzylic alcohols. Actually, the process began with the *in situ* formation of the corresponding benzylic mesylate, which was subsequently treated with an arylzinc reagent, LiI, and a chiral nickel catalyst generated *in situ* from $NiCl_2$·glyme and chiral bisoxazoline **5** (Scheme 3.10). The corresponding Negishi products were achieved in good to excellent yields (84–98%) and with enantioselectivities of 83–95% ee. A wide array of functional groups (aryl iodide, thiophene, *N*-Boc-indole) were shown to be compatible with the mild reaction conditions. The utility of this novel methodology was demonstrated in the synthesis of (*S*)-sertraline tetralone, which is a precursor of Zoloft, a leading antidepressant drug.

Motivated by potential applications in biomedical research, substantial efforts have been dedicated to the development of methods for the preparation of organofluorine compounds.[24] In the case of alkyl fluorides, advances have been described in the catalytic enantioselective synthesis of stereogenic centres that bear a fluorine substituent, particularly α to a carbonyl group. Although most studies have addressed the generation of secondary stereocentres, a few reports have examined the establishment of tertiary centres. For example, Fu and Liang have developed the synthesis of tertiary alkyl fluorides on the basis of an enantioselective nickel-catalysed Negishi cross-coupling reaction occurring between racemic α,α-dihalo ketones and arylzinc reagents.[25] As shown in Scheme 3.11, a range of chiral tertiary alkyl fluorides were produced in moderate to good yields (43–78%) and with high enantioselectivities of 82–99% ee, starting from the corresponding α-bromo-α-fluoro ketones in the presence of a combination of chiral bisoxazoline ligand **6** and $NiCl_2$·glyme. This ligand was

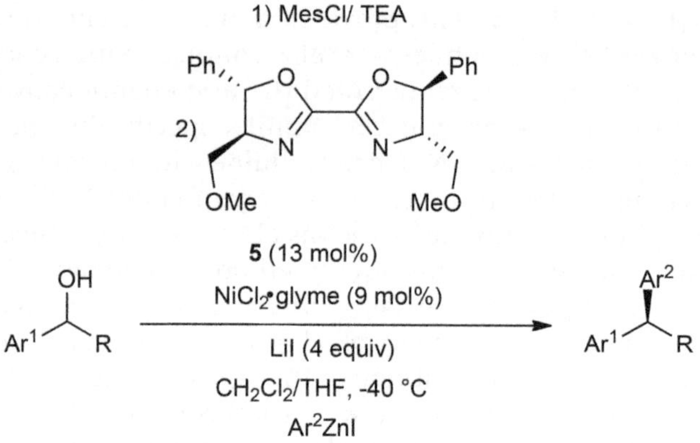

Ar^1 = Ph, Ar^2 = *p*-anisyl, R = Me: 88%, ee = 88%
Ar^1 = Ph, Ar^2 = *p*-anisyl, R = Et: 92%, ee = 94%
Ar^1 = Ph, Ar^2 = *p*-anisyl, R = *n*-Pr: 90%, ee = 94%
Ar^1 = Ph, Ar^2 = *p*-anisyl, R = CH_2Bn: 90%, ee = 88%
Ar^1 = *p*-$MeOC_6H_4$, Ar^2 = *p*-anisyl, R = *n*-Pr: 94%, ee = 95%
Ar^1 = *p*-$PivOC_6H_4$, Ar^2 = *p*-anisyl, R = *n*-Pr: 84%, ee = 84%
Ar^1 = *o*-$MeOC_6H_4$, Ar^2 = *p*-anisyl, R = Et: 94%, ee = 94%
Ar^1 = 2-furyl, Ar^2 = *p*-anisyl, R = *n*-Bu: 88%, ee = 93%
Ar^1 = 1-thienyl, Ar^2 = *p*-anisyl, R = *n*-Pr: 54%, ee = 83%
Ar^1 = 2-thienyl, Ar^2 = *p*-anisyl, R = *n*-Pr: 93%, ee = 94%
Ar^1 = *p*-$MeOC_6H_4$, Ar^2 = Ph, R = Et: 96%, ee = 94%
Ar^1 = *p*-$MeOC_6H_4$, Ar^2 = *p*-Tol, R = Et: 91%, ee = 93%
Ar^1 = *p*-$MeOC_6H_4$, Ar^2 = *p*-FC_6H_4, R = Et: 95%, ee = 94%
Ar^1 = *p*-$MeOC_6H_4$, Ar^2 = *p*-ClC_6H_4, R = Et: 95%, ee = 94%
Ar^1 = *p*-$MeOC_6H_4$, Ar^2 = *p*-$CF_3C_6H_4$, R = Et: 98%, ee = 94%
Ar^1 = *p*-$MeOC_6H_4$, Ar^2 = 2-Naph, R = Et: 93%, ee = 92%

Scheme 3.10 Cross-coupling of benzylic alcohols with arylzinc reagents.

selected among a range of bisoxazolines, including a PYBOX moiety and also a chiral diamine ligand. The R group of the ketone could vary in size, although a lower enantiomeric excess of 82% ee was obtained with a bulky isopropyl substituent. General high enantioselectivities were obtained whether the aromatic group (Ar^1) was *para*, *meta*, or *ortho* substituted, and whether it was electron-rich or electron-poor. Moreover, functional groups such as an alkene, alkyl chloride, aryl methyl ether, and aryl fluoride were found compatible with the reaction conditions. The scope of the process was also broad with respect to the nucleophile. Thus, *para* and *meta* (but not *ortho*) substituted arylzinc reagents with electron-poor or electron-rich substituents were suitable partners, furnishing the desired α-fluoro ketones in high enantioselectivities (90–99% ee). In addition, it was found that this methodology could be extended to the reaction of an α-chloro-α-fluoro ketone, providing comparable excellent results (68% yield, 98% ee with Ar^1 = Ar^2 = Ph, R = Bn). It must be noted that this remarkable work consisted of the first catalytic asymmetric cross-coupling

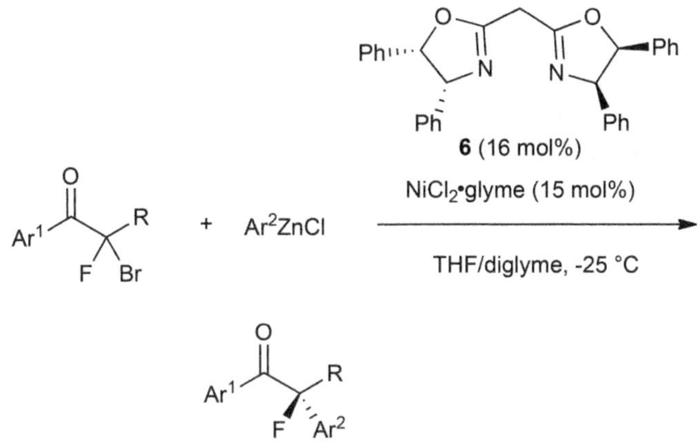

$Ar^1 = Ar^2 = Ph$, R = Et: 62%, ee = 97%
$Ar^1 = Ar^2 = Ph$, R = Bn: 73%, ee = 98%
$Ar^1 = Ar^2 = Ph$, R = allyl: 43%, ee = 97%
$Ar^1 = Ar^2 = Ph$, R = *i*-Pr: 55%, ee = 82%
Ar^1 = *p*-(*t*-Bu)C_6H_4, Ar^2 = Ph, R = $(CH_2)_2Cl$: 76%, ee = 98%
Ar^1 = *p*-$MeOC_6H_4$, Ar^2 = Ph, R = Bn: 62%, ee = 96%
Ar^1 = *p*-PhC_6H_4, Ar^2 = Ph, R = Et: 54%, ee = 96%
Ar^1 = *p*-FC_6H_4, Ar^2 = Ph, R = $(CH_2)_2Cl$: 63%, ee = 97%
Ar^1 = *m*-$MeOC_6H_4$, Ar^2 = Ph, R = Et: 54%, ee = 97%
Ar^1 = *m*-FC_6H_4, Ar^2 = Ph, R = Bn: 59%, ee = 97%
Ar^1 = *o*-$MeOC_6H_4$, Ar^2 = Ph, R = Et: 62%, ee = 92%
Ar^1 = *o*-FC_6H_4, Ar^2 = Ph, R = Bn: 59%, ee = 97%
$Ar^1 = Ar^2 = Ph$, R = $(CH_2)_2Cl$: 62%, ee = 98%
Ar^1 = Ph, Ar^2 = *p*-(OTBS)C_6H_4, R = Bn: 67%, ee = 97%
Ar^1 = Ph, Ar^2 = *p*-(*i*-Pr)C_6H_4, R = Et: 71%, ee = 97%
Ar^1 = Ph, Ar^2 = *p*-FC_6H_4, R = Bn: 65%, ee = 94%
Ar^1 = Ph, Ar^2 = *m*-(OTBS)C_6H_4, R = Bn: 78%, ee = 99%
Ar^1 = Ph, Ar^2 = *m*-$(CF_3)C_6H_4$, R = Bn: 65%, ee = 90%
Ar^1 = Ph, Ar^2 = *m*-(*i*-Pr)C_6H_4, R = Et: 73%, ee = 97%
Ar^1 = 3,5-$Me_2C_6H_3$, Ar^2 = Ph, R = Et: 65%, ee = 97%
Ar^1 = 2-Naph, Ar^2 = Ph, R = Et: 51%, ee = 97%

Scheme 3.11 Cross-coupling of α,α-dihalo ketones with arylzinc reagents.

method that employed germinal dihalides as electrophiles. Undoubtedly, these fluorinated products will be easily transformed into an array of interesting chiral organofluorine target compounds.

The development of efficient methods for the generation of enantioenriched sulfonamides and sulfones is an important objective for fields such as organic synthesis and medicinal chemistry; however, there have been relatively few reports of direct catalytic asymmetric approaches to controlling the stereochemistry of the sulfur-bearing carbon of such targets. In 2014, Fu *et al.* described nickel-catalysed stereoconvergent Negishi arylations of racemic α-bromo sulfonamides and α-bromo sulfones that furnished the corresponding cross-coupling products in very good yields and enantioselectivities, as shown in Scheme 3.12.[26] Bisoxazoline 7 was selected as optimal ligand in this study, providing very high enantioselectivities of up to 99% ee in the reaction of various α-bromo sulfonamides in which the nitrogen could bear either an alkyl or an aryl substituent. The same conditions applied to the reaction of α-bromo sulfones, provided the corresponding products with

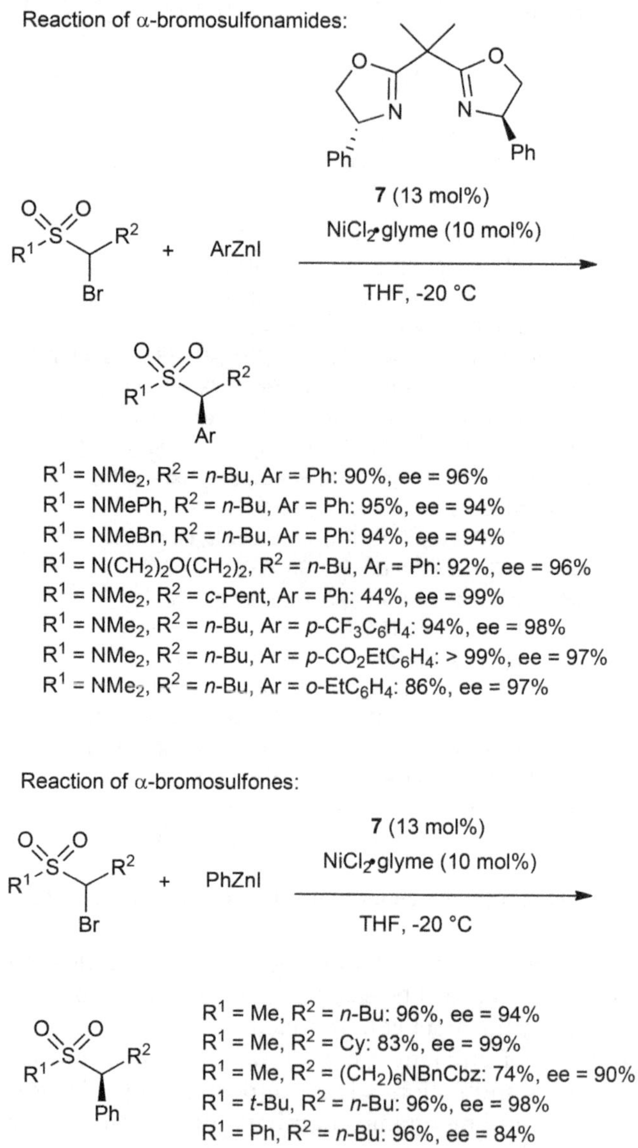

Scheme 3.12 Cross-couplings of α-bromo sulfonamides and α-bromo sulfones with arylzinc reagents.

comparable high enantioselectivities of up to 99% ee, as shown in Scheme 3.12. The scope was broad since the R^1 substituent of the sulfone could range in steric demand from methyl to *t*-butyl, and could also be aromatic. Furthermore, the R^2 substituent of the sulfone could be linear or branched. The authors also demonstrated that the reactions could be performed with alkenylzirconium reagents in place of arylzinc reagents as nucleophiles, providing enantioselectivities of 80–96% ee.

Finally, Doyle *et al.* developed nickel-catalysed asymmetric arylations of an *in situ* generated *N*-acyl-4-methoxypyridinium ion to give a broad range of chiral 2,3-dihydropyrid-4-ones in good yields and with enantioselectivities of up to >99% ee, as shown in Scheme 3.13.[27] This Negishi reaction occurred

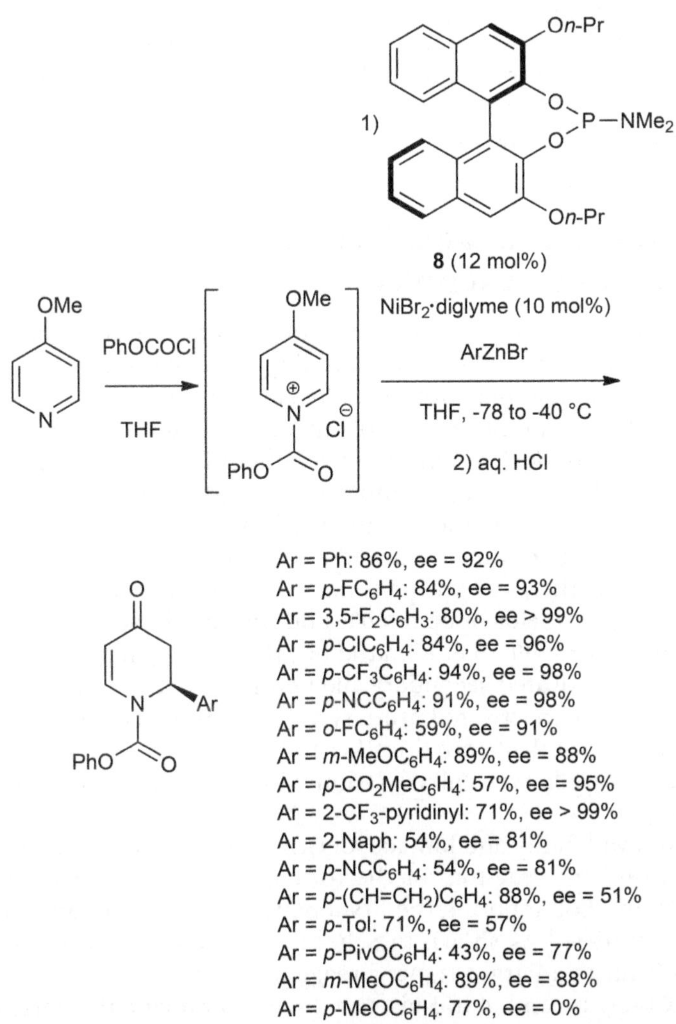

Scheme 3.13 Cross-coupling of a methoxypyridinium salt with arylzinc bromides.

between the methoxypyridinium salt derived from commercially available *p*-methoxypyridine and arylzinc bromides under catalysis with a nickel catalyst generated *in situ* from $NiBr_2$·diglyme and chiral phosphoramidite **8**, the latter selected as optimal ligand among several other ligands of the same family. Generally, *ortho*-, *meta*-, and *para*-substituted zinc nucleophiles were well tolerated. Reactions with electron-withdrawing zinc reagents fared exceptionally well, delivering the products bearing various functions in enantioselectivities of 96–99% ee. Notably, zinc reagents substituted with methyl ester and pivalate groups afforded the corresponding coupled products, which are otherwise inaccessible by standard Grignard methods. Moreover, a reaction with a heteroaromatic nucleophile, such as 2-(trifluoromethyl) pyridine, furnished the corresponding enantiopure product. Whereas electron-neutral arylzinc nucleophiles, such as 4-vinyl-, 4-methyl-, and 2-naphthylphenyl compounds, performed modestly in terms of enantioselectivity (51–81% ee), electron-rich nucleophiles, such as the 4-methoxyphenylzinc reagent, underwent reaction with no stereoinduction owing to a competitive racemic background reaction.

3.3 Hiyama, Kumada, Suzuki, and Related Cross-Coupling Reactions

Organosilicon reagents have many advantages, such as availability, low toxicity, and high functional-group compatibility.[28] Further increasing the scope of cross-coupling reactions with secondary alkyl electrophiles, Fu *et al.* developed the first asymmetric version of the nickel-catalysed Hiyama coupling with racemic secondary α-bromo esters and chiral diamine ligand (*S*,*S*)-**9** (Scheme 3.14).[29] The ligand, organosilane, and fluoride activator all played a critical role in the enantioselectivity of the reaction. Under optimised conditions, a variety of functionalised α-bromo esters, which could contain additional ester, ether, and alkene functional groups, could be coupled with arylsilanes in good yields of up to 80% and with moderate to excellent enantioselectivities of 68–99% ee. Remarkably, the authors have shown that the coupling of the activated secondary alkyl bromide occurred preferentially in the presence of an unactivated primary alkyl bromide. The reaction was sensitive to the steric bulk of both the alkyl and ester moieties. Furthermore, the α-bromo esters underwent alkenylation with moderate enantioselectivities (66–72% ee).

The ready availability and low cost of Grignard reagents make the Kumada coupling a valuable reaction for the formation of carbon–carbon bonds.[30] In 2010, Fu and Lou reported the first enantioselective Kumada coupling of alkyl electrophiles.[31] As shown in Scheme 3.15, reaction occurred between α-bromo ketones and aryl Grignard reagents in the presence of a combination of $NiCl_2$·glyme and chiral bisoxazoline 7 to provide the corresponding Kumada products in good yields of up to 91% and with enantioselectivities of up to 92% ee. The method was compatible with a diverse spectrum of

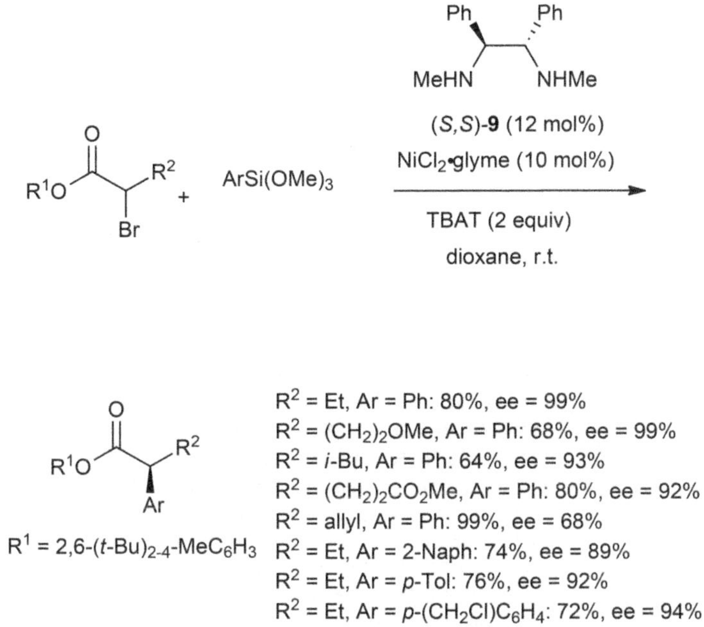

Scheme 3.14 Hiyama reaction of α-bromo esters.

functional groups, including esters, halides, nitriles, ethers, and heteroaromatic rings. Regardless of the electron-withdrawing or electron-donating nature of the substituent on the aromatic ring of the nucleophile, consistently good enantioselectivities and yields were reached. Moreover, a variety of α-bromo ketones were suitable electrophilic partners. In the case of alkyl aryl ketones, the aromatic group could be electron-rich or electron-poor, and it could bear a variety of substitution patterns. Furthermore, the coupling proceeded smoothly with a heteroaromatic substituent (R^1 = thienyl, 87% ee), as well as with an array of functionalised alkyl groups. When the same conditions were applied to dialkyl ketones, more modest enantioselectivities were observed. However, by modifying the structure of the bisoxazoline ligand **10** and raising the reaction temperature to −40 °C, promising enantioselectivities (73–90% ee) could be achieved for a variety of reaction partners, as shown in Scheme 3.15. In addition to being a remarkable novel process, this methodology has the advantages of being applicable to dialkyl ketones, and to employ readily available chiral bisoxazolines for the first time in cross-couplings of alkyl electrophiles.

The Suzuki reaction is one of the most versatile and widely used cross-coupling reactions.[32] Among the reasons for its appeal are the commercial availability of a large range of boronic acids, the ease with which these reagents can be handled, and their high functional-group compatibility. On the basis of the pioneering work of Suzuki and co-workers reported in 1992,[8] efforts by Fu's group since 2001 [33] led to the catalytic coupling of primary alkyl halides.

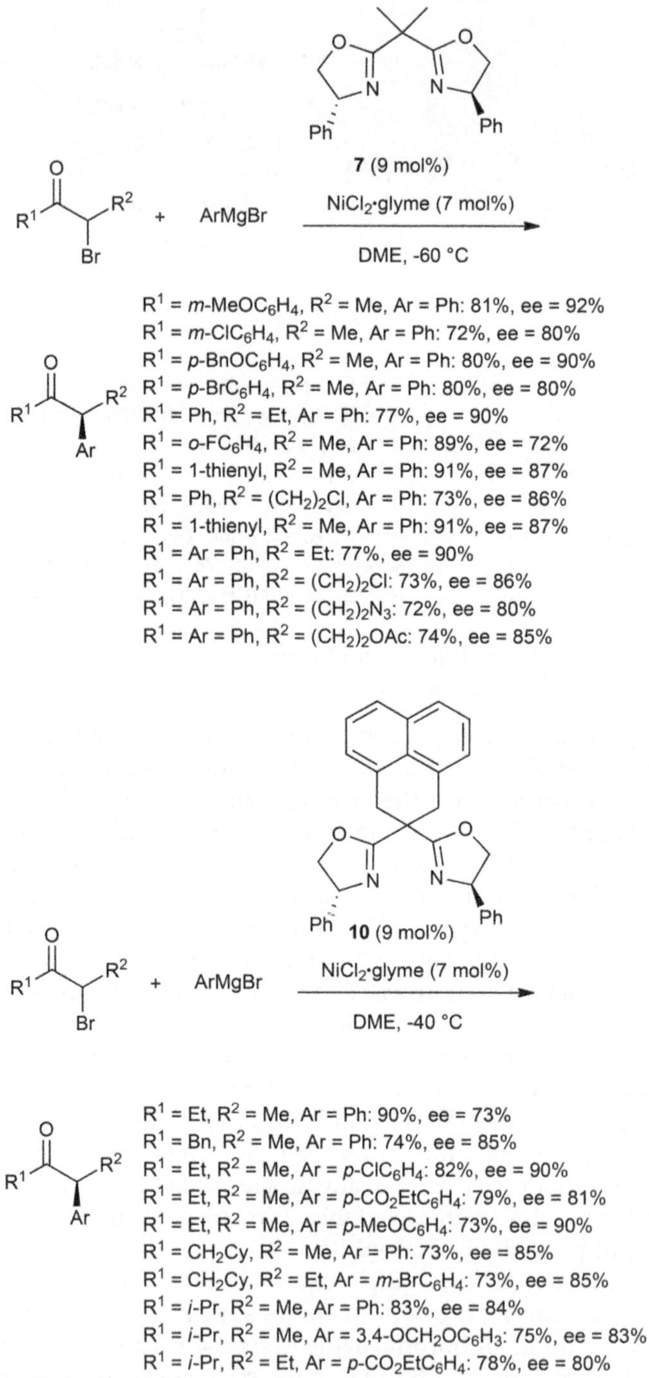

Scheme 3.15 Kumada reactions of α-bromo ketones.

Extending their process in this field, Zhou and Fu in 2004 developed the first Suzuki coupling of unactivated secondary alkyl bromides and iodides catalysed by nickel.[34] In 2008, Saito and Fu discovered that the combination of Ni(cod)$_2$ with chiral ligand (*R*,*R*)-**11** (Scheme 3.16) could be used to couple racemic secondary unactivated alkyl bromides with alkylboranes to give the corresponding chiral coupling products.[35] Fine-tuning of the nickel source, ligand, solvent, and reaction temperature led to the formation of these products in good yields of up to 86% and with moderate to high enantioselectivities of 40–94% ee. The process required the presence of *t*-BuOK and i-BuOH for the asymmetric Suzuki reaction of racemic acyclic secondary homobenzylic bromides with alkylboranes. Proper positioning of the aromatic group was essential for achieving good enantioselectivity, as the catalyst system seemed to differentiate between the CH$_2$R^3 group and the alkyl group R^2 of the homobenzylic bromide. The enantioselectivity was somewhat

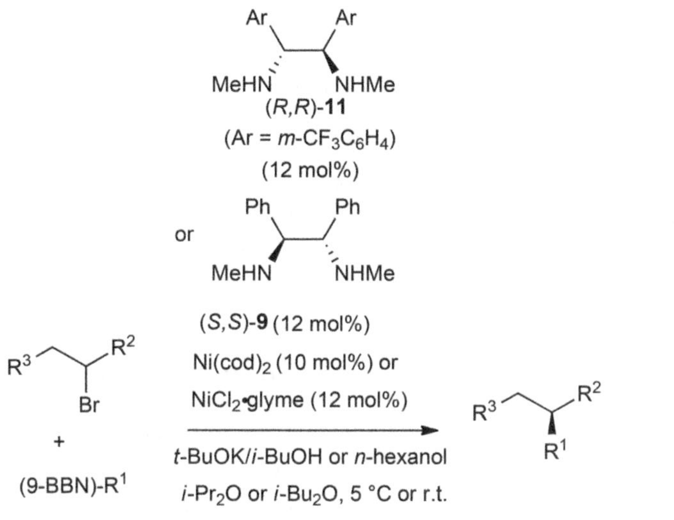

with ligand **11**:
R^1 = (CH$_2$)$_2$Bn, R^2 = Me, R^3 = Ph: 78%, ee = 90%
R^1 = (CH$_2$)$_2$Bn, R^2 = *n*-Bu, R^3 = Ph: 84%, ee = 94%
R^1 = (CH$_2$)$_2$Bn, R^2 = CH$_2$Bn, R^3 = Ph: 68%, ee = 90%
R^1 = (CH$_2$)$_2$Bn, R^2 = Me, R^3 = *o*-Tol: 86%, ee = 86%
R^1 = (CH$_2$)$_2$Bn, R^2 = Me, R^3 = *p*-CF$_3$C$_6$H$_4$: 82%, ee = 70%
R^1 = *n*-Hex, R^2 = CH$_2$OBn, R^3 = Ph: 69%, ee = 40%
R^1 = (CH$_2$)$_3$Bn, R^2 = Me, R^3 = Ph: 68%, ee = 80%
R^1 = Me, R^2 = (CH$_2$)$_2$Bn, R^3 = Bn: ee = 14%
with ligand **9**:
R^1 = (CH$_2$)$_3$-*p*-FC$_6$H$_4$, R^2 = *i*-Bu, R^3 = OCONPhBn: 66%, ee = 98%
R^1 = (CH$_2$)$_3$-*p*-FC$_6$H$_4$, R^2 = *n*-Hex, R^3 = OCONPhBn: 70%, ee = 98%
R^1 = *n*-Hex, R^2 = Et, R^3 = OCONPhBn: 74%, ee = 96%

Scheme 3.16 Suzuki reaction of unactivated secondary bromides.

diminished when the aryl group contained an electron-withdrawing substituent (70% ee). Heteroatom-containing electrophiles and alkylboranes were good coupling partners, although the products were generally isolated with lower enantioselectivities than those observed when unfunctionalised substrates were used. Later in 2010, these authors found that the closely related ligand (*S*,*S*)-**9** allowed, under almost the same reaction conditions, the products to be achieved in better enantioselectivities of up to 98% ee, as shown in Scheme 3.16.[36]

In 2011, a related methodology using chiral diamine (*R*,*R*)-**4** as ligand was extended to a highly enantioselective cross-coupling reaction of secondary alkyl chlorides bearing a variety of pendant arylamines (Scheme 3.17).[37] The corresponding chiral products were achieved in good yields of up to 86% and with moderate to excellent enantioselectivities of 71–96% ee. Mechanistic experiments were consistent with coordination of the nitrogen to the catalyst to promote high enantioselectivity.

The scope of the nickel-catalysed Suzuki cross-coupling reaction was extended by the same authors to the reaction of racemic α-chloroamides.[38] As shown in Scheme 3.18, the chiral nickel catalyst generated *in situ* from chiral diamine ligand (*S*,*S*)-**11** and $NiBr_2$·glyme was shown to promote the Suzuki reaction of α-chloro indolinylamides with arylboron reagents in the presence of *t*-BuOK and i-BuOH to give the corresponding coupled products in good yields and with enantioselectivities of 84–94% ee. A variety of acyclic α-chloroamides were found less suitable cross-coupling partners than the

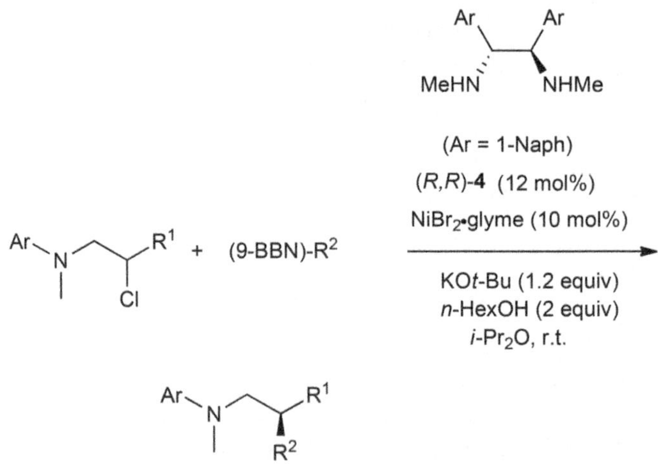

R^1 = *i*-Pr, R^2 = $(CH_2)_3$-*o*-$MeOC_6H_4$, Ar = Ph: 52%, ee = 96%
R^1 = *n*-Bu, R^2 = $(CH_2)_4$-C(Me)(OCH_2CH_2O), Ar = *o*-$MeOC_6H_4$: 86%, ee = 94%
R^1 = *n*-Bu, R^2 = $(CH_2)_4$-C(Me)(OCH_2CH_2O), Ar = *m*-FC_6H_4: 70%, ee = 92%
R^1 = CH_2Cy, R^2 = $(CH_2)_4$-C(Me)(OCH_2CH_2O), Ar = *o*-FC_6H_4: 82%, ee = 71%
R^1 = *n*-Bu, R^2 = $(CH_2)_3OTBS$, Ar = *m*-Tol: 63%, ee = 91%

Scheme 3.17 Suzuki reaction of unactivated secondary alkyl chlorides.

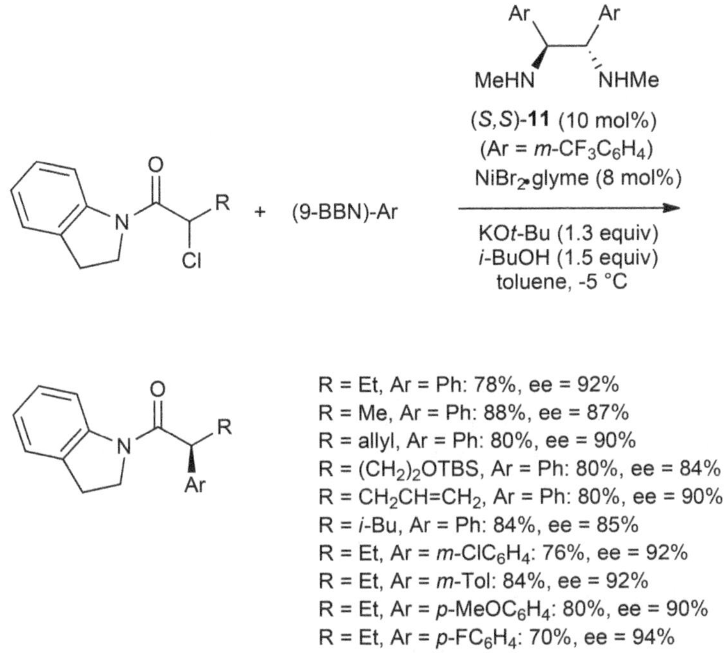

Scheme 3.18 Suzuki reaction of α-chloroamides.

indolinylamides (5–76% ee). Functional groups, such as an alkene and a silyl ether, as well as β-branching, were tolerated in the alkyl chain of the electrophile. For the nucleophile, a *meta* or *para* substituent could be present, and it could be electron-withdrawing or electron-donating. Moreover, the method was also applicable to the corresponding enantioselective cross-coupling of α-bromoamides (91% ee). It must be noted that this beautiful process represented the first example of an asymmetric arylation of an α-haloamide, an enantioselective arylation of a α-chloro carbonyl compound, and an asymmetric Suzuki reaction with an activated alkyl electrophile or an arylboron reagent. Furthermore, the coupling products could be transformed without racemisation into useful enantioenriched α-aryl carboxylic acids and primary alcohols.

In comparison with α- and β-alkylation reactions, the range of useful methods for the catalytic enantioselective incorporation of alkyl substituents γ to a carbonyl group is rather limited. One unexplored approach to this objective is the asymmetric coupling of a γ-halo carbonyl compound with an alkynylmetal reagent. In this context, Fu and Zultanski have developed asymmetric Suzuki reactions of γ-chloroamides with primary alkylborane reagents catalysed by a chiral nickel catalyst generated *in situ* from $NiBr_2$·glyme and chiral diamine (*S*,*S*)-**9** as ligand in the presence of *t*-BuOK and *n*-hexanol.[39] As shown in Scheme 3.19, a range of chiral γ-alkylated *N*,*N*-diphenylamides was achieved in moderate yields (51–74%) and with good enantioselectivities of

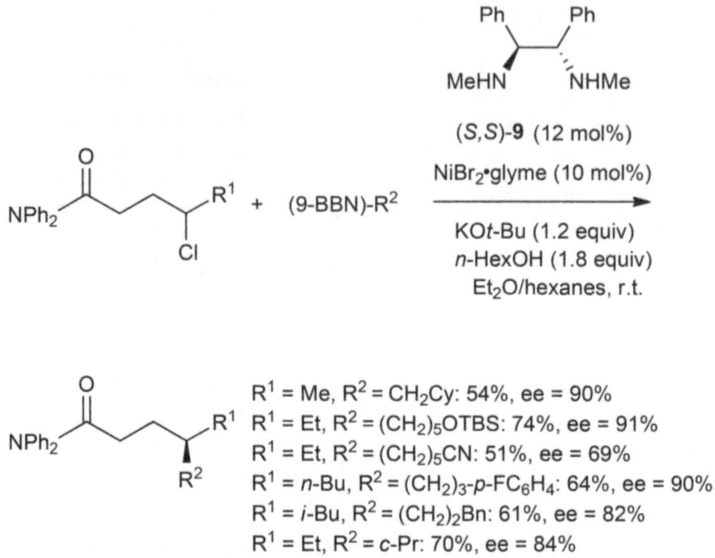

Scheme 3.19 Suzuki reaction of γ-chloroamides.

69–91% ee. A variety of functional groups was compatible with the reaction conditions. Furthermore, the process was not limited to γ-chloro diphenylamides since the corresponding bromides were also suitable electrophiles (69% ee), and the γ-chloro Weinreb amide also proceeded with promising enantioselectivity (86% ee). In addition, a secondary alkyl(cyclopropyl) borane reagent could be used, providing 84% ee. This result represented the first example of an asymmetric cross-coupling of an unactivated alkyl electrophile with a secondary alkynylmetal reagent.

Later, the same reaction conditions were applied to the enantioselective nickel-catalysed Suzuki reaction of racemic β-bromo-protected amines with primary alkylborane reagents.[40] As shown in Scheme 3.20, various β-bromo secondary unactivated carbamates and sulfonamides could serve as directing groups in nickel-catalysed cross-coupling reactions with unactivated primary alkylboranes to give the Suzuki products in good enantioselectivities of 72–91% ee combined with good yields. The method was compatible with an aryl carbamate and an aryl methyl ether. In the case of sulfonamides as substrates, both tosyl- and mesyl-protected secondary dialkylamines were suitable cross-coupling partners, undergoing stereoconvergent C–C bond formation in good enantioselectivities (72–90% ee). This nice work represented the first example of using sulfonamides as effective directing groups in metal-catalysed asymmetric C–C bond-forming reactions.

In 2014, Doyle *et al.* described enantioselective nickel-catalysed Suzuki cross-coupling reactions of quinolinium ions generated *in situ* from 2-ethoxy-1-(ethoxycarbonyl)-1,2-dihydroquinolines in the presence of a combination of [(methallyl)Ni]$_2$ and TADDOL-derived chiral ligand **12**.[41] This ligand was selected as optimal among a range of other ligands, including

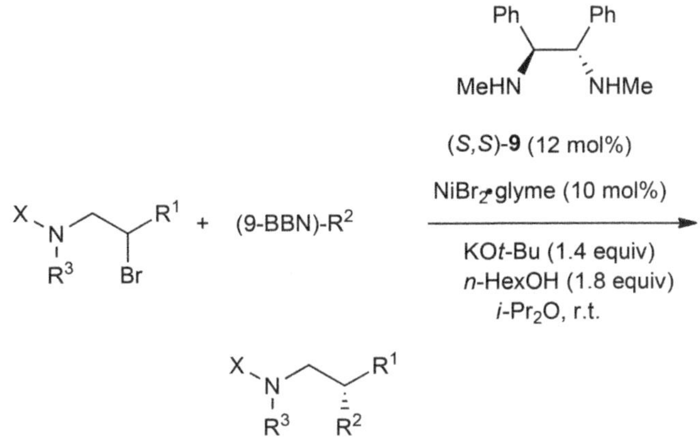

R^1 = Et, R^2 = $(CH_2)_4$-C(Me)(OCH_2CH_2O), R^3 = PMB, X = CO_2Ph: 56%, ee = 91%
R^1 = R^3 = *n*-Bu, R^2 = $(CH_2)_3$-*p*-$MeOC_6H_4$, X = CO_2Ph: 83%, ee = 90%
R^1 = CH_2Cy, R^2 = $(CH_2)_3$-*o*-Tol, R^3 = Me, X = CO_2Ph: 74%, ee = 90%
R^1 = Et, R^2 = *n*-Hex, R^3 = Ph, X = CO_2Ph: 56%, ee = 90%
R^1 = Et, R^2 = $(CH_2)_2Cy$, R^3 = Ph, X = CO_2Bn: 66%, ee = 80%
R^1 = *n*-Hex, R^2 = $(CH_2)_3$-*o*-$MeOC_6H_4$, R^3 = *n*-Bu, X = Mes: 76%, ee = 90%
R^1 = CH_2Cy, R^2 = $(CH_2)_3$-*o*-$MeOC_6H_4$, R^3 = *n*-Bu, X = Ts: 54%, ee = 90%
R^1 = Et, R^2 = $(CH_2)_3$-*p*-FC_6H_4, R^3 = PMB, X = Ts: 58%, ee = 90%
R^1 = Et, R^2 = $(CH_2)_2Cy$, R^3 = Ph, X = Ts: 68%, ee = 72%

Scheme 3.20 Suzuki reaction of β-bromo-protected amines.

phosphoramidites. The reactions of (substituted) 2-ethoxy-1-(ethoxycarbonyl)-1,2-dihydroquinolines with arylboroxines were performed in the presence of NaOPh·$3H_2O$ instead of commonly used *t*-BuOK, and provided the corresponding chiral 2-aryl- and 2-heteroaryl-1,2-dihydroquinolines in moderate to high yields (33–99%) and with enantioselectivities of up to 90% ee, as shown in Scheme 3.21. Substitution at the *meta* and *para* positions of the nucleophile was well tolerated in the reaction, whereas substitution at the *ortho* position required an increase in catalyst loading and caused a slight decrease in enantioselectivity (77% ee). A variety of electron-neutral nucleophiles performed well; electron-deficient nucleophiles provided low enantioselectivity (25% ee), while electron-rich nucleophiles were the most selective. Notably, a thioether was tolerated, suggesting that the catalyst was resistant to common poisons. The scope of the electrophile was also investigated, showing that both yield and enantioselectivity suffered relative to reactions with unsubstituted 2-ethoxy-1-(ethoxycarbonyl)-1,2-dihydroquinolines. Substitution on the quinoline ring was tolerated to a moderate extent at the 5-, 6-, and 7-positions, albeit with moderate enantioselectivities of 58–77% ee.

In addition to organozinc, -silyl, -magnesium, and -borane reagents employed respectively in Negishi, Hiyama, Kumada, and Suzuki cross-coupling reactions with various electrophiles, organozirconium reagents were

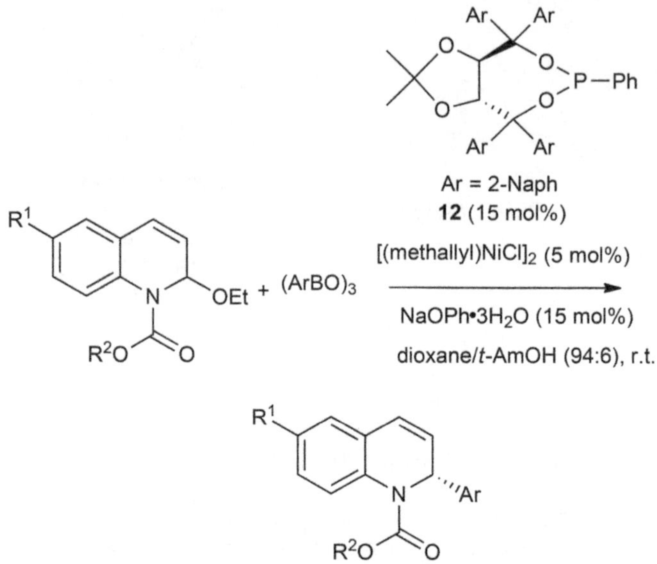

R^1 = H, R^2 = Et, Ar = Ph: 99%, ee = 85%
R^1 = H, R^2 = Et, Ar = *p*-Tol: 85%, ee = 87%
R^1 = H, R^2 = Et, Ar = 2-Naph: 76%, ee = 87%
R^1 = H, R^2 = Et, Ar = *p*-$CF_3C_6H_4$: 42%, ee = 25%
R^1 = H, R^2 = Et, Ar = *p*-$MeOC_6H_4$: 82%, ee = 87%
R^1 = H, R^2 = Et, Ar = *m*-$MeSC_6H_4$: > 99%, ee = 90%
R^1 = CO_2Me, R^2 = Et, Ar = *p*-Tol: 51%, ee = 88%
R^1 = CF_3, R^2 = Et, Ar = *p*-Tol: 33%, ee = 78%
R^1 = H, R^2 = Me, Ar = *p*-Tol: 74%, ee = 73%

Scheme 3.21 Suzuki reaction of quinolinium ions.

involved for the first time in 2010 by Fu and Lou in enantioselective nickel-catalysed cross-coupling reactions with racemic α-bromo ketones.[42] As shown in Scheme 3.22, these secondary alkyl electrophiles reacted with alkenylzirconium compounds to give the corresponding chiral β,γ-unsaturated ketones in good yields (82–95%) and with enantioselectivities of up to 98% ee when the reaction was catalysed with a combination of $NiCl_2$·glyme and tetrasubstituted chiral bisoxazoline ligand **13**. The C–C bond formation occurred below ambient temperature (10 °C) without the need for any additives. A broad array of alkenylzirconium reagents were suitable partners. Thus, the R^2 group could range in steric demand from hydrogen to *t*-butyl. Moreover, a diverse set of α-bromo ketones were suitable electrophiles. Different electron-withdrawing as well as electron-donating substituents could be present on the aromatic ring of aryl ketones (R^3 = aryl). Furthermore, a heteroaryl ketone (R^3 = thienyl) provided a very high enantioselectivity of 94% ee. In addition, an array of alkyl groups (R^3 = alkyl) on the ketone were tolerated (91–98% ee). The fact that the process could be applied to dialkyl

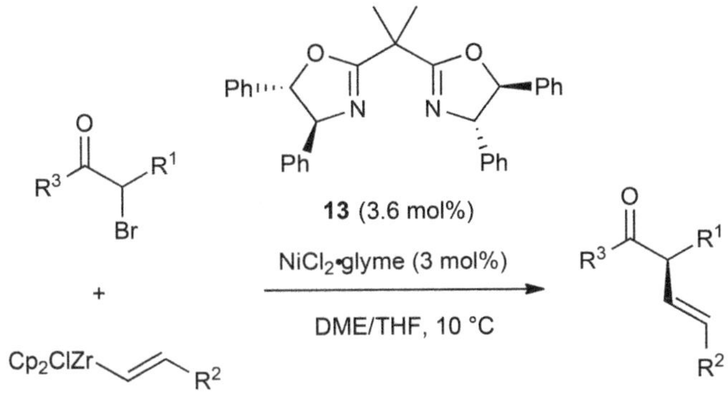

R^1 = Et, R^2 = $(CH_2)_4Cl$, R^3 = Ph: 83%, ee = 92%
R^1 = $(CH_2)_2Cl$, R^2 = *m*-Tol, R^3 = Ph: 95%, ee = 80%
R^1 = *i*-Bu, R^2 = Bn, R^3 = Ph: 81%, ee = 82%
R^1 = Me, R^2 = Bn, R^3 = *p*-$BnOC_6H_4$: 89%, ee = 95%
R^1 = Me, R^2 = *c*-Pr, R^3 = *p*-$CF_3C_6H_4$: 80%, ee = 92%
R^1 = Me, R^2 = CH_2Cy, R^3 = 2-thienyl: 85%, ee = 94%
R^1 = Me, R^2 = Bn, R^3 = Et: 86%, ee = 90%
R^1 = Et, R^2 = Bn, R^3 = *i*-Pr: 82%, ee = 98%
R^1 = Me, R^2 = Bn, R^3 = Cy: 87%, ee = 91%
R^1 = Me, R^2 = H, R^3 = Ph: 92%, ee = 90%
R^1 = Me, R^2 = *n*-Bu, R^3 = Ph: 92%, ee = 92%
R^1 = Me, R^2 = Cy, R^3 = Ph: 75%, ee = 81%
R^1 = Me, R^2 = Ph, R^3 = Ph: 89%, ee = 91%
R^1 = Me, R^2 = *t*-Bu, R^3 = Ph: 86%, ee = 84%

Scheme 3.22 Cross-coupling of α-bromo ketones with organozirconium reagents.

ketones was in contrast to a previous study reported by these authors dealing with Kumada reactions of ketones with aryl Grignard reagents, wherein different coupling conditions (ligand and temperature) were necessary for alkyl aryl ketones *versus* dialkyl ketones.[31]

Finally, the first enantioselective sp–sp^3 cross-coupling of alkynyl organometallic reagents with racemic secondary benzyl bromides was described by Caeiro *et al.* in 2008.[43] A substoichiometric amount of a trialkynylindium reagent was used as the nucleophilic component in the reaction, as the indium reagent was able to transfer all three organic groups to the electrophile (Scheme 3.23). The coupling was catalysed by a $NiBr_2$/(*S*)-i-Pr-PYBOX catalyst system and proceeded at room temperature to give the corresponding chiral products in moderate to good yields (40–82%) and with enantioselectivities of up to 87% ee. In this study the authors found that palladium was also able to catalyse this transformation, although the products were formed with less than 10% ee. The reaction was stereoconvergent and enabled the coupling of a variety of functionalised alkyl groups.

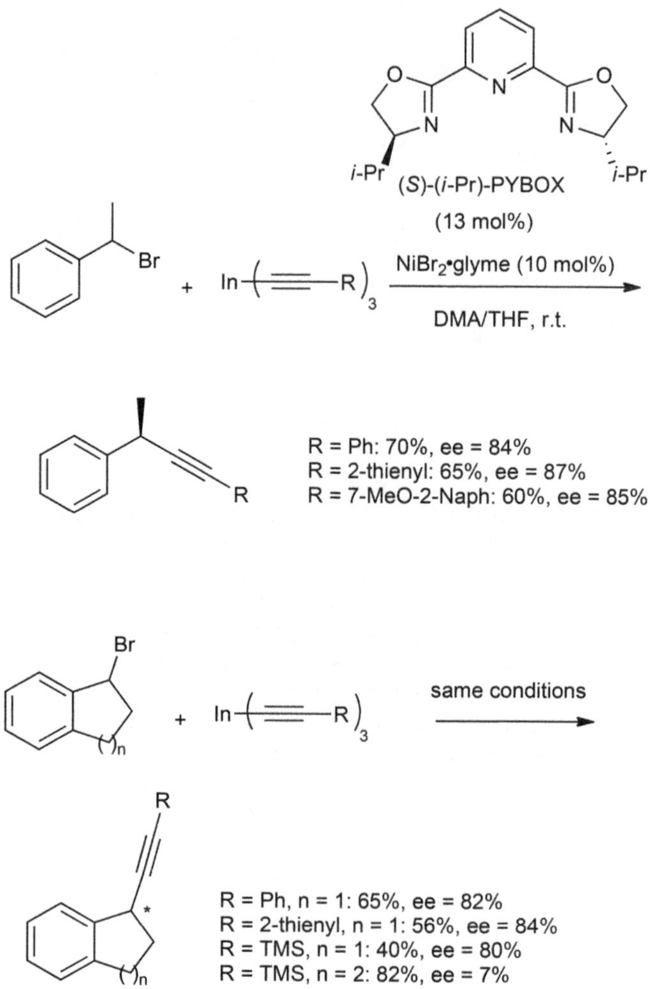

Scheme 3.23 Cross-couplings of secondary benzyl bromides with trialkynylindium reagents.

3.4 Other Coupling Reactions

Nickel-catalysed reductive coupling reactions of alkynes have emerged as powerful synthetic tools for the selective preparation of functionalised alkenes.[1a,e,h,j] In 1997, Montgomery and co-workers reported the first example of a nickel-catalysed reductive coupling reaction of alkynes and aldehydes using diethylzinc as the stoichiometric reducing agent.[44] This cross-coupling reaction afforded allylic alcohols, which are found in a variety of natural products.[45] One of the greatest challenges associated with this type of transformation is control of the regioselectivity. In 2003, Jamison *et al.* reported the first examples of catalytic, enantioselective reductive couplings of alkynes and aldehydes, using chiral nickel complexes of organophosphines as ligands.[46] These reactions provided the corresponding chiral allylic alcohols in enantioselectivities of up to 96% ee, along with regioselectivities of >95 : 5. Later, several classes of alkynes were shown to afford excellent regioselectivity in nickel-catalysed coupling reactions, including aryl-substituted alkynes (Ar–C≡C–alkyl), alkynylsilanes (R–C≡C–SiR_3), and terminal alkynes (R–C≡C–H).[46,47] However, alkynes substituted with two sterically and electronically similar groups, such as dialkylalkynes (alkyl–C≡C–alkyl′), typically afforded poor regioselectivity. To address this deficiency, and based on a hypothesis that the high regioselectivity observed with aryl-substituted alkynes was likely due to an electronic differentiation between the alkyl and the aryl substituents, Jamison *et al.* considered the possibility that another conjugating group, such as an alkene, could provide similar regiocontrol.[48] Indeed, these authors found that coupling reactions of 1,3-enynes with aldehydes were highly regioselective when trialkylphosphines were used as ligands. This nickel-catalysed reductive coupling reaction of 1,3-enynes with aldehydes was employed by these authors and others as a novel strategy for the synthesis of 1,3-dienes.[49] The first asymmetric version of this methodology was developed by Jamison *et al.* in 2005.[50] As shown in Scheme 3.24, it used chiral ferrocenylphosphine ligand (*R*)-**14** in combination with $Ni(cod)_2$ as the catalyst system, with BEt_3 as the reducing agent. Performed in the presence of triethylborane, the reaction of 1,3-enynes with aromatic aldehydes afforded the corresponding chiral 1,3-dienes in moderate yields (47–77%) and with moderate enantioselectivities of up to 58% ee. Both electron-donating and electron-withdrawing substituents were tolerated on the phenyl ring of the aldehydes. Better stereoselectivities were reached by these authors by using chiral aldehydes as substrates in addition to the presence of another chiral phosphine ligand.[51] In these conditions, the double induction allowed a diastereoselectivity of 80% de to be achieved. The same reaction conditions as those used for aldehydes in Scheme 3.24 were applied to the reductive coupling of 1,3-enynes with various ketones by using chiral ferrocenylphosphine (*S*)-**14** as ligand instead of (*R*)-**14**. The process afforded regioselectively (>95 : 5) the corresponding 1,3-dienes bearing a quaternary stereocentre with moderate yields (39–71%) and enantioselectivities of 40–70% ee, as shown in Scheme 3.24. Aromatic and heteroaromatic ketones were compatible with the reaction conditions, as well as an α,β-unsaturated ketone, such as 1-acetylcyclohex-1-ene, which provided the best enantioselectivity (70% ee). It must be noted that this work represented the first catalytic asymmetric reductive coupling of alkynes and ketones, which afforded synthetically useful chiral 1,3-dienes bearing a quaternary carbinol stereocentre.

In 2007, Montgomery *et al.* investigated the use of a chiral N-heterocyclic carbene ligand in enantioselective nickel-catalysed reductive coupling of alkynes with aldehydes in the presence of Et_3SiH as reducing agent.[52] The authors selected **15** as most efficient ligand among a range of variously substituted N-heterocyclic carbenes. The reaction catalysed by a combination of this ligand and $Ni(cod)_2$ yielded chiral protected allylic alcohols, as shown in

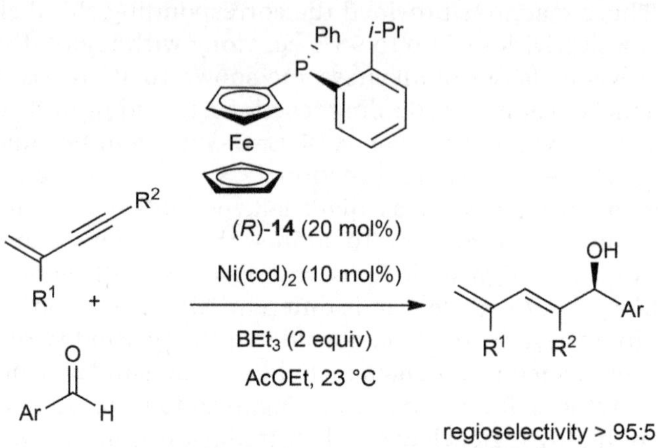

R^1 = Me, R^2 = Et, Ar = Ph: 66%, ee = 56%
R^1 = Me, R^2 = Et, Ar = *p*-Tol: 73%, ee = 56%
R^1 = Me, R^2 = Et, Ar = *p*-$AcOC_6H_4$: 55%, ee = 56%
R^1 = Me, R^2 = Et, Ar = *p*-$MeOC_6H_4$: 63%, ee = 56%
R^1 = Me, R^2 = Et, Ar = 1-Naph: 52%, ee = 48%
R^1 = Me, R^2 = Et, Ar = 2-Naph: 47%, ee = 58%
R^1 = Me, R^2 = Cy, Ar = Ph: 77%, ee = 55%
R^1 = H, R^2 = Ar = Ph: 58%, ee = 44%
R^1 = H, R^2 = Cy, Ar = Ph: 77%, ee = 55%

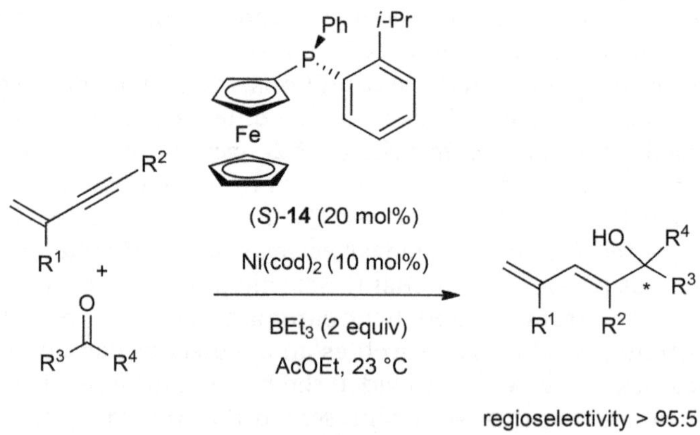

R^1 = H, R^2 = *n*-Hex, R^3 = Me, R^4 = Ph: 69%, ee = 64%
R^1 = R^3 = Me, R^2 = Et, R^4 = *p*-Tol: 71%, ee = 64%
R^1 = R^3 = Me, R^2 = Et, R^4 = 2-furyl: 58%, ee = 40%
R^1 = R^3 = Me, R^2 = Et, R^4 = 1-cyclohexenyl: 39%, ee = 70%

Scheme 3.24 Reductive couplings of 1,3-enynes with aldehydes and ketones.

Scheme 3.25. A range of substrates was tolerated in the process, providing relatively uniform yields and with enantioselectivities of up to 85% ee. Indeed, key functional groups cleanly tolerated in the procedure included aromatic as well as branched and unbranched aldehydes, internal alkynes that either possess or lack an aromatic substituent, terminal alkynes, and unprotected alcohols, wherein the trialkylsilyl group was regioselectively installed on the newly formed hydroxyl group. The regioselection of the alkyne insertion was high, with the exception of internal alkynes that possessed two aliphatic substituents (75 : 25).

In 2008, these reactions were investigated by Zhou *et al.* using another type of reductant, such as $ZnMe_2$, and another type of ligand, such as phosphoramidite **16** (Scheme 3.26).[53] This ligand was selected by the authors among a range of variously substituted phosphoramidites in addition to bidentate ligands, such as BINAP and PHOX. Under the optimised reaction conditions, the reactions of a range of alkynes with aldehydes afforded the corresponding allylic alcohols **17** containing tetrasubstituted alkenes with high regioselectivity (**17** : **18** > 95 : 5) in good yields and with excellent enantioselectivities

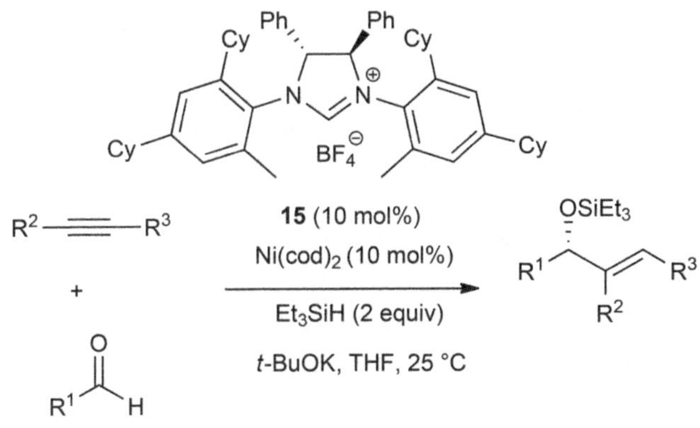

$R^1 = R^3$ = Ph, R^2 = Me: 98%, regioselectivity = 91:9, ee = 78%
R^1 = Ph, $R^2 = R^3$ = Et: 82%, ee = 70%
R^1 = *i*-Pr, R^2 = Me, R^3 = Ph: 86%, regioselectivity > 95:5, ee = 70%
R^1 = *i*-Pr, R^2 = $(CH_2)_2Bn$, R^3 = Me: 86%, regioselectivity > 75:25, ee = 75%
R^1 = Cy, $R^2 = R^3$ = Et: 84%, ee = 85%
R^1 = CH_2Bn, $R^2 = R^3$ = Et: 75%, ee = 78%
R^1 = Cy, R^2 = Me, R^3 = Ph: 78%, regioselectivity > 95:5, ee = 81%
R^1 = Cy, R^2 = H, R^3 = *n*-Hex: 64%, regioselectivity > 95:5, ee = 65%
R^1 = *n*-Hex, R^2 = Me, R^3 = Ph: 70%, regioselectivity = 91:9, ee = 73%
R^1 = Cy, R^2 = $(CH_2)_4OH$, R^3 = Ph: 99%, regioselectivity = 90:10, ee = 79%
R^1 = Cy, R^2 = *n*-Pent, R^3 = Me: 79%, regioselectivity = 75:25, ee = 76%
R^1 = Cy, R^2 = Me, R^3 = *n*-Pent: 47%, regioselectivity = 86:14, ee = 79%

Scheme 3.25 Reductive coupling of alkynes and aldehydes with an N-heterocyclic carbene ligand.

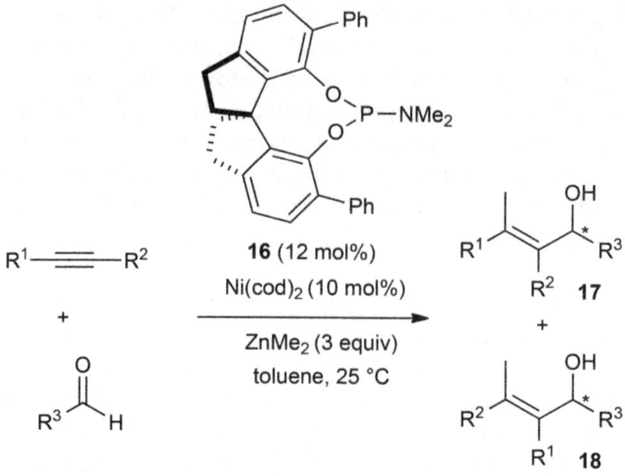

$R^1 = R^3 = Ph$, $R^2 = Me$: 88%, **17**:**18** > 95:5, ee (**17**) = 95% (*R*)
$R^1 = Ph$, $R^2 = Me$, R^3 = *m*-$MeOC_6H_4$: 92%, **17**:**18** > 95:5, ee (**17**) = 95%
$R^1 = Ph$, $R^2 = Me$, R^3 = *p*-Tol: 95%, **17**:**18** > 95:5, ee (**17**) = 94%
$R^1 = Ph$, $R^2 = Me$, R^3 = *p*-$MeOC_6H_4$: 94%, **17**:**18** = 93:7, ee (**17**) = 93%
$R^1 = Ph$, $R^2 = Me$, R^3 = *p*-FC_6H_4: 89%, **17**:**18** > 95:5, ee (**17**) = 96%
$R^1 = Ph$, $R^2 = Me$, R^3 = *p*-$CF_3C_6H_4$: 89%, **17**:**18** = 85:15, ee (**17**) = 95%
$R^1 = Ph$, $R^2 = Me$, R^3 = 2-Naph: 70%, **17**:**18** > 95:5, ee (**17**) = 86%
$R^1 = Ph$, $R^2 = Me$, R^3 = 2-thienyl: 82%, **17**:**18** > 95:5, ee (**17**) = 92%
$R^1 = Ph$, $R^2 = Me$, R^3 = *n*-Bu: 88%, **17**:**18** > 95:5, ee (**17**) = 88%
$R^1 = R^2 = Et$, R^3 = *p*-FC_6H_4: 90%, ee = 98%
$R^1 = R^2$ = *n*-Pr, R^3 = *p*-FC_6H_4: 92%, ee = 99%
$R^1 = R^2 = Ph$, R^3 = *p*-FC_6H_4: 92%, ee = 98%
$R^1 = Ph$, $R^2 = Et$, R^3 = *p*-FC_6H_4: 92%, **17**:**18** = 86:14, ee (**17**) = 98%
$R^1 = Ph$, R^2 = *n*-Bu, R^3 = *p*-FC_6H_4: 91%, **17**:**18** = 86:14, ee (**17**) = 99%

Scheme 3.26 Reductive coupling of alkynes and aldehydes with a phosphoramidite ligand.

of up to 99% ee. In addition to benzaldehyde and its derivatives, naphthaldehyde, thiophene-2-carbaldehyde, as well as aliphatic *n*-butyraldehyde could also be coupled with 1-phenylprop-1-yne to produce the corresponding alcohols in good enantioselectivities (86–92% ee). Moreover, various disubstituted alkynes other than 1-phenylprop-1-yne were also investigated in the coupling reactions with *p*-fluorobenzaldehyde and the desired alcohols were obtained in high yields and excellent enantioselectivities (98–99% ee); however, lower regioselectivities (86 : 14) were observed in the reactions with 1-phenylbut-1-yne and 1-phenylhex-1-yne. The decreased regioselectivity could be attributed to the fact that as the difference between the sizes of the two substituents of the alkyne became smaller; distinguishing the two ends of the alkyne became more difficult. The authors have also investigated different organozinc reagents other than $ZnMe_2$, such as $ZnEt_2$ and $ZnPh_2$, and found that the use of $ZnMe_2$ led to the formation of a mixture of the reductive coupling product as the minor product along with the alkylative coupling product as the major product in 57 and 71% ee, respectively, whereas the use of $ZnPh_2$ gave no reaction. It must be noted that this nice work constituted the first highly enantioselective alkylative coupling of alkynes and aldehydes catalysed by a nickel complex of a chiral spiro-phosphoramidite.

While nickel-catalysed three-component coupling reactions between alkynes, carbonyl compounds, and organometallics have been widely developed for the preparation of allylic alcohols, nickel-catalysed coupling reactions of alkynes with imines are more difficult to perform because imines are weaker electrophiles than aldehydes and, furthermore, the allylic amine products may deactivate or decompose the nickel catalyst. The first breakthrough in the nickel-catalysed coupling reaction of alkynes with imines was made by Jamison and co-workers.[54] They achieved the alkylative coupling of disubstituted alkynes, *N*-alkylimines, and organoboron reagents to produce allylic amines with a tetrasubstituted alkene moiety in good to excellent yields, with high regioselectivities and moderate to high enantioselectivities. However, the reductive coupling of alkynes with imines, which produces allylic amines with a trisubstituted alkene moiety, was only a side reaction of Jamison's system. In 2007, Krische *et al.* reported highly enantioselective iridium-catalysed reductive coupling of dialkyl-substituted alkynes with *N*-sulfonylimines, and the rhodium-catalysed couplings of acetylene and imines.[55] More recently, Zhou *et al.* developed an enantioselective nickel-catalysed reductive coupling of aromatic alkynes with imines using $ZnEt_2$ as reductant and a chiral nickel catalyst generated *in situ* from $Ni(cod)_2$ and chiral spiro-phosphine ligand **19**.[56] As shown in Scheme 3.27, the reaction afforded the expected reductive coupling **20** as the major product along with the alkylative coupling product **21** in a ratio ranging from 92:8 to 50:50. In particular, aromatic imines smoothly underwent the coupling reaction to produce the corresponding allylic amines with good yields (62–80%), good chemoselectivity ratios of 92:8 to 80:20 (for **20**:**21**), and good to high enantioselectivities of 76–94% ee. Except for *o*-methoxyphenylimine and *o*-chlorophenylimine, which interestingly afforded the highest (94% ee) and lowest (76% ee) enantioselectivities, respectively, the other aromatic imines, including naphthyl-derived substrates, gave almost the same level of enantioselectivities. In contrast, imines derived from aliphatic aldehydes gave much lower enantioselectivities (9–11% ee). Changing the aromatic alkyne (R^1 = Ph) into a dialkyl-substituted alkyne (R^1 = R^2 = Et) led to the formation of products **20** and **21** in a 50:50 ratio. An isotope-labelling experiment showed that the transferred hydrogen was most likely from the ethyl group of $ZnEt_2$. It must be noted that this novel methodology favouring aromatic alkynes provided a complement to Krische's iridium catalyst which preferred aliphatic alkynes.[55]

Earlier in 2007, the same authors reported the first example of intermolecular enantioselective nickel-catalysed reductive coupling of 1,3-dienes with carbonyl compounds.[57] As shown in Scheme 3.28, the reaction of

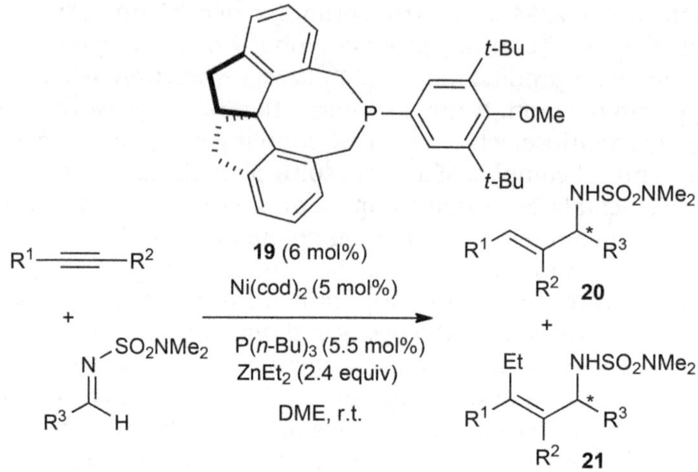

$R^1 = R^3$ = Ph, R^2 = Me: 74%, **20**:**21** = 90:10, ee (**20**) = 87%
R^1 = Ph, R^2 = Me, R^3 = *o*-MeOC_6H_4: 63%, **20**:**21** = 87:13, ee (**20**) = 94%
R^1 = Ph, R^2 = Me, R^3 = *o*-ClC_6H_4: 80%, **20**:**21** = 92:8, ee (**20**) = 76%
R^1 = Ph, R^2 = Me, R^3 = *m*-Tol: 77%, **20**:**21** = 92:8, ee (**20**) = 86%
R^1 = Ph, R^2 = Me, R^3 = *p*-MeOC_6H_4: 68%, **20**:**21** = 93:7, ee (**20**) = 89%
R^1 = Ph, R^2 = Me, R^3 = *p*-ClC_6H_4: 78%, **20**:**21** = 87:13, ee (**20**) = 85%
R^1 = Ph, R^2 = Me, R^3 = 3,4-$Cl_2C_6H_4$: 68%, **20**:**21** = 90:10, ee (**20**) = 87%
R^1 = Ph, R^2 = Me, R^3 = 1-Naph: 62%, **20**:**21** = 80:20, ee (**20**) = 88%
R^1 = Ph, R^2 = Me, R^3 = 2-Naph: 80%, **20**:**21** = 89:11, ee (**20**) = 87%
R^1 = Ph, R^2 = Me, R^3 = *n*-Pr: 65%, **20**:**21** = 86:14, ee (**20**) = 11%
R^1 = Ph, R^2 = Me, R^3 = Cy: 51%, **20**:**21** = 86:14, ee (**20**) = 9%
$R^1 = R^3$ = Ph, R^2 = Et: 81%, **20**:**21** = 90:10, ee (**20**) = 88%
$R^1 = R^2 = R^3$ = Ph: 62%, **20**:**21** = 87:13, ee (**20**) = 86%
$R^1 = R^2$ = Et, R^3 = Ph: 41%, **20**:**21** = 50:50, ee (**20**) = 67%

Scheme 3.27 Reductive coupling of alkynes with imines.

1,4-diphenylbuta-1,3-diene with aromatic aldehydes in the presence of $ZnEt_2$ as the reductant was catalysed by a chiral complex generated *in situ* from Ni$(cod)_2$ and chiral phosphoramidite ligand **22**. It afforded the corresponding chiral bishomoallylic alcohols in excellent *anti*/*syn* diastereoselectivity ratios always >98 : 2, along with high yields (85–99%) and enantioselectivities of 86–96% ee. Bulky ligand **22** was selected as optimal among several other phosphoramidites and phosphine ligands. A range of aromatic aldehydes underwent reductive coupling with the alkyne, and it was found that the presence of an electron-donating substituent, such as OMe or NMe_2, at the *para* position of the phenyl ring slightly enhanced the enantioselectivity of the reaction, while an electron-withdrawing substituent, such as Cl or CF_3, at the *para* position diminished the enantioselectivity. In addition to benzaldehydes, naphthaldehyde, furan-2-carbaldehyde, and thiophene-2-carbaldehyde could also be coupled with the 1,3-diene to afford the corresponding

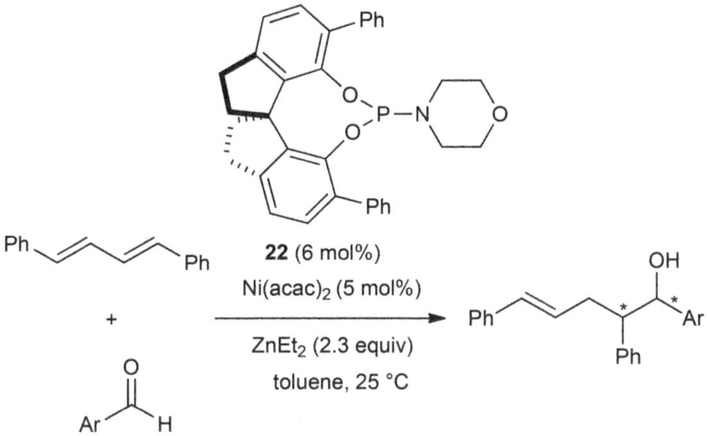

Ar = Ph: 99%, *anti/syn* > 99:1, ee = 96%
Ar = *o*-MeOC$_6$H$_4$: 95%, *anti/syn* > 99:1, ee = 93%
Ar = *m*-Tol: 94%, *anti/syn* > 99:1, ee = 94%
Ar = *m*-MeOC$_6$H$_4$: 99%, *anti/syn* > 99:1, ee = 92%
Ar = *p*-ClC$_6$H$_4$: 92%, *anti/syn* > 99:1, ee = 90%
Ar = *p*-CF$_3$C$_6$H$_4$: 98%, *anti/syn* > 99:1, ee = 85%
Ar = *p*-Tol: 95%, *anti/syn* > 99:1, ee = 95%
Ar = *p*-MeOC$_6$H$_4$: 94%, *anti/syn* > 99:1, ee = 96%
Ar = *p*-NMe$_2$C$_6$H$_4$: 85%, *anti/syn* = 99:1, ee = 96%
Ar = 1-Naph: 94%, *anti/syn* = 98:2, ee = 93%
Ar = 2-Naph: 96%, *anti/syn* > 99:1, ee = 86%
Ar = 2-furyl: 96%, *anti/syn* > 99:1, ee = 92%
Ar = 2-thienyl: 98%, *anti/syn* > 99:1, ee = 91%

Scheme 3.28 Reductive coupling of a diene with aromatic aldehydes.

products in high enantioselectivities (86–93% ee). However, butyraldehyde gave the desired bishomoallylic alcohol in only 72% ee, showing that aliphatic aldehydes were less enantioselective in the process. Other examples of enantioselective nickel-catalysed reductive coupling of 1,3-dienes as well as allenes with carbonyl compounds are described in Chapter 4, Section 4.3.1, dealing with multicomponent reactions.

On the other hand, the first enantioselective nickel-catalysed reductive acyl cross-coupling reaction was reported by Reisman and co-workers in 2013.[58] As shown in Scheme 3.29, the reaction occurred between an acid chloride and a secondary benzyl chloride under catalysis with a combination of NiCl$_2$·(dme) and chiral bisoxazoline 7 in the presence of dimethylbenzoic acid (DBMA), with molecular sieves as additives and manganese (3 equivalents) as the reducing agent. It afforded the corresponding chiral α-alkyl-α-aryl ketones in good yields and with good to high enantioselectivities of up to 94% ee. A screen of various bisoxazoline ligands, including PYBOX derivatives, demonstrated that ligand 7 was optimal. Investigating the substrate scope of the reaction showed that benzoyl chlorides bearing

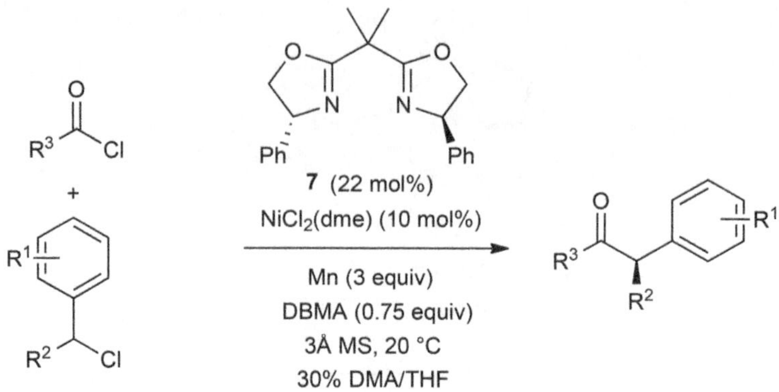

R^1 = H, R^2 = Me, R^3 = *p*-$MeOC_6H_4(CH_2)_2$: 79%, ee = 93%
R^1 = 4-Me, R^2 = Me, R^3 = *p*-$MeOC_6H_4(CH_2)_2$: 74%, ee = 93%
R^1 = 3-Me, R^2 = Me, R^3 = *p*-$MeOC_6H_4(CH_2)_2$: 75%, ee = 93%
R^1 = 2-Me, R^2 = Me, R^3 = *p*-$MeOC_6H_4(CH_2)_2$: 35%, ee = 72%
R^1 = 4-OMe, R^2 = Me, R^3 = *p*-$MeOC_6H_4(CH_2)_2$: 56%, ee = 86%
R^1 = 4-Cl, R^2 = Me, R^3 = *p*-$MeOC_6H_4(CH_2)_2$: 76%, ee = 91%
R^1 = 4-Br, R^2 = Me, R^3 = *p*-$MeOC_6H_4(CH_2)_2$: 73%, ee = 86%
R^1 = 4-CF_3, R^2 = Me, R^3 = *p*-$MeOC_6H_4(CH_2)_2$: 64%, ee = 82%
R^1 = 2-Naph, R^2 = Me, R^3 = *p*-$MeOC_6H_4(CH_2)_2$: 65%, ee = 91%
R^1 = H, R^2 = Et, R^3 = *p*-$MeOC_6H_4(CH_2)_2$: 50%, ee = 94%
R^1 = H, R^2 = Bn, R^3 = *p*-$MeOC_6H_4(CH_2)_2$: 79%, ee = 92%
R^1 = H, R^2 = CH_2OTBS, R^3 = *p*-$MeOC_6H_4(CH_2)_2$: 51%, ee = 89%
R^1 = H, R^2 = Me, R^3 = Et: 60%, ee = 91%
R^1 = H, R^2 = Me, R^3 = $(CH_2)_4CO_2Et$: 64%, ee = 92%
R^1 = H, R^2 = Me, R^3 = $(CH_2)_5Cl$: 76%, ee = 92%

Scheme 3.29 Reductive coupling of acid chlorides with secondary benzyl chlorides.

electron-releasing substituents furnished the corresponding ketones in high enantioselectivities; however, these substrates reacted slowly. In contrast, benzyl chlorides bearing electron-withdrawing substituents reacted rapidly and proceeded to full conversion. It was found that *ortho*-substituted benzyl chlorides were poor substrates, providing the products in low yields (35%) and moderate enantioselectivities (72% ee). Concerning the scope of the acid chloride coupling partner, alkyl halide and ester functionalities were well tolerated.

More recently, these same authors have developed a highly enantioselective nickel-catalysed reductive cross-coupling reaction between vinyl bromides and benzyl chlorides.[59] As shown in Scheme 3.30, the process employed a combination of $NiCl_2$·(dme) and chiral bisoxazoline **23** as the catalyst system in the presence of manganese as the reducing agent, NaI as an additive, and DMA as solvent. Remarkably, the coupling products from various vinyl bromides and benzyl chlorides were achieved in high yields and with enantioselectivities of 85–97% ee. Whereas *meta* and *para* substitution

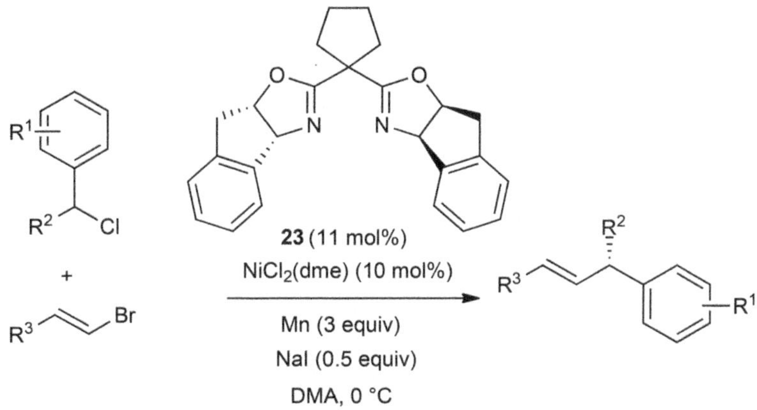

R^1 = H, R^2 = Me, R^3 = *p*-MeOC$_6$H$_4$: 91%, ee = 93%
R^1 = 4-Me, R^2 = Me, R^3 = *p*-MeOC$_6$H$_4$: 82%, ee = 94%
R^1 = 3-Me, R^2 = Me, R^3 = *p*-MeOC$_6$H$_4$: 88%, ee = 93%
R^1 = 2-Me, R^2 = Me, R^3 = *p*-MeOC$_6$H$_4$: 44%, ee = 85%
R^1 = 4-OMe, R^2 = Me, R^3 = *p*-MeOC$_6$H$_4$: 64%, ee = 93%
R^1 = 4-F, R^2 = Me, R^3 = *p*-MeOC$_6$H$_4$: 81%, ee = 89%
R^1 = 4-Cl, R^2 = Me, R^3 = *p*-MeOC$_6$H$_4$: 75%, ee = 88%
R^1 = 4-Br, R^2 = Me, R^3 = *p*-MeOC$_6$H$_4$: 59%, ee = 90%
R^1 = H, R^2 = Et, R^3 = *p*-MeOC$_6$H$_4$(CH$_2$)$_2$: 80%, ee = 97%
R^1 = H, R^2 = Bn, R^3 = *p*-MeOC$_6$H$_4$: 82%, ee = 93%
R^1 = H, R^2 = (CH$_2$)$_2$OH, R^3 = *p*-MeOC$_6$H$_4$: 81%, ee = 96%
R^1 = H, R^2 = Me, R^3 = *p*-Tol: 83%, ee = 96%
R^1 = H, R^2 = Me, R^3 = *p*-FC$_6$H$_4$: 74%, ee = 94%
R^1 = H, R^2 = Me, R^3 = *p*-ClC$_6$H$_4$: 66%, ee = 95%
R^1 = H, R^2 = Me, R^3 = *p*-CF$_3$C$_6$H$_4$: 49%, ee = 94%
R^1 = H, R^2 = Me, R^3 = *p*-OCF$_3$C$_6$H$_4$: 81%, ee = 94%
R^1 = H, R^2 = Me, R^3 = *p*-TBSOC$_6$H$_4$: 82%, ee = 96%
R^1 = H, R^2 = Me, R^3 = *p*-BpinC$_6$H$_4$: 59%, ee = 94%
R^1 = H, R^2 = Me, R^3 = *p*-HOC$_6$H$_4$: 86%, ee = 93%
R^1 = H, R^2 = Me, R^3 = (CH$_2$)$_2$Cl: 60%, ee = 94%
R^1 = H, R^2 = Me, R^3 = (CH$_2$)$_2$OH: 81%, ee = 96%
R^1 = H, R^2 = Me, R^3 = furyl: 80%, ee = 91%
R^1 = H, R^2 = Me, R^3 = (*E*)-PhCH=CH: 82%, ee = 92%

Scheme 3.30 Reductive coupling of vinyl bromides with benzyl chlorides.

at the phenyl ring of the benzyl chlorides were well tolerated, *ortho* substituents resulted in a lower yield and enantioselectivity (44% yield, 35% ee). Functional groups, such as methoxide, fluoride, chloride, bromide, and trifluoromethoxide, were compatible with the procedure. Remarkably, a benzyl chloride bearing a free alcohol could also be coupled in high yield (81%) and enantioselectivity (96% ee). Moreover, a broad scope of styryl bromides underwent the cross-coupling. Notably, a pinacol boronate and a free alcohol derivative were compatible with the reaction, but substrates possessing an aryl ester or nitrile reacted poorly (<50% yield). Furthermore, the reductive cross-coupling could be extended beyond styryl systems; for example, furan and diene derivatives were prepared in high yields (80–82%) and enantioselectivities (91–92% ee).

Axially chiral biaryls, which are common scaffolds in a large number of natural products and biologically active molecules,[60] have found extensive use in chiral auxiliaries and chiral ligands for a variety of asymmetric reactions.[61] Although considerable attention has been paid to their preparation by asymmetric synthesis, direct synthetic methods giving the enantiomerically enriched biaryls from achiral substrates are still rare. The asymmetric synthesis of axially chiral biaryls can be grouped into two categories according to the mode of generation of the axial chirality. One is cross- or homo-coupling of two aryl units where the axial chirality is generated at the formation of the biaryl skeleton. The best examples are nickel-[62] or palladium-catalysed[63] asymmetric cross-coupling of aryl halides with aryl Grignard reagents or arylboronic acids. A third type of asymmetric catalytic synthesis of axially chiral 1,1′-binaphthyls was described by Hayashi *et al.* on the basis of an asymmetric cross-coupling reaction of dinaphthothiophene with a Grignard reagent in the presence of a chiral nickel catalyst.[64] As shown in Scheme 3.31, the reaction performed with a combination of Ni(cod)$_2$ and chiral oxazoline **24** as the catalyst system afforded the corresponding cross-coupling products in excellent yields (92–97%) and with very high enantioselectivities of 93–95% ee when using an aryl Grignard reagent. Much lower enantioselectivities were achieved when using alkyl Grignard reagents. The cross-coupling of 1,9-disubstituted dibenzothiophenes was also studied, albeit providing lower enantioselectivities (up to 82% ee) along with high yields (up to 90%). By further reactions of the thiol group in the coupling products, they could

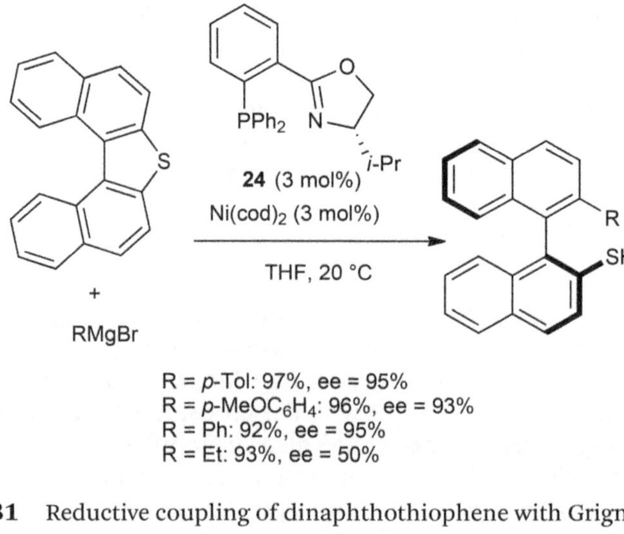

Scheme 3.31 Reductive coupling of dinaphthothiophene with Grignard reagents.

be converted into useful chiral building blocks, demonstrating the utility of this novel synthesis of chiral biaryls.

Among biaryls, axially chiral biaryl dials with two *o*-carbaldehyde functionalities represent a promising class of compounds since they are extremely useful precursors to a range of important biaryl compounds. In 2010, Xu and Lin reported the first nickel-catalysed asymmetric Ullmann coupling of bis-*ortho*-substituted aryl halides, which afforded these products.[65] As shown in Scheme 3.32, the coupling reaction of various 3-alkoxy-2-bromobenzaldehydes performed in the presence of a nickel complex of chiral BINOL-derived phosphoramidite **25**, zinc, and Bu_4NI in DMA as solvent yielded the corresponding chiral biaryl dials in moderate yields and with enantioselectivities of up to 68% ee. Substrates with substitution variations at the *meta* and *para* positions of the benzene ring all gave the expected biaryl products in comparable yields and enantioselectivities. A significant ee drop was found when changing the *o*-alkoxy to carbonyl containing acetoxy (15% ee), suggesting a coordination pattern difference in the reaction transition state. Even if this method did not provide high enantioselectivities, its utility was demonstrated in a total synthesis of (+)-isoschizandrin. It must be noted that this work deals with a homo- and not a cross-coupling process; however, it was decided to situate it in this section dealing with cross-coupling reactions since it is a single example of enantioselective nickel-catalysed homo-coupling reported in recent years.

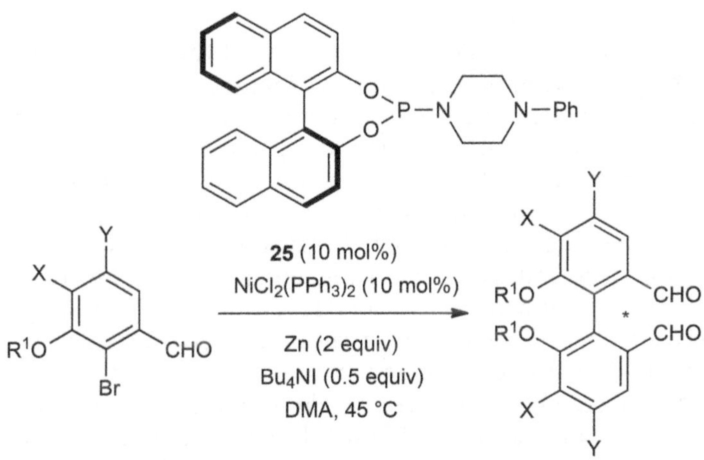

R^1 = Me, X = Y = OMe: 67%, ee = 68%
R^1 = Me, X = OMe, Y = H: 55%, ee = 58%
R^1 = Me, X = H, Y = OMe: 65%, ee = 44%
R^1 = Me, X = OBn, Y = H: 62%, ee = 61%
R^1 = Me, X = OBn, Y = OMe: 60%, ee = 58%
R^1 = Bn, X = OMe, Y = H: 62%, ee = 58%
R^1 = Ac, X = OMe, Y = H: 31%, ee = 15%

Scheme 3.32 Ullmann coupling of bis-*ortho*-substituted aryl bromides.

The development of multimetallic catalytic systems and their application to asymmetric catalysis is an emerging area in modern organic synthesis.[66] The use of a multimetallic entity in catalyst design is a viable approach to the construction of multifunctional catalysts, which can activate multiple substrates simultaneously. The bimetallic cooperative catalysts use many different types of metals, such as alkali metals, transition metals, and lanthanides. The proper arrangement of those metals in close proximity is probably the key to the success for efficient catalysis. From a mechanistic point of view, one metal generally plays a role as a Lewis acid for activating electrophiles, while the other metal ion serves as the counterion for nucleophiles. In the last few years, cooperative bimetallic catalysts have emerged as a new strategy in asymmetric catalysis to achieve high efficiency and selectivity. Moreover, the dual activation of reactants by multiple metal centres can render the reaction conditions milder. Although there are challenges to rationally design efficient bimetallic catalysts, a number of outstanding chiral bimetallic catalysts have been developed to showcase the unique power of such a bio-inspired approach. Cooperative work of multiple catalytically active sites in the asymmetric multimetallic catalysts allows for enhanced catalytic activity and stereoselectivity over monometallic catalysts. In 2013, Feng *et al.* reported an example of a cooperative bimetallic catalyst system which was successfully applied in the asymmetric oxidative cross-dehydrogenative coupling of xanthene and α-substituted β-keto esters.[67] Under oxidative conditions, two C–H bonds could be directly coupled to form a new C–C bond without prior installation of functional groups. For the activation of the sp^3 C–H bond in this reaction, the generation of a carbocation mediated by an oxidant, such as TBHP, was the key step. The catalyst system of the process was composed of a mixture of $NiBr_2$, $FeBF_4{\cdot}6H_2O$, and L-proline-derived *N*,*N'*-dioxide **26** as ligand (Scheme 3.33). Under these reaction conditions, the coupling of xanthene and β-keto esters led to the corresponding chiral xanthene derivatives bearing a quaternary stereogenic carbon centre in moderate to good yields of up to 90% and with remarkable enantioselectivities of 97–99% ee. Indeed, excellent uniform results were achieved for a range of substituted indanone pronucleophiles bearing *tert*-butyl ester substituents. The electronic effect had no influence on the enantioselectivity, and the yields were only slightly affected by either the electron-donating or electron-withdrawing substituent. On the basis of control experiments to understand the activation model of the reaction, the authors showed that the chiral nickel complex of **26** played an important role in the chiral induction of the reaction, and at the same time the chiral iron complex of **26** accelerated the reaction rate for the generation of the coupling product. The importance of this work is related to the fact that asymmetric oxidative coupling of an unreactive benzylic C–H bond remains a challenge due to the instability of the generated carbocation, and that xanthene, which contains a benzylic C–H bond, is pharmaceutically and biochemically active and has been widely used as dyes and fluorescent materials.

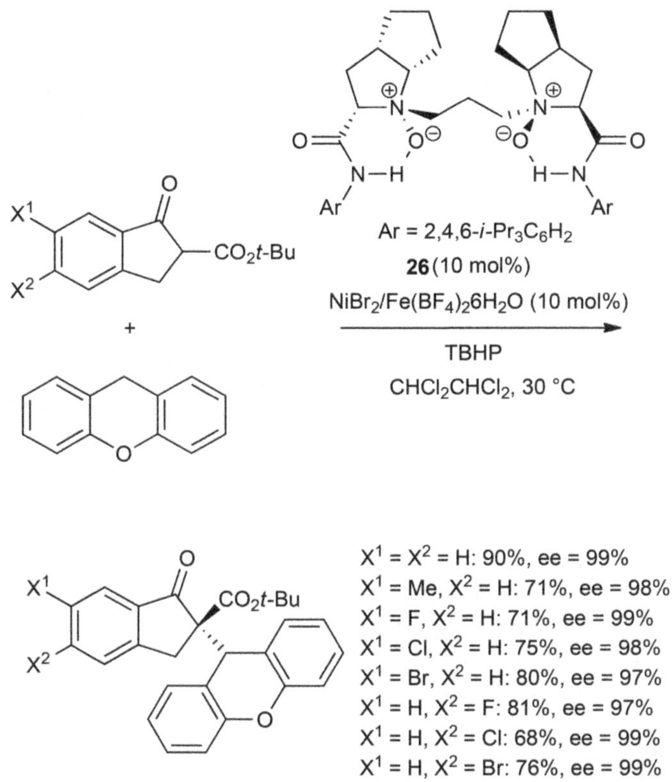

Scheme 3.33 Coupling of xanthene with β-keto esters catalysed by a combination of nickel and iron.

3.5 Conclusions

In the last decade, significant advances have been made in enantioselective nickel-catalysed cross-coupling reactions, making the C–C bond formation through these methodologies a strategy of choice in the design for the total synthesis of biologically active and natural products. In particular, an impressive number of highly efficient asymmetric versions of the Negishi reaction involving a range of activated and non-activated secondary halides has been successfully developed under catalysis with various chiral nickel complexes, providing excellent enantioselectivities in almost all cases. Indeed, the use of a broad variety of secondary electrophiles, including aryl bromides, allylic chlorides, alkyl iodides, propargylic bromides, bromoindanes, propargylic carbonates, benzylic alcohols, α-bromo ketones, α-bromo nitriles, α-zincated *N*-Boc-pyrrolidines, α,α-dihalo ketones, α-bromo sulfonamides and α,α-dibromo sulfones, as well as methoxypyridinium salts, has considerably expanded the scope of asymmetric Negishi reactions. In addition, the first enantioselective versions of the Kumada, Hiyama, and Suzuki reactions were recently achieved with excellent enantioselectivities. For the last reaction in particular, many remarkable results have been reported using various unactivated secondary halides. In addition to organozinc, organosilyl, -magnesium, and -borane reagents have been employed respectively in nickel-catalysed asymmetric Negishi, Hiyama, Kumada, and Suzuki cross-coupling reactions with various electrophiles, and organozirconium and trialkynylindium reagents have been recently successfully involved for the first time in this type of enantioselective reaction. In the last decade, important advances have also been made in the area of enantioselective nickel-catalysed reductive coupling reactions. For example, excellent regio- and enantioselectivities were achieved in reductive couplings of alkynes and aldehydes. Moreover, the first asymmetric reductive coupling of dienes with aldehydes was described with success, as well as the first reductive coupling of acid chlorides with secondary benzyl chlorides, and the first highly enantio- and regioselective reductive coupling of vinyl bromides with benzyl chlorides. Other interesting couplings have generated excellent enantioselectivities, such as that of dinaphthothiophene with Grignard reagents to generate chiral biaryls, and that of xanthene with β-keto esters which was multicatalysed by a combination of nickel and iron.

As demonstrated in this chapter, nickel is emerging as an extraordinarily versatile catalyst for asymmetric cross-coupling reactions with a wide range of coupling partners. Although nickel has been somewhat overlooked for a long time in favour of more popularly studied and therefore well-understood palladium, nickel is now back in the limelight, and hopefully its full potential will be unlocked in the future. Whereas palladium-catalysed methods are often limited to primary alkyl halide substrates, cheaper nickel complexes have been found to be uniquely suited to the catalysis of the cross-coupling of secondary alkyl halides. Indeed, the door has been opened to the development of general methods for the asymmetric coupling of any type of secondary electrophiles. Such methodologies will certainly have a large impact on the synthesis of complex molecules and natural products. While progress has been considerable over the past decade in this field, a better mechanistic and stereochemical understanding is needed in the future. Furthermore, the development of more active catalysts is still necessary to enable lower catalyst loadings than those currently used.

References

1. (a) *Metal-Catalyzed Cross-Coupling Reactions*, ed. F. Diederich and P. J. Stang, Wiley-VCH, Weinheim, 1998; (b) J. Montgomery, *Angew. Chem., Int. Ed.*, 2004, **43**, 3890–3908; (c) *Metal-Catalyzed Cross-Coupling Reactions*, ed. A. de Meijere and F. Diederich, Wiley-VCH, New York, 2004; (d) G. C. Fu, *J. Org. Chem.*, 2004, **69**, 3245–3249; (e) R. M. Moslin, K. Miller-Moslin and T. F. Jamison, *Chem. Commun.*, 2007, 4441–4449; (f) F. Glorius, *Angew. Chem., Int. Ed.*, 2008, **47**, 8347–8349; (g) A. Rudolph and M. Lautens, *Angew. Chem., Int. Ed.*, 2009, **48**, 2656–2670; (h) B. M. Rosen, K. W. Quasdorf, D. A. Wilson, N. Zhang, A.-M. Resmerita, N. K. Garg and V. Percec, *Chem. Rev.*, 2011, **111**, 1346–1416; (i) R. Jana, T. P. Pathak and M. S. Sigman, *Chem. Rev.*, 2011, **111**, 1417–1492; (j) B. L. H. Taylor and E. R. Jarvo, *Synlett*, 2011, **19**, 2761–2765; (k) J. Magano and J. R. Dunetz, *Chem.*

Rev., 2011, **111**, 2177–2250; (l) Y. Uozumi, *Comprehensive Chirality*, Elsevier, Amsterdam, 2012, vol. 4, pp. 18–32.
2. *Handbook of Functionalized Organometallics: Applications in Synthesis*, ed. P. Knochel, Wiley-VCH, Weinheim, 2005, vol. 1 and 2.
3. (a) *Handbook of Organopalladium Chemistry for Organic Synthesis*, ed. E. Negishi, Wiley-Interscience, New York, 2002, vol. 1 and 2; (b) K. Tamao, T. Hiyama and E. Niegishi, *J. Organomet. Chem.*, 2002, **653**, 1–4, (Special Issue: 30 Years of the Cross-Coupling Reaction); (c) *Cross-Coupling Reactions*, ed. N. Miyaura, 2002, vol. 219, Topics in Current Chemistry, Springer, Heidelberg; (d) A. F. Littke and G. C. Fu, *Angew. Chem., Int. Ed.*, 2002, **41**, 4176–4211; (e) M. R. Netherton and G. C. Fu, *Adv. Synth. Catal.*, 2004, **346**, 1525–1532; (f) R. Shintani and T. Hayashi, in *Modern Organonickel Chemistry*, ed. Y. Tamaru, Wiley-VCH, Weinheim, 2005, pp. 240–272; (g) A. C. Frisch and M. Beller, *Angew. Chem., Int. Ed.*, 2005, **44**, 674–688.
4. (a) T. Hayashi, in *Handbook of Organopalladium Chemistry for Organic Synthesis*, ed. E. Negishi, Wiley, New York, 2002, vol. 1, pp. 791–806; (b) T. Hayashi, *J. Organomet. Chem.*, 2002, **653**, 41–45; (c) M. Ogasawara and T. Hayashi, in *Catalytic Asymmetric Synthesis*, ed. I. Ojima, Wiley-VCH, New York, 2nd edn, 2000, pp. 651–674.
5. E. Negishi, A. O. King and N. Okukado, *J. Org. Chem.*, 1977, **42**, 1821–1823.
6. (a) T. Hayashi, M. Tajika, K. Tamao and M. Kumada, *J. Am. Chem. Soc.*, 1976, **98**, 3718–3719; (b) T. Hayashi, M. Konishi, M. Fukushima, K. Kanehira, T. Hioki and M. Kumada, *J. Org. Chem.*, 1983, **48**, 2195–2202.
7. (a) J. K. Kochi and M. Tamura, *J. Am. Chem. Soc.*, 1971, **93**, 1485–1487; (b) M. Tamura and J. K. Kochi, *J. Organomet. Chem.*, 1972, **42**, 205–228.
8. T. Ishiyama, S. Abe, N. Miyaura and A. Suzuki, *Chem. Lett.*, 1992, 691–694.
9. A. Devasagayaraj, T. Stüdemann and P. Knochel, *Angew. Chem., Int. Ed.*, 1995, **34**, 2723–2725.
10. (a) E. Negishi, Q. Hu, Z. Huang, G. Wang and N. Yin, in *The Chemistry of organozinc compounds*, ed. Z. Rappoport and I. Marek, Wiley, New York, 2006, ch. 11; (b) E. I. Negishi, *Bull. Chem. Soc. Jpn.*, 2007, **80**, 233–257; (c) E. I. Negishi, *Bull. Chem. Soc. Jpn.*, 2007, **80**, 1598–1607; (d) M. M. Heravi, E. Hashemi and N. Nazari, *Mol. Diversity*, 2014, **18**, 441–472.
11. J. Zhou and G. C. Fu, *J. Am. Chem. Soc.*, 2003, **125**, 14726–14727.
12. C. Fischer and G. C. Fu, *J. Am. Chem. Soc.*, 2005, **127**, 4594–4595.
13. F. O. Arp and G. C. Fu, *J. Am. Chem. Soc.*, 2005, **127**, 10482–10483.
14. S. Son and G. C. Fu, *J. Am. Chem. Soc.*, 2008, **130**, 2756–2757.
15. S. W. Smith and G. C. Fu, *J. Am. Chem. Soc.*, 2008, **130**, 12645–12647.
16. P. M. Lundin, J. Esquivias and G. C. Fu, *Angew. Chem., Int. Ed.*, 2009, **48**, 154–156.
17. A. J. Oelke, J. Sun and G. C. Fu, *J. Am. Chem. Soc.*, 2012, **134**, 2966–2969.
18. S.-I. Murahashi, *Science of Synthesis*, Thieme Verlag, Stuttgart, 2004, vol. 19.
19. J. Choi and G. C. Fu, *J. Am. Chem. Soc.*, 2012, **134**, 9102–9105.
20. J. T. Binder, C. J. Cordier and G. C. Fu, *J. Am. Chem. Soc.*, 2012, **134**, 17003–17006.
21. C. J. Cordier, R. J. Lundgren and G. C. Fu, *J. Am. Chem. Soc.*, 2013, **135**, 10946–10949.

22. *Privileged Chiral Ligands and Catalysts*, ed. Q.-L. Zhou, Wiley-VCH, Weinheim, 2011.
23. H.-Q. Do, E. R. R. Chandrashekar and G. C. Fu, *J. Am. Chem. Soc.*, 2013, **135**, 16288–16291.
24. *Efficient Preparations of Fluorine Compounds*, ed. H. W. Roesky, Wiley, Hoboken, 2013.
25. Y. Liang and G. C. Fu, *J. Am. Chem. Soc.*, 2014, **136**, 5520–5524.
26. J. Choi, P. Martin-Gago and G. C. Fu, *J. Am. Chem. Soc.*, 2014, **136**, 12161–12165.
27. S. T. Chau, J. P. Lutz, K. Wu and A. G. Doyle, *Angew. Chem., Int. Ed.*, 2013, **52**, 9153–9156.
28. T. Hiyama and E. Shirakawa, *Top. Curr. Chem.*, 2002, **219**, 61–85.
29. X. Dai, N. A. Strotman and G. C. Fu, *J. Am. Chem. Soc.*, 2008, **130**, 3302–3303.
30. J. Terao and N. Kambe, *Acc. Chem. Res.*, 2008, **41**, 1545–1554.
31. S. Lou and G. C. Fu, *J. Am. Chem. Soc.*, 2010, **132**, 1264–1266.
32. (a) N. Miyaura and A. Suzuki, *Chem. Rev.*, 1995, **95**, 2457–2483; (b) N. Miyaura, *Top. Curr. Chem.*, 2002, **219**, 11–59; (c) A. Suzuki, *Angew. Chem., Int. Ed.*, 2004, **50**, 6722–6737.
33. M. R. Netherton and G. C. Fu, in *Topics in Organometallic Chemistry: Palladium in Organic Synthesis*, ed. J. Tsuji, Springer, New York, 2005, pp. 85–108.
34. J. Zhou and G. C. Fu, *J. Am. Chem. Soc.*, 2004, **126**, 1340–1341.
35. B. Saito and G. C. Fu, *J. Am. Chem. Soc.*, 2008, **130**, 6694–6695.
36. N. A. Owston and G. C. Fu, *J. Am. Chem. Soc.*, 2010, **132**, 11908–11909.
37. Z. Lu, A. Wilsily and G. C. Fu, *J. Am. Chem. Soc.*, 2011, **133**, 8154–8157.
38. P. M. Lundin and G. C. Fu, *J. Am. Chem. Soc.*, 2010, **132**, 11027–11029.
39. S. L. Zultanski and G. C. Fu, *J. Am. Chem. Soc.*, 2011, **133**, 15362–15364.
40. A. Wilsily, F. Tramutola, N. A. Owston and G. C. Fu, *J. Am. Chem. Soc.*, 2012, **134**, 5794–5797.
41. J. D. Shields, D. T. Ahneman, T. J. A. Graham and A. G. Doyle, *Org. Lett.*, 2014, **16**, 142–145.
42. S. Lou and G. C. Fu, *J. Am. Chem. Soc.*, 2010, **132**, 5010–5011.
43. J. Caeiro, J. Pérez Sestelo and L. A. Sarandeses, *Chem.-Eur. J.*, 2008, **14**, 741–746.
44. E. Oblinger and J. Montgomery, *J. Am. Chem. Soc.*, 1997, **119**, 9065–9066.
45. S. Ikeda, *Angew. Chem., Int. Ed.*, 2003, **42**, 5120–5122.
46. K. M. Miller, W.-S. Huang and T. F. Jamison, *J. Am. Chem. Soc.*, 2003, **125**, 3442–3443.
47. (a) J. Chan and T. F. Jamison, *J. Am. Chem. Soc.*, 2003, **125**, 11514–11515; (b) E. A. Colby, K. C. O'Brien and T. F. Jamison, *J. Am. Chem. Soc.*, 2004, **126**, 998–999.
48. K. M. Miller, T. Luanphaisarnnont, C. Molinaro and T. F. Jamison, *J. Am. Chem. Soc.*, 2004, **126**, 4130–4131.
49. (a) G. M. Mahandru, G. Liu and J. Montgomery, *J. Am. Chem. Soc.*, 2004, **126**, 3698–3699; (b) H. Y. Jang, R. R. Huddleston and M. J. Krische, *J. Am. Chem. Soc.*, 2004, **126**, 4664–4668.

50. K. M. Miller, E. A. Colby, K. S. Woodin and T. F. Jamison, *Adv. Synth. Catal.*, 2005, **347**, 1533–1536.
51. T. Luanphaisarnnont, C. O. Ndubaku and T. F. Jamison, *Org. Lett.*, 2005, **7**, 2937–2940.
52. M. R. Chaulagain, G. J. Sormunen and J. Montgomery, *J. Am. Chem. Soc.*, 2007, **129**, 9568–9569.
53. Y. Yang, S.-F. Zhu, C.-Y. Zhou and Q.-L. Zhou, *J. Am. Chem. Soc.*, 2008, **130**, 14052–14053.
54. (a) S. J. Patel and T. F. Jamison, *Angew. Chem., Int. Ed.*, 2003, **42**, 1364–1367; (b) S. J. Patel and T. F. Jamison, *Angew. Chem., Int. Ed.*, 2004, **43**, 3941–3944.
55. (a) A. Barchuk, M.-Y. Ngai and M. J. Krische, *J. Am. Chem. Soc.*, 2005, **127**, 11269–11276; (b) M.-Y. Ngai, A. Barchuk and M. J. Krische, *J. Am. Chem. Soc.*, 2007, **129**, 12644–12645; (c) E. Skucas, J. R. Kong and M. J. Krische, *J. Am. Chem. Soc.*, 2007, **129**, 7242–7243.
56. C.-Y. Zhou, S.-F. Zhu, L.-X. Wang and Q.-L. Zhou, *J. Am. Chem. Soc.*, 2010, **132**, 10955–10957.
57. Y. Yang, S.-F. Zhu, H.-F. Duan, C.-Y. Zhou, L.-X. Wang and Q.-L. Zhou, *J. Am. Chem. Soc.*, 2007, **129**, 2248–2249.
58. A. H. Cherney, N. T. Kadunce and S. E. Reisman, *J. Am. Chem. Soc.*, 2013, **135**, 7442–7445.
59. A. H. Cherney and S. E. Reisman, *J. Am. Chem. Soc.*, 2014, **136**, 14365–14368.
60. F. Bracher, W. J. Eisenreich, J. Mühlbacher, M. Dreyer and G. Bringmann, *J. Org. Chem.*, 2004, **69**, 8602–8608.
61. P. Kocovsky, S. Vyskocil and M. Smrcina, *Chem. Rev.*, 2003, **103**, 3213–3245.
62. T. Hayashi, K. Hayashizaki, T. Kiyoi and Y. Ito, *J. Am. Chem. Soc.*, 1988, **110**, 8153–8156.
63. (a) J. Yin and S. L. Buchwald, *J. Am. Chem. Soc.*, 2000, **122**, 12051–12052; (b) A. Herrbach, A. Marinetti, O. Baudoin, D. Guénard and F. Guéritte, *J. Org. Chem.*, 2003, **68**, 4897–4905.
64. (a) Y.-H. Cho, A. Kina, T. Shimada and T. Hayashi, *J. Org. Chem.*, 2004, **69**, 3811–3823; (b) T. Shimada, Y.-H. Cho and T. Hayashi, *J. Am. Chem. Soc.*, 2002, **124**, 13396–13397.
65. W.-W. Chen, Q. Zhao, M.-H. Xu and G.-Q. Lin, *Org. Lett.*, 2010, **12**, 1072–1075.
66. (a) D. E. Fogg and E. N. dos Santos, *Coord. Chem. Rev.*, 2004, **248**, 2365–2379; (b) M. Shibasaki, M. Kanai, S. Matsunaga and N. Kumagai, *Acc. Chem. Res.*, 2009, **42**, 1117–1127; (c) M. Shibasaki, M. Kanai, S. Matsunaga and N. Kumagai, *Top. Organomet. Chem.*, 2011, **37**, 1–30; (d) Z.-Y. Han, C. Wang and L.-Z. Gong, in *Science of Synthesis, Asymmetric Organocatalysis*, ed. B. List and K. Maruoka, Georg Thieme Verlag, Stuttgart, 2011, section 2.3.6; (e) J. Park and S. Hong, *Chem. Soc. Rev.*, 2012, **41**, 6931–6943; (f) H. Pellissier, *Tetrahedron*, 2013, **69**, 7171–7210; (g) H. Pellissier, *Enantioselective Multicatalysed Tandem Reactions*, Royal Society of Chemistry, Cambridge, 2014.
67. W. Cao, X. Liu, R. Peng, P. He, L. Lin and X. Feng, *Chem. Commun.*, 2013, **49**, 3470–3472.

CHAPTER 4

Enantioselective Nickel-Catalysed Domino and Tandem Reactions

4.1 Introduction

The economic interest in combinations of asymmetric (metal) catalysis with these one-pot processes is obvious.[1] A domino reaction has been defined by Tietze as a reaction which involves two or more bond-forming transformations, taking place under the same reaction conditions, without adding additional reagents, catalysts, or additives, and in which the subsequent reactions result as a consequence of the functionality formed by bond formation or fragmentation in the previous step.[2] It must be recognised that a relatively narrow distinction exists between domino and tandem sequences, and they are often inappropriately used indistinguishably from one another in the literature.[3] From the point of view of an operator, the only difference between the two lies in the point along the sequence at which one or more catalysts or reagents had to be added to effect either the initiation of a sequence (that is, a domino reaction) or propagation to the next step (that is, a tandem sequence). The quality and importance of a domino reaction can be correlated to the number of bonds generated in such a process and the increase in molecular complexity. Its goal is the resembling of nature in its highly selective sequential transformations. The domino reactions can be performed as single-, two-, and multicomponent transformations. Multicomponent reactions are defined as domino reactions involving at least three substrates and, consequently, constitute a subgroup of domino reactions.[4,5] It must be noted that there are also some

RSC Catalysis Series No. 26
Enantioselective Nickel-Catalysed Transformations
By Hélène Pellissier

Published by the Royal Society of Chemistry, www.rsc.org

confusing ideas among chemists about the definition of a multicomponent reaction. According to Yus,[5c] this type of reaction should be clearly differentiated from other one-pot processes, such as tandem or cascade reactions, and in general from all processes that involve the reaction between two reagents to yield an intermediate which is captured by the successive addition of a new reagent (sequential component reactions). The use of (multicomponent) domino reactions in organic synthesis is increasing constantly,[6] since they allow the synthesis of a wide range of complex molecules, including natural products and biologically active compounds, in an economically favourable way by using processes that avoid the use of costly and time-consuming protection–deprotection processes, as well as purification procedures of intermediates.[7] Indeed, decreasing the number of laboratory operations required and the quantities of chemicals and solvents used have made domino and multicomponent reactions unavoidable processes.[8] The proliferation of these reactions is evidenced by the number of reviews covering the literature through 1992.[2,9] The goal of this chapter is to cover the advances in enantioselective nickel-catalysed domino reactions and tandem sequences reported in the last decade. This area was previously reviewed in 2005 by Hayashi and Shintani in a book chapter dealing with asymmetric synthesis based on the use of organonickel chemistry.[10] This present chapter is subdivided into three parts, dealing successively with enantioselective nickel-catalysed two-component domino reactions, enantioselective nickel-catalysed multicomponent reactions, and enantioselective nickel-catalysed tandem sequences. The first part is subdivided into two sections, concerning domino reactions initiated by the Michael reaction, and miscellaneous domino reactions. The second part is also subdivided into two sections, dealing successively with three-component couplings of unsaturated hydrocarbons, carbonyl compounds and derivatives, and reducing agents, and miscellaneous multicomponent reactions.

4.2 Two-Component Domino Reactions

4.2.1 Domino Reactions Initiated by the Michael Reaction

The nucleophilic 1,4-addition of stabilised carbon nucleophiles to electron-poor alkenes, generally α,β-unsaturated carbonyl compounds, is known as the Michael addition, although it was first reported by Komnenos in 1883.[11] Michael-type reactions can be considered as one of the most powerful and reliable tools for the stereocontrolled formation of carbon–carbon and carbon–heteroatom bonds,[12] as has been demonstrated by the huge number of examples in which it has been applied as a key strategic transformation in total synthesis. Since the first catalytic domino Michael–aldol reaction reported by Noyori *et al.* in 1996,[13] there have been numerous examples of domino reactions using this methodology. Among recent examples using chiral nickel catalysts, a range of enantioselective Michael-initiated domino reactions involving nitroalkenes as acceptors have been successfully developed. For example, Arai *et al.* have investigated a catalytic asymmetric *exo'*-selective virtual [3 + 2] cycloaddition of aromatic imino esters with *trans*-nitroalkenes, evolving through a domino Michael/Mannich reaction mechanism.[14] Usually, when a *trans*-nitroalkene is used in a [3 + 2] cycloaddition, the stereoconjunction between the 3- and 4-positions is fixed in a *trans* conformation, and four diastereomers are possible, classified as *endo*, *exo*, *endo'*, and *exo'* isomers. Screenings of the metal salts to study the *exo'* adduct ratio have found that nickel salts facilitated the selective production of the *exo'* products. Thus, these authors have performed the *exo'*-selective reaction of imino esters and *trans*-nitroalkenes by using a combination of $Ni(OAc)_2$ with chiral imidazoline-aminophenol **1** as the catalyst system, which provided the corresponding pyrrolidines in good yields and with diastereoselectivities of up to 84% de combined with high enantioselectivities of up to 99% ee for the major isomers, as shown in Scheme 4.1. Indeed, a range of chiral products were obtained from the reaction of aromatic imino esters with various aromatic nitroalkenes bearing electron-deficient or electron-rich substituents. Moreover, aliphatic nitroalkenes were also converted into the corresponding *exo'* products in slightly lower yields (64–68%), diastereoselectivities (36–58% de), and enantioselectivities (92–94% ee). This novel methodology represented the first general success in the catalytic asymmetric *exo'*-selective reaction of imino esters and nitroalkenes. In order to explain the results, the authors proposed that the products were generated from a domino Michael addition of the imino esters onto the *trans*-nitroalkenes, which was then followed by a Mannich reaction, as depicted in Scheme 4.1.

The Henry reaction has been often associated with the Michael reaction in a number of successful asymmetric domino sequences.[15] In this context, Wang *et al.* have developed enantioselective nickel-catalysed domino Michael/Henry reactions of 1,2-diones with nitroalkenes to afford chiral polyfunctionalised bicyclo[3.2.1]octane derivatives containing four stereogenic centres.[16] These remarkable processes, which provided good to quantitative yields, good to high diastereoselectivities of up to 96% de, along with good to excellent enantioselectivities of up to >99% ee, were induced for the first time by a combination of $Ni(OAc)_2$ with chiral cyclohexanediamine ligand **2** (Scheme 4.2). For cyclic 1,2-diones, the best results were achieved with cyclohexane-1,2-dione, while cyclopentane-1,2-dione gave lower yield, diastereo- and enantioselectivity (76% yield, 82% de, and 51% ee with R^1 = Ph, R^2 = H). In the cases of aromatic nitroalkenes in reaction with cyclohexane-1,2-dione, the aromatic ring of these substrates tolerated both electron-donating and electron-withdrawing functionalities at any positions, although only 10% de was obtained when there was an *ortho*-bromine substituent at the aromatic ring. The additions of heteroaromatic nitroalkenes derived from furan to cyclohexane-1,2-dione proceeded equally with excellent enantio- and diastereoselectivity. The more sterically hindered nitroalkene bearing a 2-naphthyl group was also a suitable substrate. Moreover, the domino product arising from the reaction of α-bromo(phenyl)nitroalkene with cyclohexane-1,2-dione

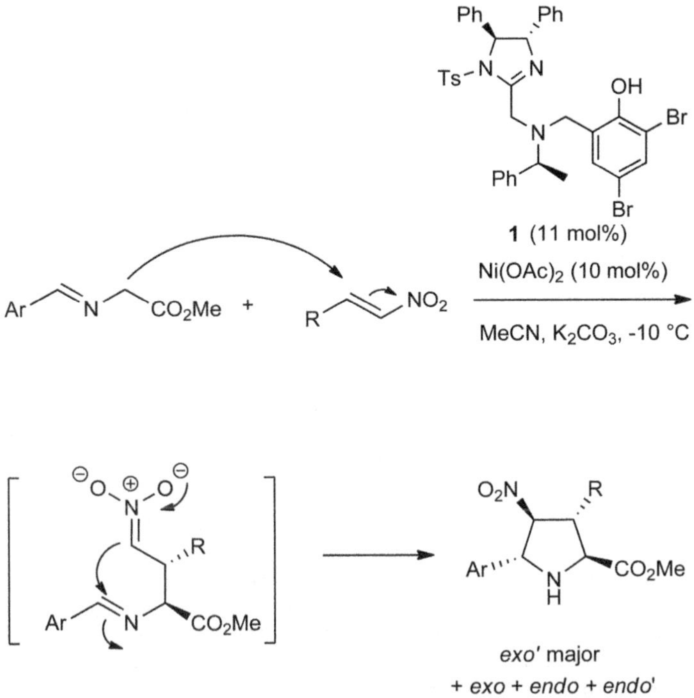

Ar = R = Ph: 99%, *exo'/endo/exo/endo'* =
82:16:1:1, ee (*exo'*) = 97%
Ar = *p*-MeOC$_6$H$_4$, R = Ph: 79%, *exo'/endo/exo/endo'* =
90:10:0:0, ee (*exo'*) = 93%
Ar = *m*-MeOC$_6$H$_4$, R = Ph: 67%, *exo'/endo/exo/endo'* =
89:11:0:0, ee (*exo'*) = 96%
Ar = *p*-ClC$_6$H$_4$, R = Ph: 84%, *exo'/endo/exo/endo'* =
87:9:0:4, ee (*exo'*) = 97%
Ar = 2-Naph, R = Ph: 70%, *exo'/endo/exo/endo'* =
92:7:0:1, ee (*exo'*) = 99%
Ar = Ph, R = *p*-MeOC$_6$H$_4$: 93%, *exo'/endo/exo/endo'* =
86:13:0:1, ee (*exo'*) = 96%
Ar = Ph, R = *p*-BrC$_6$H$_4$: 85%, *exo'/endo/exo/endo'* =
85:14:0:1, ee (*exo'*) = 97%
Ar = Ph, R = *m*-BrC$_6$H$_4$: 94%, *exo'/endo/exo/endo'* =
80:17:0:3, ee (*exo'*) = 96%
Ar = Ph, R = *p*-NO$_2$C$_6$H$_4$: 87%, *exo'/endo/exo/endo'* =
88:11:0:1, ee (*exo'*) = 96%
Ar = Ph, R = *n*-Pent: 68%, *exo'/endo/exo/endo'* =
68:28:0:4, ee (*exo'*) = 92%
Ar = Ph, R = BnCH$_2$: 64%, *exo'/endo/exo/endo'* =
79:21:0:0, ee (*exo'*) = 94%

Scheme 4.1 Domino Michael/Mannich reaction of imino esters with nitroalkenes.

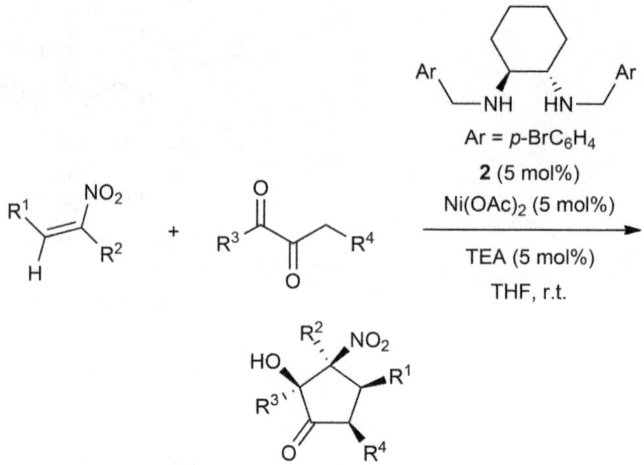

R^1 = Ph, R^2 = H, R^3,R^4 = $(CH_2)_3$: 98%, de = 82%, ee = 98%
R^1 = *p*-$MeOC_6H_4$, R^2 = H, R^3,R^4 = $(CH_2)_3$: 74%, de = 96%, ee = 98%
R^1 = 2-furyl, R^2 = H, R^3,R^4 = $(CH_2)_3$: 70%, de = 94%, ee > 99%
R^1 = *p*-FC_6H_4, R^2 = H, R^3,R^4 = $(CH_2)_3$: 98%, de = 96%, ee > 99%
R^1 = *m*-$MeOC_6H_4$, R^2 = H, R^3,R^4 = $(CH_2)_3$: 80%, de = 96%, ee = 98%
R^1 = 2-Naph, R^2 = H, R^3,R^4 = $(CH_2)_3$: > 99%, de = 78%, ee = 92%
R^1 = *p*-BrC_6H_4, R^2 = H, R^3,R^4 = $(CH_2)_3$: 77%, de = 96%, ee = 97%
R^1 = *p*-Tol, R^2 = H, R^3,R^4 = $(CH_2)_3$: 80%, de = 72%, ee > 99%
R^1 = Ph, R^2 = Br, R^3,R^4 = $(CH_2)_3$: > 99%, de = 90%, ee = 83%
R^1 = *o*-BrC_6H_4, R^2 = H, R^3,R^4 = $(CH_2)_3$: > 99%, de = 10%, ee = 90-91%
R^1 = 3,4-$Me_2C_6H_3$, R^2 = H, R^3,R^4 = $(CH_2)_3$: 91%, de = 75%, ee > 97%
R^1 = Ph, R^2 = H, R^3,R^4 = $(CH_2)_2$: 76%, de = 82%, ee = 51%
R^1 = Ph, R^2 = H, R^3 = Me, R^4 = Et: 86%, de = 60%, ee = 90%
R^1 = Ph, R^2 = H, R^3 = Et, R^4 = Me: 91%, de = 94%, ee = 97%
R^1 = Cy, R^2 = H, R^3,R^4 = $(CH_2)_3$: 51%, de = 78%, ee = 97%
R^1 = *n*-Bu, R^2 = H, R^3,R^4 = $(CH_2)_3$: 87%, de = 96%, ee = 98%

proposed transition state:

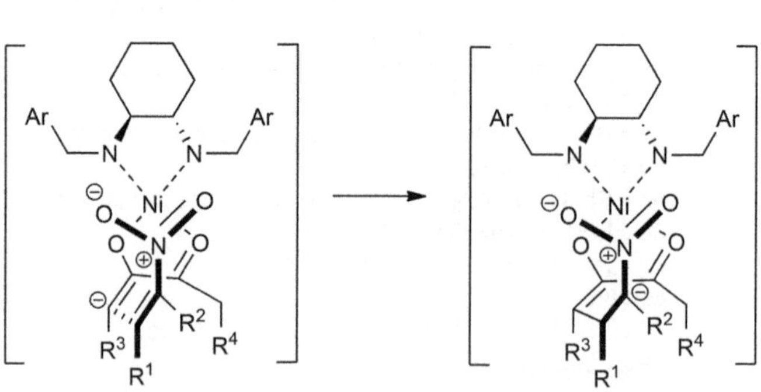

Scheme 4.2 Domino Michael/Henry reaction of 1,2-diones with nitroalkenes.

was obtained in 83% ee. Furthermore, several alkyl-substituted nitroalkenes provided the corresponding domino products in excellent enantioselectivities (97–98% ee) and moderate to good diastereoselectivities (78–96% de) by reaction with cyclohexane-1,2-dione. In addition to cyclic 1,2-diones such as cyclohexane-1,2-dione, several acyclic 1,2-diones also gave good results (90–97% ee). The authors have proposed the transition state depicted in Scheme 4.2 to explain the results. At the beginning of the transition state, the nitronate anion is stabilised by interaction at the open apical position on the nickel ion. A Michael addition then occurs with the activated metal enolate structure of the 1,2-dione. In the favoured transition state, the substituted group of the nitroalkene is oriented away from the bulky aromatic group of the chiral ligand. The π–π stacking interaction between the enolate structure of the 1,2-dione and the double bond of the nitroalkene probably has a favourable effect on the stereoselectivity of the reaction. The resulting adduct subsequently undergoes an intramolecular Henry reaction to form the final product.

Later, asymmetric domino Michael/Henry reactions of cyclohexane-1,2-dione with nitroalkenes were also developed by Ni *et al.* using an *in situ* generated nickel catalyst derived from chiral bisoxazolidine **3** and $Ni(acac)_2$ (Scheme 4.3).[17] Performing the reactions in isopropanol at room temperature with 5 mol% of catalyst loading allowed the corresponding bicyclo[3.2.1]octane derivatives bearing four stereogenic centres to be achieved in high yields (76–99%) and with moderate to good diastereoselectivities of up to 80% de and excellent enantioselectivities (90–99% ee). The results showed that the nature of the substituents on the aryl group of the nitrostyrenes influenced both yields and stereoselectivities. For aromatic nitroalkenes with a 4-halo substitution, both diastereo- and enantioselectivities increased by changing Br to F. In the case of the 2-bromo-substituted nitrostyrene, excellent enantioselectivity was also observed (98% ee), but the diastereoselectivity was relatively low (34% de). The addition of nitroalkenes bearing electron-donating, heteroaromatic, and naphthyl groups also proceeded smoothly and afforded the corresponding products in excellent yields and with high diastereo- and enantioselectivities (92–98% yield, 60–75% de, 90–91% ee). Furthermore, the catalyst was also effective for the domino reaction of cyclohexane-1,2-dione with an aliphatic nitroalkene (R = *n*-Bu) under standard reaction conditions, providing the corresponding product in 76% yield, with good diastereoselectivity of 75% de and high enantioselectivity of 95% ee. Importantly, the authors demonstrated that it was possible to decrease the catalyst loading to 3 mol% without significantly affecting the stereoselectivity.

In 2014, Arai and Yamamoto described asymmetric nickel-catalysed domino Michael/Henry reactions between 2-sulfanylbenzaldehydes and aromatic nitroalkenes to give the corresponding chiral 2-aryl-3-nitrochroman-4-ols, in most cases in almost quantitative yields and with good to high diastereo- and enantioselectivities of up to >98% de and 95% ee, respectively (R^2 = H, Scheme 4.4).[18] These reactions were promoted by an *in situ* generated catalyst from 10–11 mol% of chiral imidazoline-aminophenol ligand **1** and

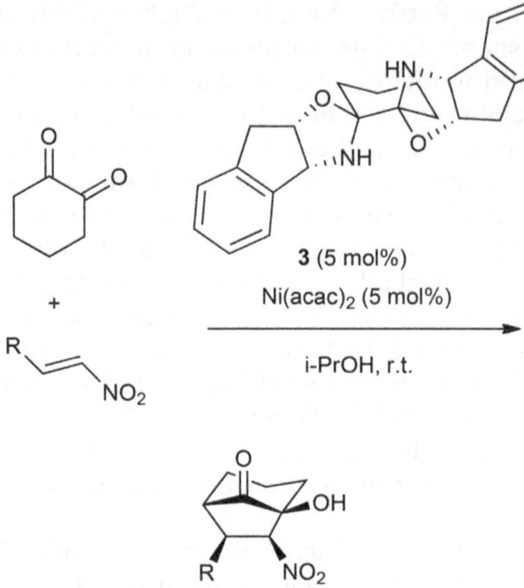

R = Ph: 99%, de = 75%, ee = 99%
R = *p*-BrC$_6$H$_4$: 88%, de = 60%, ee = 92%
R = *p*-ClC$_6$H$_4$: 80%, de = 75%, ee = 96%
R = *p*-FC$_6$H$_4$: 90%, de = 80%, ee = 99%
R = *o*-BrC$_6$H$_4$: 90%, de = 34%, ee = 98%
R = *p*-Tol: 98%, de = 75%, ee = 91%
R = *p*-MeOC$_6$H$_4$: 95%, de = 75%, ee = 90%
R = 2-furyl: 92%, de = 60%, ee = 91%
R = 1-Naph: 94%, de = 75%, ee = 91%
R = *n*-Bu: 76%, de = 75%, ee = 95%

Scheme 4.3 Domino Michael/Henry reaction of cyclohexane-1,2-dione with nitroalkenes.

$Ni(OAc)_2·4H_2O$ in chloroform at −20 °C. The substrate scope of the process showed that various electron-withdrawing and electron-donating substituents were tolerated on the benzene ring of the nitroalkene, providing the products in enantioselectivities of up to 95% ee. In addition, (*E*)-2-(2-nitrovinyl)thiophene reacted with 2-sulfanylbenzaldehyde to give the corresponding product in 94% ee and 90% de. Two substituted 2-sulfanylbenzaldehydes (R^2 = Cl, *t*-Bu) also reacted smoothly with phenylnitroethylene in 80–87% ee. Unfortunately, the use of an aliphatic nitroalkene (R^1 = *n*-Pent) resulted in a reduced yield (74%) of the desired product, as well as a low enantioselectivity of 4% ee. A proposed mechanism explaining the manner in which the chiral (2*S*,3*R*,4*R*)-thiochromanes were synthesised through domino Michael/Henry reactions is depicted in Scheme 4.4. The process begins with the formation of an L*–Ni–thiolate intermediate due to the high affinity of thiol for the nickel centre, which reacts with the nitrostyrene through a Michael addition to give the corresponding L*–Ni–nitronate intermediate. The latter leads to the formation of the six-membered ring of the thiochromane to

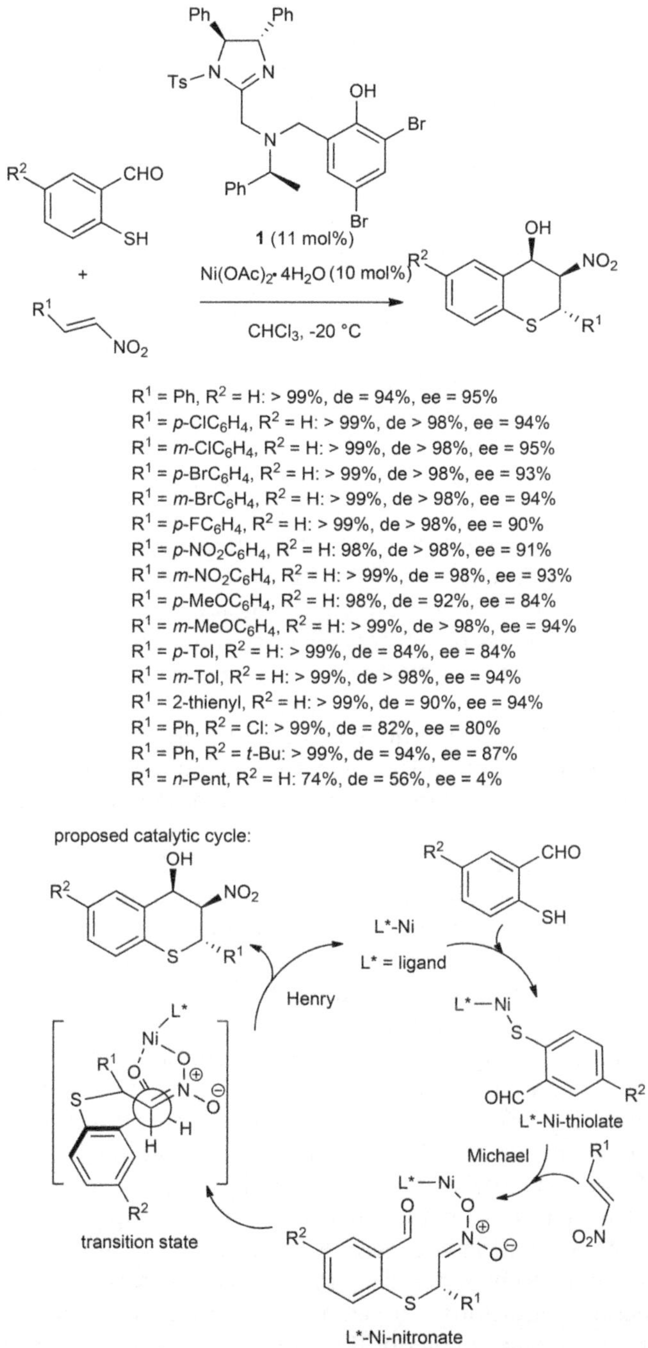

Scheme 4.4 Domino Michael/Henry reaction of sulfanylbenzaldehydes with nitroalkenes.

adopt a strained half-boat-like conformation in which the eclipsed interaction between the carbonyl and nitro groups is increased if the carbonyl group remains in the equatorial position. The Henry reaction then proceeds from the transition-state complex, in which the C4–O–Ni bond is in the pseudoaxial position, to give the final (2*S*,3*R*,4*R*)-thiochromane.

On the other hand, enantioselective nickel-catalysed Michael-initiated domino reactions involving α,β-unsaturated carbonyl compounds have also been successfully developed in the last decade. For example, Kanemasa *et al.* have reported the enantioselective domino Michael/cyclisation reaction of dimedone to 1-(2-crotonoyl)-3,5-dimethylpyrazoles **4** catalysed by an *in situ* generated bisoxazoline chiral nickel catalyst of (*R*,*R*)-DBFOX-Ph (Scheme 4.5).[19] The reaction begins with the Michael addition of the nickel enolate of dimedone to 1-(2-crotonoyl)-3,5-dimethylpyrazole **4** to give nickel intermediate **A**, which further undergoes cyclisation to provide intermediate **B**. The latter is then submitted to elimination of the pyrazole moiety, followed by dehydration to finally afford the corresponding chiral enol lactone **5** in good yields and with moderate to high enantioselectivities of up to 95% ee. The reaction employed one equivalent of acetic anhydride which trapped the pyrazole through *N*-acetylation, allowing yield and enantioselectivity to be improved. The role of the pyrazole chelating auxiliary of acceptor **4a** determined the reactivity of **4a** not only as electrophile but also as leaving group. Accordingly, the authors examined another acceptor **4b** (X = Br), which was expected to be more reactive than **4a** because of its less basic property as chelating auxiliary. Therefore, the reaction of **4b** with dimedone in isopropanol provided the corresponding enol lactone **5** in quantitative yield and 95% ee, whereas **5a** was obtained in both lower yield and enantioselectivity (80% yield, 76% ee).

Earlier, these authors also performed this type of reaction with the same catalyst, albeit employed in the presence of 2,2,6,6-tetramethylpiperidine (TMP) as an additive and with two equivalents of acetic anhydride in THF at room temperature, providing the corresponding enol lactones through the same domino Michael/cyclisation reactions as 1-(alk-2-enoyl)-4-halo-3,5-dimethylpyrazoles.[20] The chiral polyfunctionalised products were obtained in good to high yields (up to 99%) and with generally high enantioselectivities of 89–99% ee, as shown in Scheme 4.6. The use of acetic anhydride as additive was highly effective through the acetylation trapping of the liberated pyrazole, giving 1-acetyl-3,5-dimethylpyrazole. As shown in Scheme 4.6, a variety of 1-(alk-2-enoyl)-4-halo-3,5-dimethylpyrazoles having alkyl, aromatic, as well as heteroaromatic substituents were compatible with the reaction conditions, providing comparable results. The authors have proposed the structure depicted in Scheme 4.6 for the active catalyst.

These reaction conditions were also applied to the domino Michael/cyclisation reactions of other nucleophiles, such as 4-hydroxy-6-methyl-2-pyrone **6**, 4-hydroxycoumarin **7**, and 3-hydroxyperinaphthenone **8**, as shown in Scheme 4.7.[20] By reaction with variously substituted 1-(alk-2-enoyl)-4-bromo-3,5-dimethylpyrazoles, these compounds afforded the corresponding chiral

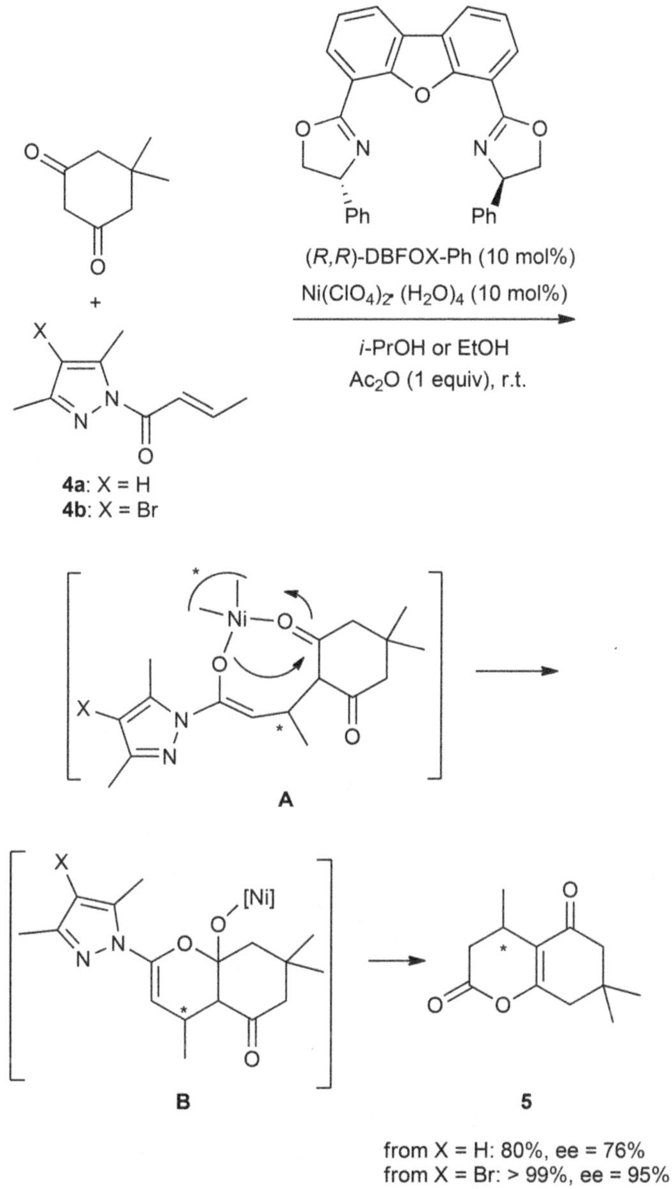

Scheme 4.5 Domino Michael/cyclisation reaction of dimedone with 1-(2-crotonoyl)-3,5-dimethylpyrazoles.

domino products **9a–c, 9d–h,** and **10,** respectively. These products, arising from domino Michael/cyclisation reactions, were obtained in moderate to good yields and with good to high enantioselectivities of up to 98% ee.

Coumarin derivatives are probably one of the most common skeletons found in natural products. Owing to their extensive array of biological

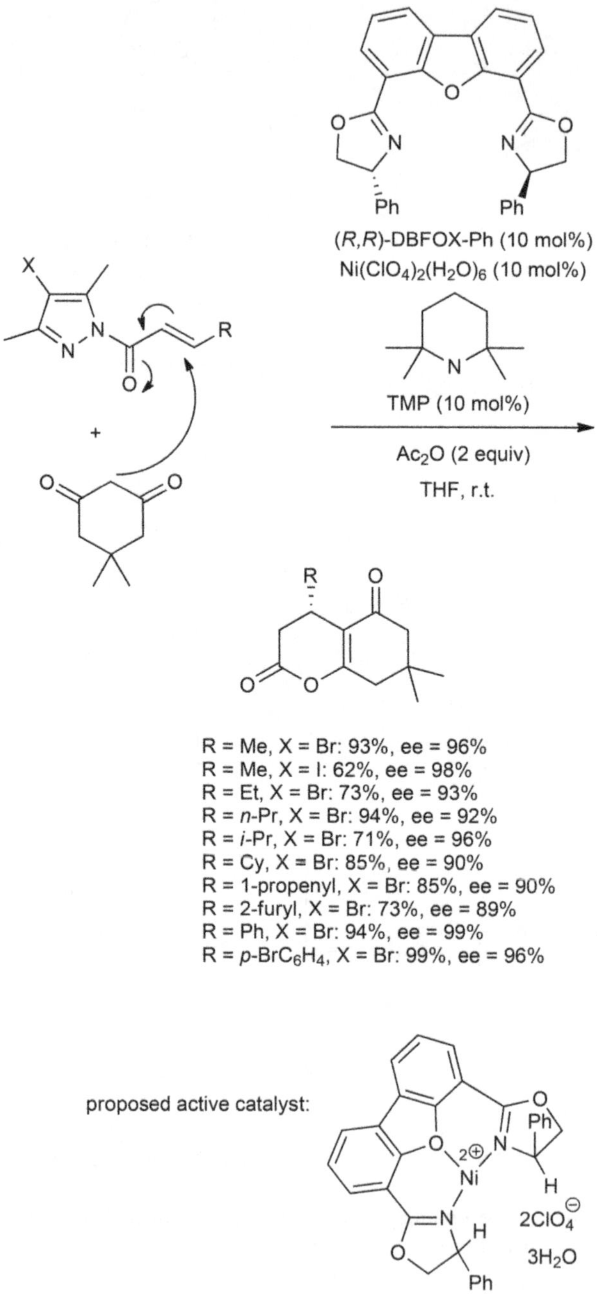

Scheme 4.6 Domino Michael/cyclisation reaction of dimedone with 1-(alk-2-enoyl)-4-halo-3,5-dimethylpyrazoles.

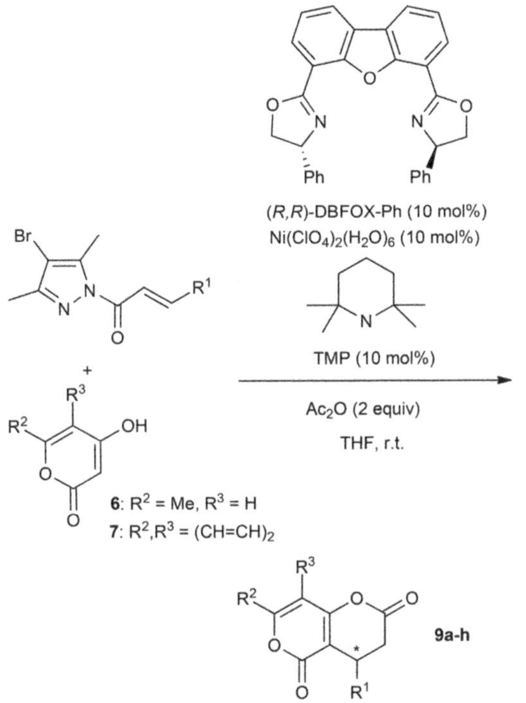

9a: $R^1 = R^2 = Me$, $R^3 = H$: 85%, ee = 98%
9b: R^1 = 5-Br-2-MeOC_6H_3, R^2 = Me, R^3 = H: 23%, ee = 84%
9c: R^1 = 3,5-$Br_2C_6H_3$, R^2 = Me, R^3 = H: 60%, ee = 96%
9d: R^1 = Me, R^2,R^3 = $(CH{=}CH)_2$: 66%, ee = 93%
9e: R^1 = Ph, R^2,R^3 = $(CH{=}CH)_2$: 64%, ee = 96%
9f: R^1 = *p*-BrC_6H_4, R^2,R^3 = $(CH{=}CH)_2$: 65%, ee = 95%
9g: R^1 = 5-Br-2-MeOC_6H_3, R^2,R^3 = $(CH{=}CH)_2$: 53%, ee = 84%
9h: R^1 = 3,5-$Br_2C_6H_3$, R^2,R^3 = $(CH{=}CH)_2$: 80%, ee = 97%

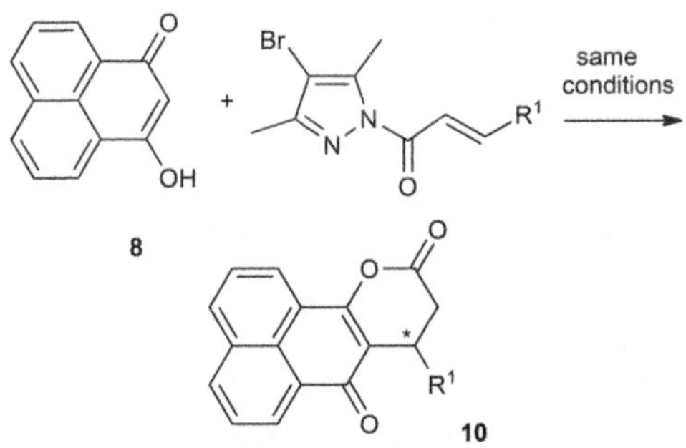

R^1 = Me: 81%, ee = 91%
R^1 = 5-Br-2-MeOC_6H_3: 31%, ee = 89%

Scheme 4.7 Domino Michael/cyclisation reactions of enols with 1-(alk-2-enoyl)-4-bromo-3,5-dimethylpyrazoles.

activities and pharmacological properties, the synthesis of this class of compound has been a long-standing challenge in organic chemistry. Among the coumarin family members, the 4-hydroxycoumarin core, such as in the anticoagulant warfarin, is especially important. Among methodologies developed to achieve diversely structured chiral warfarins, the asymmetric domino Michael/cyclisation reaction of 4-hydroxycoumarin with α,β-unsaturated systems has become a very attractive methodology in recent years. Initially, Jørgensen *et al.* developed a bisoxazoline–copper(II) chiral complex to catalyse the reaction of a cyclic 1,3-dicarbonyl compound to a β,γ-unsaturated α-keto ester, which afforded a warfarin analogue.[21] Later, several groups reported the successful use of various organocatalysts in these reactions. In 2011, Lin and Feng applied an *N,N′*-dioxide chiral nickel catalyst to induce the enantioselective domino Michael/cyclisation reaction of cyclic 1,3-dicarbonyl compounds to β,γ-unsaturated α-keto esters.[22] The catalyst was generated *in situ* from $Ni(acac)_2$ and chiral *N,N′*-dioxide **11**, both employed at 5 mol% of catalyst loading in 1,2-dichloroethane in the presence of 4 Å MS at 0 °C (Scheme 4.8). The reactions remarkably afforded the corresponding chiral warfarin analogues in nearly quantitative yields and with high enantioselectivities of up to 90% ee. Interestingly, neither the steric hindrance nor the electronic nature of the aromatic ring (R^1) of the β,γ-unsaturated α-keto ester had any obvious effect on the enantioselectivity (87–90% ee). It is worth noting that the substrates with condensed-ring, heteroaromatic, and a cinnamyl group performed well, giving the corresponding products in excellent yields and with high enantioselectivities. Moreover, the α-keto ester with an ethyl substrate (R^2 = Et) also provided an excellent yield (98%) and a good enantioselectivity (85% ee), as well as the 4-hydroxycoumarin containing a 6-methyl group which gave 98% yield combined with 87% ee. To explain these results, the authors proposed a transition state which is depicted in Scheme 4.8. They speculated that *N,N′*-dioxide ligand **11** and the β,γ-unsaturated α-keto ester coordinated with $Ni(acac)_2$ to form a complex. Then, the 4-hydroxycoumarin could only attack the *Re* face of the double bond, since the *Si* face of the double bond was hindered by the sterically bulky group. The corresponding domino product was afforded with the *S* configuration.

4.2.2 Miscellaneous Domino Reactions

Spirocyclic oxindole units are structural motifs found in natural and non-natural compounds with diverse and important biological activities.[23] While various synthetic methods are available for the synthesis of chiral spirooxindoles bearing an all-carbon quaternary stereocenter at the C-3′ position of the oxindole unit, those affording spirooxindoles bearing a nitrogen atom at this position are still rather limited. In 2013, Matsunaga and Kanai reported the first catalytic asymmetric addition of isothiocyanatooxindoles to aldehydes by using a chiral dinuclear nickel Schiff base catalyst.[24] As shown in Scheme 4.9, the reaction of a range of aliphatic

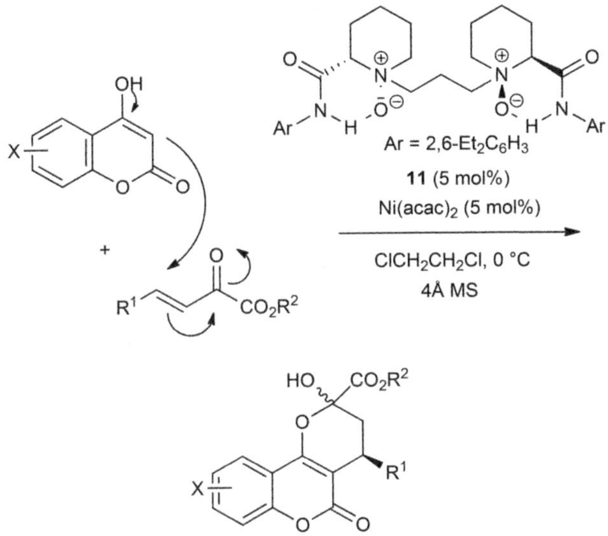

R^1 = Ph, R^2 = Me, X = H: 99%, ee = 89%
R^1 = *m*-Tol, R^2 = Me, X = H: 99%, ee = 89%
R^1 = *p*-Tol, R^2 = Me, X = H: 99%, ee = 89%
R^1 = *m*-$MeOC_6H_4$, R^2 = Me, X = H: 99%, ee = 87%
R^1 = *p*-$MeOC_6H_4$, R^2 = Me, X = H: 99%, ee = 88%
R^1 = *p*-PhC_6H_4, R^2 = Me, X = H: 98%, ee = 87%
R^1 = *m*-ClC_6H_4, R^2 = Me, X = H: 78%, ee = 89%
R^1 = *p*-ClC_6H_4, R^2 = Me, X = H: 99%, ee = 87%
R^1 = *m*-BrC_6H_4, R^2 = Me, X = H: 89%, ee = 87%
R^1 = *p*-BrC_6H_4, R^2 = Me, X = H: 99%, ee = 90%
R^1 = *p*-FC_6H_4, R^2 = Me, X = H: 99%, ee = 90%
R^1 = 2-Naph, R^2 = Me, X = H: 97%, ee = 89%
R^1 = 2-thienyl, R^2 = Me, X = H: 98%, ee = 87%
R^1 = (*E*)-PhCH=CH, R^2 = Me, X = H: 96%, ee = 89%
R^1 = Ph, R^2 = Et, X = H: 98%, ee = 85%
R^1 = Ph, R^2 = Me, X = 6-Me: 98%, ee = 87%

proposed transition state:

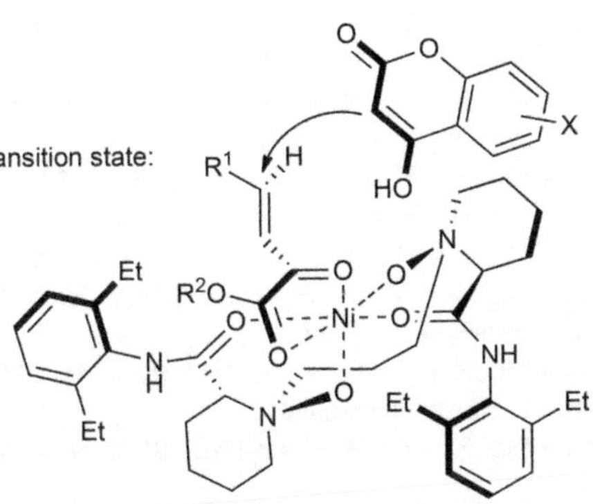

Scheme 4.8 Domino Michael/cyclisation reaction of cyclic 1,3-dicarbonyl compounds with β,γ-unsaturated α-keto esters.

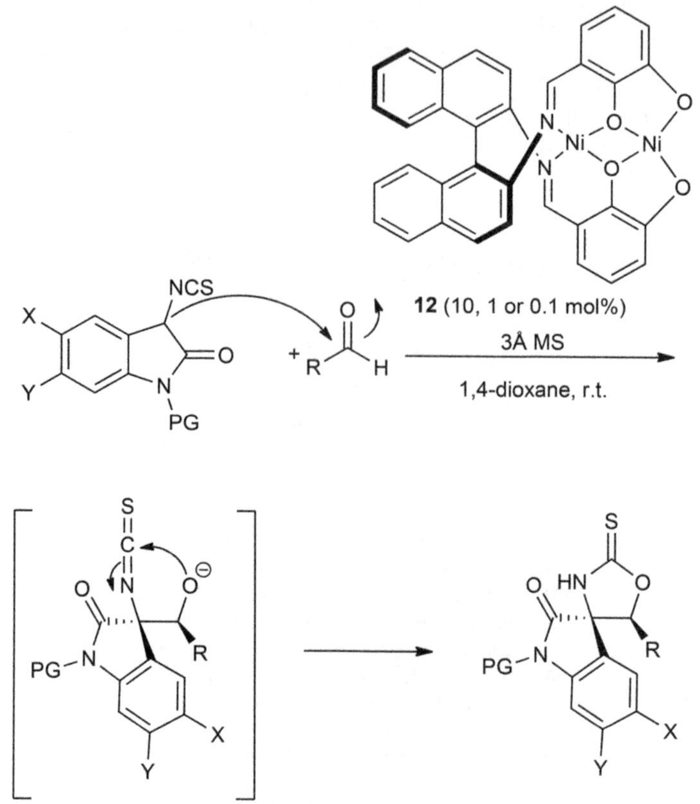

with 10 mol% of catalyst:
PG = Me, X = Y = H, R = Cy: 91%, de = 78%, ee = 91%
PG = Me, X = Y = H, R = *c*-Pent: 90%, de = 74%, ee = 92%
PG = Me, X = Y = H, R = *i*-Pr: 96%, de = 78%, ee = 90%
PG = Me, X = Y = H, R = *i*-Pent: 92%, de = 66%, ee = 80%
PG = Me, X = Y = H, R = *n*-Pent: 86%, de = 82%, ee = 99%
PG = Me, X = Y = H, R = CH_2Bn: 85%, de = 80%, ee = 97%
PG = Me, X = Y = H, R = $(CH_2)_2$-CH=CH-$(CH_2)_4$Me:
82%, de = 82%, ee = 88%
PG = Me, X = Y = H, R = $(CH_2)_3$OTBS: 99%, de = 62%, ee = 96%
PG = X = Me, Y = H, R = *n*-Pent: 90%, de = 78%, ee = 98%
PG = Me, X = H, Y = Cl, R = *n*-Pent: 84%, de = 64%, ee = 92%
PG = allyl, X = Y = H, R = *n*-Pent: 96%, de = 76%, ee = 89%
with 1 mol% of catalyst:
PG = Me, X = Y = H, R = *n*-Pent: 90%, de = 78%, ee = 99%
with 0.1 mol% of catalyst:
PG = Me, X = Y = H, R = *n*-Pent: 85%, de = 42%, ee = 98%

Scheme 4.9 Domino aldol-type/cyclisation reaction of aldehydes with isothiocyanatooxindoles.

aldehydes with isothiocyanatooxindoles afforded the corresponding spirooxindoles in high yields, diastereo-, and enantioselectivities of up to 99%, 82% de, and 99% ee, respectively. The process, evolving through a domino aldol-type/cyclisation reaction, was generally promoted by 10 mol% of chiral dinuclear Ni_2–Schiff base **12** in the presence of molecular sieves at room temperature and in 1,4-dioxane as solvent. It was demonstrated that the catalyst loading could be reduced to 0.1 mol% and still provide remarkable enantioselectivity of up to 98% ee (PG = Me, X = Y = H, R = *n*-Pent). The authors found that catalyst **12** was much more efficient than the corresponding dinuclear copper and cobalt complexes, which gave low enantioselectivities (2–21% ee). The substrate scope of the domino aldol-type/cyclisation reaction showed that α-branched aliphatic aldehydes gave the corresponding chiral tricyclic products in 66–78% de and 80–92% ee, while linear aliphatic aldehydes exhibited slightly higher enantioselectivity than the α-branched ones, providing products in 88–99% ee and up to 82% de. An aldehyde bearing a silyl ether moiety also led to the corresponding product in high enantioselectivity of 96% ee and yield of 99%, albeit with a moderate diastereoselectivity of 62% de. Moreover, oxindole donors with either a methyl or chloro substituent were applicable, providing enantioselectivities of 92–98% ee. In addition to an *N*-methyl protecting group, the reaction conditions were compatible with an oxindole bearing a removable *N*-allyl protecting group, providing an enantioselectivity of 89% ee. In contrast to aliphatic aldehydes, the system afforded poor results for aromatic aldehydes. For instance, an enantioselectivity of only 33% ee was obtained for the reaction of benzaldehyde with unsubstituted *N*-methyloxindole (X = Y = H, PG = Me), in combination with a low diastereoselectivity of just 10% de.

Catalytic methods encompassing metal carbene intermediates constitute a vast array of transformations that offer the synthetic chemist great scope in the synthesis of many complex molecules.[25] Of these processes, the catalytic domino carbonyl ylide formation/1,3-dipolar cycloaddition reaction offers an elegant route to highly substituted oxygen-containing heterocycles.[25b,c,26] This powerful methodology has been extensively advanced by the Padwa group in particular.[26,27] The development of a catalytic enantioselective version of this domino reaction has become a challenging objective. The primary work in this area was reported by Hodgson *et al.* in 1997, dealing with intramolecular enantioselective catalytic domino carbonyl-ylide formation/cyclisation reactions of α-diazo-β-keto esters in enantioselectivities of up to 53% ee by using Davies' prolinate catalyst, $Rh_2((S)\text{-DOSP})_4$.[28] Ever since, the formation of keto carbenoids by treatment of diazo keto compounds with rhodium(II) salts has been broadly employed in enantioselective domino processes as the primary step. This is then followed by the generation of a 1,3-dipole through an intramolecular cyclisation of the keto carbenoid onto an oxygen atom of a neighbouring keto group and an inter- or intramolecular 1,3-dipolar cycloaddition. Chiral catalysts of metals other than rhodium, such as nickel or silver, and chiral catalysts of ytterbium have also been employed to induce enantioselective domino carbonyl-ylide formation/1,3-dipolar cycloaddition reactions.[29] For example, Suga *et al.* have developed the first successful example of reverse-electron-demand dipole-LUMO/dipolarophile-HOMO controlled cycloaddition reactions between carbonyl ylides, which were generated *in situ* from α-diazo-*o*-(methoxycarbonyl)acetophenone and their acyl derivatives in the presence of $Rh_2(OAc)_4$, and vinyl ether derivatives activated by chiral nickel Lewis acids.[30] As shown in Scheme 4.10, when cyclohexyl vinyl ether was activated by the (*R*)-BINIM-4Me-2QN–Ni(II) complex as a chiral Lewis acid, it reacted with α,α′-dicarbonyl diazo compounds to yield the corresponding *endo* cycloadducts in high yields, complete diastereoselectivity, and high enantioselectivities of up to 97% ee. Among diazo compounds investigated, those bearing bulky acyl substituents (R = Cy, i-Pr) were shown to exhibit relatively higher enantioselectivities (96–97% ee).

The first examples of chiral Lewis-acid catalysis in the formation of chiral tetrahydro-1,2-oxazines with very high enantioselectivity were reported by Sibi *et al.* in 2005, by way of an enantioselective formal cycloaddition of nitrones to activated cyclopropanes.[31] A highly effective chiral Lewis-acid system, derived from nickel perchlorate and a chiral ligand such as (*S*,*S*)-DBFOX-Ph, allowed excellent yields of up to 99% and good to high enantioselectivities of up to 95% ee to be achieved, as shown in Scheme 4.11. The substrate scope demonstrated that phenyl-substituted nitrone derivatives (R^2 = Ph, *p*-BrC_6H_4, or *p*-$MeOC_6H_4$) gave the highest enantioselectivities (89–95% ee). For example, lower enantioselectivities of 71% and 79% ee were respectively obtained for nitrones derived from cinnamaldehyde and furfural. The authors have postulated that the domino reaction could begin with ring opening of the activated cyclopropane to a dipolar species which was then trapped by the nitrone, affording the final product through cyclisation.

In another context, Murakami *et al.* have demonstrated that highly reactive azanickelacycles could be generated from 2*H*-1,2,3,4-benzothiatriazine 1,1-dioxides through extrusion of N_2.[32] These azanickelacycles further incorporated a variety of allenes through a regio- and enantioselective manner, providing a new synthetic route to chiral biologically interesting substituted 3,4-dihydro-2*H*-1,2-benzothiazine 1,1-dioxides **13**. As shown in Scheme 4.12, this domino process was promoted by a combination of 10 mol% of (*R*)-QUINAP with 10 mol% of $Ni(cod)_2$ by heating in 1,4-dioxane or 1 : 1 mixture of THF/MeCN as solvent. The reaction of cyclohexylpropa-1,2-diene with various allenes was performed in 1,4-dioxane at 100 °C, while those of other monosubstituted allenes were carried out at a lower temperature of 60 °C in 1 : 1 mixture of THF/MeCN as solvent. (*R*)-QUINAP was selected as the most efficient ligand among a range of C_2-symmetric bidentate biphosphine ligands, such as (*S*)-BINAP, (*S*,*S*′,*R*,*R*′)-TANGPHOS, and (*R*,*R*)-Me-DUPHOS, as well as unsymmetrical bidentate P,N-type ligands, such as (*R*)-(*S*)-ppfa, and (*R*,*R*)-i-Pr-FOXAP. In the case of reaction of cyclohexylpropa-1,2-diene, primary and secondary alkyl groups on the nitrogen atom of the 2*H*-1,2,3,4-benzothiatriazine 1,1-dioxides were suitable, giving

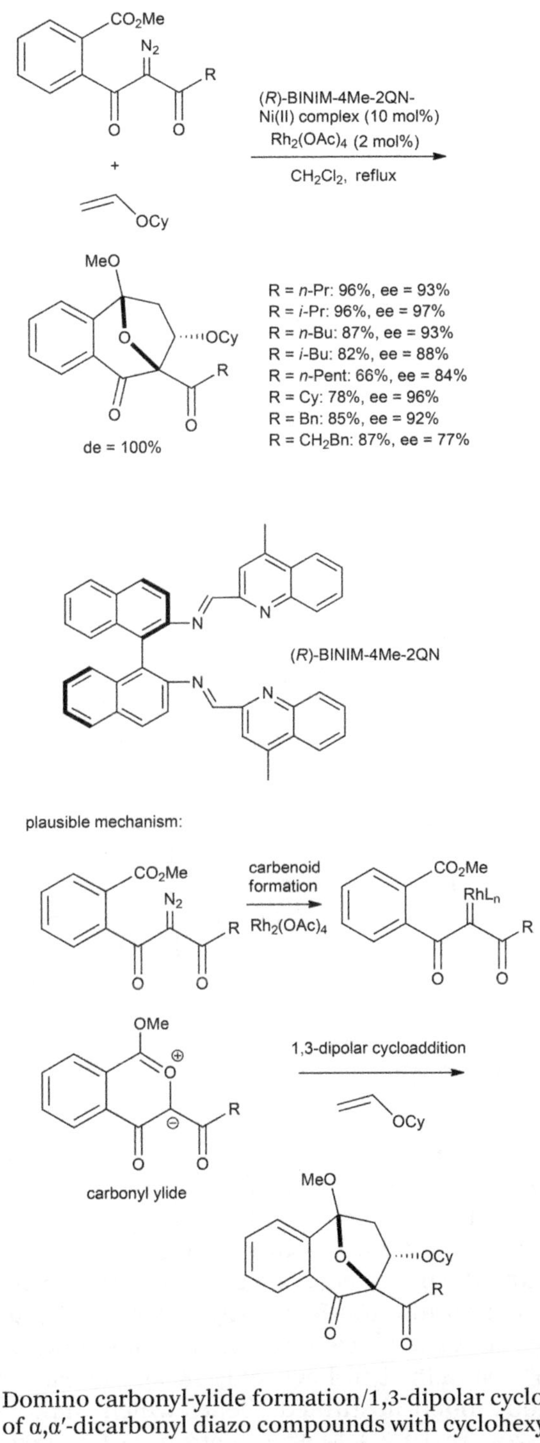

Scheme 4.10 Domino carbonyl-ylide formation/1,3-dipolar cycloaddition reaction of α,α′-dicarbonyl diazo compounds with cyclohexyl vinyl ether.

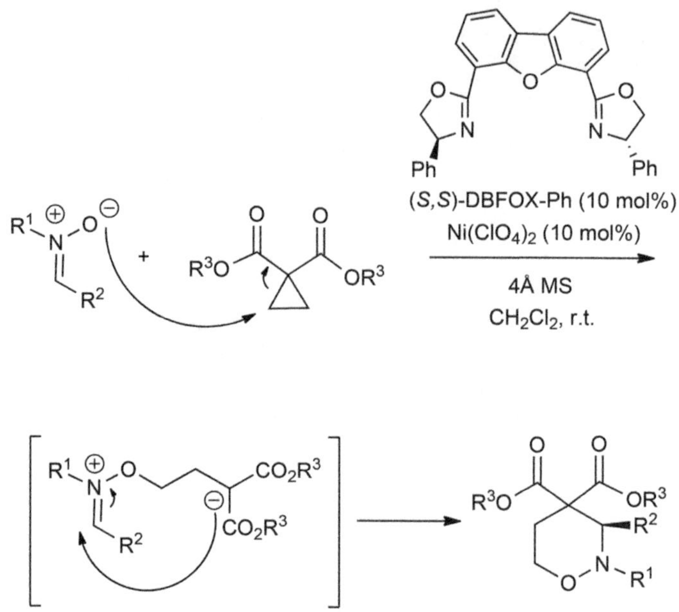

$R^1 = R^3 = Me$, $R^2 = Ph$: 96%; ee = 89%
$R^1 = Me$, $R^2 = Ph$, $R^3 = Et$: 99%; ee = 92%
$R^1 = R^2 = Ph$, $R^3 = Me$: 97%; ee = 91%
$R^1 = R^2 = Ph$, $R^3 = Et$: 99%; ee = 94%
$R^1 = R^2 = Ph$, $R^3 = t\text{-Bu}$: 39%; ee = 95%
$R^1 = Me$, $R^2 = Ph$, $R^3 = Et$: 99%; ee = 92%
$R^1 = Me$, $R^2 = p\text{-}BrC_6H_4$, $R^3 = Et$: 99%; ee = 95%
$R^1 = R^2 = Ph$, $R^3 = Et$: 99%; ee = 94%
$R^1 = Bn$, $R^2 = Ph$, $R^3 = Et$: 99%; ee = 93%
$R^1 = Bn$, $R^2 = p\text{-}MeOC_6H_4$, $R^3 = Et$: 99%; ee = 90%
$R^1 = Me$, $R^2 = \text{cinnamyl}$, $R^3 = Et$: 99%; ee = 71%
$R^1 = Me$, $R^2 = \text{2-furyl}$, $R^3 = Et$: 95%; ee = 79%

Scheme 4.11 Domino ring opening/cyclisation reaction of nitrones with activated cyclopropanes.

enantioselectivities of 88–97% ee. It should be noted that a small amount (2–9%) of regioisomers **14** was observed in all cases of substrates. On the other hand, the *t*-butyl-substituted substrate favoured the formation of undesired side-product **14** [$R^1 = t\text{-Bu}$, $R^2 = Cy$: 67% (**13**/**14** = 13 : 87)], related to steric repulsion around the bulky *tert*-butyl group which changed the preferred site of allylic amidation to the primary allylic carbon. A *p*-tolyl-substituted substrate was also converted into the corresponding product in good enantioselectivity (86% ee), albeit in low yield (28%). The substrate scope was extended to various monosubstituted allenes which smoothly afforded, by reaction with 2*H*-2-methyl-1,2,3,4-benzothiatriazine 1,1-dioxide at 60 °C, the

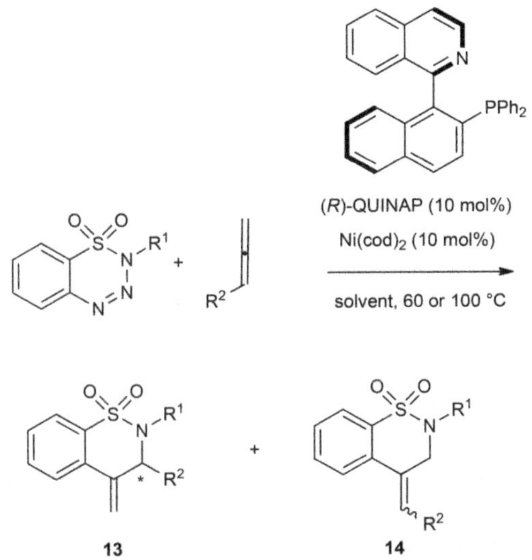

in 1,4-dioxane at 100 °C:
R^1 = Et, R^2 = Cy: 84% (**13**/**14** = 92:8), ee = 97%
R^1 = Bn, R^2 = Cy: 74% (**13**/**14** = 94:6), ee = 97%
R^1 = PMB, R^2 = Cy: 69% (**13**/**14** = 98:2), ee = 91%
R^1 = *i*-Pr, R^2 = Cy: 77% (**13**/**14** = 91:9), ee = 88%
R^1 = *p*-Tol, R^2 = Cy: 28% (**13**/**14** = 88:12), ee = 86%
R^1 = *t*-Bu, R^2 = Cy: 67% (**13**/**14** = 13:87)
in THF/MeCN (1:1) at 60 °C:
R^1 = Me, R^2 = *n*-Hex: 87% (**13**/**14** = 96:4), ee = 85%
R^1 = Me, R^2 = CH_2Cy: 92% (**13**/**14** = 98:2), ee = 81%
R^1 = Me, R^2 = *n*-Pent: 97% (**13**/**14** = 97:3), ee = 85%
R^1 = Me, R^2 = *t*-Bu: 92% (**13**/**14** = 87:13), ee = 84%
R^1 = Me, R^2 = Ph: 99% (**13**/**14** = 86:14), ee = 85%
R^1 = Me, R^2 = $(CH_2)_2OTBS$: 98% (**13**/**14** = 91:9), ee = 72%
R^1 = Me, R^2 = $(CH_2)_2OBn$: 91% (**13**/**14** = 93:7), ee = 73%
R^1 = Me, R^2 = $(CH_2)_2N(Phth)$: 95% (**13**/**14** = 93:7), ee = 76%

proposed mechanism:

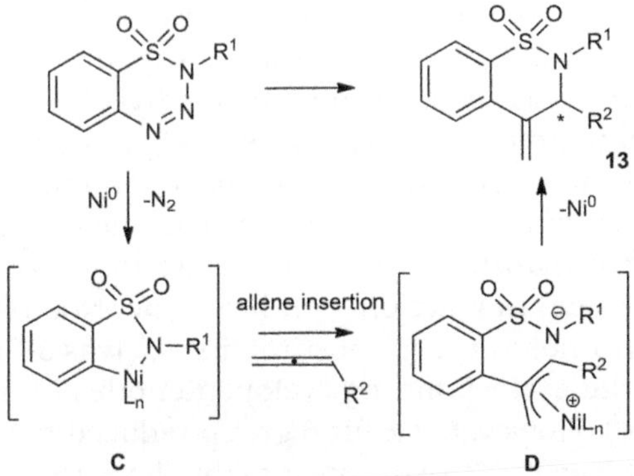

Scheme 4.12 Domino denitrogenative annulation of 2*H*-1,2,3,4-benzothiatriazine 1,1-dioxides with allenes by using (*R*)-QUINAP as ligand.

corresponding substituted 3,4-dihydro-2*H*-1,2-benzothiazine 1,1-dioxides **13** in high yields (87–99%). Enantioselectivities of 81–85% ee were observed with simple allenes that contained a primary, secondary, tertiary, or phenyl substituent. Functional groups, such as siloxy, benzyloxy, or *N*-phthalimidoyl groups, on the alkyl chains of the allenes were tolerated, although providing lower enantioselectivities of 72–76% ee. To explain these results, the authors have proposed the mechanism depicted in Scheme 4.12, beginning with an oxidative addition of the N=N bond to nickel(0), followed by extrusion of N_2 to give five-membered ring azanickelacycle **C**, and then insertion of an allene to form π-allylnickel intermediate **D**, which finally undergoes allylic amidation at the more-substituted carbon to release the product and the chiral nickel catalyst.

These reactions were also investigated by the same authors using (*S*,*S*)-i-Pr-FOXAP as the chiral ligand.[33] Therefore, nona-1,2-diene (R^2 = *n*-Hex) reacted with various *N*-aryl-substituted 2*H*-1,2,3,4-benzothiatriazine 1,1-dioxides under catalysis with a combination of 20 mol% of (*S*,*S*)-i-Pr-FOXAP and 10 mol% of $Ni(cod)_2$ to give the corresponding chiral substituted 3,4-dihydro-2*H*-1,2-benzothiazine 1,1-dioxides **13** in both high yields (81–99%) and with enantioselectivities of 90–97% ee. The regioselectivity of the reactions was good since only 1–7% of side-products **14** were produced in all cases of substrates studied, except with cyclohexylallene which gave a **13**/**14** product ratio of 73 : 27. Moreover, the process was found to tolerate the presence of a variety of functional groups on the 2*H*-1,2,3,4-benzothiatriazine 1,1-dioxide moiety as well as the allene moiety, as shown in Scheme 4.13, leading to the corresponding products in comparable enantioselectivities (91–97% ee) and high yields (91–99%).

In 2014, Cramer *et al.* reported another type of enantioselective nickel-catalysed annulation reaction occurring between α,β-unsaturated aromatic esters and alkynes, which afforded chiral cyclopentenones, constituting versatile structural motifs in many natural products and bioactive compounds as well as key synthetic intermediates.[34] This formal [3 + 2] cycloaddition was induced by a combination of $Ni(cod)_2$ and chiral bulky C_1-symmetric N-heterocyclic carbene ligand **15** in the presence of superstoichiometric amounts of BEt_3 and *tert*-butyl alcohol in cyclopentyl methyl ether (CPME) as solvent. A wide range of cinnamic esters were tolerated, as shown in Scheme 4.14. The aromatic portion accommodated the most common electron-donating and electron-withdrawing groups. Irrespective of the position of the substituents (*ortho*, *meta*, or *para*), the yields and enantioselectivities were consistently high (81–97% ee) in reaction with hex-3-yne. Notably, substrates with condensed arenes or heterocyclic substituents, such as 3-furyl and 3-thienyl, reliably provided the corresponding cyclopentenones in high enantioselectivities (85–90% ee). However, a 2-furyl group reduced both yield and enantioselectivity (48% yield, 55% ee), presumably because of chelation of the nickel centre with the oxygen atom of the furan. Alkyl-substituted acrylates also underwent the annulation reaction, but with lower enantioselectivities (59–77% ee). In addition to hex-3-yne, a variety of other symmetrical

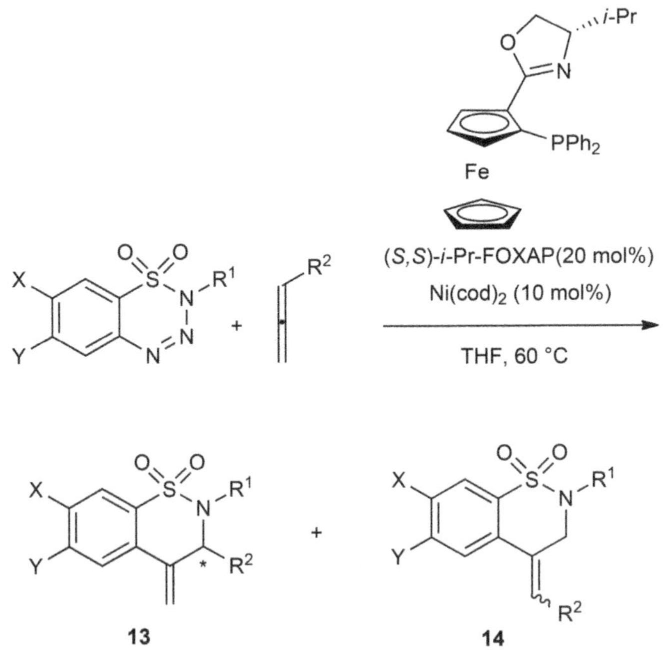

R^1 = *p*-Tol, X = Y = H, R^2 = *n*-Hex: 96% (**13**/**14** = 98:2), ee = 90%
R^1 = Ph, X = Y = H, R^2 = *n*-Hex: 99% (**13**/**14** = 97:3), ee = 91%
R^1 = *p*-$MeOC_6H_4$, X = Y = H, R^2 = *n*-Hex: 99% (**13**/**14** = 96:4), ee = 92%
R^1 = *p*-ClC_6H_4, X = Y = H, R^2 = *n*-Hex: 94% (**13**/**14** = 95:5), ee = 93%
R^1 = *o*-$MeOC_6H_4$, X = Y = H, R^2 = *n*-Hex: 98% (**13**/**14** = 98:2), ee = 91%
R^1 = Ph, X = Y = OMe, R^2 = *n*-Hex: 99% (**13**/**14** = 94:6), ee = 92%
R^1 = Ph, X = H, Y = CO_2Me, R^2 = *n*-Hex: 95% (**13**/**14** = 95:5), ee = 97%
R^1 = *p*-$CF_3C_6H_4$, X = Y = H, R^2 = *n*-Hex: 92% (**13**/**14** = 93:7), ee = 93%
R^1 = $CONPh_2$, X = Y = H, R^2 = *n*-Hex: 81% (**13**/**14** = 99:1), ee = 95%
R^1 = *p*-Tol, X = Y = H, R^2 = $(CH_2)_2OBn$: 98% (**13**/**14** = 94:6), ee = 91%
R^1 = *p*-Tol, X = Y = H, R^2 = $(CH_2)_2OSi$*t*-$BuMe_2$:
92% (**13**/**14** = 95:5), ee = 91%
R^1 = *p*-Tol, X = Y = H, R^2 = $(CH_2)_2OH$: 91% (**13**/**14** = 92:8), ee = 97%
R^1 = *p*-Tol, X = Y = H, R^2 = $(CH_2)_3CN$: 99% (**13**/**14** = 94:6), ee = 93%
R^1 = *p*-Tol, X = Y = H, R^2 = Cy: 76% (**13**/**14** = 73:27), ee = 96%
R^1 = *p*-Tol, X = Y = H, R^2 = $(CH_2)_2N(Phth)$: 99% (**13**/**14** = 96:4), ee = 97%

Scheme 4.13 Domino denitrogenative annulation of 2*H*-1,2,3,4-benzothiatriazine 1,1-dioxides with allenes by using (*R*,*R*)-i-Pr-FOXAP as ligand.

dialkylalkynes provided enantioselectivities of up to 96% ee, with enantioselectivities of 90–92% ee achieved for those bearing functional groups. On the other hand, it was shown that diarylalkynes did not provide the desired cyclopentenones. Moreover, the authors also evaluated nonsymmetrical alkynes, which led to the corresponding products in high enantioselectivities of 95–97% ee combined with high yields (84–94%), except in the case of an

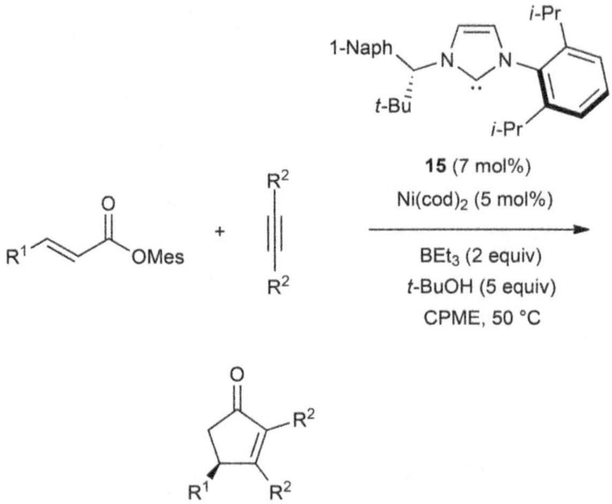

R^1 = *p*-Tol, R^2 = Et: 89%, ee = 93%
R^1 = *m*-Tol, R^2 = Et: 90%, ee = 94%
R^1 = *p*-FC_6H_4, R^2 = Et: 67%, ee = 93%
R^1 = *p*-$CF_3C_6H_4$, R^2 = Et: 63%, ee = 81%
R^1 = *p*-$MeOC_6H_4$, R^2 = Et: 75%, ee = 95%
R^1 = *m*-$MeOC_6H_4$, R^2 = Et: 90%, ee = 90%
R^1 = *o*-$MeOC_6H_4$, R^2 = Et: 88%, ee = 97%
R^1 = 3,4-(OCH_2O)C_6H_3, R^2 = Et: 83%, ee = 96%
R^1 = 1-Naph, R^2 = Et: 83%, ee = 95%
R^1 = 2-Naph, R^2 = Et: 92%, ee = 90%
R^1 = ferrocenyl, R^2 = Et: 34%, ee = 90%
R^1 = 2-furyl, R^2 = Et: 48%, ee = 55%
R^1 = 3-furyl, R^2 = Et: 61%, ee = 85%
R^1 = 3-thienyl, R^2 = Et: 80%, ee = 90%
R^1 = *n*-Pr, R^2 = Et: 77%, ee = 59%
R^1 = *i*-Pr, R^2 = Et: 28%, ee = 77%
R^1 = Ph, R^2 = *n*-Pr: 83%, ee = 96%
R^1 = Ph, R^2 = *n*-Bu: 84%, ee = 96%
R^1 = Ph, R^2 = $(CH_2)_3OMe$: 88%, ee = 92%
R^1 = Ph, R^2 = $(CH_2)_3OBn$: 88%, ee = 90%
R^1 = Ph, R^2 = $(CH_2)_3N(Phth)$: 75%, ee = 90%

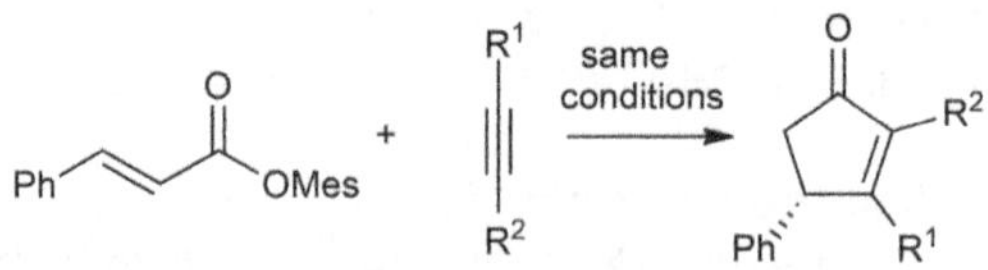

R^1 = *n*-Pr, R^2 = Me: 94%, rs = 10:1, ee = 96%
R^1 = $(CH_2)_2OBn$, R^2 = Me: 84%, rs = 1.5:1, ee = 96%
R^1 = $(CH_2)_2OTIPS$, R^2 = Me: 94%, rs = 10:1, ee = 95%
R^1 = $(CH_2)_2OTIPS$, R^2 = Et: 84%, rs = 1.5:1, ee = 96%
R^1 = Ph, R^2 = Me: 48%, rs = 3:1, ee = 97%

Scheme 4.14 Formal [3 + 2] cycloadditions of α,β-unsaturated esters with alkynes.

alkyl(aryl)alkyne which turned out to be less reactive (48% yield with R^1 = Ph, R^2 = Me), as shown in the second equation of Scheme 4.14.

To explain the preceding results, the authors proposed the mechanism depicted in Scheme 4.15. First, both the enoate and alkyne substrates coordinate to the Ni(0) catalyst, which bears the chiral N-heterocyclic carbene ligand. Next, the enantioselectivity-determining step of the process is the oxidative cyclisation, giving metallocyclic intermediate **E**. Then, cyclisation and subsequent β-alkoxide elimination releases the enone. Transmetallation with BEt_3, followed by β-hydride elimination, gives a nickel hydride. Reductive elimination then closes the catalytic cycle.

Earlier, Kurahashi and Matsubara described regio- and enantioselective nickel-catalysed decarbonylative formal cycloadditions of phthalic anhydrides with allenes to give in a single step the corresponding chiral δ-lactones.[35] The process represented an unprecedented insertion reaction of a carbon–carbon double bond into a carbon–oxygen bond. It was performed in the presence of a chiral nickel catalyst generated *in situ* from $Ni(cod)_2$ and (*S*,*S*)-i-Pr-FOXAP in pyridine at reflux. The chiral δ-lactones were achieved in both moderate to good yields (64–73%) and with enantioselectivities of 59–81% ee, as shown in Scheme 4.16. Better yields (73–90%) and enantioselectivities

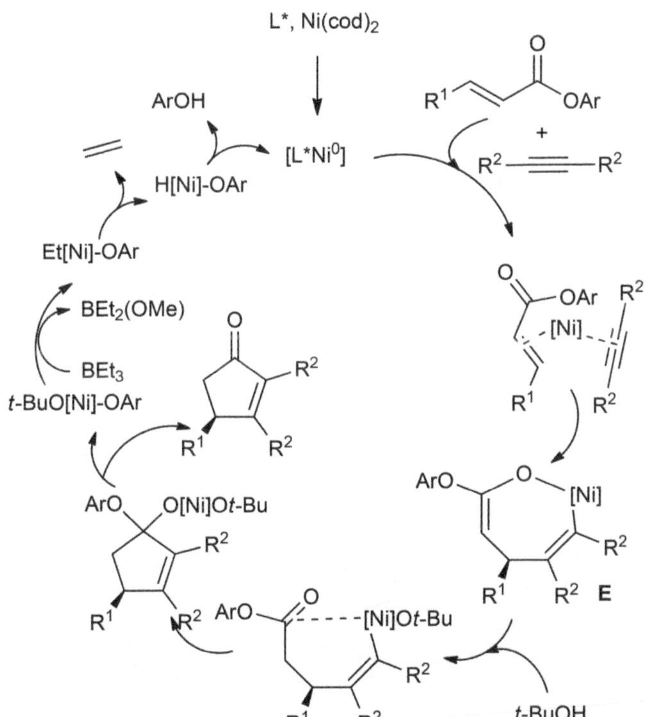

Scheme 4.15 Proposed mechanism for formal [3 + 2] cycloaddition of α,β-unsaturated esters with alkynes.

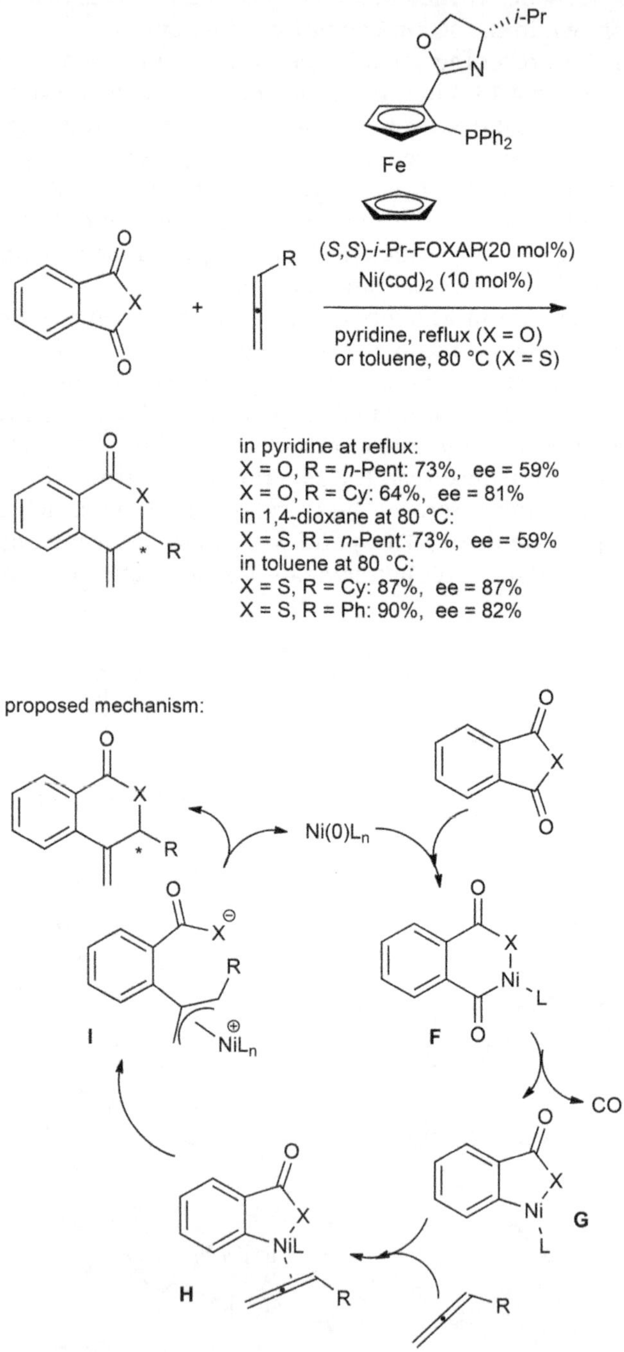

Scheme 4.16 Domino decarbonylative cycloaddition reaction of (thio)phthalic anhydrides with allenes.

(59–87% ee) were achieved in the reactions of thiophthalic anhydrides with allenes, which were performed in toluene or 1,4-dioxane as solvent. A plausible reaction pathway to account for the formation of these products is proposed in Scheme 4.16. The catalytic cycle of the reaction could consist of an oxidative addition of a CO–X bond (X = O or S) to a Ni(0) complex to give intermediate **F**. Subsequent decarbonylation to provide **G** and coordination of the allene takes place to give nickel(II) intermediate **H**. The allene could then insert into the C–Ni bond to give the more stable acyclic π-allylnickel intermediate **I**. Nucleophilic addition of a heteroatom onto the π-allylnickel at the more substituted carbon takes place to afford the final cycloadduct and regenerate the starting Ni(0) complex.

In 2013, Jiang *et al.* reported the synthesis of the novel chiral mixed metal–organic framework CMOF **16** from new enantiopure dicarboxyl-functionalised Ni(salen) metalloligand **17** and $CdCl_2$ (Scheme 4.17).[36] This novel catalyst was characterised by X-ray diffraction, showing that each tetranuclear cadmium cluster was linked by eight Ni–ligand groups, and each Ni–ligand was linked by two tetranuclear cadmium clusters to generate a 3D framework. This novel catalyst, in which the Ni(salen) units function as Lewis acid sites, was applied to promote the asymmetric synthesis of chiral propylene carbonate through formal cycloaddition of CO_2 with racemic propylene oxide, performed in the presence of NBu_4Br as cocatalyst. As shown in Scheme 4.17, the product was obtained in low yield (28%) and with moderate enantioselectivity of 52% ee by using a remarkably low catalyst loading of 2×10^{-4} mol%. In addition to its considerable activity, this catalyst was shown to be recyclable and reusable, retaining its framework after being used three times. The authors have proposed a plausible Lewis acidic activation mechanism for the domino reaction, which is depicted in Scheme 4.17. The coordinatively unsaturated Ni^{2+} in chiral channels of catalyst CMOF **16** could selectively complex one enantiomer of the racemic propylene oxide. Then, Br^- generated from NBu_4Br attacks the less-substituted carbon of the coordinated propylene oxide regioselectively, leading to its enantioselective ring-opening; the intermediate then reacts with CO_2 adsorbed into the channels of the catalyst, and further forms optically active propylene carbonate through intramolecular cyclic elimination.

In order to expand the scope of cross-coupling reactions of alkyl electrophiles, Cong and Fu recently developed enantioselective nickel-catalysed couplings of arylboron reagents bearing a pendant alkene with unactivated alkyl bromides, providing the corresponding 2,3-dihydrobenzofurans through a domino cyclisation/cross-coupling reaction.[37] When the domino process was promoted by a chiral catalyst generated *in situ* from $NiBr_2$·glyme and chiral diamine ligand **18** in the presence of superstoichiometric amounts of KO*t*-Bu and i-BuOH, a range of chiral 2,3-dihydrobenzofurans was achieved in moderate to good yields (45–82%) and with good to high enantioselectivities of up to 97% ee (Scheme 4.18). Among various alkyl electrophiles used, those including a silane, an acetal, and an imide function were compatible with the reaction conditions. Moreover, the method was not limited to unhindered primary alkyl bromides, since β-branched primary and secondary bromides

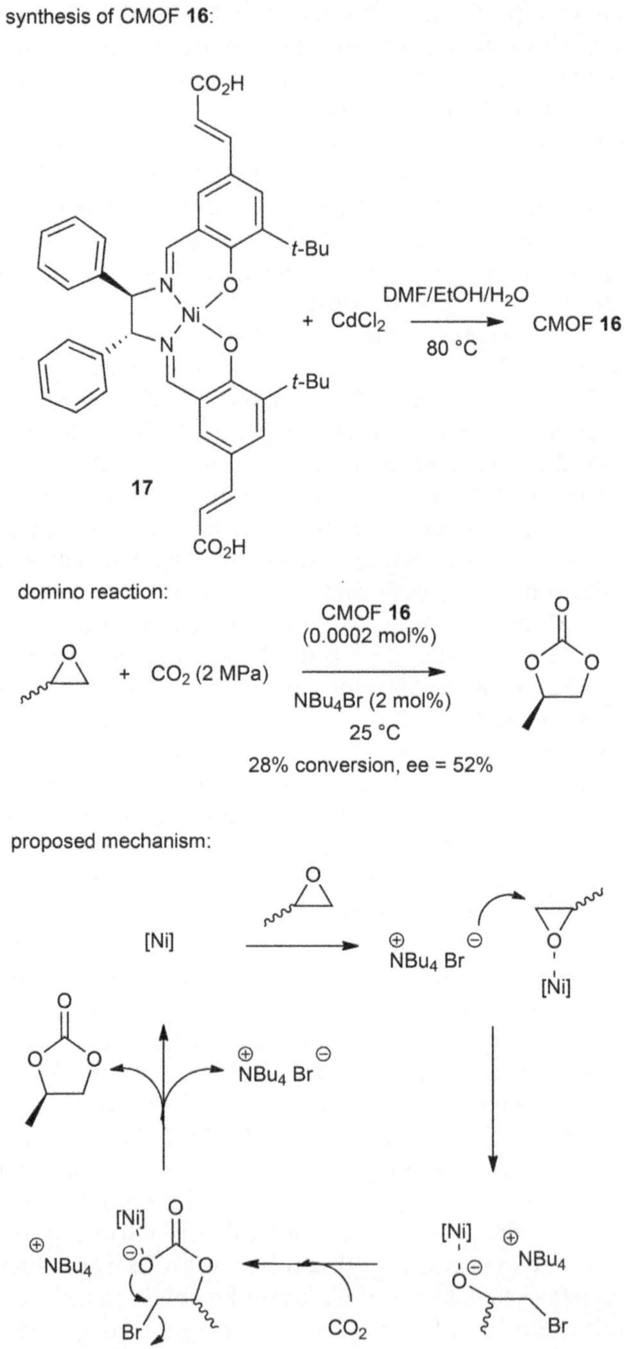

Scheme 4.17 Domino formal cycloaddition of CO_2 with propylene oxide.

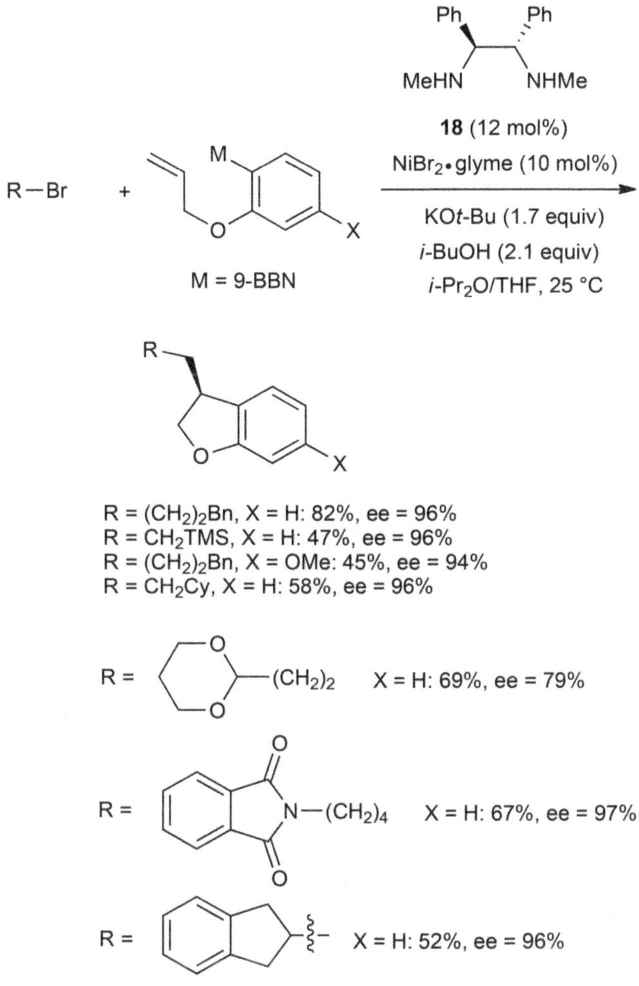

Scheme 4.18 Domino cyclisation/cross-coupling reaction.

also underwent the process, with enantioselectivities of 96–97% ee. The utility of this novel methodology, allowing the formation of two carbon–carbon bonds and a stereogenic centre in one step, has been demonstrated by its application to the synthesis of the dihydrobenzofuran core of the pharmaceutical fasiglifam.

4.3 Multicomponent Reactions

Even though the history of multicomponent reactions dates back to the second half of the 19th century with the reactions of Strecker, Hantzsch, and Biginelli, it was only in recent decades with the work of Ugi that the concept of the multicomponent reaction has emerged as a powerful tool in synthetic chemistry, allowing researchers to easily reach high molecular complexity in an economically favourable way with advantages of savings in solvent, time, energy, and costs by avoiding costly protecting groups and time-consuming purification procedures after each step. In the last decade, a variety of highly efficient enantioselective multicomponent reactions have been catalysed by chiral nickel catalysts.

4.3.1 Three-Component Couplings of Unsaturated Hydrocarbons, Carbonyl Compounds and Derivatives, and Reducing Agents

4.3.1.1 Reactions of 1,3-Dienes, Carbonyl Compounds, and Reducing Agents

Carbon dioxide is regarded as an important source of C_1 due to its abundant reserve and low toxicity. Hence, the development of transition metal-catalysed reactions for CO_2 incorporation into organic molecules is of great importance. It is especially quite challenging to develop efficient catalytic protocols that enable carbon–carbon bond formation between CO_2 and substrates in an enantioselective manner. However, such catalytic asymmetric CO_2 incorporation processes have rarely been explored due to the limited number of available catalytic carbon–carbon bond-forming CO_2 incorporation reactions that can be efficiently carried out under mild conditions. In this context, Mori *et al.* have reported enantioselective nickel-catalysed carboxylative cyclisation reactions of bis-1,3-dienes performed under mild conditions.[38] These multicomponent reactions occur between a diene, CO_2 as the carbonyl component, and a dialkyl- or diarylzinc reagent in the presence of a chiral nickel catalyst generated *in situ* from $Ni(acac)_2$ and the chiral phosphine ligand (*S*)-MeO-MOP employed at 10 and 20 mol% catalyst loadings, respectively. Both excellent yields and enantioselectivities of up to >99% and 96% ee, respectively, were obtained for a range of chiral products having a carbocyclic five-membered ring skeleton arising from the reactions of dimethyl- and diphenylzincs with various functionalised bis-1,3-dienes, as shown in Scheme 4.19. It was notable that an unsymmetrical diene (Y = Me) could also be used in the process, with high enantioselectivity (up to 96% ee) and with regioselective introduction of CO_2 into the less-substituted 1,3-diene moiety.

Later, the same authors investigated enantioselective nickel-catalysed three-component reactions of bis-1,3-dienes, aldehydes, and dimethylzinc, evolving through intramolecular cyclodimerisation of the bis-1,3-diene moiety.[39] The domino reaction was promoted at room temperature by a chiral nickel catalyst generated *in situ* from $Ni(acac)_2$ and the chiral phosphine ligand (*S*)-H-MOP, affording the corresponding coupling products **19** as inseparable mixtures of two isomers in moderate to good yields (51–93%; Scheme 4.20). These intermediate alcohol domino products **19** were subsequently converted into the corresponding carbonyl compounds **20** by treatment with PCC. These ketones were obtained as sole products in good enantioselectivities of up to 85% ee.

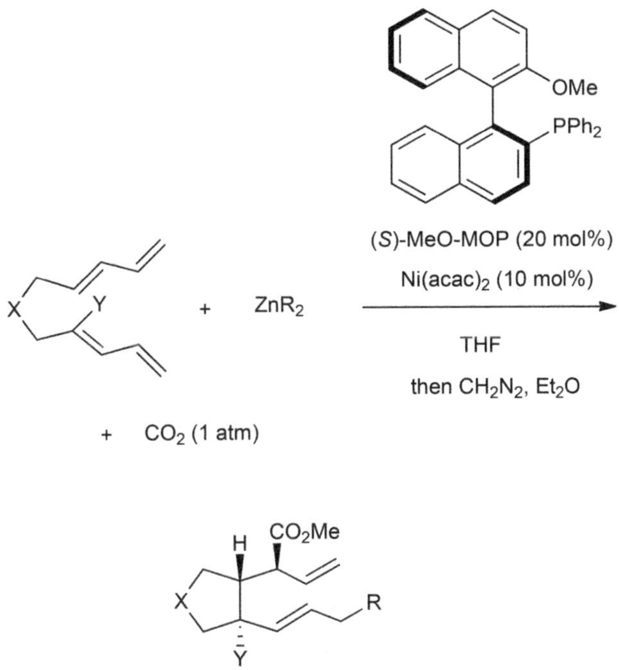

X = NTs, Y = H, R = Ph: 81%, ee = 95%
X = NTs, Y = H, R = Et: 57%, ee = 94%
X = $C(CO_2Me)_2$, Y = H, R = Me: > 99%, ee = 94%
X = $C(CO_2Me)_2$, Y = H, R = Ph: 89%, ee = 92%
X = $C(CH_2OBn)_2$, Y = H, R = Me: 95%, ee = 95%
X = $C(CH_2OBn)_2$, Y = H, R = Ph: 80%, ee = 90%
X = $CH_2OCOOCH_2$, Y = H, R = Me: 90%, ee = 94%
X = $CH_2OCOOCH_2$, Y = H, R = Ph: 83%, ee = 95%
X = NTs, Y = R = Me: 88%, ee = 96%
X = NTs, Y = Me, R = Ph: 87%, ee = 91%

Scheme 4.19 Three-component reaction of bis-1,3-dienes, CO_2, and organozinc reagents.

The authors proposed a possible mechanism (shown in Scheme 4.20) which starts with the oxidative cycloaddition of the bis-1,3-diene to the Ni(0) complex to produce bis-allylnickel complex **J**, and subsequent insertion of the aldehyde into the nickel–carbon bond affords oxanickelacycle **K**. Transmetallation of **K** with $ZnMe_2$ provides methylnickel complex **L**, which can easily undergo reductive elimination to afford zinc alkoxide **M**. Hydrolysis of **M** in an aqueous workup procedure provides the final alcohol product **19**.

In addition, Zhou *et al.* have developed diastereo- and enantioselective reductive coupling of 1,3-dienes, such as 1,4-diphenylbuta-1,3-diene, with aromatic and heteroaromatic aldehydes by using $ZnEt_2$ as the reducing agent and nickel complexes of chiral spiro phosphoramidites such as **21** (Scheme 4.21).[40] In this case, the three-component reaction provided the

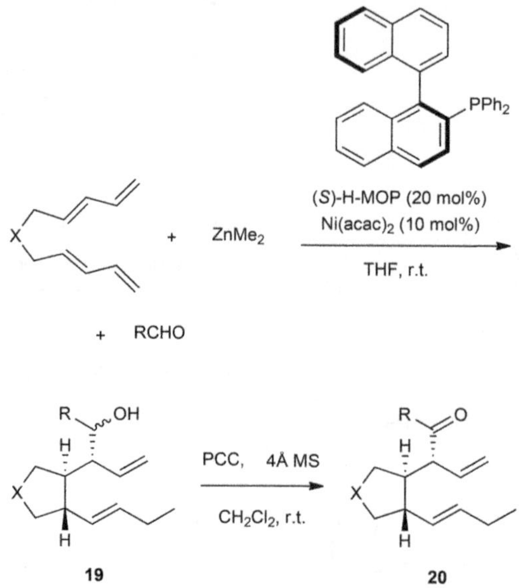

X = NTs, R = Ph: 58% (**19**), ee (**20**) = 81%
X = NTs, R = *n*-Pr: 51% (**19**), ee (**20**) = 78%
X = NTs, R = H: 93% (**19**), ee (**20**) = 85%
X = $C(CO_2Me)_2$, R = Ph: 75% (**19**), ee (**20**) = 79%
X = $C(CO_2Me)_2$, R = H: 86% (**19**), ee (**20**) = 85%

proposed mechanism:

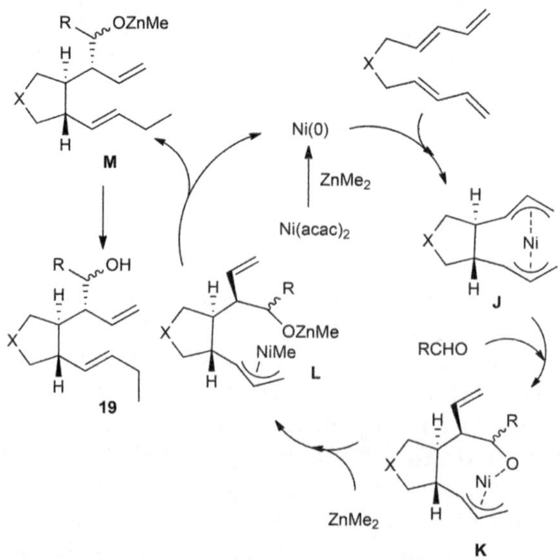

Scheme 4.20 Three-component reaction of bis-1,3-dienes, aldehydes, and dimethylzinc.

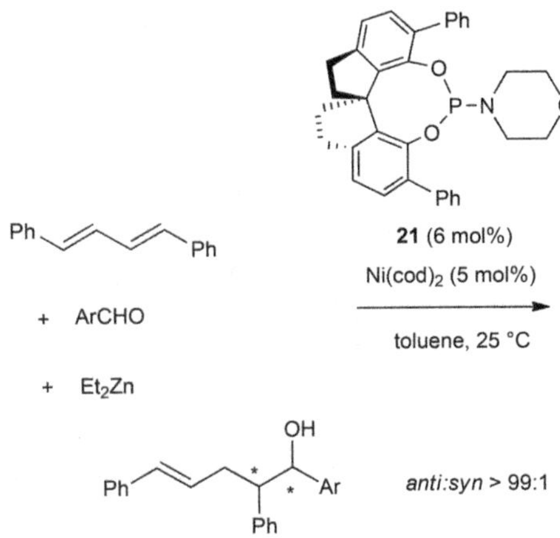

Ar = Ph: 99%, ee = 96%
Ar = *o*-Tol: 95%, ee = 93%
Ar = *o*-$MeOC_6H_4$: 98%, ee = 91%
Ar = *m*-Tol: 94%, ee = 94%
Ar = *m*-$MeOC_6H_4$: 99%, ee = 92%
Ar = *p*-Tol: 95%, ee = 95%
Ar = *p*-$MeOC_6H_4$: 94%, ee = 96%
Ar = *p*-$Me_2NC_6H_4$: 85%, ee = 96%
Ar = *p*-ClC_6H_4: 92%, ee = 90%
Ar = *p*-$CF_3C_6H_4$: 98%, ee = 85%
Ar = 1-Naph: 94%, ee = 93%
Ar = 2-Naph: 96%, ee = 86%
Ar = 2-furyl: 96%, ee = 92%

possible mechanism:

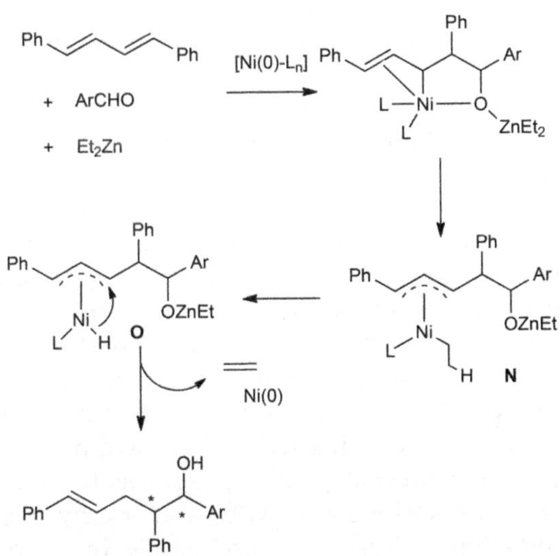

Scheme 4.21 Three-component reaction of 1,4-diphenylbuta-1,3-diene, aldehydes, and diethylzinc.

corresponding chiral bishomoallylic alcohols in excellent yields and diastereoselectivities (*anti*:*syn* up to >99:1) in almost all cases of substrates studied, combined with general high enantioselectivities of up to 96% ee. The substrate scope showed that, generally, an electron-donating substituent such as OMe and NMe_2 at the *para* position slightly enhanced the enantioselectivity (96% ee), while an electron-withdrawing substituent such as Cl and CF_3 at the *para* position diminished the enantioselectivity (85–90% ee). In addition to benzaldehyde derivatives, heteroaromatic aldehydes afforded the corresponding bishomoallylic alcohols in high enantioselectivities of up to 92% ee. However, the authors found that aliphatic aldehydes were much less enantioselective in this process, since butyraldehyde led to the corresponding coupling product in only 72% ee. It is important to note that this nice work represented the first example of intermolecular catalytic asymmetric reductive coupling of 1,3-dienes with carbonyl compounds. A possible mechanism can be proposed (Scheme 4.21), beginning with cooperative oxidative addition of the diene and the aldehyde moieties onto Ni(0), followed by ethyl group transfer from $ZnEt_2$ to Ni(II) to give intermediate **N**, which then undergoes β-H elimination of the Ni–Et bond and reductive elimination of the thus-formed NiH to generate intermediate **O**, Ni(0) complexes, and ethylene.

Later, Gade *et al.* performed the same reaction between 1,4-diphenylbuta-1,3-diene, benzaldehyde, and diethylzinc in the presence of another chiral nickel catalyst generated *in situ* from $NiBr_2$·(DME) and novel chiral bulky cyclophosphazane **22**.[41] Using 3 mol% of this robust and readily available ligand, the three-component reaction afforded the corresponding bishomoallylic alcohol in quantitative yield, with a high *anti* diastereoselectivity (*anti*/*syn* = 92:8) and a good enantioselectivity of 84% ee, as shown in Scheme 4.22. In spite of this moderate enantioselectivity, this work demonstrated that chiral cyclophosphazanes constitute a promising family of chiral ligands.

On the other hand, Sato *et al.* have developed nickel-catalysed three-component coupling of 1,3-dienes, aldehydes, and triethylsilane as reducing agent to give regio-, diastereo-, and enantioselectively the corresponding β-triethylsilyloxy (*Z*)-alkenes exclusively.[42] On the basis of the screening of various chiral N-heterocyclic carbene precursors, chiral imidazolium salt **23**, having 1-(mesitylphenyl)propyl groups on the nitrogen, was selected as the most efficient ligand for nickel to induce chirality in the process, allowing various coupling products to be synthesised in good to quantitative yields and with good to high enantioselectivities of up to 97% ee (Scheme 4.23). The substrate scope showed that aliphatic as well as aromatic aldehydes were tolerated as substrates in the coupling reaction, providing comparable results. Moreover, the reaction of internal 1,3-dienes, such as 1,4-diphenylbuta-1,3-diene among others, with various aldehydes and triethylsilane provided the corresponding *anti*- and (*Z*)-products exclusively in good to quantitative yields (81% to quantitative) and with high enantioselectivities of up to 97% ee (Scheme 4.23, second equation). Interestingly, using an unsymmetrical internal diene also proceeded in a regio- and diastereoselective manner, but

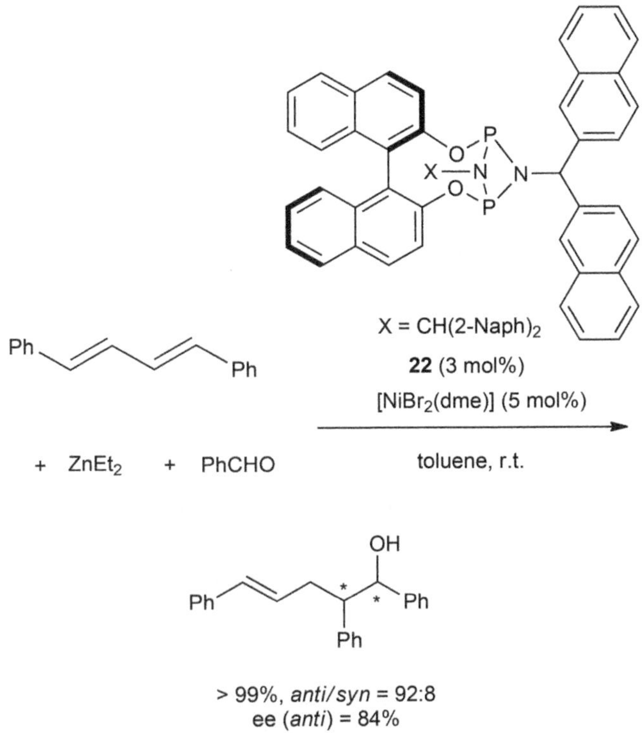

Scheme 4.22 Three-component reaction of 1,4-diphenylbuta-1,3-diene, benzaldehyde, and diethylzinc.

the corresponding product was obtained in lower enantioselectivity (50% ee). The catalyst was generated *in situ* from $Ni(cod)_2$, the chiral imidazolium salt ligand, and Cs_2CO_3 as a base. The mechanism of the process (see Scheme 4.23) proceeded *via* σ-bond metathesis between nickelacycle **P** and triethylsilane, exclusively affording (*Z*)-alkenes as sole products.

More recently, the same authors have investigated a closely related coupling reaction of 1,3-dienes with aldehydes using [dimethyl(phenyl)silyl] pinacolborane as the reductant instead of triethylsilane in the previous work.[43] In this case, chiral phosphoramidite **24** was selected as the most efficient ligand for nickel, providing the corresponding (*E*)-silanes as single diastereomers in moderate to high yields and with enantioselectivities of up to 97% ee (Scheme 4.24). It must be noted that the coupling of internal 1,3-dienes gave generally lower yields (22–51%). The substrate scope showed that aromatic as well as aliphatic aldehydes were tolerated, providing comparable results. A low enantioselectivity (20% ee) and moderate yield (51%) were obtained in the reaction of 1,4-diphenylbuta-1,3-diene with benzaldehyde and [dimethyl(phenyl)silyl]pinacolborane. Remarkably, in each case of substrate studied, a single diastereomer was isolated. It is important to highlight that this novel three-component reaction represented the first

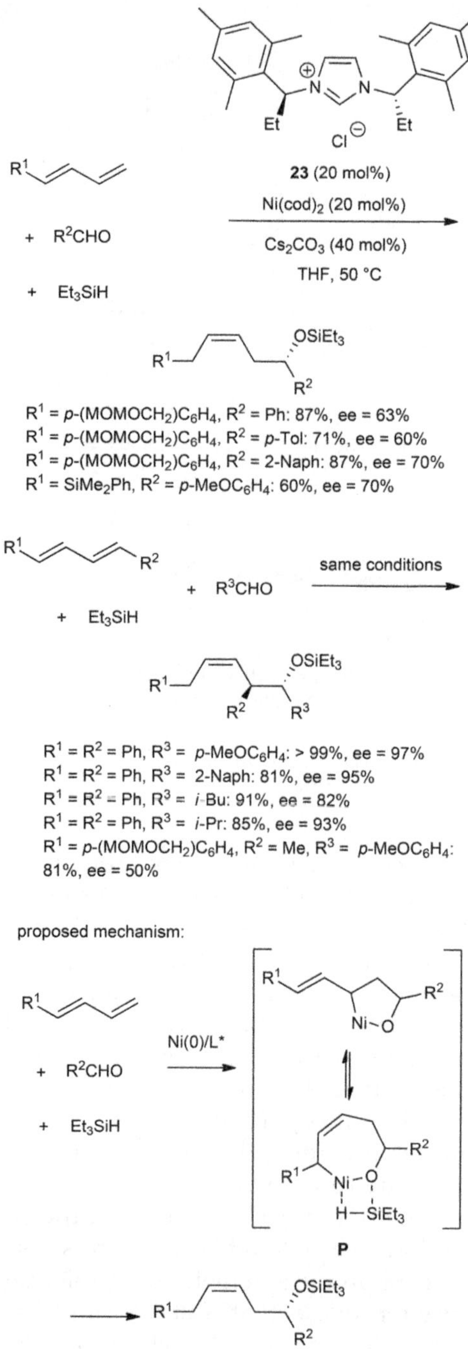

Scheme 4.23 Three-component reaction of 1,3-dienes, aldehydes, and triethylsilane.

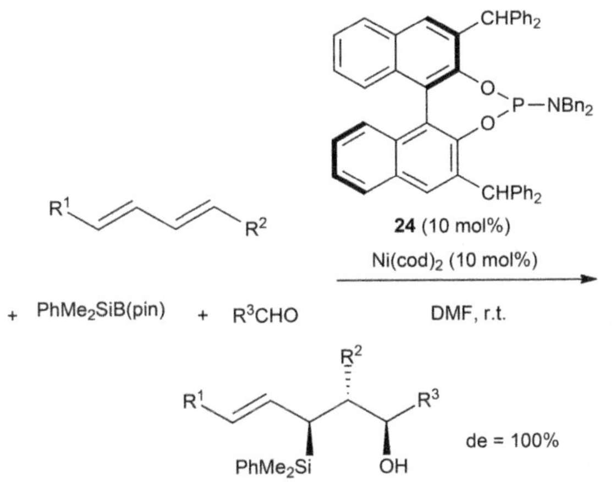

R^1 = *p*-(MOMCH$_2$)-C$_6$H$_4$, R^2 = H, R^3 = *p*-Tol: 89%, ee = 96%
R^1 = *p*-(MOMCH$_2$)-C$_6$H$_4$, R^2 = H, R^3 = *p*-MeO-C$_6$H$_4$: 92%, ee = 92%
R^1 = *p*-(MOMCH$_2$)-C$_6$H$_4$, R^2 = H, R^3 = *p*-CF$_3$-C$_6$H$_4$: 29%, ee = 85%
R^1 = *p*-(MOMCH$_2$)-C$_6$H$_4$, R^2 = H, R^3 = 2-Naph: 72%, ee = 92%
R^1 = *p*-(MOMCH$_2$)-C$_6$H$_4$, R^2 = H, R^3 = *i*-Pr: 68%, ee = 96%
R^1 = *p*-(MOMCH$_2$)-C$_6$H$_4$, R^2 = H, R^3 = *i*-Bu: 56%, ee = 97%
R^1 = *p*-(MOMCH$_2$)-C$_6$H$_4$, R^2 = H, R^3 = Cy: 74%, ee = 94%
R^1 = PhSiMe$_2$, R^2 = H, R^3 = Ph: 80%, ee = 82%
R^1 = *p*-(MOMCH$_2$)-C$_6$H$_4$, R^2 = Me, R^3 = Ph: 22%, ee = 92%
R^1 = R^2 = R^3 = Ph: 51%, ee = 20%

proposed mechanism:

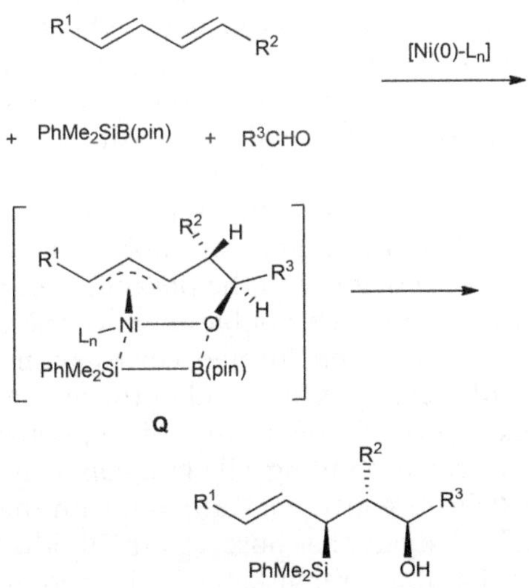

Scheme 4.24 Three-component reaction of 1,3-dienes, aldehydes, and a silylborane.

example of an asymmetric coupling of two different types of unsaturated compounds with a bimetallic reagent, and constituted a new synthetic approach to α-chiral allylsilane derivatives. The authors have proposed the mechanism depicted in Scheme 4.24 in which the key oxanickelacycle **Q** intermediate is generated by oxidative cycloaddition of the diene and the aldehyde with a nickel(0) complex. In the presence of a heterobimetallic compound, such as a silylborane, the reaction of **Q** and the silylborane proceeds by a highly oxophilic boron atom interacting with an oxygen atom, giving the final coupling product through different bond-forming reactions (C–C, C–Si, and O–B) in one pot.

4.3.1.2 Reactions of Allenes, Aldehydes, and Reducing Agents

In 2005, Jamison *et al.* reported the first highly enantioselective coupling reaction of chiral allenes with aldehydes and silanes.[44] This novel three-component reaction was promoted by a achiral nickel catalyst generated *in situ* from $Ni(cod)_2$ and achiral imidazolinyl carbene ligand **25** in THF at room temperature, providing the corresponding chiral protected allylic alcohols in moderate to good yields and with generally remarkable enantioselectivities of up to 98% ee, as shown in Scheme 4.25. The axial chirality of the allene was transferred completely to the product, providing a trisubstituted (*Z*)-allylic alcohol protected as a silyl ether. Indeed, in all cases examined the degree of chirality transfer was 100%, and the *Z*/*E* selectivity was uniformly >95 : 5. A definite preference for sp rather than sp^2 coupling was observed, and differentially substituted allenes underwent highly selective coupling. Indeed, the ratio of allylic and homoallylic products (*i.e.*, coupling at the sp *vs.* sp^2 carbon) was >95 : 5. The substrate scope showed that the process was compatible with aldehydes bearing Lewis-base ethers, esters, and aryl chloride functions. An electron-donating MeO group in the *para* position of benzaldehyde had little effect on the transformation, while an electron-withdrawing CO_2Me substituent on the benzaldehyde reduced the yield to 56% and the allylic : homoallylic selectivity to 90 : 10. Moreover, the authors have extended the scope of the reaction to chiral 1,3-allenes in which the two allene substituents were different, adding yet another selectivity variable in the process, namely site selectivity, with the possibility of producing two allylic alcohols depending upon which double bond of the allene reacted. Remarkably, single allylic products were formed, corresponding to the exclusive reaction of the more hindered double bond of the allene. Interestingly, the multicomponent reaction of *t*-butylbuta-1,2-diene, benzaldehyde, and triethylsilane afforded the corresponding allylic alcohol in reduced yield (40%) and allylic : homoallylic selectivity (85 : 15), but with the same level of *Z*/*E*, enantio-, and site selectivity as that observed in all other cases of substrates. It must be noted that these remarkable regio- and enantioselective transformations constituted the first enantioselective multicomponent coupling processes of allenes. The authors proposed the mechanism depicted in Scheme 4.25 to explain the results. It begins with the formation of nickel

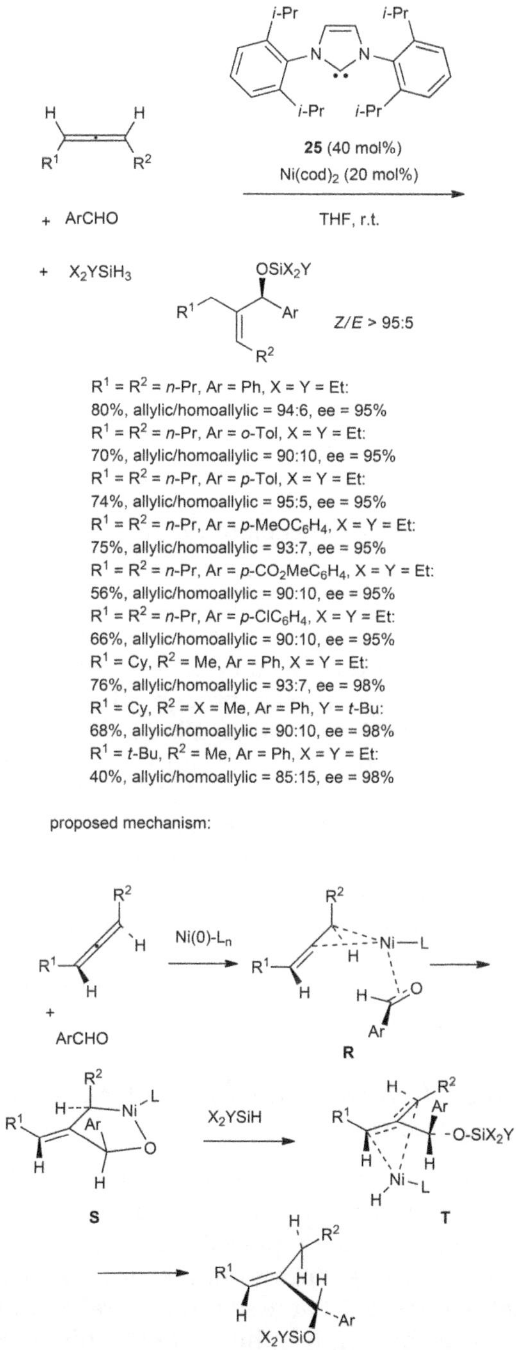

Scheme 4.25 Three-component reaction of chiral allenes, aldehydes, and silanes evolving through chirality transfer.

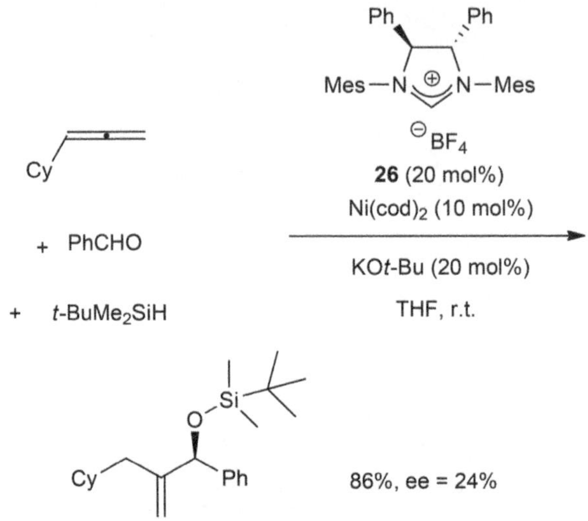

Scheme 4.26 Three-component reaction of cyclohexylallene, benzaldehyde, and *t*-BuMe$_2$SiH.

intermediate **R** with the aldehyde coordinated away from the R^2 group of the allene. Oxidative addition of **R** leads to metallacycle **S** which reacts with the silane to afford η^3-allyl–Ni complex **T** through σ-bond metathesis. Further reductive elimination leads to the final (*Z*)-alkene.

Even more interesting, the same authors later investigated related three-component reactions involving achiral terminal allenes to afford other chiral allylic alcohols by using chiral N-heterocyclic carbene ligands as chirality promotors.[45] Among a range of chiral N-heterocyclic carbene ligands evaluated, ligand **26** was selected as the most efficient in the reaction of cyclohexylallene with benzaldehyde and *t*-BuMe$_2$SiH, selectively providing the corresponding enantio-enriched 1,1-disubstituted allylic alcohol protected as a silyl ether in good yield (86%), albeit with a low enantioselectivity of 24% ee, as shown in Scheme 4.26.

4.3.1.3 *Reactions of Alkynes, Aldehydes or Aldimines, and Reducing Agents*

Chiral allylic amines constitute key synthetic intermediates, auxiliaries, and resolving agents in the synthesis of both natural and nonnatural products. In 2004, Jamison and Patel reported the first highly enantioselective catalytic synthesis of allylic amines from alkynes, imines, and organoboranes such as triethylborane.[46] Catalysed by a chiral complex derived from Ni(cod)$_2$ and chiral ferrocenylphosphine (*R*)-**27**, this novel three-component process provided chiral tetrasubstituted allylic amines in good yields in one-pot (Scheme 4.27). These products were obtained in moderate to very good enantioselectivities of up to 89% ee. Both symmetrical and

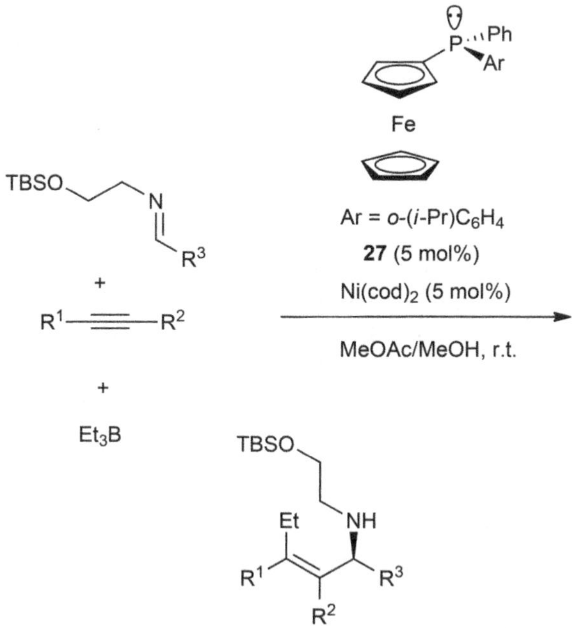

$R^1 = R^2$ = *n*-Pr, R^3 = Ph: 85%, ee = 89%
$R^1 = R^2$ = *n*-Bu, R^3 = Ph: 83%, ee = 89%
$R^1 = R^2$ = Et, R^3 = Ph: 89%, ee = 83%
$R^1 = R^2$ = *n*-Pr, R^3 = *o*-Tol: 74%, ee = 85%
$R^1 = R^2$ = *n*-Pr, R^3 = *p*-anisyl: 75%, ee = 82%
$R^1 = R^2$ = *n*-Pr, R^3 = *p*-$CF_3C_6H_4$: 91%, ee = 85%
$R^1 = R^2$ = *n*-Pr, R^3 = 2-Naph: 90%, ee = 73%
$R^1 = R^2$ = Et, R^3 = Cy: 53%, ee = 51%
$R^1 = R^3$ = Ph, R^2 = Me: 45% (80:20), ee = 84%
$R^1 = R^3$ = Ph, R^2 = Et: 62% (98:2), ee = 71%
R^1 = 2-Naph, R^2 = Me, R^3 = Ph: 42% (85:15), ee = 70%

Scheme 4.27 Three-component reaction of alkynes, imines, and triethylborane.

unsymmetrical alkyl(aryl)alkynes were effective in these reactions, the latter allowing the preparation of chiral tetrasubstituted allylic amines with four different substituents on the alkene with good to complete control of the alkene geometry and regioselectivity. Furthermore, several aromatic imines underwent the process to give the corresponding products in both high enantioselectivity and yield. No significant difference in enantioselectivity was observed with electron-donating or electron-withdrawing substituents on the phenyl group of the imine; however, the imine derived from 2-naphthaldehyde was less selective (70% ee). Notably, even an enolisable aliphatic imine (R^3 = Cy) underwent the reaction in moderate enantioselectivity (51% ee). It was shown that the *tert*-butyldimethylsilyl group on the imine nitrogen not only maximised the reactivity and selectivity in these reactions, but also presented the advantage to be easily removed after the coupling process, thus providing a direct access to versatile primary allylic chiral amines.

The nickel-catalysed three-component reaction of 1,3-enynes, aldehydes, and triethylborane was employed by several groups as a novel strategy for the synthesis of 1,3-dienes.[47] The first asymmetric version of this methodology was developed by Jamison *et al.* in 2005.[48] As shown in Scheme 4.28, it used chiral ferrocenylphosphine ligand (*R*)-**27** in combination with $Ni(cod)_2$ as the catalyst system. Performed in the presence of triethylborane as reducing agent, the reaction of 1,3-enynes with aromatic aldehydes afforded the corresponding chiral 1,3-dienes in moderate yields (47–77%) and moderate enantioselectivities of up to 58% ee. Both electron-donating and electron-withdrawing substituents were tolerated on the phenyl ring of the aldehydes. Better stereoselectivities were reached by these authors by using chiral aldehydes as substrates in addition to the presence of another chiral phosphine ligand.[49] Under these conditions, the double induction allowed a diastereoselectivity of 80% de to be achieved. The same reaction conditions as those used for aldehydes in Scheme 4.28 were applied to the three-component reaction of 1,3-enynes and triethylborane with various ketones by using chiral ferrocenylphosphine (*S*)-**27** as ligand instead of (*R*)-**27**. The process afforded regioselectively (>95 : 5) the corresponding 1,3-dienes bearing a quaternary stereocentre with moderate yields (39–71%) and enantioselectivities of 40–70% ee (Scheme 4.28). Aromatic and heteroaromatic ketones were compatible with the reaction conditions, as well as an α,β-unsaturated ketone, 1-acetylcyclohex-1-ene, which provided the best enantioselectivity (70% ee). It must be noted that this work represented the first catalytic asymmetric reductive coupling of alkynes and ketones, which afforded synthetically useful chiral 1,3-dienes bearing a tertiary alcohol.

In the same area, Montgomery *et al.* have developed asymmetric nickel-catalysed reductive couplings of aldehydes and alkynes, using triethylsilane as the reducing agent.[50] When the process was catalysed by nickel complexes of chiral N-heterocyclic carbene ligands derived from C_2-symmetric diamines, such as **28**, in the presence of KO*t*-Bu as a base, it provided the corresponding chiral silyl ethers in moderate to high yields (47–98%) associated with moderate to good enantioselectivities (65–85% ee), as shown in Scheme 4.29. The authors found that yields and enantioselectivities were relatively uniform across a broad range of substrates. Key functional groups cleanly tolerated in the procedure included aromatics as well as branched and linear aliphatic aldehydes, internal alkynes that either possess or lack an aromatic substituent, terminal alkynes, and unprotected alcohols, wherein the trialkylsilyl group was regioselectively installed on the newly formed hydroxyl. Regioselection of the alkyne insertion was high, with the exception of internal alkynes possessing two aliphatic substituents. To explain the results, the authors have proposed the mechanism depicted in Scheme 4.29 in which the reaction proceeds through the generation of a three-coordinate complex **U**.

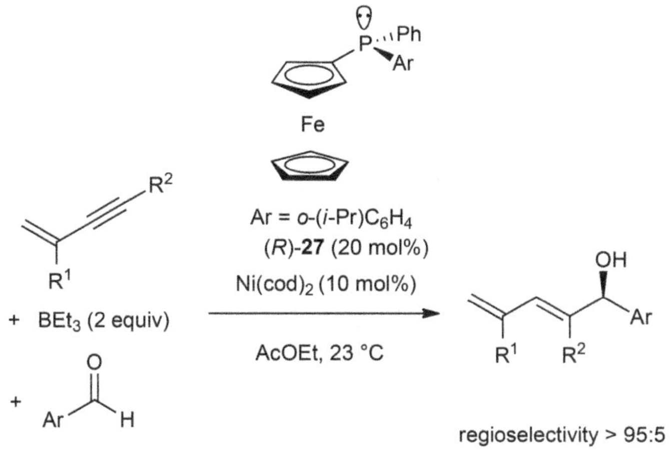

R^1 = Me, R^2 = Et, Ar = Ph: 66%, ee = 56%
R^1 = Me, R^2 = Et, Ar = *p*-Tol: 73%, ee = 56%
R^1 = Me, R^2 = Et, Ar = *p*-AcOC_6H_4: 55%, ee = 56%
R^1 = Me, R^2 = Et, Ar = *p*-MeOC_6H_4: 63%, ee = 56%
R^1 = Me, R^2 = Et, Ar = 1-Naph: 52%, ee = 48%
R^1 = Me, R^2 = Et, Ar = 2-Naph: 47%, ee = 58%
R^1 = Me, R^2 = Cy, Ar = Ph: 77%, ee = 55%
R^1 = H, R^2 = Ar = Ph: 58%, ee = 44%
R^1 = H, R^2 = Cy, Ar = Ph: 77%, ee = 55%

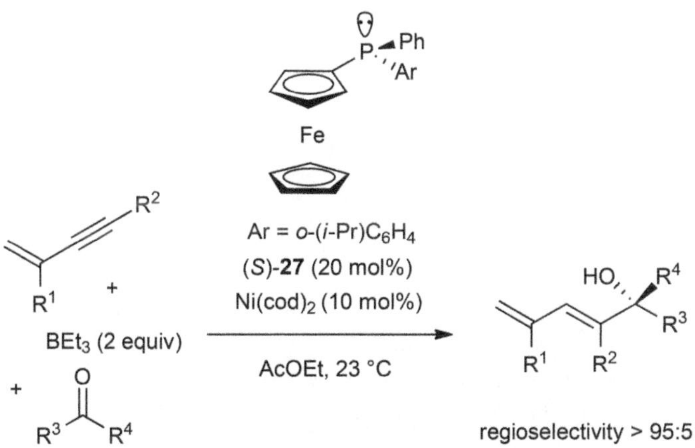

R^1 = H, R^2 = *n*-Hex, R^3 = Me, R^4 = Ph: 69%, ee = 64%
R^1 = R^3 = Me, R^2 = Et, R^4 = *p*-Tol: 71%, ee = 64%
R^1 = R^3 = Me, R^2 = Et, R^4 = 2-furyl: 58%, ee = 40%
R^1 = R^3 = Me, R^2 = Et, R^4 = 1-cyclohexenyl: 39%, ee = 70%

Scheme 4.28 Three-component reactions of 1,3-enynes, aldehydes or ketones, and triethylborane.

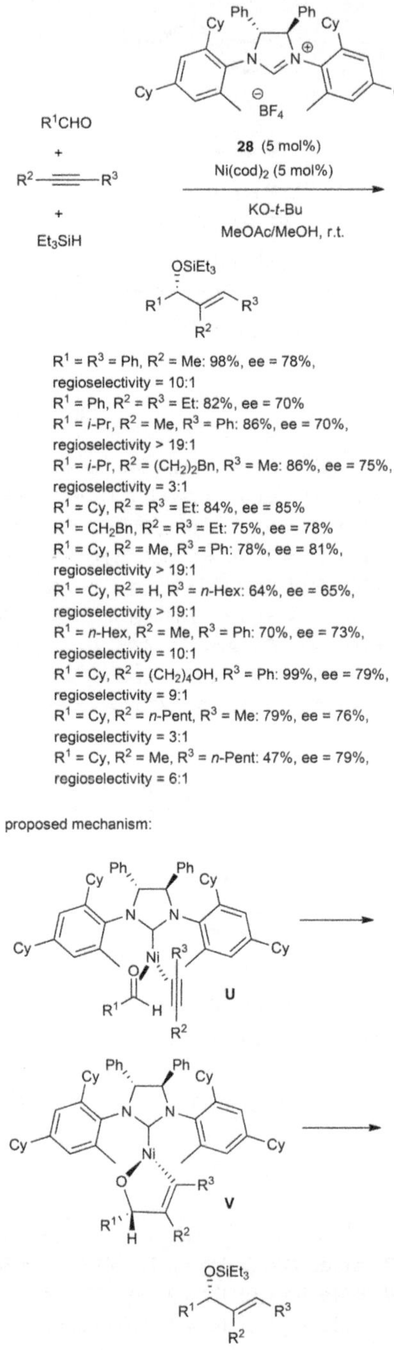

Scheme 4.29 Three-component reaction of alkynes, aldehydes, and triethylsilane.

Tilting of the *N*-aryl ring of **U** relative to the imidazolidine ring positioned the *ortho*-cyclohexyl substituent *anti* to the backbone phenyl group and distal to nickel. This orientation then positioned the *ortho*-methyl substituent *syn* to the backbone phenyl group and proximal to nickel. It was the *ortho*-methyl substituent that thus dictated the selectivity of aldehyde binding according to this model. Oxidative cyclisation of **U** to metallacycle **V** then led to the formation of the final product.

In 2008, these reactions were investigated by Zhou *et al.* by using another type of reductant, such as $ZnMe_2$, and another type of ligand, such as phosphoramidite **29** (Scheme 4.30).[51] This ligand was selected by the authors among a range of variously substituted phosphoramidites in addition to bidentate ligands, such as BINAP and PHOX. Under the optimised reaction conditions, the reactions of a range of alkynes with aldehydes and $ZnMe_2$ afforded the corresponding allylic alcohols **30** containing tetrasubstituted alkenes with

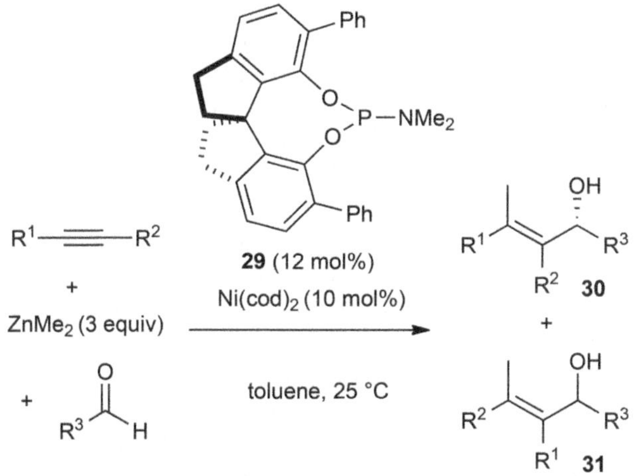

$R^1 = R^3$ = Ph, R^2 = Me: 88%, **30**:**31** > 95:5, ee (**30**) = 95% (*R*)
R^1 = Ph, R^2 = Me, R^3 = *m*-$MeOC_6H_4$: 92%, **30**:**31** > 95:5, ee (**30**) = 95%
R^1 = Ph, R^2 = Me, R^3 = *p*-Tol: 95%, **30**:**31** > 95:5, ee (**30**) = 94%
R^1 = Ph, R^2 = Me, R^3 = *p*-$MeOC_6H_4$: 94%, **30**:**31** = 93:7, ee (**30**) = 93%
R^1 = Ph, R^2 = Me, R^3 = *p*-FC_6H_4: 89%, **30**:**31** > 95:5, ee (**30**) = 96%
R^1 = Ph, R^2 = Me, R^3 = *p*-$CF_3C_6H_4$: 89%, **30**:**31** = 85:15, ee (**30**) = 95%
R^1 = Ph, R^2 = Me, R^3 = 2-Naph: 70%, **30**:**31** > 95:5, ee (**30**) = 86%
R^1 = Ph, R^2 = Me, R^3 = 2-thienyl: 82%, **30**:**31** > 95:5, ee (**30**) = 92%
R^1 = Ph, R^2 = Me, R^3 = *n*-Bu: 88%, **30**:**31** > 95:5, ee (**30**) = 88%
$R^1 = R^2$ = Et, R^3 = *p*-FC_6H_4: 90%, ee = 98%
$R^1 = R^2$ = *n*-Pr, R^3 = *p*-FC_6H_4: 92%, ee = 99%
$R^1 = R^2$ = Ph, R^3 = *p*-FC_6H_4: 92%, ee = 98%
R^1 = Ph, R^2 = Et, R^3 = *p*-FC_6H_4: 92%, **30**:**31** = 86:14, ee (**30**) = 98%
R^1 = Ph, R^2 = *n*-Bu, R^3 = *p*-FC_6H_4: 91%, **30**:**31** = 86:14, ee (**30**) = 99%

Scheme 4.30 Three-component reaction of alkynes, aldehydes, and dimethylzinc.

high regioselectivity (**30**:**31** > 95:5), in good yields and with excellent enantioselectivities of up to 99% ee. In addition to benzaldehyde and its derivatives, naphthaldehyde, thiophene-2-carbaldehyde, as well as the aliphatic *n*-butyraldehyde, could also be coupled with 1-phenylprop-1-yne to produce the corresponding alcohols in good enantioselectivities (86–92% ee). Moreover, various disubstituted alkynes other than 1-phenylprop-1-yne were also investigated in the coupling reactions with *p*-fluorobenzaldehyde and the desired alcohols were obtained in high yields and with excellent enantioselectivities (98–99% ee); however, lower regioselectivities (86:14) were observed in the reactions with 1-phenylbut-1-yne and 1-phenylhex-1-yne. The decreased regioselectivity could be attributed to the fact that as the difference between the sizes of the two substituents of the alkyne becomes smaller, distinguishing the two ends of the alkyne becomes more difficult. The authors have also investigated different organozinc reagents other than $ZnMe_2$, such as $ZnEt_2$ and $ZnPh_2$, and found that the use of $ZnEt_2$ led to the formation of a mixture of the reductive coupling product as a minor product along with the alkylative coupling product as the major product in 57 and 71% ee, respectively, whereas the use of $ZnPh_2$ gave no reaction. It must be noted that this nice work constituted the first highly enantioselective alkylative coupling of alkynes and aldehydes catalysed by a nickel complex of a chiral spiro phosphoramidite.

In 2010, Zhou *et al.* described a highly efficient nickel-catalysed reductive coupling of alkynes and imines using Et_2Zn as reductant, affording a range of chiral allylic amines with high yields and moderate to good chemoselectivities ranging from 1:1 to 16:1.[52] Indeed, the reaction also produced a by-product in 4–41% yields arising from alkylative coupling of the product with $ZnEt_2$ (Scheme 4.31). Chiral induction was achieved by employing a nickel catalyst containing chiral spiro phosphine ligand **32**, which allowed enantioselectivities of up to 94% ee to be reached. It must be noted that homogeneous results were obtained in the case of aromatic imines, while imines derived from aliphatic aldehydes gave low enantioselectivities of 9–11% ee.

4.3.2 Miscellaneous Multicomponent Reactions

The Reformatsky reaction, discovered more than 125 years ago,[53] is the well-recognised carbon–carbon bond-forming reaction of α-halo esters with aldehydes or ketones in the presence of Zn metal to give β-hydroxy esters. High functional-group tolerance and the *in situ* preparation of the reagent have contributed to its success. Imines are suitable substrates for the Reformatsky reaction, and the so-called imino-Reformatsky reaction has great potential in synthesis. In 2005, Cozzi and Rivalta described the first practical one-pot three-component enantioselective imino-Reformatsky reaction, which was based on the use of *N*-methylephedrine **33** as a cheap and recoverable chiral ligand (Scheme 4.32).[54] The nickel-catalysed three-component reaction occurred between aldehydes, α-bromo esters, and *o*-anisidine, in which $ZnMe_2$ played multiple roles as dehydrating agent for the formation of

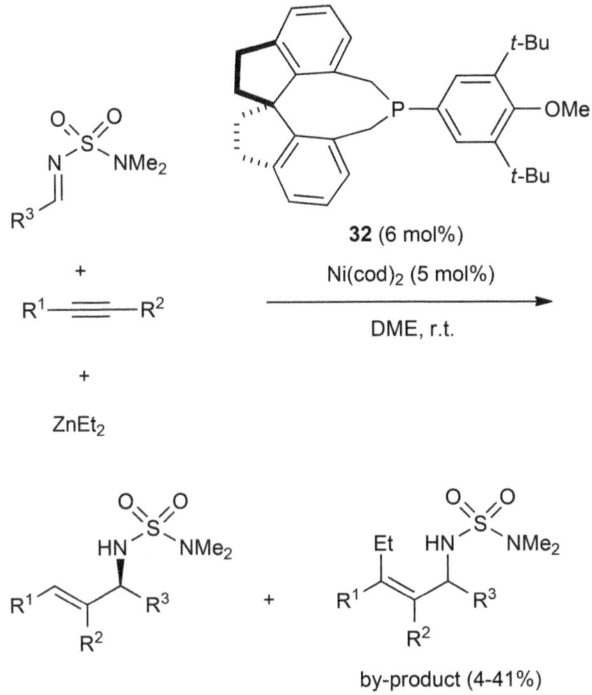

$R^1 = R^3 = Ph$, $R^2 = Me$: 74%, ee = 64%
$R^1 = Ph$, $R^2 = Me$, $R^3 = o\text{-}MeOC_6H_4$: 63%, ee = 94%
$R^1 = Ph$, $R^2 = Me$, $R^3 = o\text{-}ClC_6H_4$: 80%, ee = 76%
$R^1 = Ph$, $R^2 = Me$, $R^3 = m\text{-}Tol$: 77%, ee = 86%
$R^1 = Ph$, $R^2 = Me$, $R^3 = m\text{-}ClC_6H_4$: 75%, ee = 85%
$R^1 = Ph$, $R^2 = Me$, $R^3 = p\text{-}MeOC_6H_4$: 68%, ee = 89%
$R^1 = Ph$, $R^2 = Me$, $R^3 = p\text{-}Tol$: 71%, ee = 86%
$R^1 = Ph$, $R^2 = Me$, $R^3 = p\text{-}FC_6H_4$: 72%, ee = 86%
$R^1 = Ph$, $R^2 = Me$, R^3 = 1-Naph : 62%, ee = 88%
$R^1 = Ph$, $R^2 = Me$, R^3 = 2-Naph : 80%, ee = 87%
$R^1 = Ph$, $R^2 = Me$, $R^3 = n\text{-}Pr$: 65%, ee = 11%
$R^1 = Ph$, $R^2 = Me$, $R^3 = Cy$: 51%, ee = 9%
$R^1 = R^3 = Ph$, $R^2 = Et$: 81%, ee = 88%
$R^1 = R^2 = R^3 = Ph$: 62%, ee = 86%
$R^1 = R^2 = Et$, $R^3 = Ph$: 41%, ee = 67%

Scheme 4.31 Three-component reaction of alkynes, imines, and diethylzinc.

the imine, reductant of the Ni(II) salt to Ni(0), and coordinating the metal. In this context, a range of chiral β-amino esters could be synthesised in enantioselectivities of up to 92% ee on the basis of this domino reaction. The reaction scope was broad since aromatic, aliphatic, unsaturated, and heterocyclic aldehydes were reactive and resulted in good to excellent enantioselectivities, but with only moderate yields (30–80%). The proposed mechanism of

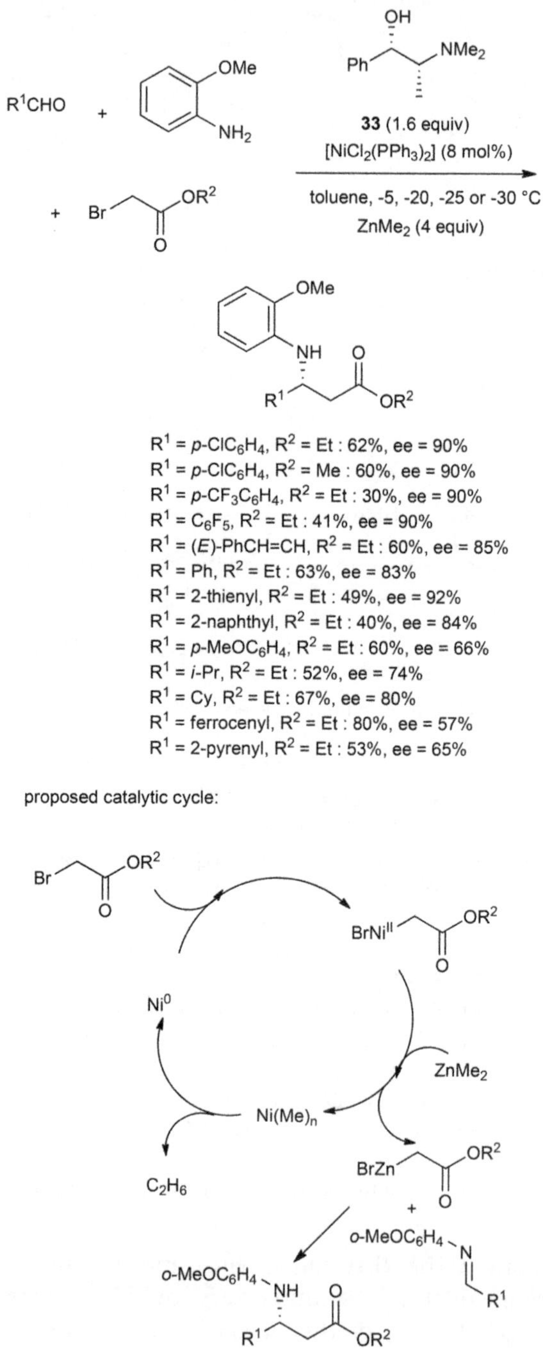

Scheme 4.32 Three-component imino-Reformatsky reaction.

the process (Scheme 4.32) involved reduction of the Ni(II) complex to a Ni(0) complex, and a Ni(II)/Zn(II) exchange, which led to an organozinc Reformatsky reagent.

Transition metal-catalysed [2 + 2 + 2] cycloaddition of unsaturated motifs, such as alkyne and alkene, constitutes the most atom-economical and facile protocol for the construction of a six-membered ring system.[55] In particular, enantioselective [2 + 2 + 2] cycloaddition is a fascinating protocol for the construction of chiral cyclic skeletons.[56] In this context, remarkable levels of regio- and enantioselectivities of up to >95 : 5, and 99% ee, respectively, were reported by Murakami *et al.* in a novel intermolecular formal [2 + 2 + 2] cycloaddition of two molecules of isocyanates with allenes.[57] This unprecedented pseudo-three-component reaction was catalysed by a combination of $Ni(cod)_2$ with the unsymmetrical phosphino-oxazoline chiral ligand (*S*,*S*)-i-Pr-FOXAP. The latter has been selected among a range of various chiral ligands, such as the C_2-symmetric biphosphine ligands (*S*,*S*)-CHIRAPHOS, (*S*,*S*)-NORPHOS, and (*S*)-BINAP, which gave lower regioselectivities. This process provided an efficient access to chiral dihydropyrimidine-2,4-diones in moderate to good yields, as shown in Scheme 4.33. Various combinations of monosubstituted allenes and isocyanates were investigated, demonstrating that allenes possessing a primary alkyl group readily reacted with high regio- and enantioselectivities, whereas the reaction of cyclohexylallene was sluggish to give the corresponding product in only 26% yield. Functional groups, such as benzyloxy, siloxy, and alkenyl groups, were tolerated, providing excellent enantioselectivities of up to 99% ee. Generally, higher regioselectivity was observed with electron-rich aryl isocyanates than with electron-deficient aryl isocyanates. On the other hand, other alkyl isocyanates, including hexyl isocyanate, cyclohexyl isocyanate, and *tert*-butyl isocyanate, all failed to undergo the reaction. A plausible mechanism for the production of the dihydropyrimidine-2,4-dione from the corresponding allene and isocyanate is depicted in Scheme 4.33. Initially, the intermolecular oxidative cyclisation of a heteropair of the allene and isocyanate occurs on nickel(0) to give the five-membered ring azanickelacyclic intermediate **W**. Subsequent insertion of another molecule of isocyanate into the nickel–nitrogen bond expands **W** to the seven-membered ring azanickelacycle **X**, which is in equilibrium with the zwitterionic π-allylnickel species **Y**. Finally, an intramolecular recombination occurs at the more substituted carbon of the allyl moiety to afford the final product, along with nickel(0).

In the last few years, an explosive number of multiple-catalyst systems for various organic transformations have been developed.[58] In particular, the combination of organocatalysts and transition metal catalysts has evolved as a new strategy to carry out enantioselective transformations that could not be performed in a traditional way by simply employing one of the two catalysts. These transformations not only demonstrate the potential of this merged catalytic approach, but they also show that there are more options to render a reaction highly enantioselective than testing different chiral metal–ligand complexes, organocatalysts, or additives. By using appropriate combinations

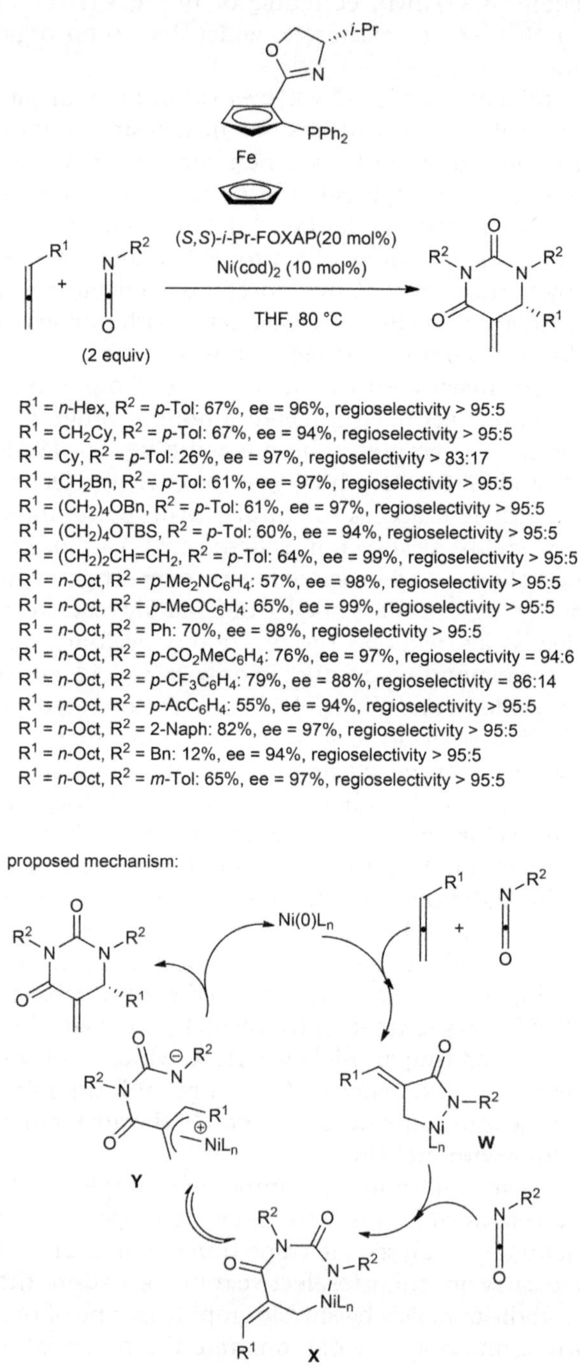

Scheme 4.33 Pseudo-three-component reaction of isocyanates and allenes.

of an organocatalyst and an achiral or chiral transition metal catalyst, facile ways for reaction optimisation can be achieved by simply varying one of the two existing catalysts. The first example of combining a transition metal and an organocatalyst was reported by Ito *et al.* in 1986, dealing with a remarkable enantioselective domino aldol/cyclisation reaction of aldehydes with methyl isocyanoacetate catalysed by a combination of a gold complex and a chiral tertiary amine as organocatalyst, allowing diastereo- and enantioselectivities of up to >99% de and 97% ee, respectively, to be achieved in combination with yields of 83–100%.[59] Although the combination of transition metal catalysis with organocatalysis has allowed a range of novel and useful reactions to be achieved,[58d,60] the development of domino reactions induced by a combination of two types of catalysts still remains a challenge. While organocatalysis is dominated by Lewis base catalysts, such as amines, carbenes, and tertiary phosphines, a metal catalyst usually has an empty coordination site to interact and activate a substrate. The challenge in combining an organocatalyst and a metal catalyst is in part to avoid the deactivation of catalyst by Lewis acid/base interaction. Even in the absence of a catalyst poison, the presence of a Lewis base can erode the chiral environment of a chiral metal complex. Consequently, the success of tandem catalysis will need fine tuning of the hardness and softness of the metal catalyst and the organocatalyst to increase their compatibility. The combination of relay nickel catalysis with organocatalysis has been recently applied to develop highly efficient asymmetric multicomponent reactions. As an example, McQuade and co-workers have developed an original one-pot multicomponent reaction catalysed by a microencapsulated amine catalyst **34** and chiral nickel complex **35** (Scheme 4.34).[61] Although the enantioselectivity of this process was not so high (72% ee), the site-isolation of two otherwise incompatible catalysts provided by microencapsulation brought new insight into the development of amine–Lewis acid domino sequences. The encapsulation of the amine catalyst was the key for the success of the reaction for the following reasons: (1) the use of a soluble amine catalyst led to catalyst deactivation by complexation with the nickel catalyst; (2) the silica MCM-41 or polystyrene supported amine catalyst failed to catalyse nitroalkene formation at room temperature, but the encapsulated poly(ethyleneimine) could; (3) the microencapsulated amines swollen in methanol retained their catalytic potency when in toluene, which allowed the one-pot reaction to be run in a mixture of two different solvents, and the microencapsulated amine and nickel catalyst could work under their respective ideal solvents of methanol and toluene.

4.4 Tandem Sequences

The cyclopropane ring is an important structural motif in a great number of natural products and biologically active agents.[62] In addition, cyclopropyl derivatives also constitute valuable synthetic building blocks in organic synthesis.[63] Consequently, a number of efforts have been made to develop efficient synthetic methods of chiral cyclopropanes.[64] Among them, asymmetric

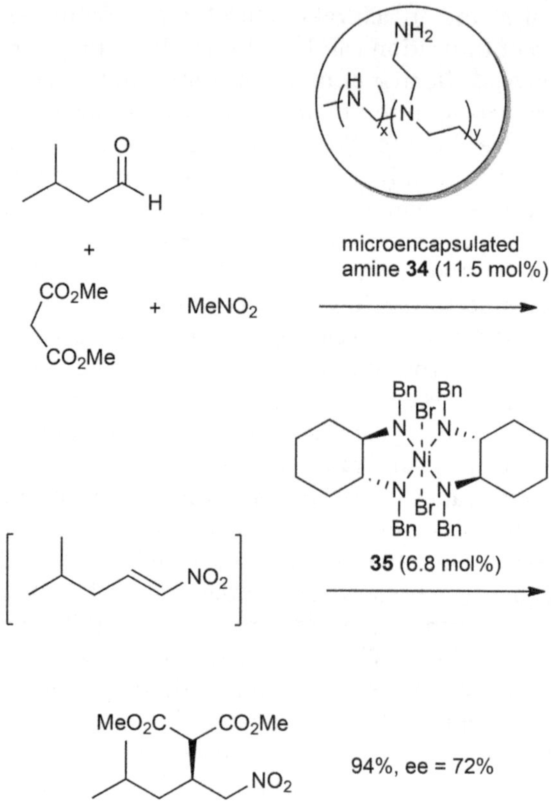

Scheme 4.34 Multicatalysed three-component domino Henry/Michael reaction.

domino reactions and tandem sequences involving a Michael addition followed by an intramolecular alkylation have been developed by several groups for the synthesis of chiral functionalised cyclopropanes.[65] The reactions can involve domino or tandem Michael/intramolecular alkylation reactions of bromonitromethanes with α,β-unsaturated carbonyl derivatives, as well as additions of bromomalonates to nitroalkenes. So far, most of these reactions have been catalysed by organocatalysts. In 2012, Kim *et al.* reported the first example of an enantioselective tandem Michael/intramolecular alkylation sequence between bromomalonates and nitroalkenes promoted by a chiral nickel catalyst (Scheme 4.35).[66] The process begins with the Michael addition of the bromomalonate to the nitroalkene in the presence of 5 mol% of chiral preformed diamine nickel catalyst **36** in dibromomethane at room temperature, which is followed by an intramolecular alkylation induced by addition of DBU to the reaction mixture to afford the final chiral cyclopropane. A range of aromatic and heteroaromatic nitroalkenes reacted smoothly with ethyl as well as methyl bromomalonates to give the corresponding cyclopropanes as single *trans*-diastereoisomers in uniformly high yields and enantioselectivities (85–99% ee). To explain the stereoselectivity of the process, the

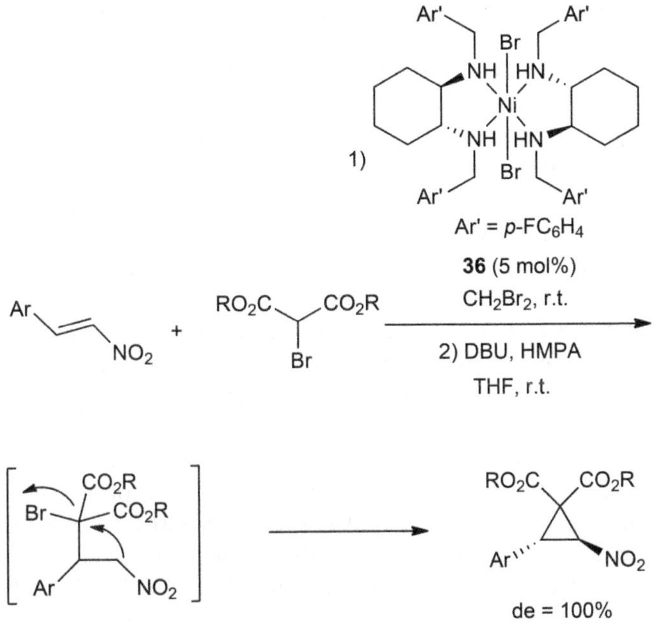

Ar = Ph, R = Me: 87%, ee = 94%
Ar = Ph, R = Et: 99%, ee = 97%
Ar = *p*-MeOC_6H_4, R = Me: 89%, ee = 96%
Ar = *p*-MeOC_6H_4, R = Et: 96%, ee = 99%
Ar = *p*-Tol, R = Me: 86%, ee = 97%
Ar = *p*-Tol, R = Et: 97%, ee = 99%
Ar = *p*-BrC_6H_4, R = Me: 85%, ee = 91%
Ar = *p*-ClC_6H_4, R = Et: 92%, ee = 98%
Ar = *p*-FC_6H_4, R = Et: 95%, ee = 98%
Ar = *o*-$NO_2C_6H_4$, R = Et: 93%, ee = 91%
Ar = *o*-BrC_6H_4, R = Et: 90%, ee = 93%
Ar = 2-thienyl, R = Et: 89%, ee = 99%
Ar = 2-furyl, R = Et: 91%, ee = 91%
Ar = 1-Naph, R = Et: 70%, ee = 85%

proposed transition state:

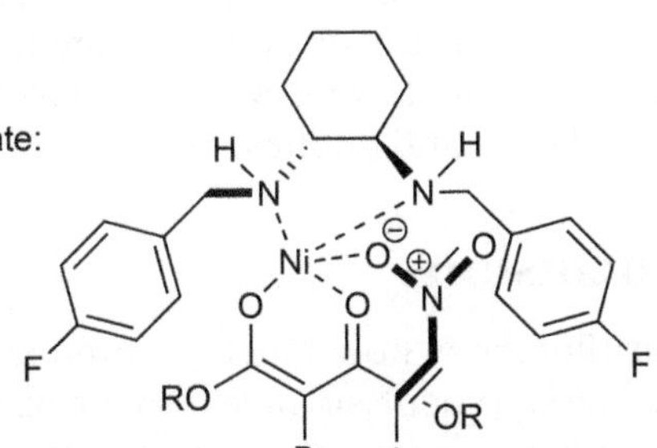

Scheme 4.35 Tandem Michael/intramolecular alkylation sequence between bromomalonates and nitroalkenes.

authors proposed that bromomalonate was activated by the nickel catalyst in a bidentate fashion. Then, the bromomalonate anion attacked the *Si* face of the double bond of the nitroalkene, as shown in the transition state depicted in Scheme 4.35.

While progress has been made in the development of efficient enantioselective catalytic conjugate addition protocols, notable shortcomings have remained unaddressed for a long time. Among them are approaches involving the conjugate addition of alkenyl groups to less reactive acyclic trisubstituted α,β-unsaturated carbonyl compounds, which provided chiral products with all-carbon-substituted quaternary stereogenic centres. Very recently, this goal was remarkably achieved by Hoveyda and McGrath, who disclosed the first examples of enantioselective Michael-type additions of alkenylmetal reagents to acyclic trisubstituted enones evolving through multicatalytic tandem sequences.[67] The requisite β-substituted alkenylaluminum reagents **37** were synthesised *in situ* with exceptional stereoselectivity by a nickel-catalysed hydroalumination process between the corresponding alkynes and DIBAL, using $Ni(PPh_3)_2Cl_2$ as catalyst (Scheme 4.36). These β-substituted alkenylaluminum reagents reacted with enones, prepared through a site- and (*E*)-stereoselective zirconocene-catalysed carboalumination/acylation reaction, in the presence of a combination of Ag(I)-based N-heterocyclic carbene complex **38** and $CuCl_2{\cdot}2H_2O$ in THF at room temperature. The one-pot tandem sequence afforded the corresponding chiral β-alkenyl ketones **39** in moderate to good yields and with high enantioselectivities of up to 96% ee. The substrate scope of the multicomponent tandem sequence was broad since aryl-, heteroaryl-, and alkyl-substituted alkynes could be used. Moreover, enones bearing an electron-donating or an electron-withdrawing aryl unit as well as an alkyl substituent were effective substrates. Similarly, α-substituted alkenylaluminum reagents **40**, derived from hydroaluminations catalysed by $Ni(dppp)Cl_2$, reacted under closely related reaction conditions (−30 °C instead of r.t.) with a range of enones to give the corresponding chiral β-alkenyl ketones **41** in even better yields and enantioselectivities (up to 98% ee) than the analogous reactions of the less encumbered β-alkenylmetal variants. These unprecedented multicatalytic one-pot tandem sequences were complete within four hours, and present a number of advantages including generating all-carbon-substituted stereogenic centres, using alkenyl-based nucleophiles for the first time, and related to the variety of functionalisation feasible with alkenes and enones.

4.5 Conclusions

This chapter illustrates how much asymmetric nickel catalysis has contributed to the development of novel enantioselective domino, multicomponent, and tandem sequential reactions. It updates the major progress in the field of enantioselective two- and multicomponent domino reactions as well as tandem sequences promoted by chiral nickel catalysts, covering the literature since the beginning of 2004. It well illustrates the power of these

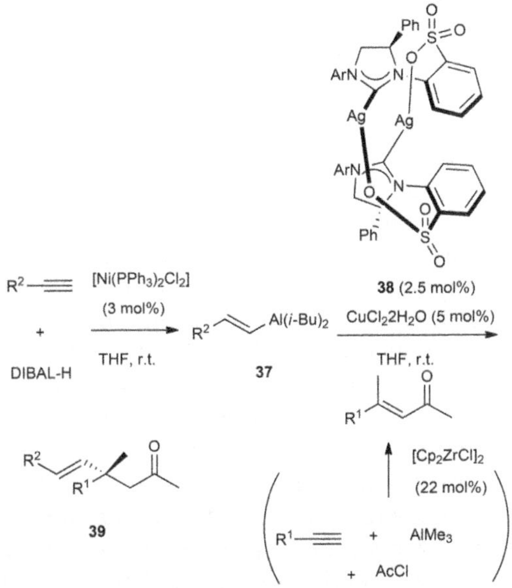

R^1 = *p*-$MeOC_6H_4$, R^2 = Ph: 75%, ee = 96%
R^1 = *p*-BrC_6H_4, R^2 = Ph: 60%, ee = 92%
R^1 = Ph, R^2 = *p*-$MeOC_6H_4$: 60%, ee = 93%
R^1 = Ph, R^2 = *p*-$CF_3C_6H_4$: 72%, ee = 96%
R^1 = *o*-$MeOC_6H_4$, R^2 = Ph: 60%, ee = 96%
R^1 = Ph, R^2 = 3-thienyl: 54%, ee = 93%
R^1 = Cy, R^2 = Ph: 51%, ee = 94%
R^1 = Bn, R^2 = Ph: 55%, ee = 81%
R^1 = Ph, R^2 = Cy: 60%, ee = 89%

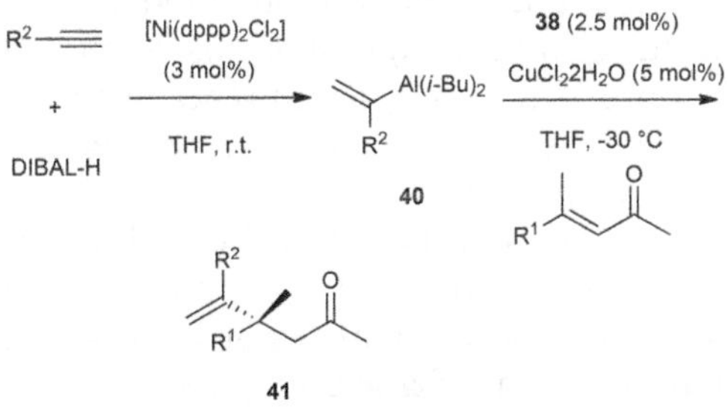

R^1 = R^2 = Ph: 82%, ee = 98%
R^1 = *p*-$MeOC_6H_4$, R^2 = Ph: 95%, ee = 95%
R^1 = *m*-FC_6H_4, R^2 = Ph: 88%, ee = 98%
R^1 = *o*-FC_6H_4, R^2 = Ph: 48%, ee = 98%
R^1 = Ph, R^2 = *p*-$MeOC_6H_4$: 95%, ee = 96%
R^1 = Ph, R^2 = *m*-$CF_3C_6H_4$: 97%, ee = 96%
R^1 = Ph, R^2 = *o*-FC_6H_4: 71%, ee = 98%
R^1 = *p*-$MeOC_6H_4$, R^2 = 3-thienyl: 81%, ee = 96%
R^1 = *p*-BrC_6H_4, R^2 = Ph: 95%, ee = 98%
R^1 = Bn, R^2 = Ph: 85%, ee = 95%
R^1 = Cy, R^2 = Ph: 83%, ee = 84%

Scheme 4.36 Multicatalysed tandem Michael-type sequences between alkynes, DIBAL, and trisubstituted enones.

elegant one-pot processes of two or more bond-forming reactions, evolving under identical conditions in which the subsequent transformation takes place at the functionalities obtained in the former transformation, following the same principles that are found in biosynthesis from the nature. These fascinating reactions have rapidly become one of the most current fields in organic chemistry. During the last 10 years, an impressive number of novel powerful asymmetric domino and multicomponent processes have been developed on the basis of asymmetric nickel catalysis. In particular, a number of enantioselective Michael-initiated domino reactions have been described, involving nitroalkenes as well as various α,β-unsaturated carbonyl compounds as acceptor-activated alkenes, which provided a wide variety of chiral functionalised (poly)cyclic products in enantioselectivities uniformly excellent (often up to 99% ee). Moreover, other types of enantioselective novel two-component domino reactions have been successfully catalysed by chiral nickel complexes, such as the first domino aldol-type/cyclisation reactions between aldehydes and isothiocyanatooxindoles, the first domino carbonyl-ylide formation/reverse-electron-demand 1,3-dipolar cycloaddition reactions between α,α′-dicarbonyl diazo compounds and cyclohexyl vinyl ether, domino denitrogenative annulation reactions of 2*H*-1,2,3,4-benzothiatriazine 1,1-dioxides with allenes, and domino cyclisation/cross-coupling reactions of alkylboron reagents bearing a pendant alkene with unactivated alkyl bromides, all providing excellent enantioselectivities of up to 97–99% ee.

In the context of enantioselective nickel-catalysed multicomponent reactions, many excellent results have also been achieved. For example, three-component reactions between 1,3-dienes, carbonyl compounds such as aldehydes or carbon dioxide, and various reducing agents such as organozinc reagents, silanes, or silaboranes have provided a variety of cyclic as well as acyclic chiral products in very high enantioselectivities of up to 97% ee in all cases. Furthermore, three-component reactions between allenes, aldehydes, and silanes have allowed chiral allylic alcohols to be easily produced with enantioselectivities of up to 98% ee along with *Z*/*E* ratios of >95 : 5. These chiral products along with chiral allylic amines were also generated with high enantioselectivities (up to 94% ee) on the basis of three-component reactions between alkynes, aldehydes or imines, and reducing agents such as boranes, silanes, or dialkylzincs.

Other types of multicomponent reactions have also been successfully developed, such as the first practical three-component imino-Reformatsky reaction, and a pseudo-three-component reaction between allenes and isocyanates, providing enantioselectivities of up to 92% and 99% ee, respectively. Finally, excellent results were described for several novel enantioselective tandem sequences. For example, remarkable enantioselectivities of up to 99% ee were reached in tandem Michael/intramolecular cyclisation sequences, as well as in a remarkable multicatalytic Michael sequence occurring between enones, alkynes, and DIBAL, which stereoselectively afforded a range of chiral β-alkenyl ketones bearing an all-carbon-substituted quaternary stereogenic centre in enantioselectivities of up to 98% ee.

The economical interest in combinations of asymmetric nickel catalytic processes with the concept of domino, multicomponent, and sequential reactions is obvious, and has allowed reaching easily high molecular complexity with often excellent levels of stereocontrol with simple operational one-pot procedures, and advantages of savings in solvent, time, energy, and costs by avoiding costly protecting groups and time-consuming purification procedures after each step. Undoubtedly, the future direction in this field is to continue expanding the scope of these enantioselective one-pot reactions through the combination of different types of reactions, the employment of novel chiral nickel catalysts, and to apply these powerful strategies to the synthesis of biologically interesting molecules, including natural products, and that of novel chiral ligands, and functional materials.

References

1. H. Clavier and H. Pellissier, *Adv. Synth. Catal.*, 2012, **18**, 3347–3403.
2. (a) L. F. Tietze and U. Beifuss, *Angew. Chem., Int. Ed.*, 1993, **32**, 131–163; (b) L. F. Tietze, *Chem. Rev.*, 1996, **96**, 115–136; (c) L. F. Tietze, G. Brasche and K. Gericke, *Domino Reactions in Organic Synthesis*, Wiley-VCH, Weinheim, 2006; (d) L. F. Tietze, *Domino Reactions-Concepts for Efficient Organic Synthesis*, Wiley-VCH, Weinheim, 2014.
3. K. C. Nicolaou, D. J. Edmonds and P. G. Bulger, *Angew. Chem., Int. Ed.*, 2006, **45**, 7134–7186.
4. C. J. Chapman and C. G. Frost, *Synthesis*, 2007, 1–21.
5. (a) L. F. Tietze and A. Modi, *Med. Res. Rev.*, 2000, **20**, 304–322; (b) *Multicomponent Reactions*, ed. J. Zhu and H. Bienaymé, Wiley-VCH, Weinheim, 2005; (c) D. J. Ramon and M. Yus, *Angew. Chem., Int. Ed.*, 2005, **44**, 1602–1634; (d) *Synthesis of Heterocycles via Multicomponent Reactions*, ed. R. V. A. Orru and E. Ruijter, Springer, Berlin, 2010, vol. I and II, Topics in Heterocyclic Chemistry.
6. (a) I. Ugi, R. Meyr, U. Fetzer and C. Steinbrückner, *Angew. Chem.*, 1959, **71**, 386; (b) R. V. A. Orru and M. de Greef, *Synthesis*, 2003, 1471–1499; (c) E. Ruijter, R. Scheffelaar and R. V. A. Orru, *Angew. Chem., Int. Ed.*, 2011, **50**, 6234–6246.
7. (a) C. Grondal, M. Jeanty and D. Enders, *Nat. Chem.*, 2010, **2**, 167–178; (b) K. C. Nicolaou and J. S. Chen, *Chem. Soc. Rev.*, 2009, **38**, 2993–3009; (c) B. B. Touré and D. G. Hall, *Chem. Rev.*, 2009, **109**, 4439–4486; (d) M. Colombo and I. Peretto, *Drug Discovery Today*, 2008, **13**, 677–684; (e) A. Padwa and S. K. Bur, *Tetrahedron*, 2007, **63**, 5341–5378; (f) C. Hulme and V. Gore, *Curr. Med. Chem.*, 2003, **10**, 51–80.
8. B. J. Cowen and S. J. Miller, *Chem. Soc. Rev.*, 2009, **38**, 3102–3116, Special Issue on Rapid formation of molecular complexity in organic synthesis.
9. (a) G. H. Posner, *Chem. Rev.*, 1986, **86**, 831–844; (b) T.-L. Ho, *Tandem Organic Reactions*, Wiley, New York, 1992; (c) R. A. Bunce, *Tetrahedron*, 1995, **51**, 13103–13159; (d) P. J. Parsons, C. S. Penkett and A. J. Shell, *Chem. Rev.*, 1996, **96**, 195–206; (e) L. F. Tietze and N. Rackelmann, *Pure Appl. Chem.*,

2004, **76**, 1967–1983; (f) P. I. Dalko and L. Moisan, *Angew. Chem., Int. Ed.*, 2004, **43**, 5138–5175; (g) H. Pellissier, *Tetrahedron*, 2006, **62**, 2143–2173; (h) H. Pellissier, *Tetrahedron*, 2006, **62**, 1619–1665; (i) D. Enders, C. Grondal and M. R. M. Hüttl, *Angew. Chem., Int. Ed.*, 2007, **46**, 1570–1581; (j) G. Guillena, D. J. Ramon and M. Yus, *Tetrahedron: Asymmetry*, 2007, **18**, 693–700; (k) D. M. D'Souza and T. J. J. Müller, *Chem. Soc. Rev.*, 2007, **36**, 1095–1108; (l) A.-N. Alba, X. Companyo, M. Viciano and R. Rios, *Curr. Org. Chem.*, 2009, **13**, 1432–1474; (m) M. Ruiz, P. Lopez-Alvarado, G. Giorgi and J. C. Menéndez, *Chem. Soc. Rev.*, 2011, **40**, 3445–3454; (n) L. Albrecht, H. Jiang and K. A. A. Jørgensen, *Angew. Chem., Int. Ed.*, 2011, **50**, 8492–8509; (o) H. Pellissier, *Adv. Synth. Catal.*, 2012, **354**, 237–294; (p) C. De Graaff, E. Ruijter and R. V. A. Orru, *Chem. Soc. Rev.*, 2012, **41**, 3969–4009.

10. R. Shintani and T. Hayashi, in *Modern Organonickel Chemistry*, ed. Y. Tamaru, Wiley-VCH, Weinheim, 2005, pp. 240–272.
11. T. Komnenos, *Justus Liebigs Ann. Chem.*, 1883, **218**, 145–169.
12. (a) P. Perlmutter, in *Conjugate Addition Reactions in Organic Synthesis*, Pergamon Press, Oxford, 1992; (b) M. P. Sibi and S. Manyem, *Tetrahedron*, 2000, **56**, 8033–8061; (c) M. Kanai and M. Shibasaki, in *Catalytic Asymmetric Synthesis*, 2nd edn, Wiley, New York, 2000, p. 569; (d) N. Krause and A. Hoffmann-Roder, *Synthesis*, 2001, 171–196.
13. M. Kitamura, T. Miki, K. Nakano and R. Noyori, *Tetrahedron Lett.*, 1996, **37**, 5141–5144.
14. T. Arai, N. Yokoyama, A. Mishiro and H. Sato, *Angew. Chem., Int. Ed.*, 2010, **49**, 7895–7898.
15. (a) J. Boruwa, N. Gogoi, P. P. Saikia and N. C. Barua, *Tetrahedron: Asymmetry*, 2006, **17**, 3315–3326; (b) C. Palomo, M. Oiarbide and A. Laso, *Eur. J. Org. Chem.*, 2007, 2561–2574; (c) E. Marquès-Lopez, P. Merino, T. Tejero and R. P. Herrera, *Eur. J. Org. Chem.*, 2009, 2401–2420; (d) Y. Alvarez-Casao, E. Marques-Lopez and R. P. Herrera, *Symmetry*, 2011, **3**, 220–245.
16. W. Li, X. Liu, Z. Mao, Q. Chen and R. Wang, *Org. Biomol. Chem.*, 2012, **10**, 4767–4773.
17. Q. Liang, J. He and B. Ni, *Tetrahedron: Asymmetry*, 2014, **25**, 1146–1149.
18. T. Arai and Y. Yamamoto, *Org. Lett.*, 2014, **16**, 1700–1703.
19. (a) H. Yanagita, K. Kodama and S. Kanemasa, *Tetrahedron Lett.*, 2006, **47**, 9353–9357; (b) F. Ono, M. Hasegawa, S. Kanemasa and J. Tanaka, *Tetrahedron Lett.*, 2008, **49**, 5105–5107.
20. K. Itoh, M. Hasegawa, J. Tanaka and S. Kanemasa, *Org. Lett.*, 2005, **7**, 979–981.
21. N. Halland, T. Velgaard and K. A. Jørgensen, *J. Org. Chem.*, 2003, **68**, 5067–5074.
22. Z. Dong, J. Feng, W. Cao, X. Liu, L. Lin and X. Feng, *Tetrahedron Lett.*, 2011, **52**, 3433–3436.
23. C. V. Galliford and K. A. Scheidt, *Angew. Chem., Int. Ed.*, 2007, **46**, 8748–8758.
24. S. Kato, M. Kanai and S. Matsunaga, *Chem.–Asian J.*, 2013, **8**, 1768–1771.

25. (a) A. Padwa and M. D. Weingarten, *Chem. Rev.*, 1996, **96**, 223–269; (b) M. P. Doyle and D. C. Forbes, *Chem. Rev.*, 1998, **98**, 911–935; (c) K. V. Gothelf and K. A. Jørgensen, *Chem. Rev.*, 1998, **98**, 863–909; (d) S. Karlsson and H.-E. Högberg, *Org. Prep. Proced. Int.*, 2001, **33**, 103–172; (e) I. N. N. Namboothiri and A. Hassner, *Top. Curr. Chem.*, 2001, **216**, 1–49; (f) S. Kanemasa, *Synlett*, 2002, 1371–1387; (g) K. V. Gothelf, *Synthesis*, 2002, 211–247; (h) G. Broggini, G. Molteni, A. Terraneo and G. Zecchi, *Heterocycles*, 2003, **59**, 823–858; (i) I. Coldham and R. Hufton, *Chem. Rev.*, 2005, **105**, 2765–2809; (j) H. Pellissier, *Tetrahedron*, 2007, **63**, 3235–3285; (k) S. Kanemasa, *Heterocycles*, 2010, **82**, 87–200.
26. A. Padwa, *Acc. Chem. Res.*, 1991, **24**, 22–28.
27. (a) A. Padwa and S. F. Hornbuckle, *Chem. Rev.*, 1991, **91**, 263–309; (b) A. Padwa, *Helv. Chim. Acta*, 2005, **88**, 1357–1374; (c) A. Padwa, *J. Organomet. Chem.*, 2005, **690**, 5533–5540; (d) A. Padwa, *Chem. Soc. Rev.*, 2009, **38**, 3072–3081.
28. D. M. Hodgson, P. A. Stupple and C. Johnstone, *Tetrahedron Lett.*, 1997, **38**, 6471–6472.
29. H. Suga, T. Suzuki, K. Inoue and A. Kakehi, *Tetrahedron*, 2006, **62**, 9218–9225.
30. (a) H. Suga, D. Hishimoto, S. Higuchi, M. Ohtsuka, T. Arikawa, T. Tsuchida, A. Kakehi and T. Baba, *Org. Lett.*, 2007, **9**, 4359–4362; (b) H. Suga, S. Higuchi, M. Ohtsuka, D. Ishimoto, T. Airkawa, Y. Hashimoto, S. Misawa, T. Tsuchida, A. Kakehi and T. Baba, *Tetrahedron*, 2010, **66**, 3070–3089.
31. M. P. Sibi, Z. Ma and C. P. Jasperse, *J. Am. Chem. Soc.*, 2005, **127**, 5764–5765.
32. T. Miura, M. Yamauchi, A. Kosaka and M. Murakami, *Angew. Chem., Int. Ed.*, 2010, **49**, 4955–4957.
33. M. Yamauchi, M. Morimoto, T. Miura and M. Murakami, *J. Am. Chem. Soc.*, 2010, **132**, 54–55.
34. J. S. E. Ahlin, P. A. Donets and N. Cramer, *Angew. Chem., Int. Ed.*, 2014, **53**, 13229–13233.
35. Y. Ochi, T. Kurahashi and S. Matsubara, *Org. Lett.*, 2011, **13**, 1374–1377.
36. Y. Ren, X. Cheng, S. Yang, C. Qi, H. Jiang and Q. A. Mao, *Dalton Trans.*, 2013, **42**, 9930–9937.
37. H. Cong and G. C. Fu, *J. Am. Chem. Soc.*, 2014, **136**, 3788–3791.
38. M. Takimoto, Y. Nakamura, K. Kimura and M. Mori, *J. Am. Chem. Soc.*, 2004, **126**, 5956–5957.
39. M. Takimoto, Y. Kajima, Y. Sato and M. Mori, *J. Org. Chem.*, 2005, **70**, 8605–8608.
40. Y. Yang, S.-F. Zhu, H.-F. Duan, C.-Y. Zhou, L.-X. Wang and Q.-L. Zhou, *J. Am. Chem. Soc.*, 2007, **129**, 2248–2249.
41. T. Roth, H. Wadepohl, D. S. Wright and L. H. Gade, *Chem.–Eur. J.*, 2013, **19**, 13823–13837.
42. Y. Sato, Y. Hinata, R. Seki, Y. Oonishi and N. Saito, *Org. Lett.*, 2007, **9**, 5597–5599.
43. N. Saito, A. Kobayashi and Y. Sato, *Angew. Chem., Int. Ed.*, 2012, **51**, 1228–1231.

44. (a) S.-S. Ng and T. F. Jamison, *J. Am. Chem. Soc.*, 2005, **127**, 7320–7321; (b) S.-S. Ng and T. F. Jamison, *Tetrahedron*, 2005, **61**, 11405–11417.
45. S.-S. Ng and T. F. Jamison, *Tetrahedron*, 2006, **62**, 11350–11359.
46. S. J. Patel and T. F. Jamison, *Angew. Chem., Int. Ed.*, 2004, **43**, 3941–3944.
47. (a) G. M. Mahandru, G. Liu and J. Montgomery, *J. Am. Chem. Soc.*, 2004, **126**, 3698–3699; (b) H. Y. Jang, R. R. Huddleston and M. J. Krische, *J. Am. Chem. Soc.*, 2004, **126**, 4664–4668; (c) K. M. Miller, T. Luanphaisarnnont, C. Molinaro and T. F. Jamison, *J. Am. Chem. Soc.*, 2004, **126**, 4130–4131.
48. K. M. Miller, E. A. Colby, K. S. Woodin and T. F. Jamison, *Adv. Synth. Catal.*, 2005, **347**, 1533–1536.
49. T. Luanphaisarnnont, C. O. Ndubaku and T. F. Jamison, *Org. Lett.*, 2005, **7**, 2937–2940.
50. M. R. Chaulagain, G. J. Sormunen and J. Montgomery, *J. Am. Chem. Soc.*, 2007, **129**, 9568–9569.
51. Y. Yang, S.-F. Zhu, C.-Y. Zhou and Q.-L. Zhou, *J. Am. Chem. Soc.*, 2008, **130**, 14052–14053.
52. C.-Y. Zhou, S.-F. Zhu, L.-X. Wang and Q.-L. Zhou, *J. Am. Chem. Soc.*, 2010, **132**, 10955–10957.
53. S. Reformatsky, *Ber. Dtsch. Chem. Ges.*, 1887, **20**, 1210–1211.
54. (a) P. G. Cozzi and E. Rivalta, *Angew. Chem., Int. Ed.*, 2005, **44**, 3600–3603; (b) P. G. Cozzi and E. Rivalta, *Pure Appl. Chem.*, 2006, **78**, 287–291.
55. (a) Y. Yamamoto, *Curr. Org. Chem.*, 2005, **9**, 503–509; (b) S. Kotha, E. Brahmachary and K. Lahiri, *Eur. J. Org. Chem.*, 2005, 4741–4761; (c) P. R. Chopade and J. Louie, *Adv. Synth. Catal.*, 2006, **348**, 2307–2327; (d) Y. Shibata and K. Tanaka, *Synthesis*, 2012, 323–350.
56. (a) T. Shibata, *Org. Biomol. Chem.*, 2008, 1317–1323; (b) K. Tanaka, *Chem.–Asian J.*, 2009, **4**, 508–518.
57. T. Miura, M. Morimoto and M. Murakami, *J. Am. Chem. Soc.*, 2010, **132**, 15836–15838.
58. (a) J.-A. Ma and D. Cahard, *Angew. Chem., Int. Ed.*, 2004, **43**, 4566–1508; (b) M. Kanai, N. Kato, E. Ichikawa and M. Shibasaki, *Synlett*, 2005, 1491–1508; (c) D. H. Paull, C. J. Abraham, M. T. Scerba, E. Alden-Danforth and T. Leckta, *Acc. Chem. Res.*, 2008, **41**, 655–663; (d) Z. Shao and H. Zhang, *Chem. Soc. Rev.*, 2009, **38**, 2745–2755; (e) C. Zhong and X. Shi, *Eur. J. Org. Chem.*, 2010, 2999–3025; (f) M. Rueping, R. M. Koenigs and I. Atodiresei, *Chem. Eur. J.*, 2010, **16**, 9350–9365; (g) J. Zhou, *Chem.–Asian J.*, 2010, **5**, 422–434; (h) L. M. Ambrosini and T. H. Lambert, *ChemCatChem*, 2010, **2**, 1373–1380; (i) S. Piovesana, D. M. Scarpino Schietroma and M. Bella, *Angew. Chem., Int. Ed.*, 2011, **50**, 6216–6232; (j) N. T. Patil, V. S. Shinde and B. A. Gajula, *Org. Biomol. Chem.*, 2012, **10**, 211–224; (k) A. E. Allen and D. W. C. MacMillan, *Chem. Sci.*, 2012, **3**, 633–658; (l) Z. Du and Z. Shao, *Chem. Soc. Rev.*, 2013, **42**, 1337–1378; (m) H. Pellissier, *Tetrahedron*, 2013, **69**, 7171–7210; (n) H. Pellissier, *Enantioselective Multicatalysed Tandem Reactions*, Royal Society of Chemistry, Cambridge, 2014.
59. Y. Ito, M. Sawamura and T. Hayashi, *J. Am. Chem. Soc.*, 1986, **108**, 6405–6406.

60. M. Klussmann, *Angew. Chem., Int. Ed.*, 2009, **48**, 7124–7125.
61. (a) S. L. Poe, M. Kobaslija and D. T. McQuade, *J. Am. Chem. Soc.*, 2006, **128**, 15586–15587; (b) S. L. Poe, M. Kobaslija and D. T. McQuade, *J. Am. Chem. Soc.*, 2007, **129**, 9216–9221.
62. L. A. Wessjohann, W. Brandt and T. Thiemann, *Chem. Rev.*, 2003, **103**, 1625–1648.
63. H.-U. Reissig and R. Zimmer, *Chem. Rev.*, 2003, **103**, 1151–1196.
64. H. Pellissier, *Tetrahedron*, 2008, **64**, 7041–7095.
65. (a) H. Xie, L. Zu, H. Li, J. Wang and W. Wang, *J. Am. Chem. Soc.*, 2007, **129**, 10886–10894; (b) R. Rios, H. Sunden, J. Vesely, G.-L. Zhao, P. Dziedzic and A. Cordova, *Adv. Synth. Catal.*, 2007, **349**, 1028–1032; (c) H. M. Hansen, D. A. Longbottom and S. V. Ley, *Chem. Commun.*, 2006, 4838–4840.
66. H. J. Lee, S. M. Kim and D. Y. Kim, *Tetrahedron Lett.*, 2012, **53**, 3437–3439.
67. K. P. McGrath and A. H. Hoveyda, *Angew. Chem., Int. Ed.*, 2014, **53**, 1910–1914.

CHAPTER 5

Enantioselective Nickel-Catalysed Hydrovinylation, Hydrophosphination, Hydrocyanation, and Hydroalkynylation Reactions of Alkenes

5.1 Introduction

In the last decade, several powerful protocols for the asymmetric hydrovinylation of alkenes have been described in which nearly quantitative yields of the desired products can be obtained using low catalytic amounts of nickel complexes, along with high levels of chemo-, regio-, and enantioselectivity, often >95% ee. Nickel catalysts have been by far the most used catalysts in this type of reaction. All these novel procedures have used highly versatile phosphoramidite ligands derived from Feringa's ligand, with NaBArF as catalyst activator, providing remarkable enantioselectivities in all cases. For example, enantioselectivities of up to 99% ee were achieved in the asymmetric hydrovinylation of a range of vinylarenes to give the corresponding 3-arylbutenes. Moreover, when this reaction was followed by an oxidative degradation, the sequence offered a novel route to important anti-inflammatory chiral 2-arylpropionic acids, such as naproxen, ibuprofen, fenoprofen, and flurbiprofen. Comparable excellent enantioselectivities of up to 99% ee were also achieved in the case of substrates such as (α-alkylvinyl)arenes, as

RSC Catalysis Series No. 26
Enantioselective Nickel-Catalysed Transformations
By Hélène Pellissier

Published by the Royal Society of Chemistry, www.rsc.org

well as cyclic 1,3-dienes. It is important to note that the hydrovinylation of (α-alkylvinyl)arenes provided a new efficient access to the construction of chiral all-carbon quaternary centres. Furthermore, functionalised alkenes, such as silyl-protected allylic alcohols, could also be submitted to asymmetric hydrovinylation, with high enantioselectivities of up to 95% ee. Even higher enantioselectivities of up to 99% ee were reached for the nickel-catalysed hydrovinylation of strained alkenes, such as norbornenes, while slightly lower enantioselectivities of up to 87% ee were obtained in the hydrovinylation of strained heterobicyclo[2.2.1]heptenes and cyclobutenes.

In the area of nickel-catalysed asymmetric hydrophosphination of alkenes, the first highly enantioselective reaction was developed by using the C_1-symmetric trisphosphine Pigiphos, which provided very good enantioselectivities of up to 94% ee. In addition, a highly efficient enantioselective nickel-catalysed hydrocyanation of arylalkenes was performed with enantioselectivities of up to 92% ee by employing a TADDOL-derived phosphine/phosphite ligand. Finally, the first nickel-catalysed hydroalkynylation of 1-arylbuta-1,3-dienes was achieved by using a phosphoramidite ligand, providing up to 93% ee. All the formed chiral products from hydrovinylation, hydrophosphination, hydrocyanation, and hydroalkynylation reactions of alkenes constitute useful building blocks for the total synthesis of natural products and biologically active compounds, since they can be readily transformed into a variety of other functionalised compounds.

5.2 Hydrovinylations

A regioselective, Markovnikov addition of ethylene to alk-1-enes provides an efficient way of creating stereogenic centres in the carbon–carbon bond-forming process, considering that a wide variety of alk-1-enes are readily available. Moreover, since ethylene is a cheap, abundantly available feedstock carbon source, and the resulting vinyl group in the product of the hydrovinylation reaction is readily transformed into a variety of other common functional groups, this reaction has huge potential as a scalable, environmentally benign method for the preparation of valuable chemical intermediates. The hydrovinylation reaction has a long history, and various metals (*e.g.* cobalt, nickel, ruthenium, rhodium, and palladium) have been used to catalyse this reaction.[1] Therefore, the development of its enantioselective variant is of high value in organic synthesis.[2] The first highly enantioselective example of hydrovinylation of styrene was reported in 1988 by Wilke and Monkiewicz, allowing enantioselectivity of up to 93% ee to be achieved in the hydrovinylation of styrene in the presence of a chiral phosphine–amine ligand.[3] Ever since, the development of asymmetric hydrovinylation has been mostly focused on the use of chiral nickel complexes as catalysts. Earlier in 1985, Buono *et al.* reported the first asymmetric nickel-catalysed hydrovinylation of cyclohexa-1,3-diene with comparable enantioselectivity by using other chiral phosphine–amine ligands.[4] Later, several groups successfully developed asymmetric nickel-catalysed hydrovinylations of styrene and its derivatives by using various types of phosphorus-based chiral ligands.[5] The chiral information is usually introduced through chiral monodentate phosphorus ligands, whereby some of them bear an additional hemilabile donor group. The catalytic cycle of the nickel-catalysed hydrovinylation occurs *via* cationic intermediates, and the active species is typically generated from a neutral nickel halide precursor with a suitable co-catalyst. This activation step can be achieved by replacing the halide ligand with a weakly coordinating anion. Sodium tetrakis[(3,5-trifluoromethyl)phenyl]borate (NaBArF)[5d] has been used successfully for this purpose. It is well established that the nature of this weakly coordinating anion plays a decisive role in controlling the activity, chemoselectivity, and even enantioselectivity.

In 2005, RajanBabu *et al.* investigated a range of tunable phosphines, phosphinites, phosphites, and phosphoramidites as chiral ligands for enantioselective nickel-catalysed hydrovinylation of vinylarenes.[6] A quantitative yield combined with a high enantioselectivity of 91% ee were achieved for the asymmetric hydrovinylation of a styrene by using chiral dialkylphospholane **1** bearing an acetal derived from (*R*,*R*)-butane-2,3-diol in the presence of $NaBAr'_4$ as catalyst activator (Scheme 5.1).[7] Increasing the size of the phospholane 2,5-substituents from methyl to ethyl had a small effect on the enantioselectivity of the reaction, but significantly, this resulted in a slower reaction. The stereoselectivity was demonstrated to be dictated by the chirality of the phospholane ring, with the (*R*,*R*)-phospholane favouring (*S*)-3-arylbutene.

Later, the same authors reported a catalyst tuning of the biaryl and amino moieties of Feringa's phosphoramidite ligand **3**[8] for the same reaction (Scheme 5.2).[9] From this study, ligand (R_a,S_c)-**2**, in which the (*S*)-*N*-α-methylbenzyl groups in Feringa's ligand were replaced by an achiral benzyl and a chiral (*S*)-α-methyl-1-naphthyl group, was proven to be by far the best ligand in the presence of NaBArF for the asymmetric hydrovinylation of vinylarenes,

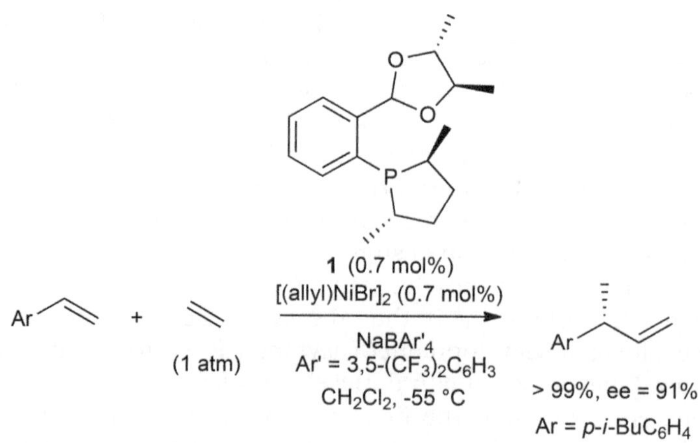

Scheme 5.1 Hydrovinylation of a vinylarene with an *in situ* generated nickel catalyst from a monophosphine ligand.

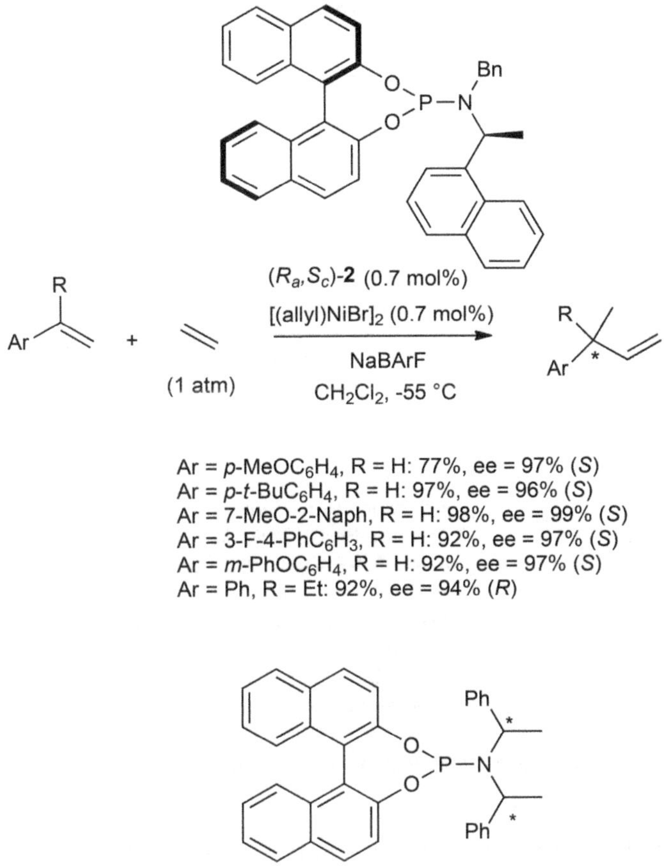

Feringa's ligand **3**

Scheme 5.2 Hydrovinylation of vinylarenes with an *in situ* generated nickel catalyst from a phosphoramidite ligand derived from Feringa's ligand.

providing high yields of up to 98% and excellent enantioselectivities of up to 99% ee. The scope of the reaction was extended to a variety of monosubstituted vinylarenes as well as to (1-alkylvinyl)arenes, such as (1-ethylvinyl) styrene, which afforded the corresponding product bearing an all-carbon quaternary stereogenic centre in 92% yield and 94% ee, as shown in Scheme 5.2. One remarkable feature of this process was its low catalyst loading of only 0.7 mol%. It must be noted that the enantioselectivities and yields observed in this asymmetric hydrovinylation of vinylarenes to give the corresponding 3-arylbutenes were among the highest for all asymmetric catalytic processes reported so far in the synthesis of intermediates for the widely used anti-inflammatory 2-arylpropionic acids, including naproxen, ibuprofen, fenoprofen, and flurbiprofen.

The utility of this methodology was demonstrated in the development of a two-step procedure for the direct synthesis of (*S*)-2-arylpropionic acids from

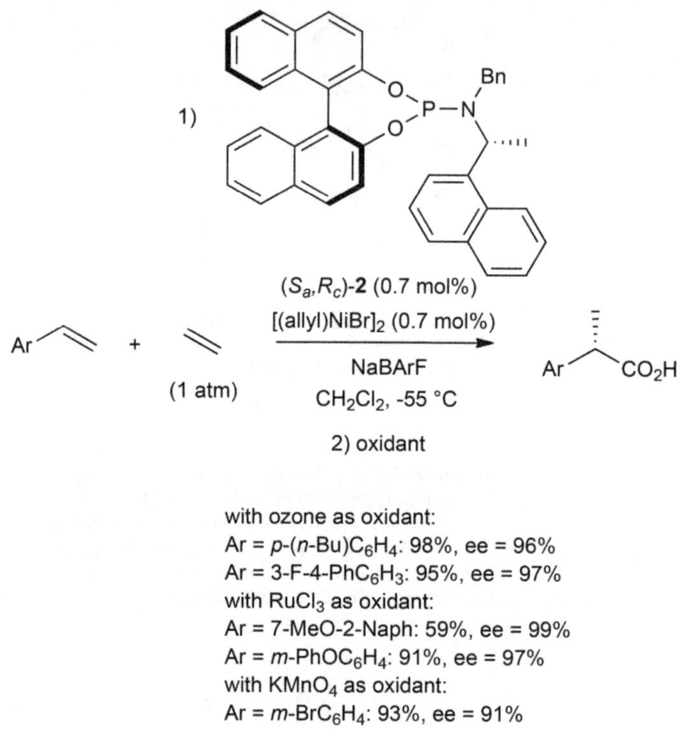

Scheme 5.3 Asymmetric synthesis of anti-inflammatory agents, 2-arylpropionic acids.

vinylarenes in nearly enantiomerically pure form.[10] As shown in Scheme 5.3, the same methodology as that described in Scheme 5.2, albeit using enantiomer ligand (S_a,R_c)-**2**, provided the corresponding (*R*)-3-arylbutenes, which were subsequently submitted to oxidative degradation by treatment with ozone or $RuCl_3/NaIO_4$ to give final (*S*)-2-arylpropionic acids, including ibuprofen, naproxen, flurbiprofen, and fenoprofen in enantioselectivities of 91–99% ee combined with high yields.

In 2007, Leitner *et al.* reported the synthesis of novel monodentate Quinaphos phosphoramidites bearing different substituents in the 2-position of the 1,2-dihydroquinoline backbone.[11] These ligands were further investigated in the enantioselective nickel-catalysed hydrovinylation of styrene in the presence of NaBArF. Among them, Cl-Quinaphos phosphoramidite (R_a,R_c)-**4**, bearing an *n*-butyl substituent in the 2-position of the 1,2-dihydroquinoline backbone, was demonstrated as optimal, providing the corresponding 3-phenylbut-1-ene in enantioselectivity of 91% ee and almost quantitative yield, as shown in Scheme 5.4.

In addition to the use of NaBArF, another method for activation of the nickel catalyst is the use of a Lewis acid (LA), which can abstract the halide ligand with formation of a corresponding adduct LAX acting as a weakly coordinating anion. In 2009, Leitner *et al.* investigated a range of available Lewis acids,

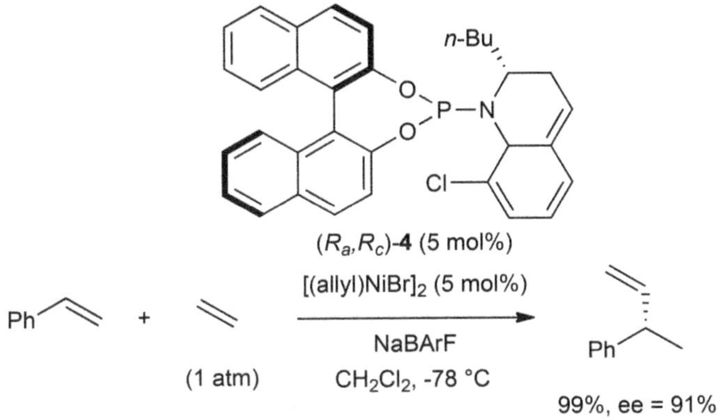

Scheme 5.4 Hydrovinylation of styrene with an *in situ* generated nickel catalyst from a Cl-Quinaphos ligand.

including chloride, bromide, and triflate salts of ytterbium, bismuth, lanthanum, indium, zinc, gadolinium, zinc, and scandium, for their ability to activate and regulate chiral nickel catalysts in asymmetric hydrovinylation processes using styrene as a model substrate and Feringa's ligand as the benchmark system.[12] The colour change during the activation step associated with the halide abstraction furnished helpful indications to adapt the activation conditions to the pre-catalyst/Lewis acid system. In general, metal halide Lewis acids led to higher activities and enantioselectivities than the corresponding triflates. In particular, the use of InI_3 as co-catalyst resulted in the same chemo- and enantioselectivity and even higher activity than the benchmark system based on NaBArF. As shown in Scheme 5.5, the asymmetric hydrovinylation of styrene with a combination of preformed chiral nickel catalyst derived from Feringa's ligand (R_a,S_c,S_c)-**3** and InI_3, both employed at a very low catalyst loading of 0.1 mol%, afforded the corresponding product in 82% yield and 92% ee. InI_3 presented the advantage to be safe to handle and cheap, thus providing a simple and practical novel protocol for an efficient hydrovinylation reaction.

These authors also investigated a range of novel chiral phosphorus triamides based on (*S*)-*N*-(pyrrolidin-2-ylmethyl)aniline as ligand in the presence of NaBArF in the same reaction; however, only moderate enantioselectivities of up to 60% ee were achieved.[13] On the other hand, much better enantioselectivities of up to >99% ee were reached by using 0.4 mol% of a preformed chiral nickel catalyst of a novel phosphoramidate (R_a,S_c,S_c)-**5** derived from Feringa's ligand and bearing four naphthyl groups.[14] As shown in Scheme 5.6, when this catalyst was activated by NaBArF, it allowed products to be obtained from the corresponding styrenes bearing electron-rich and electron-poor substituents in quantitative yields and excellent enantioselectivities of from 96 to >99% ee. The authors found that comparable excellent results were also obtained by using InI_3 instead of NaBArF as catalyst activator. Studying other ligands of the same type, the authors demonstrated that the steric bulk and hemilabile interaction of the naphthyl groups at the

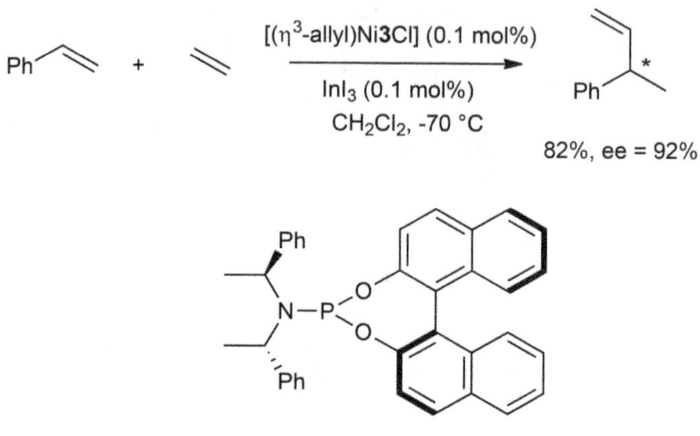

(R_a,S_c,S_c)-**3**

Scheme 5.5 Hydrovinylation of styrene with a preformed phosphoramidite chiral nickel catalyst activated by InI_3.

α-position to the nitrogen strongly controlled the efficacy of the ligands. Remarkably, these reactions could be performed for the first time at room temperature, providing slightly lower enantioselectivities of 90–95% ee with still excellent yields (Scheme 5.6).

In 2013, Francio *et al.* reported the synthesis of novel phosphoramidite and phosphorodiamidite ligands derived from (S_a)-2′-hydroxy-2-(phenylamino)-1,1′-binaphthyl (*N*-Ph-NOBIN) and bis(1-phenylethyl)amine.[15] These novel ligands were fully characterised, and the absolute configuration of the stereogenic phosphorus atoms assigned. Among these ligands, phosphoramidite ligand **6** featured three non-bridged substituents at phosphorus, comprising the bis(1-phenylethyl)amine and two NOBIN moieties. The NOBIN units were bound to the phosphorus through the oxygen atoms with two pendant nitrogen atoms. In the enantioselective nickel-catalysed hydrovinylation of styrene, no conversion was observed with the phosphorodiamidites, while the phosphoramidite ligands led to active catalysts with a marked cooperative effect on selectivities. Whereas the racemic product was obtained with the (S_a,S_a,S_c,S_c) diastereomer, the (S_a,S_a,R_c,R_c) diastereomer **6** proved to be one of the best ligands for this reaction, leading to almost perfect selectivity, with enantioselectivities of up to 91% ee and a high yield of 96%, as shown in Scheme 5.7. The striking difference in the enantioselectivity obtained with these two diastereomeric ligands showed the importance of cooperative effects between the various elements within the ligand structure. An intervention of the free NH functionalities during the catalytic cycle could not be excluded at this stage.

Despite impressive progress that has been achieved in the asymmetric hydrovinylation of styrene and its derivatives, the asymmetric hydrovinylation of (α-alkylvinyl)arenes, which has a potential for being a novel methodology for the construction of chiral all-carbon quaternary centres, has not been well documented. In this context, Zhou *et al.* reported in 2006 the first

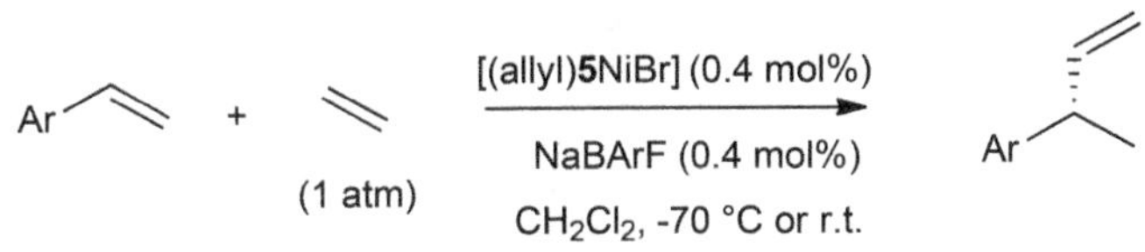

at -70 °C:
Ar = Ph: 90%, ee = 99%
Ar = 7-MeO-2-Naph: > 99%, ee > 99%
Ar = *p*-(*i*-Bu)-C_6H_4: > 99%, ee = 96%
Ar = *p*-BrC_6H_4: 99%, ee = 98%
at r.t.:
Ar = Ph: 98%, ee = 93%
Ar = 7-MeO-2-Naph: > 99%, ee = 95%
Ar = *p*-(*i*-Bu)-C_6H_4: > 99%, ee = 90%
Ar = *p*-BrC_6H_4: 93%, ee = 92%

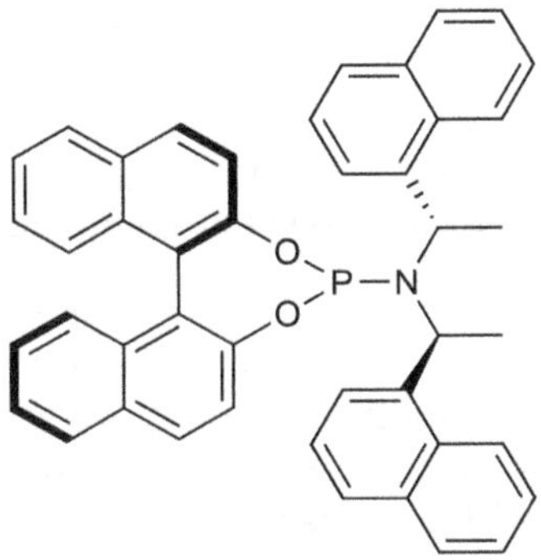

(R_a,S_c,S_c)-**5**

Scheme 5.6 Hydrovinylation of styrenes with a preformed phosphoramidite chiral nickel catalyst.

highly enantioselective hydrovinylation of (α-alkylvinyl)arenes using chiral spiro phosphoramidite ligands, providing a new efficient access to the construction of chiral all-carbon quaternary centres.[16] Among a series of chiral spiro monophosphorus ligands previously reported by these authors, ligand (S_a,R_c,R_c)-7 was found optimal in the enantioselective nickel-catalysed hydrovinylation of α-isopropylstyrene, providing the corresponding product in 75% yield and with an excellent enantioselectivity of 99% ee. As shown in Scheme 5.8, the scope of the process could be extended to a range of (α-alkylvinyl)arenes, providing new chiral arenes all bearing an all-carbon quaternary stereocentre. The reaction of α-alkylstyrenes without an electron-deficient group at a *para* or *meta* position on the phenyl ring gave almost quantitative conversion of substrates. However, in the case of 4-chloro-α-isopropylstyrene the electron-withdrawing chlorine substituent led to a lower yield (65%). The *ortho*-substituted α-alkylstyrenes, such as α-isopropyl-2-methylstyrene and α-ethyl-2-methylstyrene, were also examined under the same reaction conditions, but no reaction was observed, indicating that the steric hindrance of the substrate had a remarkable negative effect on the reactivity. Moreover, there was a high correlation between the enantioselectivity

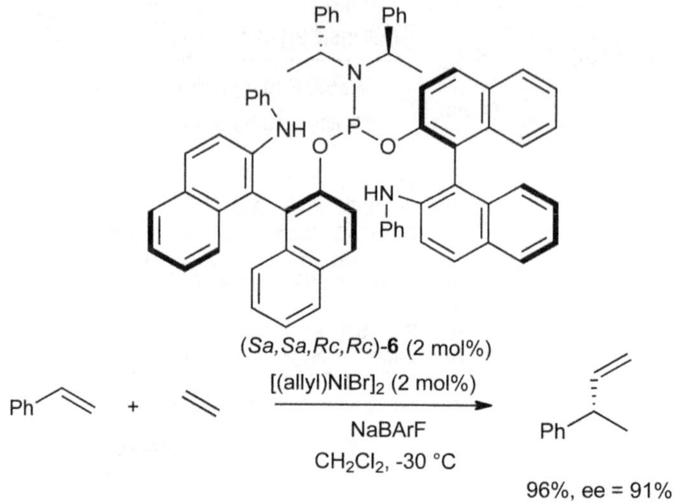

Scheme 5.7 Hydrovinylation of styrene with an *in situ* generated nickel catalyst from a NOBIN-based phosphorodiamidite ligand.

of the reaction and the size of the α-alkyl group in the substrates. Indeed, when the ethyl in the substrate was changed to *n*-propyl, isobutyl, or isopropyl, the enantioselectivity excess values of the corresponding hydrovinylation products increased successively from 70 to 82, 88, and 99% ee. Generally, all the substrates with an α-isopropyl or α-cyclohexyl group gave excellent enantioselectivities of 94–99% ee, demonstrating that a bulky alkyl group at the α-position of the vinylarenes was definitely necessary for obtaining chiral all-carbon quaternary centres in high enantioselectivity.

These reactions were also investigated by RajanBabu *et al.* by using a nickel catalyst generated *in situ* from Feringa's ligand (R_a,S_c,S_c)-**3** and [(allyl)NiBr]$_2$ in the presence of NaBArF.[17] As shown in Scheme 5.9, a range of chiral products from the corresponding (α-alkylvinyl)arenes were produced in moderate to high yields and with enantioselectivities of up to >95% ee. Interestingly, the scope of the reaction could be extended to tetralin derivatives bearing an exomethylene group, which underwent the hydrovinylation reaction with >99% ee (Scheme 5.9). Even an oxygenated derivative afforded the corresponding product in 94% ee. The hydrovinylation products which bore two versatile latent functionalities, an aryl and a vinyl group, are potentially useful for the synthesis of several important natural products containing benzylic all-carbon quaternary centres. For example, some chiral derivatives derived from the tetralin derivatives (second equation, X = CH_2, R = H, OMe) have already been used as intermediates in the syntheses of the analgesic (−)-eptazocine, the narcotic (−)-aphanorphine, and related compounds.[18]

The asymmetric hydrovinylation of functionalised alkenes remained a challenge until the work reported by Zhou *et al.* in 2010, in which chiral spiro phosphoramidite (S_a,R,R)-**7** previously reported by these authors was applied to the asymmetric hydrovinylation of silyl-protected allylic alcohols.[19] A range

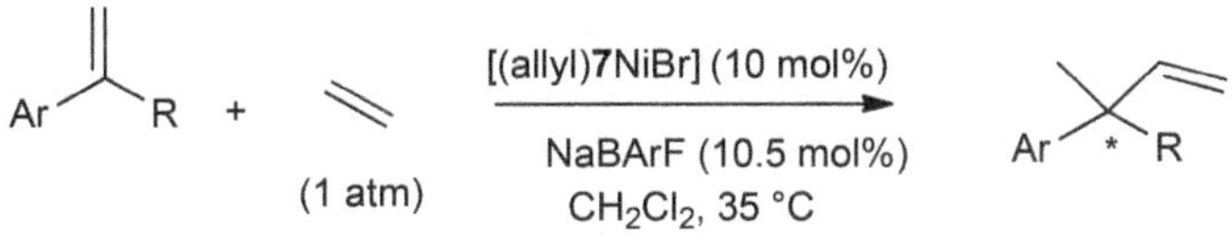

Ar = Ph, R = *i*-Pr: 75%, ee = 99% (*S*)
Ar = Ph, R = *n*-Pr: 77%, ee = 82%
Ar = Ph, R = Et: 71%, ee = 70% (*S*)
Ar = *p*-Tol, R = *i*-Pr: 85%, ee = 97%
Ar = *m*-Tol, R = *i*-Pr: 85%, ee = 98%
Ar = *m*-MeOC_6H_4, R = *i*-Pr: 80%, ee = 98%
Ar = *p*-MeOC_6H_4, R = *i*-Pr: 81%, ee = 98%
Ar = *p*-ClC_6H_4, R = *i*-Pr: 65%, ee = 94%
Ar = Ph, R = Et: 71%, ee = 70% (*S*)
Ar = Ph, R = *n*-Pr: 77%, ee = 82%
Ar = Ph, R = *i*-Bu: 79%, ee = 88%
Ar = 2-Naph, R = *i*-Pr: 75%, ee = 99%
Ar = 2-Naph, R = Cy: 79%, ee = 99%

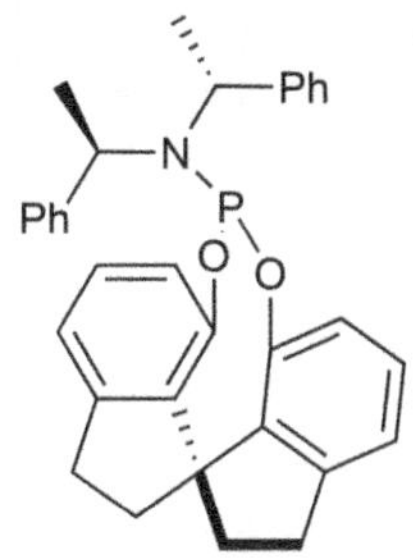

(S_a,R_c,R_c)-**7**

Scheme 5.8 Hydrovinylation of (α-alkylvinyl)arenes with a preformed spiro phosphoramidite chiral nickel catalyst.

of chiral homoallylic alcohols bearing a chiral quaternary carbon centre was synthesised in this way in moderate to high yields and with enantioselectivities of up to 95% ee, as shown in Scheme 5.10. Indeed, all substrates with a *meta* or a *para* substituent on the phenyl ring smoothly underwent the reaction and produced the products in good yields (42–97%) and with good to high enantioselectivities of 76–95% ee. On the other hand, the *ortho* substitution on the phenyl ring of the substrate fully prohibited the reaction under standard conditions. In addition to styrene derivatives, the substrates containing naphthyl and thiophenyl groups could also undergo the hydrovinylation reaction, with high yields and good enantioselectivities (73–81% ee). It must be noted that the double bond and the hydroxyl group present in the chiral products of this hydrovinylation reaction provided potential for conversion to various chiral bifunctional compounds, which are significant intermediates in the synthesis of natural products and pharmaceuticals.

Since the first enantioselective nickel-catalysed hydrovinylation of cyclohexa-1,3-diene reported by Buono *et al.* in 1985, which was based on the use

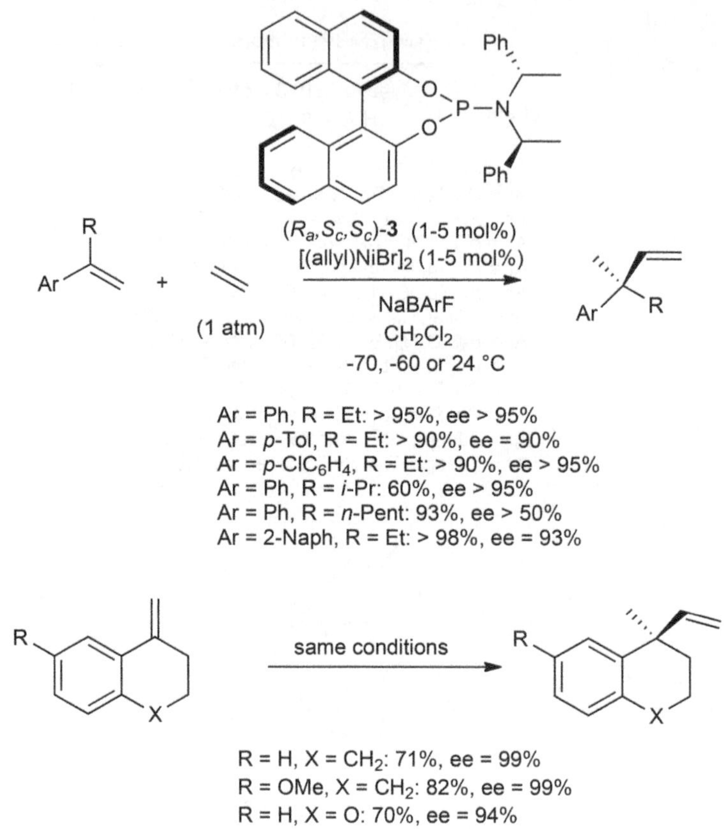

Scheme 5.9 Hydrovinylations of (α-alkylvinyl)arenes with an *in situ* generated nickel catalyst derived from Feringa's phosphoramidite ligand.

of a chiral aminophosphine/phosphinite ligand,[4] the asymmetric nickel-catalysed hydrovinylation of 1,3-dienes has been rarely investigated. As an early example, RajanBabu and Zhang reported the use of Feringa's phosphoramidite ligand (R_a,S_c,S_c)-**3** in the enantioselective nickel-catalysed hydrovinylation of various cyclic 1,3-dienes.[20] As shown in Scheme 5.11, the corresponding chiral products were regioselectively achieved in quantitative yields and with excellent enantioselectivities from 95 to >99% ee. This remarkable methodology using 3–6 mol% of catalyst loading applied to acyclic 1,3-dienes, however, provided only racemic products.

The above powerful methodology was applied to the synthesis of both steroid C_{20} (*S*) and C_{20} (*R*) derivatives from the corresponding 1,3-dienes **8** and **9** derived from two prototypical steroids, estrone and 3-epiandrosterone.[21] By the proper choice of ligands between Feringa's phosphoramidite ligand (S_a,R_c,R_c)-**3** and another phosphoramidite ligand (*S*,*S*)-**10**, the authors showed that it was possible to install with complete stereoselectivity either the C_{20} (*R*) or C_{20} (*S*) configuration. As shown in Scheme 5.12, the nickel-catalysed

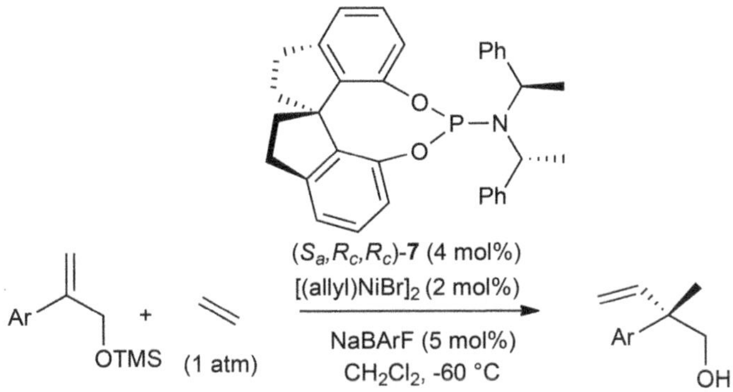

Ar = Ph: 77%, ee = 81%
Ar = *p*-Tol: 91%, ee = 78%
Ar = *m*-$MeOC_6H_4$: 97%, ee = 85%
Ar = *p*-FC_6H_4: 80%, ee = 94%
Ar = *p*-ClC_6H_4: 69%, ee = 92%
Ar = *p*-BrC_6H_4: 62%, ee = 95%
Ar = *p*-PhC_6H_4: 82%, ee = 80%
Ar = *m*-$MeOC_6H_4$: 86%, ee = 76%
Ar = *m*-FC_6H_4: 49%, ee = 91%
Ar = *m*-ClC_6H_4: 42%, ee = 88%
Ar = 2-Naph: 92%, ee = 81%
Ar = 2-thienyl: 90%, ee = 73%

Scheme 5.10 Hydrovinylation of silyl-protected allylic alcohols with an *in situ* generated nickel catalyst from a spiro phosphoramidite ligand.

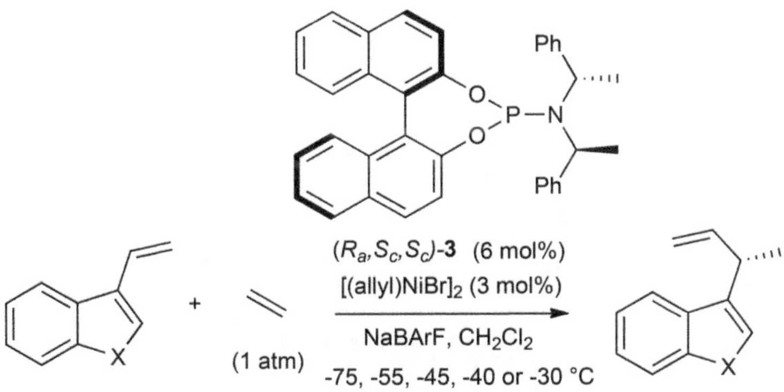

X = CH_2: > 99%, ee = 96%
X = $(CH_2)_2$: > 99%, ee > 99%
X = OCH_2: > 99%, ee = 95%

Scheme 5.11 Hydrovinylation of 1,3-dienes with an *in situ* generated nickel catalyst derived from Feringa's phosphoramidite ligand.

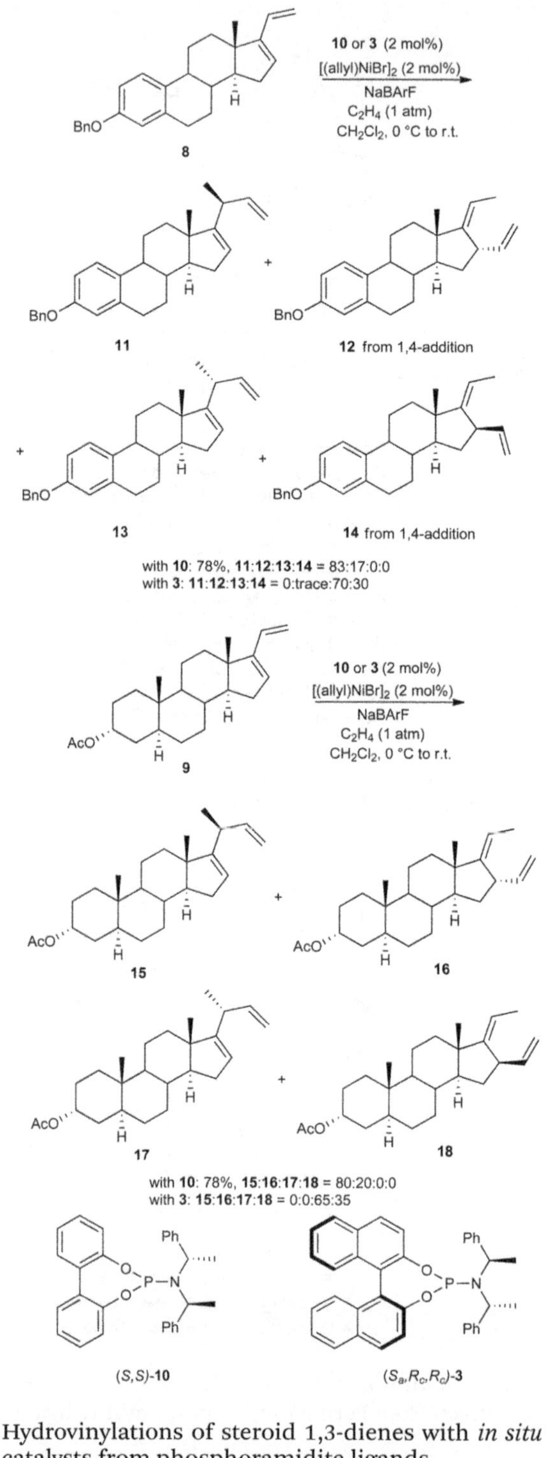

Scheme 5.12 Hydrovinylations of steroid 1,3-dienes with *in situ* generated nickel catalysts from phosphoramidite ligands.

hydrovinylation of 1,3-diene **8** derived from estrone performed in the presence of ligand (*S*,*S*)-**10** afforded the highest proportion of the 1,2-adduct **11** bearing the C_{20} (*S*) configuration (1,2-adduct **11**/1,4-adduct **12** = 83 : 17), while using Feringa's ligand gave (*R*)-1,2-adduct **13** as the major product (1,2-adduct **13**/1,4-adduct **14** = 70 : 30). Comparable regio- and stereoselectivities were obtained in the hydrovinylation of diene **15** derived from 3-acetylepiandrosterone, since the use of ligand **10** afforded an 80 : 20 mixture of 1,2-adduct **15** bearing the C_{20} (*S*) configuration as the major product along with the 1,4-adduct **16**, while using Feringa's ligand led to a 65 : 35 mixture of major 1,2-adduct **17** with the C_{20} (*R*) configuration along with 1,4-adduct **18**. This novel stereoselective ligand-dependent protocol allowed the installation of exocyclic stereocentres in a steroid D-ring *via* asymmetric hydrovinylation. Feringa's phosphoramidite (S_a,R_c,R_c)-**3** gave exclusively the (*R*)-hydrovinylation products, whereas phosphoramidite ligand (*S*,*S*)-**10** provided exclusively the (*S*)-hydrovinylation products.

In 2010, RajanBabu and Liu extended the scope of the asymmetric hydrovinylation reaction, performed with a nickel catalyst derived from Feringa's ligand, to strained alkenes.[22] For example, the reaction of a norbornene derivative was demonstrated to give the corresponding product in both high yield (93%) and with enantioselectivity of up to 99% ee, as shown in Scheme 5.13. A series of monophosphine and phosphoramidite ligands were screened for this reaction, demonstrating that Feringa's ligand was the most efficient. The scope of this process was extended to other strained alkenes, such as heterobicyclo[2.2.1]heptenes and cyclobutenes (Scheme 5.13). Reactions involving heterobicyclo[2.2.1]heptene compounds constituted rare examples for this class of substrates where the metal-catalysed C–C bond-forming reactions proceeded without a concomitant ring-opening process. While the enantioselectivity in these systems remained modest (87% ee), they led to products that could be potentially transformed into highly substituted key chiral cyclohexane derivatives. In the case of the cyclobutene derivative, the highest enantioselectivity (82% ee) was reached by using chiral phosphoramidite ligand (*S*,*S*)-**10**.

5.3 Hydrophosphinations

Chiral phosphines are valuable as ligands in numerous important catalytic asymmetric processes.[23] Intense research efforts have produced many classes of chiral phosphines with diverse steric and electronic properties, and researchers have examined how these properties influence activity and stereoselectivity in catalysis. Given the rapid development of new methods in asymmetric catalysis and the prominence of chiral phosphines in this area of chemistry, it is surprising that few catalytic asymmetric syntheses of chiral phosphines have been described.[24] Instead, enantiopure phosphines are most commonly prepared either *via* stereospecific reactions of resolved starting materials or through routes which require an additional resolution step, such as fractional crystallisation of diastereomers. On the other

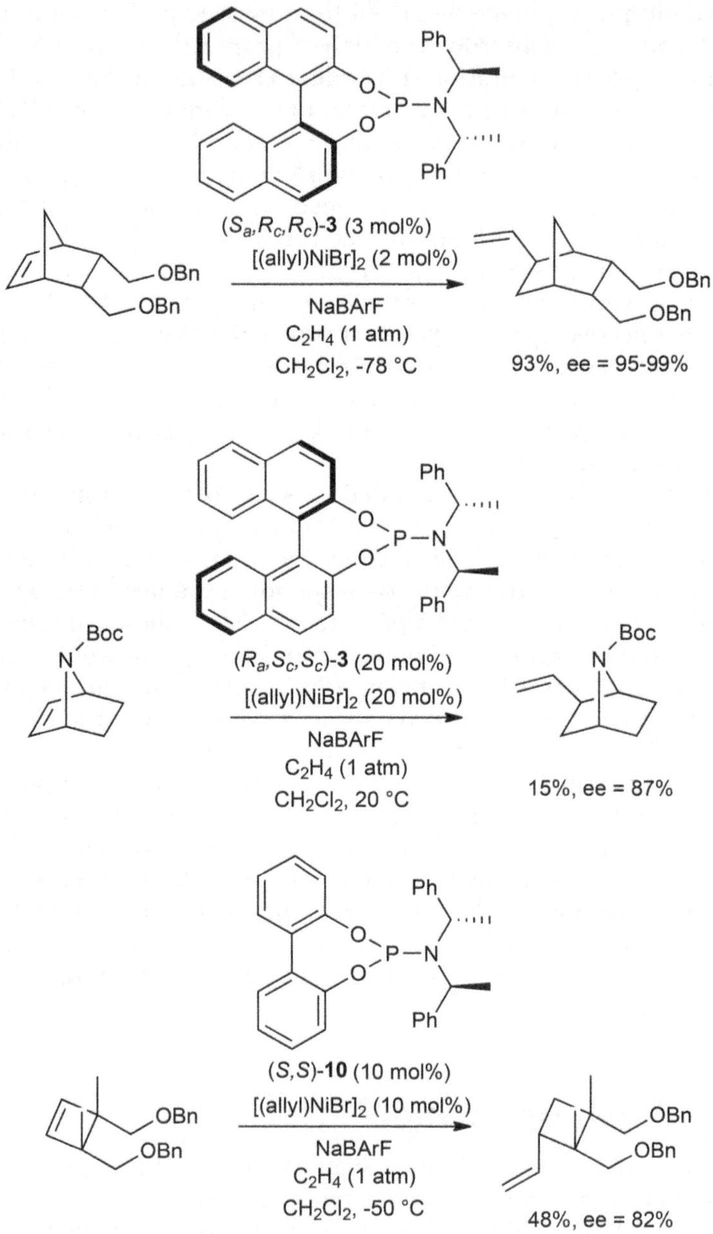

Scheme 5.13 Hydrovinylations of strained alkenes with *in situ* generated nickel catalysts from phosphoramidite ligands.

hand, catalytic asymmetric hydrophosphination of alkenes is potentially an alternative and efficient synthetic route to chiral phosphines.[25] Transition metals catalyse this reaction with high selectivities.[26] In 2000, a platinum complex of MeDUPHOS was used to catalyse the asymmetric hydrophosphination of methacrylonitrile but with low enantioselectivity of 27% ee.[27] Later,

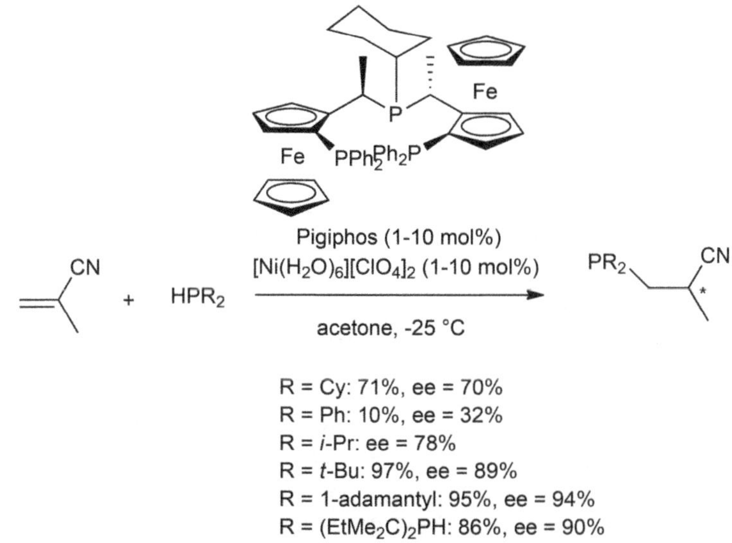

Scheme 5.14 Hydrophosphination of methacrylonitrile with an *in situ* generated nickel catalyst derived from the trisphosphine Pigiphos ligand.

better enantioselectivities were achieved in lanthanide-catalysed intramolecular phosphinations.[28] In 2004, Togni *et al.* reported a novel method for the preparation of a series of chiral (2-cyanopropyl)phosphines based on the first nickel-catalysed asymmetric hydrophosphination of methacrylonitrile with various secondary phosphines.[29] The authors employed a nickel catalyst generated *in situ* from $[Ni(H_2O)_6][ClO_4]_2$ and the C_1-symmetric trisphosphine Pigiphos. The reaction was performed at −25 °C in acetone, providing the corresponding hydrophosphination products in low to excellent yields (10–97%) and with enantioselectivities of 32–94% ee, as shown in Scheme 5.14. Actually, this process could also be situated in Chapter 2 dealing with enantioselective nickel-catalysed conjugate additions since the alkene is a Michael acceptor. The results showed that bulky dialkylphosphines gave more promising results while diphenylphosphine provided the lowest enantioselectivity (32% ee). This suggested a pathway involving the coordination of methacrylonitrile to the nickel centre followed by nucleophilic attack of the secondary phosphine.[30] On the other hand, primary phosphines, such as $CyPH_2$ and $(1\text{-}MeCy)PH_2$, did not lead to isolable hydrophosphination products. It must be noted that this work constituted the first highly enantioselective catalytic hydrophosphination reaction.

5.4 Hydrocyanations

A regioselective, Markovnikov addition of HCN to alk-1-enes constitutes an efficient way of creating stereogenic centres in the carbon–carbon bond-forming process. The products of this reaction are nitriles, which are potentially very versatile building blocks that can be used as precursors for amines, isocyanates, amides, carboxylic acids, esters, and N-heterocycles.[31] Therefore, the development of enantioselective versions of hydrocyanation reactions is of high value in organic synthesis.[32] Despite the apparent simplicity of addition reactions to alkenes, it is still an important challenge to control the regio- and stereoselectivity of such processes. Whereas branched nitriles are especially interesting for fine chemical applications, linear nitriles are the desired products for most large-scale targets, such as polymers. Owing to the relative low costs of these products, the corresponding production processes have to be highly efficient, with low catalyst costs but high yields and stereoselectivity. However, the difficulties in handling the highly toxic and volatile HCN are often regarded as a serious drawback. In fact, most of the literature on this field has been reported by industrial rather than by academic laboratories. The first report dealing with a homogeneous catalysed HCN addition to non-functionalised alkenes involved a cobalt catalyst and was reported by Arthur in 1954.[33] The general possibility to perform hydrocyanations enantioselectively in the presence of a chiral ligand was first demonstrated by Elmes and Jackson in the palladium-catalysed hydrocyanation of norbornene, achieved in 30% ee.[34] Since then, only a few reports on asymmetric hydrocyanation have appeared in the literature.[35] In 1992, RajanBabu and Casalnuovo succeeded in applying the nickel-catalysed hydrocyanation of vinylarenes in the synthesis of branched products.[35c] After optimisation, these authors achieved the nickel-catalysed hydrocyanation of 2-methoxy-6-vinylnaphthalene with high enantioselectivity of up to 95% ee, employing a carbohydrate-derived diphosphinite ligand.[35e] Several subsequent results have been reported, but they did not provide enantioselectivities >80% ee.[35h,i] For example, inspired by their early work reported in the 1990s and based on the use of glucose- and fructose-derived bidentate chiral phosphinite ligands in asymmetric nickel-catalysed hydrocyanation of styrenes with enantioselectivities of 56–95% ee,[35d,e] RajanBabu *et al.* later reported the asymmetric hydrocyanation of 1,3-dienes by using bis-1,2-diarylphosphinites **19** and **20** derived from D-glucose as nickel ligands.[36] As shown in Scheme 5.15, the reaction of 1-phenylbuta-1,3-diene led to the corresponding 1,2-addition product using ligand **20** at 3 mol% of catalyst loading in combination with the same quantity of $Ni(cod)_2$. This chiral product was achieved in good yield (87%) and with enantioselectivity of 78% ee. Other acyclic and cyclic 1,3-dienes led to the corresponding products in the presence of this ligand or ligand **19**, but with lower enantioselectivities (68–75% ee), as shown in Scheme 5.15.

The enantioselective hydrocyanation of cyclohexa-1,3-diene was investigated by Vogt *et al.* using a nickel catalyst generated *in situ* from $Ni(cod)_2$ and chiral diphosphite ligand **21** (Scheme 5.16).[37] The reaction resulted in the formation of cyclohex-2-ene-1-carbonitrile, which could arise from both 1,2- and 1,4-addition. By using DCN instead of HCN, the authors demonstrated that the product of this reaction actually resulted from an equal 1,2-/1,4-product distribution. The product was achieved in a moderate yield (45%) but with good enantioselectivity of 86% ee.

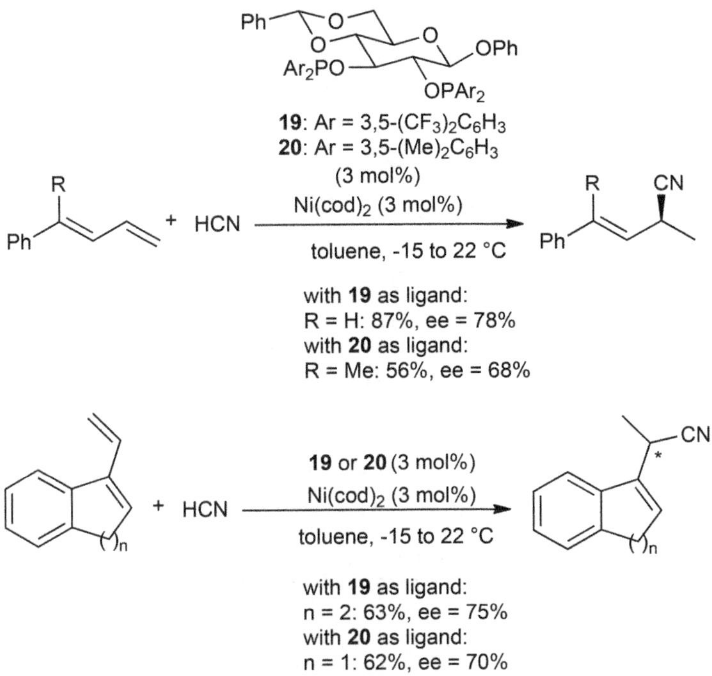

Scheme 5.15 Hydrocyanations of various 1,3-dienes with *in situ* generated nickel catalysts from D-glucose-derived diaryldiphosphinite ligands.

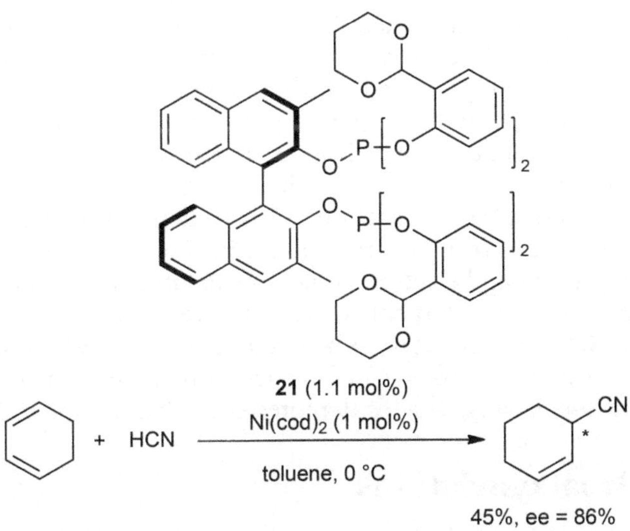

Scheme 5.16 Hydrocyanation of cyclohexa-1,3-diene with an *in situ* generated nickel catalyst derived from a diphosphite ligand.

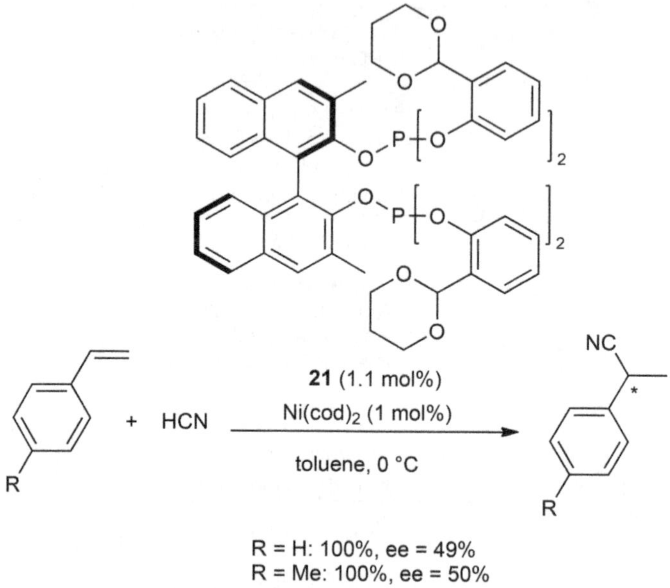

Scheme 5.17 Hydrocyanation of styrenes with an *in situ* generated nickel catalyst derived from a diphosphite ligand.

The same reaction conditions were applied by these authors to the enantioselective hydrocyanation of styrenes, which afforded the corresponding products in quantitative yields, albeit with moderate enantioselectivities (49–50% ee), as shown in Scheme 5.17.[38] The steric parameter seemed to be equally important as the electronic effect in these reactions.

More recently, Schmalz *et al.* identified a superior chiral ligand for nickel-catalysed asymmetric hydrocyanations of vinylarenes.[39] As shown in Scheme 5.18, the use of TADDOL-derived phosphine/phosphite ligand **22** allowed these reactions to be achieved in good yields and with enantioselectivities of up to 92% ee. The reaction also had the advantage that the handling of toxic HCN could be circumvented by employing TMSCN as a safe reagent in the presence of methanol. Various substituted arylalkenes were compatible with this process, since good results were obtained in almost all cases of substrates. Electron-withdrawing and electron-donating substituents were tolerated, and even styrenes with a β-substituent could be reacted, with enantioselectivities of 67–84% ee (Scheme 5.18). It must be noted that this novel protocol opened a reliable and scalable access to a broad spectrum of chiral nitriles with high levels of enantioselectivity, which still represents a challenging task.

5.5 Hydroalkynylations

Much interest has been focused on the catalytic, direct conversion of alkyne C–H bonds through C–C bond-forming reactions without the stoichiometric generation of acetylides.[40] One of the most widely used procedures for such an

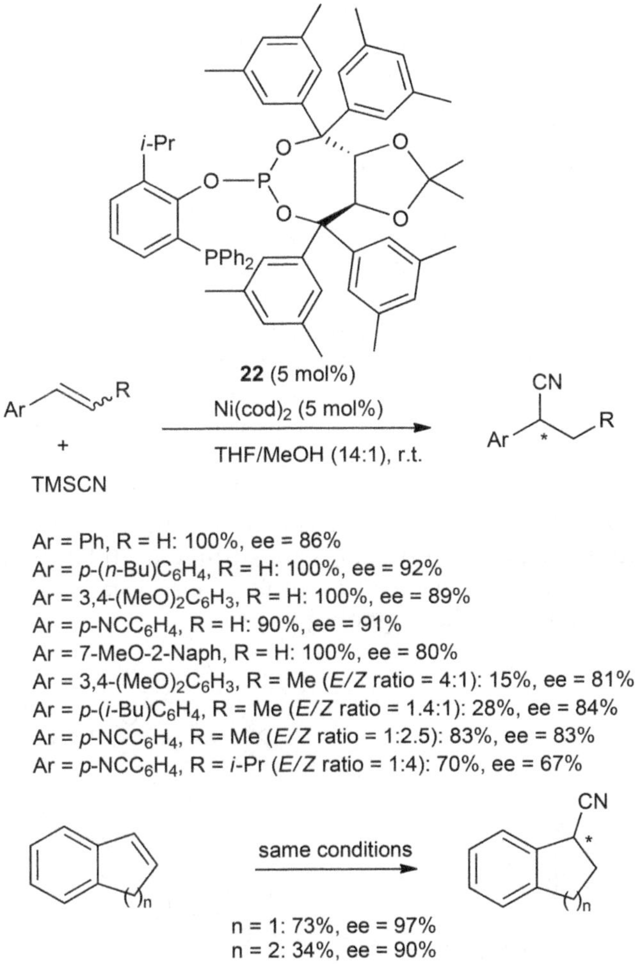

Scheme 5.18 Hydrocyanations of arylalkenes with an *in situ* generated nickel catalyst derived from a phosphine/phosphite ligand.

atom-economical process is the nucleophilic alkynylation of carbonyl compounds, α,β-unsaturated carbonyl compounds, or related electrophiles, in which catalytically generated metal acetylides often play a key role.[41] Recent attention has focused on the development of asymmetric variants of these nucleophilic alkynylation reactions for the synthesis of highly functionalised chiral alkyne derivatives.[42] Besides these nucleophilic alkynylation reactions, hydroalkynylation, *i.e.* the addition of alkyne C–H bonds across unactivated carbon–carbon multiple bonds, has attracted increasing attention.[43] After extensive studies on the homo- and cross-dimerisation reactions of alkynes using rhodium, palladium, and nickel catalysts,[44] hydroalkynylation has been extended to carbon–carbon double bonds, such as those in allenes and cyclopropenes.[45] However, the scope of the hydroalkynylation reaction is still significantly limited.[45b–d] As a consequence, no successful catalytic asymmetric hydroalkynylation reactions had been established until 2010, except for the rhodium-catalysed hydroalkynylation of allenes.[46] However, Shirakura and Suginome reported in 2010 the first nickel-catalysed asymmetric hydroalkynylation reaction of 1-arylbuta-1,3-dienes, which involved the use of TADDOL-derived phosphoramidite ligands.[47] In addition to the significance of these chiral ligands, the use of a terminal alkyne that contained an α-siloxy-*sec*-alkyl group on the alkynyl carbon was important to achieve sufficient reaction efficiency. As shown in Scheme 5.19, the reaction of alkyne **23** with *trans*-1-phenylbuta-1,3-diene (Ar = Ph) was carried out in the presence of a combination of $Ni(cod)_2$ and chiral ligand **24** selected as optimal, and afforded the corresponding hydroalkynylation product in 60% yield and with high enantioselectivity of 91% ee. A range of *trans*-1-arylbuta-1,3-dienes variously substituted on the aromatic ring, as well as 1-naphthyl derivatives, were subjected to the reaction, providing the corresponding functionalised chiral products in moderate yields (41–68%) but with very good enantioselectivities of 90–93% ee. The slow addition of the alkyne allowed minimising its dimerisation. The formed products constituted useful building blocks in asymmetric organic synthesis. For example, the siloxyalkyl group could be easily transformed into other organic groups.

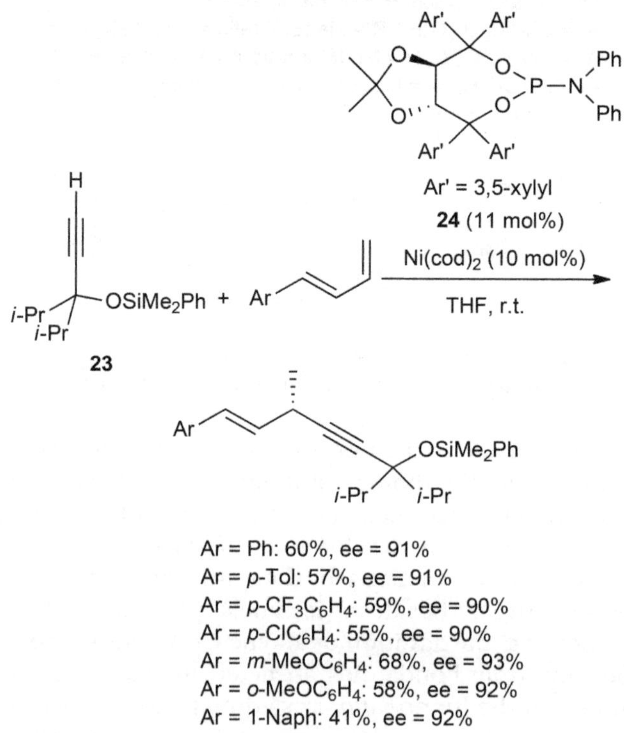

Scheme 5.19 Hydroalkynylation of 1-arylbuta-1,3-dienes with an *in situ* generated nickel catalyst derived from a phosphoramidite ligand.

5.6 Conclusions

The last decade has known many important advances in the field of asymmetric nickel-catalysed hydrovinylations of alkenes, probably related to the fact that nickel complexes have been the main catalysts in hydrovinylation reactions. All these novel procedures have used very low catalyst loadings of highly versatile phosphoramidite ligands derived from Feringa's ligand, with NaBArF as catalyst activator, providing excellent enantioselectivities in all cases. For example, enantioselectivities of up to 99% ee were achieved in the asymmetric hydrovinylation of a range of vinylarenes to give the corresponding 3-arylbutenes. Moreover, when this reaction was followed by an oxidative degradation, the sequence offered a novel route to important anti-inflammatory chiral 2-arylpropionic acids, such as naproxen, ibuprofen, fenoprofen, and flurbiprofen. Comparable excellent enantioselectivities of up to 99% ee were also achieved in the case of other substrates, such as (α-alkylvinyl)arenes, as well as cyclic 1,3-dienes. It is important to note that the hydrovinylation of (α-alkylvinyl)arenes provided a new efficient access to the construction of chiral all-carbon quaternary centres. Furthermore, functionalised alkenes, such as silyl-protected allylic alcohols, could also be submitted to asymmetric hydrovinylation with high enantioselectivities of up to 95% ee. Even higher enantioselectivities of up to 99% ee were reached for the nickel-catalysed hydrovinylation of strained alkenes, such as norbornenes, while slightly lower enantioselectivities of up to 87% ee were obtained in the hydrovinylation of strained heterobicyclo[2.2.1]heptenes and cyclobutenes.

In the area of nickel-catalysed asymmetric hydrophosphination of alkenes, the first highly enantioselective reaction was developed by using the C_1-symmetric trisphosphine Pigiphos, which provided very good enantioselectivities of up to 94% ee. In addition, a highly efficient enantioselective nickel-catalysed hydrocyanation of arylalkenes was performed with enantioselectivities of up to 92% ee by employing a TADDOL-derived phosphine/phosphite ligand. Finally, the first nickel-catalysed hydroalkynylation of 1-arylbuta-1,3-dienes was achieved by using a phosphoramidite ligand, providing up to 93% ee.

All the formed chiral products from hydrovinylation, hydrophosphination, hydrocyanation, and hydroalkynylation reactions of alkenes constitute useful building blocks in asymmetric organic synthesis. Thus, the resulting vinyl group in the products of hydrovinylation reactions can be readily transformed into a variety of other common functional groups. On the other hand, given the rapid development of new methods in asymmetric catalysis and the prominence of chiral phosphines in this area of chemistry, the utility of asymmetric nickel-catalysed hydrophosphination is obvious. Moreover, chiral nitriles derived from asymmetric nickel-catalysed hydrocyanations of alkenes are potentially very versatile building blocks which can be used as precursors for amines, isocyanates, amides, carboxylic acids, esters, and N-heterocycles. In addition, the potential applicability of the enantioselective nickel-catalysed hydroalkynylation of 1-arylbuta-1,3-dienes in asymmetric organic synthesis is obvious since it afforded multifunctionalised chiral products. One can reasonably anticipate that future studies will provide new applications of these chiral products in the total synthesis of natural products and biologically active compounds. Efforts will have to be made in developing new catalytic systems through ligand design to perform these reactions, applying the already known ligands, and expansion of the scope of the reactions.

References

1. (a) G. Francio, W. Leitner and P. L. Alsters, in *Science of Synthesis*, Thieme, Stuttgart, 2011, vol. 1, pp. 477–519; (b) D. Vogt, *Angew. Chem., Int. Ed.*, 2010, **49**, 7166–7168.
2. (a) T. V. RajanBabu, in *Asymmetric Synthesis II*, ed. M. Christmann and S. Brase, Wiley, Weinheim, 2012, pp. 293–301; (b) T. V. RajanBabu, in *Comprehensive Chirality*, Wiley, Hoboken, 2012, vol. 5, pp. 355–398; (c) G. Hilt, *Eur. J. Org. Chem.*, 2012, **24**, 4441–4451; (d) T. V. RajanBabu, *Chem. Rev.*, 2003, **103**, 2845–2860; (e) L. J. Gooβen, *Angew. Chem., Int. Ed.*, 2002, **41**, 3775–3778; (f) T. V. RajanBabu, N. Nomura, B. Radetich, H. Park and M. Nandi, *Chem.–Eur. J.*, 1999, **5**, 1963–1968.
3. G. Wilke and J. Monkiewicz, *Chem. Abstr.*, 1988, **109**, P6735.
4. G. Buono, C. Siv, G. Peiffer, C. Triantaphylides, P. Denis, A. Mortreux and F. Petit, *J. Org. Chem.*, 1985, **50**, 1781–1782.
5. (a) N. Nomura, J. Jin, H. Park and T. V. RajanBabu, *J. Am. Chem. Soc.*, 1998, **120**, 459–460; (b) A. Wegner and W. Leitner, *Chem. Commun.*, 1999, 1583–1584; (c) A. Bösmann, G. Francio, E. Janssen, M. Solinas, W. Leitner and P. Wasserscheid, *Angew. Chem., Int. Ed.*, 2001, **40**, 2697–2699; (d) G. Francio, F. Faraone and W. Leitner, *J. Am. Chem. Soc.*, 2002, **124**, 736–737; (e) H. Park and T. V. RajanBabu, *J. Am. Chem. Soc.*, 2002, **124**, 734–735; (f) R. Kumareswaran, M. Nandi and T. V. RajanBabu, *Org. Lett.*, 2003, **5**, 4345–4348.
6. H. Park, R. Kumareswaran and T. V. RajanBabu, *Tetrahedron*, 2005, **61**, 6352–6367.
7. A. Zhang and T. V. RajanBabu, *Org. Lett.*, 2004, **6**, 1515–1517.
8. B. L. Feringa, *Acc. Chem. Res.*, 2000, **33**, 346–353.
9. C. R. Smith and T. V. RajanBabu, *Org. Lett.*, 2008, **10**, 1657–1659.
10. C. R. Smith and T. V. RajanBabu, *J. Org. Chem.*, 2009, **74**, 3066–3072.
11. C. J. Diez-Holz, C. Böing, F. Giancarlo, M. Hölscher and W. Leitner, *Eur. J. Org. Chem.*, 2007, 2995–3002.
12. N. Lassauque, G. Francio and W. Leitner, *Eur. J. Org. Chem.*, 2009, 3199–3202.
13. K. Barta, M. Hölscher, G. Francio and W. Leitner, *Eur. J. Org. Chem.*, 2009, 4102–4116.
14. N. Lassauque, G. Francio and W. Leitner, *Adv. Synth. Catal.*, 2009, **351**, 3133–3138.

15. M. Schmitkamp, W. Leitner and G. Francio, *Catal. Sci. Technol.*, 2013, **3**, 589–594.
16. W.-J. Shi, Q. Zhang, J.-H. Xie, S.-F. Zhu, G.-H. Hou and Q.-L. Zhou, *J. Am. Chem. Soc.*, 2006, **128**, 2780–2781.
17. (a) A. Zhang and T. V. RajanBabu, *J. Am. Chem. Soc.*, 2006, **128**, 5620–5621; (b) C. R. Smith, H. J. Lim, A. Zhang and T. V. RajanBabu, *Synthesis*, 2009, **12**, 2089–2100.
18. A. N. Hulme, S. S. Henry and A. I. Meyers, *J. Org. Chem.*, 1995, **60**, 1265–1270.
19. Q. Zhang, S.-F. Zhu, Y. Cai, L.-X. Wang and Q.-L. Zhou, *Sci. China Chem.*, 2010, **53**, 1899–1906.
20. A. Zhang and T. V. RajanBabu, *J. Am. Chem. Soc.*, 2006, **128**, 54–55.
21. B. Saha, C. R. Smith and T. V. RajanBabu, *J. Am. Chem. Soc.*, 2008, **130**, 9000–9005.
22. W. Liu and T. V. RajanBabu, *J. Org. Chem.*, 2010, **75**, 7636–7643.
23. *Catalytic Asymmetric Synthesis*, ed. I. Ojima, VCH, New York, 1993.
24. (a) P. E. C. Corbridge, in *Studies in Inorganic Chemistry 20, Phosphorus: An outline of its Chemistry, Biochemistry and Technology*, Elsevier, Amsterdam, 5th edn, 1995; (b) C. Baillie and J. Xiao, *Curr. Org. Chem.*, 2003, **7**, 477–514; (c) D. S. Glueck, *Chem.–Eur. J.*, 2008, **14**, 7108–7117; (d) A. L. Reznichenko and K. C. Hultzsch, *Science of Synthesis*, Thieme, Stuttgart, 2011, vol. 1, pp. 689–729; (e) V. Koshti, S. Gaikwad and S. H. Chikkali, *Coord. Chem. Rev.*, 2014, **265**, 52–73.
25. M. Tanaka, *Top. Curr. Chem.*, 2004, **232**, 25–34.
26. (a) W. A. Nugent, T. V. RajanBabu and M. J. Burk, *Science*, 1993, **259**, 479–483; (b) M. Beller, A. Seayad, A. Tillack and H. Jiao, *Angew. Chem., Int. Ed.*, 2004, **43**, 3368–3398; (c) D. S. Glueck, *Synlett*, 2007, **17**, 2627–2634; (d) L. Bini, C. Müller and D. Vogt, *Chem. Commun.*, 2010, **46**, 8325–8334; (e) T. V. RajanBabu, Hydrocyanation of Alkenes and Alkynes, *Org. React.*, 2011, **75**, 1–73; (f) S. A. Pullarkat and P.-H. Leung, in *Top. Organomet. Chem.*, Springer, Heidelberg, 2013, vol. 43, pp. 145–166.
27. I. Kovacik, D. K. Wicht, N. S. Grewal, D. S. Glueck, C. D. Incarvito, I. A. Guzei and A. L. Rheingold, *Organometallics*, 2000, **19**, 950–953.
28. M. R. Douglass, M. Ogasawara, S. Hong, M. V. Metz and T. J. Marks, *Organometallics*, 2002, **21**, 283–292.
29. A. D. Sadow, I. Haller, L. Fadini and A. Togni, *J. Am. Chem. Soc.*, 2004, **126**, 14704–14705.
30. A. D. Sadow and A. Togni, *J. Am. Chem. Soc.*, 2005, **127**, 17012–17024.
31. (a) F. F. Fleming, L. Yao, P. C. Ravikumar, L. Funk and B. C. Shook, *Med. Chem.*, 2010, **53**, 7902–7917; (b) M. B. Smith and J. March, in *March's Advanced Organic Chemistry: Reactions and Mechanisms*, Wiley-Interscience, Hoboken, 6th edn, 2007.
32. L. Bini, C. Müller and D. Vogt, *ChemCatChem*, 2010, **2**, 590–608.
33. P. Arthur, D. C. England, B. C. Pratt and G. M. Whitman, *J. Am. Chem. Soc.*, 1954, **76**, 5364–5367.
34. P. S. Elmes and W. R. Jackson, *J. Am. Chem. Soc.*, 1979, **101**, 6128–6129.

35. (a) M. Hogson and D. Parker, *J. Organomet. Chem.*, 1987, **325**, C27–C30; (b) M. Hogson, D. Parker, R. J. Taylor and G. Ferguson, *Organometallics*, 1988, 7, 1761–1766; (c) M. J. Parker and P. G. Pringle, *J. Chem. Soc., Chem. Commun.*, 1991, 1292–1293; (d) T. V. RajanBabu and A. L. Casalnuovo, *J. Am. Chem. Soc.*, 1992, **114**, 6265–6266; (e) A. L. Casalnuovo, T. V. RajanBabu, T. A. Ayers and T. H. Warren, *J. Am. Chem. Soc.*, 1994, **116**, 9869–9882; (f) T. V. RajanBabu and A. L. Casalnuovo, *J. Am. Chem. Soc.*, 1996, **118**, 6325–6326; (g) T. Horiuchi, E. Shirakawa, K. Nozaki and H. Takaya, *Tetrahedron: Asymmetry*, 1997, **8**, 57–63; (h) A. L. Casalnuovo and T. V. RajanBabu, *Chirality in Industry II*, Wiley, New York, 1997, pp. 309–333; (i) T. V. RajanBabu and A. L. Casalnuovo, *Comprehensive Asymmetric Catalysis I–III*, Springer, Heidelberg, 1999, vol. 1, pp. 367–378; (j) M. Yan, Q.-Y. Xu and A. S. C. Chan, *Tetrahedron: Asymmetry*, 2000, **11**, 845–849; (k) W. Goertz, P. C. J. Kamer, P. W. N. M. van Leeuwen and D. Vogt, *Chem.–Eur. J.*, 2001, 7, 1614–1618; (l) N.-u. H. Khan, R. I. Kureshy, S. H. R. Abdi and S. Agrawal, *Coord. Chem. Rev.*, 2008, **252**, 593–623; (m) P. W. N. M. van Leewen, *Science of Synthesis*, Thieme, Stuttgart, 2011, vol. 1, pp. 409–475; (n) D. Vogt and J. Wilting, *Comprehensive Chirality*, Elsevier, Amsterdam, 2012, vol. 5, pp. 343–354; (o) P. I. Dalko, *Comprehensive Enantioselective Organocatalysis*, Wiley, Weinheim, 2013, vol. 3, pp. 873–897.
36. B. Saha and T. V. RajanBabu, *Org. Lett.*, 2006, **8**, 4657–4659.
37. J. Wilting, M. Janssen, C. Müller and D. Vogt, *J. Am. Chem. Soc.*, 2006, **128**, 11374–11375.
38. J. Wilting, M. Janssen, C. Müller, M. Lutz, A. L. Spek and D. Vogt, *Adv. Synth. Catal.*, 2007, **349**, 350–356.
39. A. Falk, A.-L. Göderz and H.-G. Schmalz, *Angew. Chem., Int. Ed.*, 2013, **52**, 1576–1580.
40. (a) D. E. Frantz, R. Fassler, C. S. Tomooka and E. M. Carreira, *Acc. Chem. Res.*, 2000, **33**, 373–381; (b) *Acetylene Chemistry*, ed. F. Diederich, P.-J. Stang and R. R. Tykwinski, Wiley-VCH, Weinheim, 2005; (c) C. Anaya de Parrodi and P. A. Walsh, *Angew. Chem., Int. Ed.*, 2009, **48**, 4679–4682.
41. (a) R. Motoki, M. Kanai and M. Shibasaki, *Org. Lett.*, 2007, **9**, 2997–3000; (b) C. Li, F. Mo, W. Li and J. Wang, *Tetrahedron Lett.*, 2009, **50**, 6053–6056.
42. (a) N. K. Anand and E. M. Carreira, *J. Am. Chem. Soc.*, 2001, **123**, 9687–9688; (b) C. Wei and C.-J. Li, *J. Am. Chem. Soc.*, 2002, **124**, 5638–5639; (c) R. Takita, K. Yakura, T. Oshima and M. Shibasaki, *J. Am. Chem. Soc.*, 2005, **127**, 13760–13761; (d) T. F. Knöpfel, P. Zarotti, T. Ichikawa and E. M. Carreira, *J. Am. Chem. Soc.*, 2005, **127**, 9682–9683; (e) Y. Asano, K. Hara, H. Ito and M. Sawamura, *Org. Lett.*, 2007, **9**, 3901–3904; (f) S. Fujimori, T. F. Knöpfel, P. Zarotti, T. Ichikawa, D. Boyall and E. M. Carreira, *Bull. Chem. Soc. Jpn.*, 2007, **80**, 1635–1657; (g) T. Nishimura, X.-X. Guo, N. Uchiyama, T. Katoh and T. Hayashi, *J. Am. Chem. Soc.*, 2008, **130**, 1576–1577; (h) B. M. Trost and A. H. Weiss, *Adv. Synth. Catal.*, 2009, **351**, 963–983.
43. *Handbook of C–H Transformations*, ed. G. Dyker, Wiley-VCH, Weinheim, 2005, vol. 1, pp. 31–78.

44. (a) H. Singer and G. Wilkinson, *J. Chem. Soc. A*, 1968, 849–853; (b) B. M. Trost, C. Chan and G. Ruhter, *J. Am. Chem. Soc.*, 1987, **109**, 3486–3487; (c) M. Ishikawa, J. Ohshita, Y. Ito and A. Minato, *J. Chem. Soc., Chem. Commun.*, 1988, 804–805; (d) S. Ogoshi, M. Ueta, M. Oka and H. Kurosawa, *Chem. Commun.*, 2004, 2732–2733; (e) T. Nishimura, X.-X. Guo, K. Ohnishi and T. Hayashi, *Adv. Synth. Catal.*, 2007, **349**, 2669–2672; (f) N. Tsukada, S. Ninomiya, Y. Aoyama and Y. Inoue, *Org. Lett.*, 2007, **9**, 2919–2921; (g) N. Matsuyama, H. Tsurugi, T. Satoh and M. Miura, *Adv. Synth. Catal.*, 2008, **350**, 2274–2278; (h) N. Matsuyama, K. Hirano, T. Satoh and M. Miura, *J. Org. Chem.*, 2009, **74**, 3576–3578.
45. (a) T. Mitsudo, Y. Nakagawa, Y. Hori, H. Misawa, H. Watanabe and Y. Watanabe, *J. Org. Chem.*, 1985, **50**, 565–571; (b) M. Shirakura and M. Suginome, *J. Am. Chem. Soc.*, 2008, **130**, 5410–5411; (c) M. Shirakura and M. Suginome, *Org. Lett.*, 2009, **11**, 523–526; (d) M. Shirakura and M. Suginome, *J. Am. Chem. Soc.*, 2009, **131**, 5060–5061.
46. T. Nishimura, X.-X. Guo and T. Hayashi, *Chem.–Asian J.*, 2008, **3**, 1505–1510.
47. M. Shirakura and M. Suginome, *Angew. Chem., Int. Ed.*, 2010, **49**, 3827–3829.

CHAPTER 6

Enantioselective Nickel-Catalysed α-Heterofunctionalisation, and α-Arylation/Alkylation Reactions of Carbonyl Compounds

6.1 Introduction

A prochiral or racemic carbonyl compound can be activated toward electrophilic substitution *via* the formation of an enol or enolate intermediate, creating a tertiary or quaternary centre at the α-carbon. The use of a non-carbon electrophile leads to heterofunctionalised products while that of carbon electrophiles affords α-arylated or α-alkylated carbonyl compounds, and the generation of a new stereogenic centre in these reactions makes them amenable to the development of asymmetric methodologies. In particular, α-heterofunctionalisation of a carbonyl compound is a highly direct and strategically simple method for the synthesis of a large number of interesting molecules and synthetic building blocks, such as amino acids, α-hydroxy acids, and α-fluorinated products. Moreover, optically active α-aryl carbonyl moieties are important structural features of many natural occurring products, pharmaceutically attractive molecules, synthetically useful intermediates, and precursors to emissive polymers. In the last decade, a number of highly enantioselective nickel-catalysed electrophilic halogenation reactions have been developed. Most of them are fluorination reactions of several types of carbonyl substrates, including cyclic as well as acyclic β-keto esters, *N*-acetylthiazolidinones, acid chlorides, and α-chloro-β-keto esters, which all provided near-perfect enantioselectivities

RSC Catalysis Series No. 26
Enantioselective Nickel-Catalysed Transformations
By Hélène Pellissier

Published by the Royal Society of Chemistry, www.rsc.org

by using various ligands. On the other hand, remarkable enantioselective nickel-catalysed α-aminations of *N*-Boc-oxindoles with azodicarboxylates have been achieved by using a chiral Schiff base dinuclear catalyst. Also in the area of asymmetric nickel-catalysed α-hydroxylations with oxaziridines, advances have been made with highly enantioselective α-hydroxylations of cyclic and acyclic β-keto esters as well as malonates performed with the DBFOX ligand. In the context of α-arylations, many excellent results have been achieved, such as nickel-catalysed arylations and heteroarylations of a wide variety of indanones and tetralones with chloroarenes or aryl triflates, which provided up to 99% ee using Difluorphos or BINAP ligands. Finally, a nice α-alkylation of *N*-acylthiazolidinethiones catalysed by a BINAP–nickel catalyst was performed with up to 99% ee.

6.2 α-Halogenations

The chemistry of organofluorine compounds is a rapidly developing area of research because of their importance in biochemical and medicinal applications and material science.[1] Introduction of a fluorine atom into biologically active compounds often leads to a significant and frequently beneficial modification of their biological characteristics due to the unique properties of the fluorine atom.[2] The use of optically active compounds containing a fluorine atom at a stereogenic centre is restricted by the limited availability of effective methods for the enantioselective construction of fluorinated quaternary carbon centres. Thus, the development of effective methodologies for the preparation of chiral organic fluorine compounds through C–F bond formation is still a highly desirable goal in synthetic organic chemistry.[3] A prochiral or racemic carbonyl compound can be activated toward electrophilic substitution *via* the formation of an enol or enolate intermediate, creating a tertiary or quaternary centre at the α-carbon. The use of a non-carbon electrophile leads to chiral α-heterofunctionalised products, and the generation of a new stereogenic centre in this reaction makes it amenable to the development of asymmetric methodologies. α-Heterofunctionalisation of a carbonyl compound is a highly direct and strategically simple method for the synthesis of a large number of interesting molecules and synthetic building blocks, such as amino acids (by amination), α-hydroxy acids (by hydroxylation), and α-halogenated products (by halogenation).[4]

The first catalytic asymmetric α-fluorination of β-keto esters was reported by Hintermann and Togni in 2000,[5] using titanium TADDOLato catalysts, which gave low to high enantioselectivities of up to 90% ee.[6] These chiral nickel catalysts were found to be also applicable to chlorination and bromination of these substrates.[7] Later, a wide range of other chiral metal catalysts were successfully investigated by various groups,[4c] including palladium,[8] titanium,[9] zinc,[10] copper,[11] and nickel complexes. It must be noted that excellent enantioselectivities were reached in the case of using chiral diphosphine palladium complexes as catalysts. In the last decade, a number of excellent results have been reported in the area of asymmetric nickel-catalysed α-halogenations and especially α-fluorinations of carbonyl compounds. As an example, in 2004, Shibata, Toru and co-workers reported the asymmetric fluorination of β-keto esters bearing bulky ester substituents using *N*-fluorobenzenesulfonimide (NFSI) as the fluorinating agent and a nickel catalyst generated *in situ* from $Ni(OTf)_2$ and the DBFOX ligand.[12] Enantioselectivities of 93–99% ee in combination with good yields (75–88%) were achieved for a range of cyclic β-keto esters, as shown in Scheme 6.1. Only one acyclic substrate was investigated and provided both lower yield (75%) and enantioselectivity (83% ee). Using trifluoromethanesulfonyl chloride as a mild chlorinating agent, highly enantioselective chlorination of 2-carboxylate indanone and tetralone substrates could also be achieved in excellent enantioselectivities of 97–98% ee and good yields (61–85%) under the same reaction conditions (Scheme 6.1).[13] In this study, the choice of the metal salt

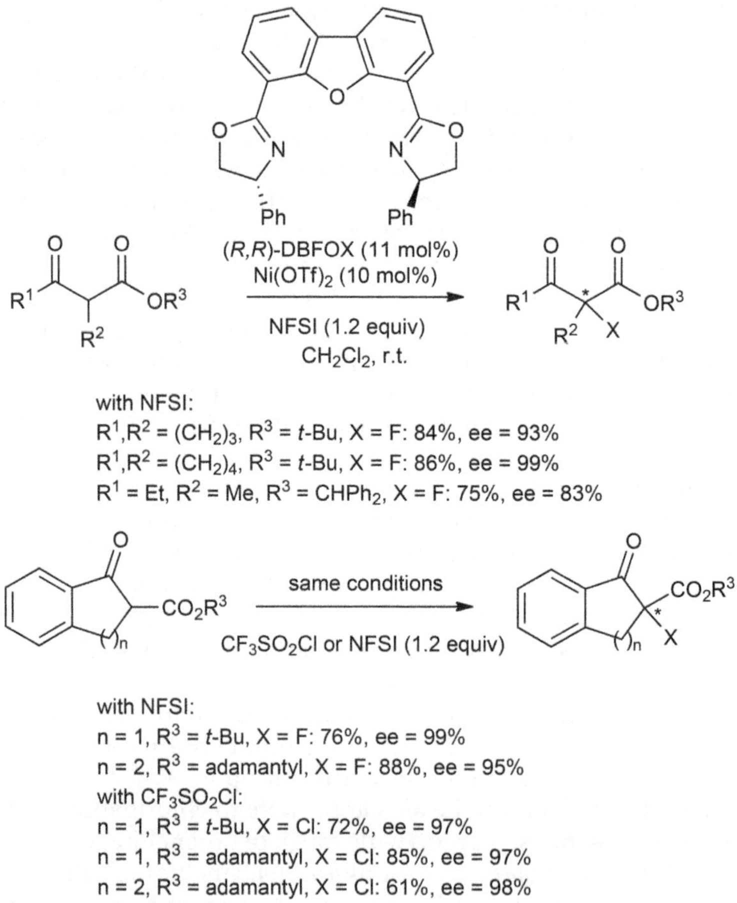

Scheme 6.1 Halogenations of β-keto esters with an *in situ* generated nickel catalyst derived from the DBFOX ligand.

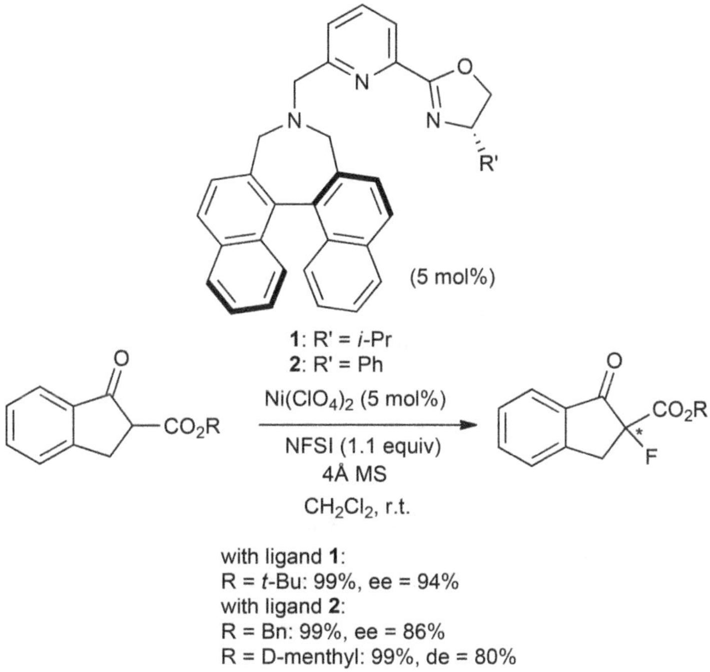

Scheme 6.2 Fluorination of β-keto esters with *in situ* generated nickel catalysts derived from cyclic amine-substituted 2-(oxazolinyl)pyridine ligands.

(Cu or Ni) was demonstrated to afford opposite enantioselectivity in the fluorination of keto esters. Moreover, employing chiral copper catalysts generated *in situ* from the *t*-Bu-BOX ligand provided a range of chiral α-chlorinated and α-brominated β-keto esters, by using CF_3SO_2Cl and NBS as respective halogenating agents, but always in lower enantioselectivities (61–80% ee) than those reached by using chiral nickel complexes of the DBFOX ligand.

In 2007, Iwasa *et al.* described the preparation and application of novel modified chiral oxazoline ligands for asymmetric fluorination reactions of β-keto esters, using $Ni(ClO_4)_2$ as metal precursor.[14] Among variously substituted *N*,*N*,*N*-tridentate ligands, possessing both binaphthyl axial chirality and carbon-centred chirality, ligands **1** and **2** were selected as optimal, allowing the fluorination of several cyclic β-keto esters to be achieved in enantioselectivities of up to 94% ee, as shown in Scheme 6.2. Even if a limited number of cyclic keto ester substrates were included in this study, nevertheless good to high enantioselectivities of up to 94% ee could be attained in combination with excellent yields.

All the reagents were mixed almost simultaneously in the above work (Scheme 6.2). To suppress unproductive binding of the electrophile to the catalyst, these authors investigated a slow introduction of the fluorinating reagent over the course of the reaction, which was supposed to be essential to obtain good results.[15] Surprisingly, under these reaction conditions the

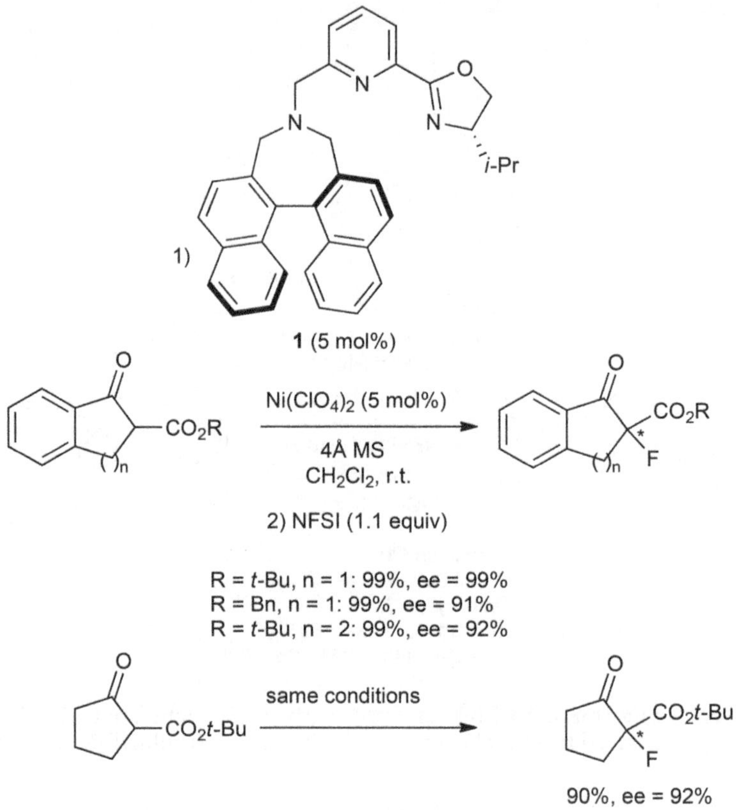

Scheme 6.3 Fluorination of β-keto esters with *in situ* generated nickel catalysts derived from cyclic amine-substituted 2-(oxazolinyl)pyridine ligands using a different operational sequence.

enantioselectivity was dramatically increased to 99% ee by using ligand **1**, as shown in Scheme 6.3. Again, the products were obtained in very good yields ranging from 90 to 99%.

In 2007, an approach for the monofluorination of a prochiral methylene group was reported by Sodeoka *et al.*, wherein the fluorination of α-arylacetic acid derivatives proceeded under a three-component catalyst system composed of an *in situ* generated catalyst from a (*R*,*R*)-derived BINAP ligand **3** and $NiCl_2$, Et_3SiOTf, and 2,6-lutidine (Scheme 6.4).[16] As the arylacetic acid moiety is found in many important medicines, such as nonsteroidal anti-inflammatory drugs, the availability of chiral α-monofluorinated arylacetic acids is expected to be useful for drug synthesis. As shown in Scheme 6.4, a range of *N*-acylthiazolidinones and *N*-acyloxazolidinones were fluorinated by treatment with NFSI to give the corresponding chiral products in moderate to excellent yields of up to 99% along with moderate to good enantioselectivities of up to 88% ee. The nature of the heteroatom in the oxazolidinone/thiazolidinone ring did not appear to affect the selectivity of the process.

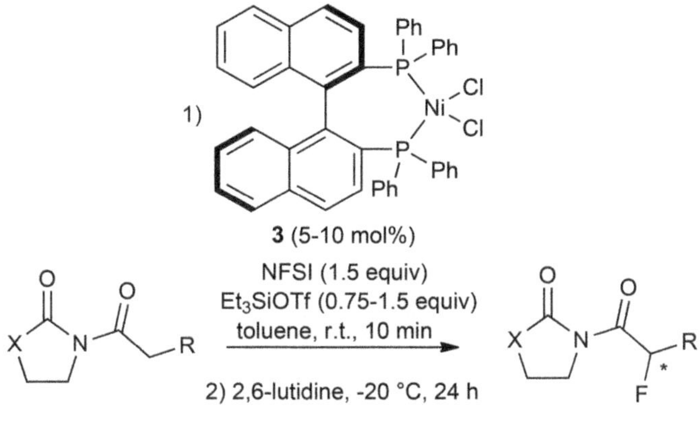

R = Ph, X = S: 99%, ee = 88%
R = *p*-FC_6H_4: 90%, ee = 83%
R = *p*-$MeOC_6H_4$: 92%, ee = 81%
Ar = *m*-$MeOC_6H_4$: 95%, ee = 82%
R = *o*-$MeOC_6H_4$: 87%, ee = 78%
R = 1-Naph: 94%, ee = 87%
R = 2-Naph: 99%, ee = 83%
R = Ph, X = O: 95%, ee = 87%
R = *n*-Pr, X = S: 15%, ee = 11%

Scheme 6.4 Fluorination of *N*-acylthiazolidinones and *N*-acyloxazolidinones with a preformed nickel catalyst derived from the BINAP ligand.

The only α-alkyl substrate (R = *n*-Pr, X = S) reacted with poor yield (15%) and enantioselectivity (11% ee). The unique combination of the catalytic components was specifically chosen to offer a dual activation of the substrate and reagent: the formation of the activated nickel–enolate complex was assisted by the presence of the noncoordinating Brønsted base, while the Lewis acidic Et_3SiOTf activated NFSI to become a stronger nucleophile, without interfering with the formation of the enolate. C–F bond formation resulted from the reaction between the activated substrate and the activated electrophile, prior to product release.

Because the chlorine atom is a good leaving group, substitution reactions of alkyl chlorides with various nucleophiles are routinely employed in organic synthesis. Such reactions occur through S_N2-type displacement with (complete) inversion of the stereochemistry. For this reason, optically active chlorinated compounds can serve as versatile intermediates for further chemical modifications. In 2011, Sodeoka *et al.* employed the chiral catalyst **3** in enantioselective nickel-catalysed chlorinations of various *N*-acetyloxazolidinones.[17] As in Scheme 6.5, the reaction was performed with Et_3SiOTf in toluene but at a lower temperature (−60 °C) and by using *N*-methylmorpholine (NMM) instead of 2,6-lutidine. Under these optimised conditions, a range of chiral chlorinated products were achieved in high yields (86–99%) along with moderate to good enantioselectivities of 75–88% ee, as shown in Scheme 6.5. It must be noted that the dichlorinated compound was not

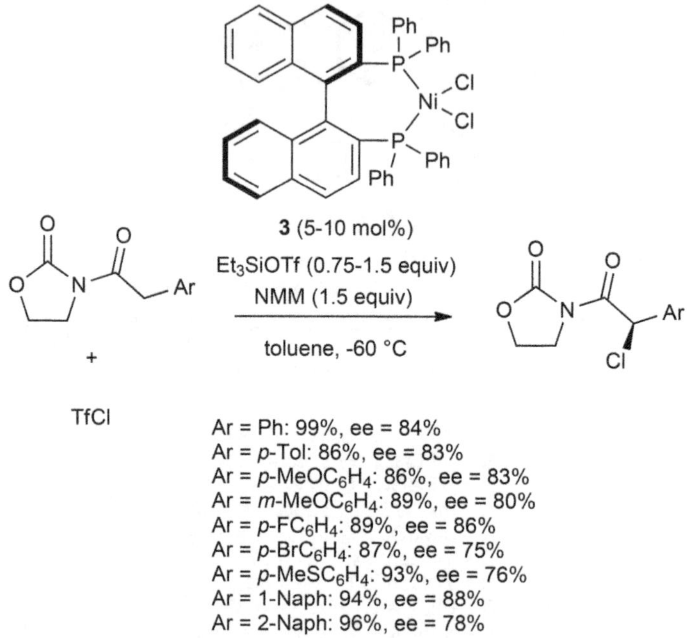

Scheme 6.5 Chlorination of *N*-acetyloxazolidinones with a preformed nickel catalyst derived from the BINAP ligand.

isolated in any of these reactions. Ether, thioether, and halogens were tolerant to the conditions, and bulkier substrates were also chlorinated smoothly. On the other hand, no reaction was observed in the case of an aliphatic substrate derived from hexanoic acid.

In 2008, α-fluorinations of 3-(2-arylacetyl)thiazolidin-2-ones were also investigated by Shibata and Toru by using $Ni(ClO_4)_2{\cdot}6H_2O$ as the Lewis acid, bisoxazoline (*R*,*R*)-DBFOX as ligand, and 2,6-lutidine as cocatalyst (Scheme 6.6).[18] Performed in dichloromethane at 0 °C, the reactions afforded the corresponding fluorinated products in good yields of up to 96% and with moderate enantioselectivities of up to 78% ee. Compared to the previous results (Scheme 6.4) obtained with the BINAP-derived ligand **3**, the enantioselectivities provided from using the nickel/(*R*,*R*)-DBFOX system were not as high, probably because of the low activity of the catalytic system which required higher reaction temperature conditions (0 °C instead of −20 °C); however, the presence of Et_3SiOTf as a supplementary Lewis acid was unnecessary.

In 2009, the same authors disclosed that the enantioselectivity of these reactions could be dramatically improved to 99% ee by the help of a catalytic amount of hexafluoroisopropanol (HFIP).[19] As shown in Scheme 6.7, the fluorinations of *N*-(arylacetyl)thiazolidinones with NFSI induced by a catalyst generated *in situ* from $Ni(ClO_4)_2{\cdot}6H_2O$ and (*R*,*R*)-DBFOX in the presence of molecular sieves, 2,6-lutidine, and 30 mol% of HFIP afforded the corresponding fluorinated chiral thiazolidinones in excellent yields (87–98%) and with

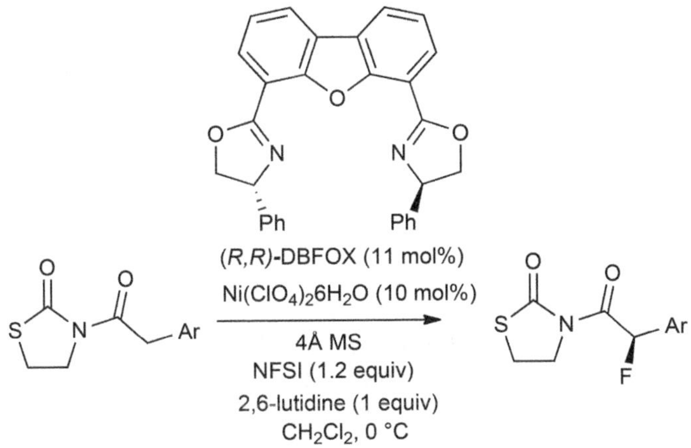

Ar = Ph: 90%, ee = 74%
Ar = *o*-MeOC_6H_4: 96%, ee = 78%
Ar = *m*-MeOC_6H_4: 94%, ee = 66%
Ar = *p*-MeOC_6H_4: 90%, ee = 65%
Ar = *o*-Tol: 69%, ee = 76%
Ar = *m*-Tol: 75%, ee = 73%
Ar = *p*-Tol: 75%, ee = 77%
Ar = *p*-FC_6H_4: 60%, ee = 62%
Ar = *p*-BrC_6H_4: 77%, ee = 56%
Ar = 1-Naph: 85%, ee = 59%
Ar = 2-Naph: 90%, ee = 60%

Scheme 6.6 Fluorination of 3-(2-arylacetyl)thiazolidin-2-ones with an *in situ* generated nickel catalyst derived from the DBFOX ligand.

enantioselectivities of 92–98% ee. Substrates having electron-donating and -withdrawing groups at the *meta* or *para* position of the benzene ring were found to be excellent substrates, affording the fluorinated products in high yields and with 94–99% ee. Aryl groups with halogen substitution also afforded high yields and high enantioselectivities (87–99% ee). The reactions of the sterically demanding naphthyl-substituted substrates also gave excellent yields and high enantioselectivities (92–99% ee). The effect of HFIP was presumably an improvement in the catalytic turnover by protonation of a metal–oxygen bond followed by release of the products in the transition state, and 2,6-lutidine could assist the enolisation of the substrates. The utility of this novel catalytic system was demonstrated by its application in the catalysis of the fluorination of *N*-(but-3-enoyl)thiazolidinones, which had no precedent in the literature. The corresponding fluorinated products were obtained in good to high yields and enantioselectivities of up to 91% ee, as shown in Scheme 6.7. Phenyl-substituted and nonsubstituted alkenes were suitable substrates in the reaction, providing high yields and enantioselectivities (78–86% ee). The substrates with methyl, chloro, and bromo substituents at the benzene ring and the bulkier naphthyl-substituted substrate were also successfully transformed into fluorinated products with good enantioselectivities (80–86% ee).

fluorination of aryl acetyl thiazolidinones:

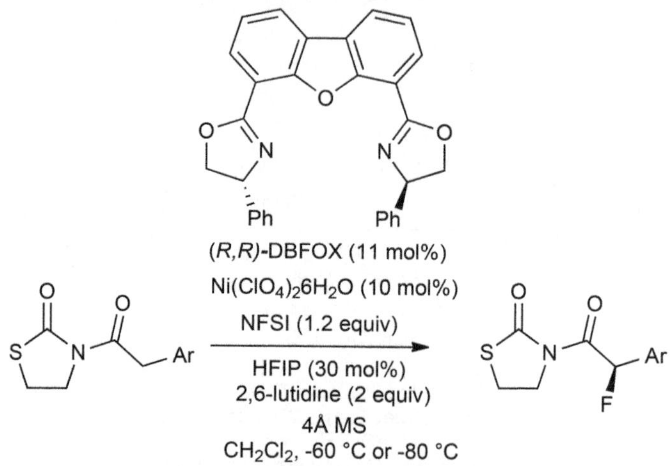

Ar = Ph: 91%, ee = 98% (at -60 °C)
Ar = *m*-MeOC_6H_4: 93%, ee = 96% (at -60 °C)
Ar = *p*-MeOC_6H_4: 93%, ee = 96% (at -60 °C)
Ar = *m*-Tol: 93%, ee = 98% (at -60 °C) (80%, ee = 99% at -80 °C)
Ar = *p*-Tol: 90%, ee = 96% (at -60 °C)
Ar = *p*-$CF_3C_6H_4$: 94%, ee = 94% (at -60 °C)
Ar = *p*-FC_6H_4: 90%, ee = 94% (at -60 °C)
Ar = *p*-BrC_6H_4: 96%, ee = 93% (at -60 °C) (87%, ee = 99% at -80 °C)
Ar = 1-Naph: 87%, ee = 92% (at -60 °C)
Ar = 2-Naph: 94%, ee = 95% (at -60 °C) (70%, ee = 99% at -80 °C)

fluorination of aryl-3-butenoyl thiazolidinones:

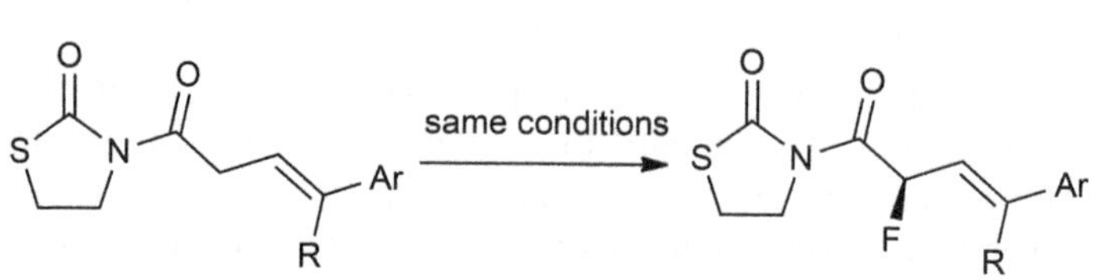

Ar = Ph, R = Me: 50%, ee = 91% (at -80 °C)
Ar = Ph, R = H: 58%, ee = 83% (at -70 °C)
Ar = R = Ph: 99%, ee = 86% (at -60 °C)
Ar = *p*-Tol, R = Me: 99%, ee = 85% (at -60 °C)
Ar = *p*-ClC_6H_4, R = Me: 99%, ee = 86% (at -70 °C)
Ar = *p*-BrC_6H_4, R = Me: 99%, ee = 80% (at -60 °C)
Ar = 2-Naph, R = Me: 97%, ee = 85% (at -60 °C)

Scheme 6.7 Fluorinations of *N*-(arylacetyl)- and *N*-(4-arylbut-3-enoyl)thiazolidinones with an *in situ* generated nickel catalyst derived from the DBFOX ligand.

A number of impressive examples of asymmetric α-fluorinations of β-keto esters, imides, and aldehydes have been successfully developed; however, the α-fluorination of ketene enolates to yield directly simple chiral α-fluorinated carboxylic acid derivatives was not reported until 2008. That year, Lectka *et al.* developed a catalytic enantioselective α-fluorination of acid chlorides to produce products of a broad scope in good yields and with excellent enantioselectivities of up to >99% ee.[20] As shown in Scheme 6.8, the process was promoted by a nickel catalyst generated *in situ* from (1,3-dppp)$NiCl_2$ and the chiral benzoylquinidine BQd in the presence of NFSI as the fluorinating agent. Isolation of the putative bis(sulfonamide) intermediate **4** was difficult owing to its lability; this fact necessitated a quenching reaction with alcohols, water, or other nucleophiles. The investigation of the substrate scope of the reaction showed that acid halides containing various aromatic substituents, including a naphthyl group, were good substrates. Along with an alcohol quench that provided esters, a water workup afforded chiral α-fluorinated carboxylic acids in 99% ee that should be of potentially broad utility. Moreover, an amine-based workup afforded enantiopure amides. One of the prime advantages of generating chiral α-fluorinated reactive intermediates "in a flask" is the ability to quench them with drugs, natural products, and other exotic nucleophiles to produce interesting and potentially useful derivatives. For example, workup of a standard fluorination of 3-phthalimidopropionyl chloride **5** with the antiprotozoal isoquinoline alkaloid natural product (+)-emetine yielded the diastereomerically pure fluorinated derivative **6** in 91% yield (Scheme 6.8). It is easy to imagine a wide range of fluorinated intermediates which will be coupled in the future with a vast array of natural nucleophiles to produce a virtually limitless number of medicinally important products. In this study, the authors demonstrated that excellent enantioselectivities were achieved upon catalysis with *trans*-(PPh_3)$PdCl_2$ instead of (1,3-dppp)$NiCl_2$ under the same reaction conditions.

A few synthetic methods for the preparation of α-chloro-α-fluoro-β-keto esters have been reported so far. In general, these compounds are prepared by the electrophilic fluorination of the corresponding α-chloro-β-keto esters using NFSI or F_2. In 2003, Togni reported the first catalytic enantioselective synthesis of these products using chiral titanium complexes with moderate enantioselectivities of up to 65% ee.[21] Later in 2007, Kim *et al.* reported enantioselectivities of up to 77% ee by using chiral palladium catalysts in these reactions.[22] In 2010, these authors reinvestigated these reactions by using chiral nickel catalysts.[23] As shown in Scheme 6.9, very good enantioselectivities of up to 99% ee combined with moderate to good yields (65–85%) were achieved by using preformed chiral nickel catalyst **7** at 10 mol% of catalyst loading in the presence of Selectfluor as the fluorinating agent. The chiral diamine ligand in catalyst **7** was selected as optimal among a range of other chiral diamine ligands bearing various substituents (Ar′). The absolute configuration of one product (Ar = Ph) was determined to be *S* by comparing chiral HPLC data with published values. The substrate scope was broad for aromatic substrates, but aliphatic substrates such as ethyl 2-chloro-3-oxobutanoate could not be fluorinated under the reaction conditions.

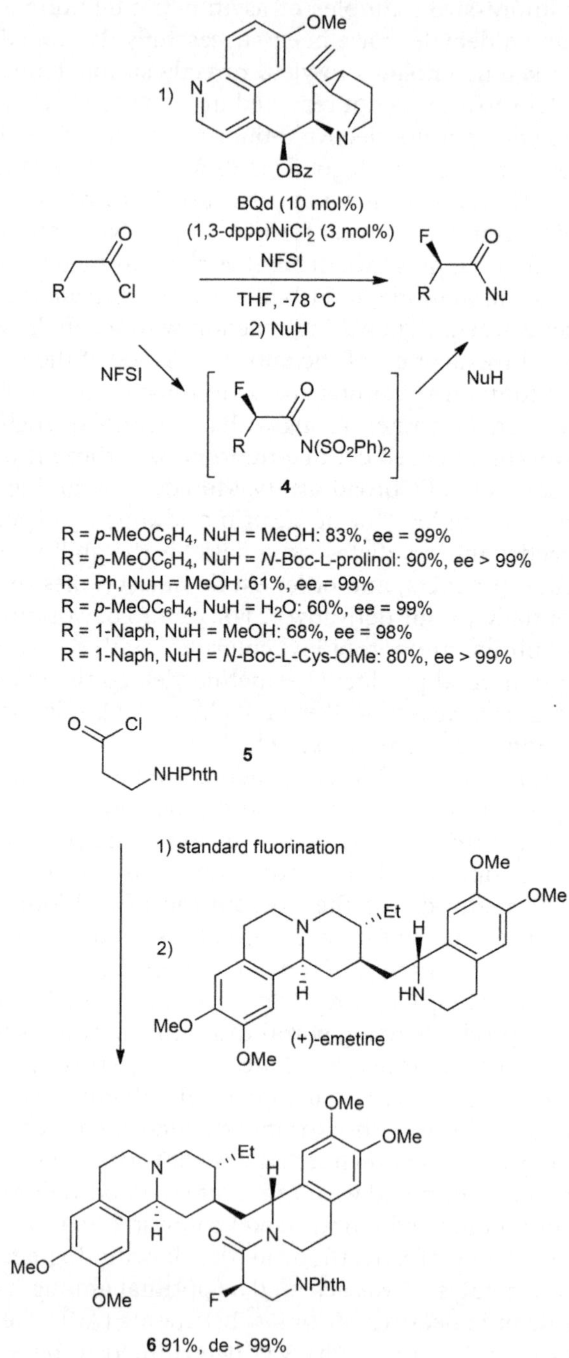

Scheme 6.8 Fluorination of acid chlorides with an *in situ* generated nickel catalyst derived from the benzoylquinidine ligand.

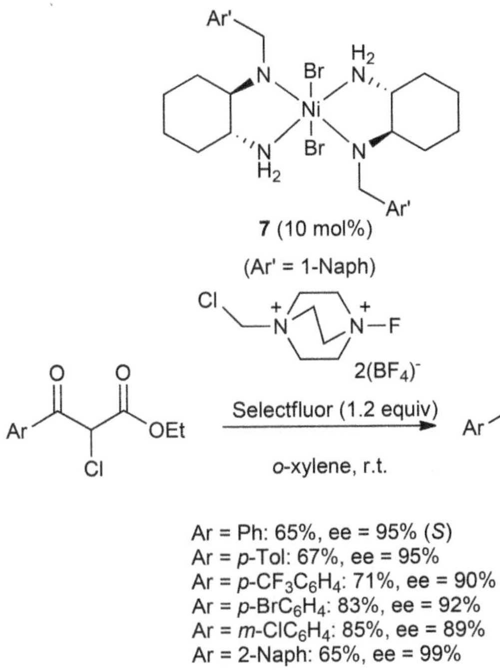

Scheme 6.9 Fluorination of α-chloro-β-keto esters with a preformed nickel catalyst derived from a diamine ligand.

6.3 α-Aminations

Not many asymmetric versions of the electrophilic α-amination of carbonyl compounds exist in spite of the importance of the formed chiral α-amino carbonyl compounds. Among them are amino acids which are used as pharmaceuticals, agrochemicals, and fundamental synthetic building blocks for the preparation of an assortment of biologically valuable molecules.[24] Indeed, the direct asymmetric electrophilic α-amination of carbonyl compounds remains a challenging synthetic transformation.[4c,25] This is largely due to the lack of "naked" sources of electrophilic nitrogen, which can afford an amine or protected amine directly in a single step, thus requiring further functional group transformations.[4c] The vast majority of results use an azodicarboxylate ($RO_2CN{=}NCO_2R$, with R = *t*-Bu, i-Pr, Et, or Me) as the electrophile, generating chiral hydrazines as products, which can be transformed into chiral amines under hydrogenating or reducing conditions. A range of metals, including magnesium,[26] scandium,[27] nickel, copper,[28] zinc,[28f,29] palladium,[30] silver,[31] lanthanum,[32] iridium,[33] and europium,[34] have been successfully applied in these reactions. The azodicarboxylate reacts with a silyl enol ether, an activated carbonyl compound, such as a β-keto ester, or an oxindole. Early in 1997, Evans *et al.* reported the enantioselective α-amination of *N*-acyloxazolidinones with azodicarboxylates by using a chiral magnesium complex of a bis(sulfonamide) ligand.[26] Enantioselectivities of 80–90% ee were achieved in this work, but the reactions required several days to reach completion even at 10 mol% of catalyst loading at low temperature. Reactions with β-keto esters were first reported by Jørgensen and co-workers by using chiral bisoxazoline copper catalysts, providing excellent results.[28b] Other chiral oxazoline-based copper catalysts,[28c,e] as well as a europium complex of the chiral PYBOX ligand, have given very high enantioselectivities in these reactions.[34] Concerning examples using chiral nickel catalysts, most of them dealt with reactions occurring between *tert*-butyl azodicarboxylate and β-keto esters or α-cyano ketones. The latter have been enantioselectively α-aminated by Kim *et al.* with *tert*-butyl azodicarboxylate in the presence of 5 mol% of a preformed nickel catalyst **8** derived from a chiral diamine ligand, affording the corresponding α-aminated α-cyano ketones in good yields and with moderate enantioselectivities of up to 83% ee, as shown in Scheme 6.10.[35] This ligand was selected as optimal among variously substituted (Ar) chiral

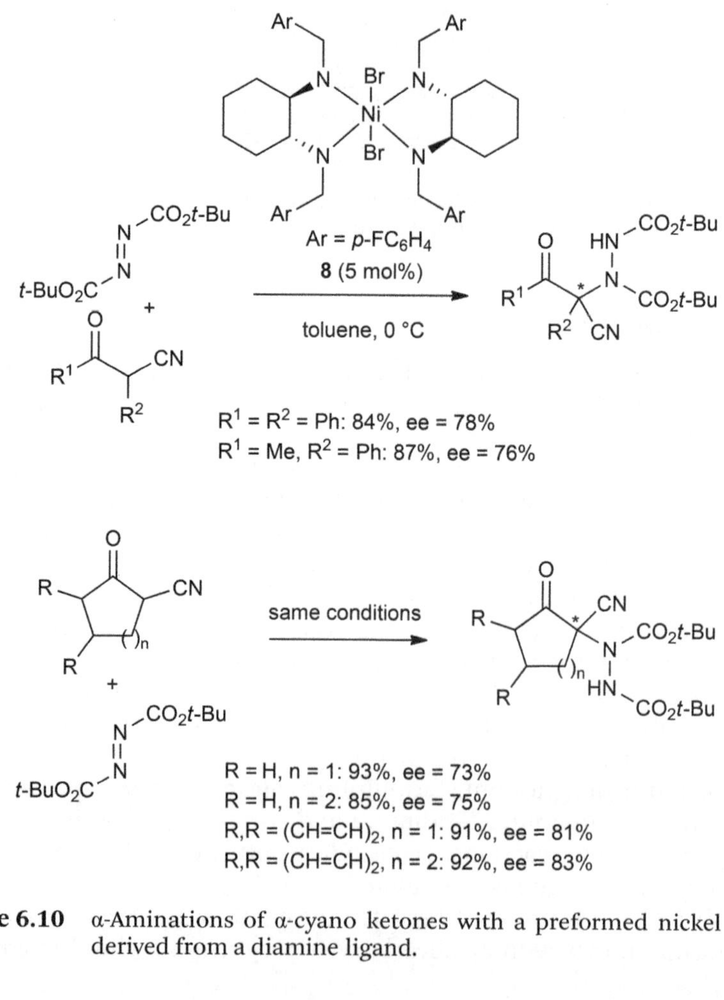

Scheme 6.10 α-Aminations of α-cyano ketones with a preformed nickel catalyst derived from a diamine ligand.

cyclohexane-1,2-diamines. Closely related enantioselectivities were obtained for cyclic and acyclic α-cyano ketones, while yields from cyclic substrates were slightly better. These reactions were also performed by these authors with chiral palladium catalysts, demonstrating that better results were achieved by employing cationic palladium(II) bisphosphine complexes which provided enantioselectivities of up to 95% ee for cyclic substrates and up to 86% ee for acyclic substrates.[30] Furthermore, enantioselectivities of 91–96% ee were reported by Ikariya *et al.* for amination of α-aryl-α-cyanoacetates by using a chiral Cp*Ir(TsDPEN) complex.[33]

In 2009, the nickel catalyst in Scheme 6.10 was applied to the enantioselective α-amination of α-fluoro-β-keto esters to give the corresponding hydrazines in good yields, albeit with moderate enantioselectivities of 20–74% ee, as shown in Scheme 6.11.[36] Even if the reactions were faster than those

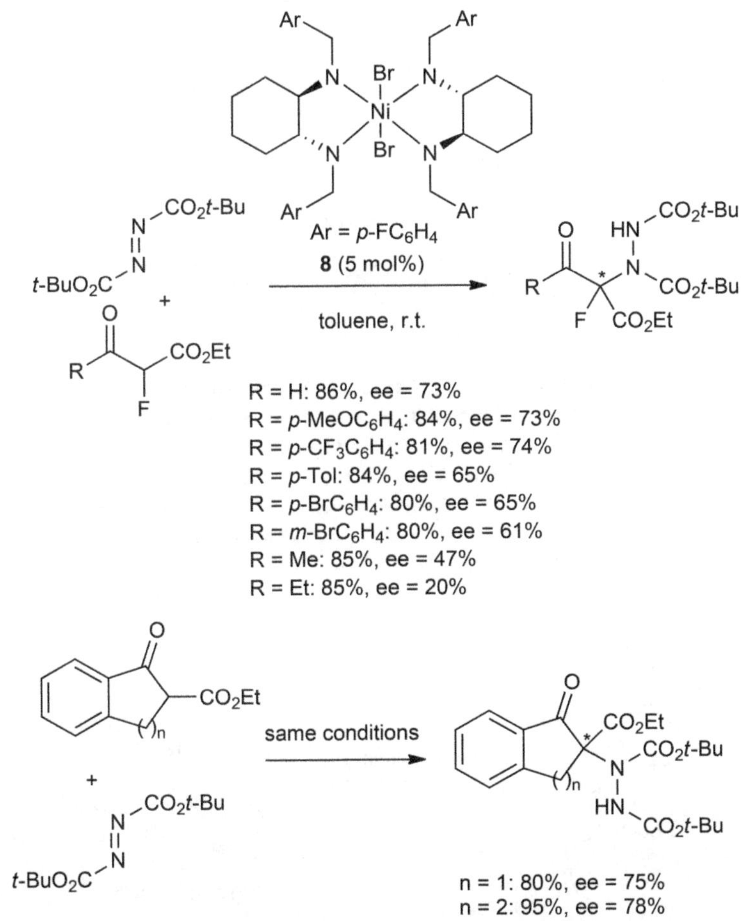

Scheme 6.11 α-Aminations of α-fluoro-β-keto esters and cyclic β-keto esters with a preformed nickel catalyst derived from a diamine ligand.

previously achieved using copper bisoxazoline catalysts,[28c] they offered lower enantioselectivities (up to 94% ee with copper). Moreover, catalyst **8** was investigated in the α-amination of cyclic β-keto esters, and provided only moderate enantioselectivities of 75–78% ee.

In 2009, these same authors applied a closely related preformed nickel catalyst **9** derived from another chiral cyclohexane-1,2-diamine in the enantioselective α-amination of cyclic β-keto esters (Scheme 6.12).[37] The reaction of various cyclic β-keto esters with *tert*-butyl azodicarboxylate provided the

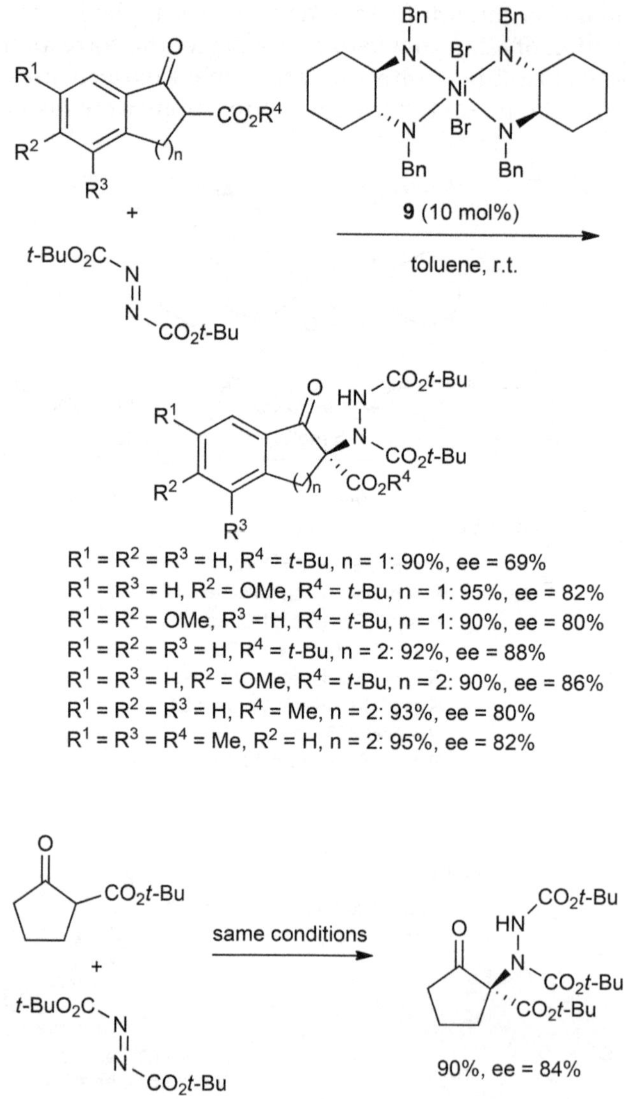

Scheme 6.12 α-Aminations of β-keto esters with a closely related preformed nickel catalyst derived from a diamine ligand.

corresponding α-aminated β-keto esters in high yields (90–95%) and with moderate to good enantioselectivities of up to 88% ee. The highest enantioselectivity was reached with the sterically hindered *tert*-butyl ester of indanonecarboxylate. It must be noted that better enantioselectivities (up to 98–99% ee) were achieved when these reactions were catalysed by copper– or zinc–oxazoline systems.[28b,c,29]

The α-amination with azodicarboxylates can also involve oxindoles. In 2010, Shibasaki *et al.* reported an interesting class of Schiff base catalyst based on chiral binaphthyldiamine backbones to be applied in these reactions.[38] A binuclear Schiff base nickel complex **10** derived from a ligand of this type was found to be highly efficient in the enantioselective α-amination of a range of *N*-Boc-oxindoles with *tert*-butyl and isopropyl azodicarboxylates to give the corresponding 3-aminooxidoles in both high yields (86–99%) and enantioselectivities (87–99% ee), as shown in Scheme 6.13. The substrate scope was broad since 3-methyl-, allyl-, (*E*)-cinnamyl-, and benzyl-substituted oxindoles all gave excellent results, as well as 5- and 6-substituted oxindoles. It was noteworthy that ester and nitrile groups were also tolerated. Furthermore, the authors also investigated mononuclear Schiff base nickel catalyst **11** in α-aminations of *N*-Boc-oxindoles with *tert*-butyl azodicarboxylate, affording the corresponding hydrazines in moderate to high enantioselectivities (80–98% ee) along with high yields (91–99%). In most cases, the enantioselectivities achieved from bimetallic catalyst **10** were higher than those derived from the use of mononuclear catalyst **11**. Importantly, a reversal of enantiofacial selectivity was observed between the bimetallic and monometallic Schiff base complexes. Indeed, the bimetallic catalyst afforded the (*R*)-products while the monometallic complex furnished the (*S*)-products. The utility of this powerful methodology was demonstrated in the transformation of the products into an optically active oxindole with a spiro-β-lactam unit and a known key intermediate for AG-041R synthesis. It must be noted that these reactions have also been achieved by Feng *et al.* using a chiral scandium catalyst derived from an *N*,*N′*-dioxide ligand.[27] The reactions required several days to complete at −20 °C, which was considerably slower than that involving nickel catalysis. However, enantioselectivities of >90% ee could be routinely achieved for a range of alkyl-substituted oxindoles, with one example of an aryl-substituted substrate (93% ee). Even more impressively, the reaction outcome was insensitive to the ester substituent on the electrophile (methyl and isopropyl azodicarboxylates), and the *N*-substituent (H, Me, or Bn). In 2012, Shibasaki reported the enantioselective α-amination of 3-aryl-substituted *N*-Boc-oxindoles with *tert*-butyl azodicarboxylate catalysed by dinuclear Schiff base nickel catalyst **10**.[39] In contrast to the reactions of 3-alkyl-substituted *N*-Boc-oxindoles, the reactions of 3-aryl-substituted *N*-Boc-oxindoles required $CHCl_3$ instead of toluene as solvent, a lower temperature of 30 °C instead of 50 °C, and the presence of 5 Å molecular sieves as additives to provide the corresponding chiral hydrazines in moderate to high enantioselectivities of 66–98% ee (Scheme 6.13).

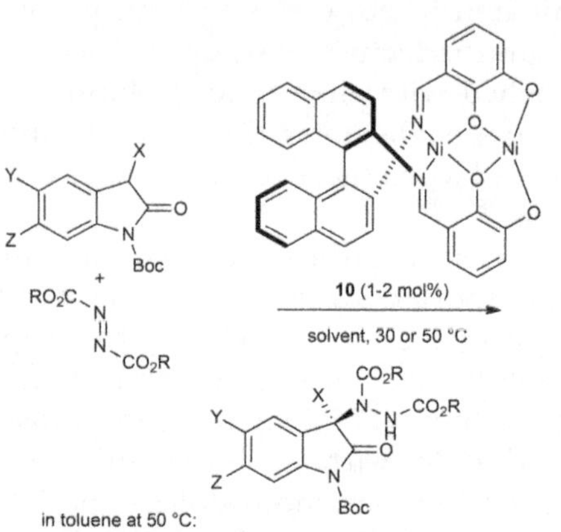

in toluene at 50 °C:
X = Me, Y = Z = H, R = *t*-Bu: 99%, ee = 99%
X = allyl, Y = Z = H, R = *t*-Bu: 99%, ee = 97%
X = (*E*)-cinnamyl, Y = Z = H, R = *t*-Bu: 86%, ee = 91%
X = Bn, Y = Z = H, R = *t*-Bu: 93%, ee = 99%
X = Me, Y = OMe, Z = H, R = *t*-Bu: 91%, ee = 94%
X = Me, Y = F, Z = H, R = *t*-Bu: 95%, ee = 96%
X = allyl, Y = F, Z = H, R = *t*-Bu: 90%, ee = 98%
X = allyl, Y = Cl, Z = H, R = *t*-Bu: 93%, ee = 95%
X = CH_2CO_2Me, Y = Z = H, R = *t*-Bu: 98%, ee = 96%
X = CH_2CN, Y = Z = H, R = *t*-Bu: 89%, ee = 87%
X = Me, Y = Z = H, R = *i*-Pr: 94%, ee = 95%
X = Me, Y = Br, Z = H, R = *i*-Pr: 92%, ee = 91%
X = Bn, Y = H, Z = Cl, R = *t*-Bu: 98%, ee = 99%
in $CHCl_3$ at 30 °C and with 5A MS as additive:
X = Ph, Y = Z = H, R = *t*-Bu: 94%, ee = 90%
X = Ph, Y = OMe, Z = H, R = *t*-Bu: 86%, ee = 82%
X = *o*-$MeOC_6H_4$, Y = Z = H, R = *t*-Bu: 75%, ee = 98%
X = *m*-$MeOC_6H_4$, Y = Z = H, R = *t*-Bu: 93%, ee = 86%
X = *p*-FC_6H_4, Y = Z = H, R = *t*-Bu: 91%, ee = 82%
X = *m*-$MeOC_6H_4$, Y = OMe, Z = H, R = *t*-Bu: 78%, ee = 66%

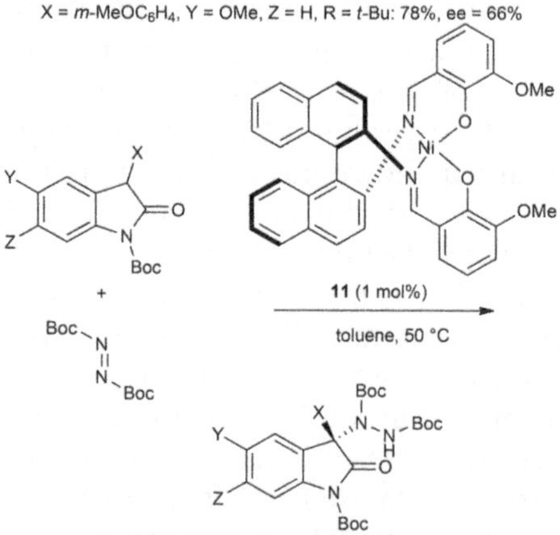

X = Me, Y = Z = H: 99%, ee = 94%
X = allyl, Y = Z = H: 94%, ee = 80%
X = (*E*)-cinnamyl, Y = Z = H: 95%, ee = 92%
X = Bn, Y = Z = H: 93%, ee = 93%
X = Me, Y = OMe, Z = H: 96%, ee = 98%
X = Me, Y = F, Z = H: 91%, ee = 87%
X = allyl, Y = F, Z = H: 93%, ee = 92%
X = allyl, Y = Cl, Z = H: 94%, ee = 87%
X = CH_2CO_2Me, Y = Z = H: 96%, ee = 91%
X = Bn, Y = H, Z = Cl: 91%, ee = 85%

Scheme 6.13 α-Aminations of *N*-Boc-oxindoles with Schiff base mononuclear and dinuclear nickel catalysts.

6.4 α-Hydroxylations

Keto–enol tautomerism generates a C=C bond, which can undergo epoxidation reactions to generate an unstable intermediate that rearranges to give an α-hydroxy ketone. An attempt to achieve this catalytically was first reported in 2002, wherein 2-hydroxy ketones were subjected to conditions of the Sharpless asymmetric epoxidation reaction.[40] Optically active hydroxylated products could be obtained with high enantioselectivities, but the reactions were extremely slow even when using an excess of catalyst, since only low conversions were obtained after several days. The operation of competitive side reactions (*e.g.*, Baeyer–Villiger oxidation) was also a significant problem in these reactions.[41] A number of metal catalysts used earlier in asymmetric α-fluorination reactions were also found to be useful for the corresponding α-hydroxylation reactions. In 2004, Togni, Mezzeti, and co-workers reported the first enantioselective α-hydroxylation of β-keto esters performed with an oxaziridine as oxidant and catalysed with a chiral TADDOL-derived titanium complex, which provided enantioselectivities reaching 94% ee.[42] Later, Toru, Shibata, and co-workers described the first example of the catalytic enantioselective α-hydroxylation of 2-oxindoles and β-keto esters with oxaziridines as oxidants using the DBFOX ligand in combination with Lewis acids based on zinc and nickel salts.[43] In the case of 2-oxindoles, the best enantioselectivities of up to 97% ee were achieved by using a zinc–DBFOX complex, while the corresponding chiral nickel catalyst provided lower enantioselectivities of up to 65% ee. On the other hand, the α-hydroxylation of β-keto esters with the same oxaziridine provided the highest enantioselectivities by using the nickel catalyst generated *in situ* from the (*S*,*S*)-DBFOX ligand and Ni $(ClO_4)_2 \cdot 6H_2O$ in the presence of molecular sieves. As shown in Scheme 6.14, a variety of cyclic β-keto esters could be hydroxylated in good to high yields (82–97%) along with excellent enantioselectivities of 94–97% ee. An acyclic β-keto ester was also compatible, giving the corresponding product in 93% ee, albeit with a lower yield of 27%. Since the α-hydroxy-β-dicarbonyl functional unit is an important structural motif found in many bioactive molecules,[44] this novel highly efficient methodology was welcome, offering higher enantioselectivities than all other methods reported earlier.[45]

In 2009, this procedure was successfully extended to the asymmetric α-hydroxylation of unsymmetrical *tert*-butyl malonates (Scheme 6.15).[46] With these less reactive substrates, prolonged reaction times were required, even at reflux. In general, the products were achieved in good to high yields (81–93%), along with high enantioselectivities of 81–98% ee. Once again, the presence of a bulky *tert*-butyl ester was found to be necessary for achieving high enantioselectivity, as almost no enantioselectivity was observed with ethyl methyl malonates (12% ee). The reaction was only catalysed by a nickel complex; the corresponding zinc complex was inactive. The utility of this interesting novel methodology was demonstrated by the syntheses of chlozolinate, an important antifungal agent, and a key intermediate for the antiandrogen bicalutamide.

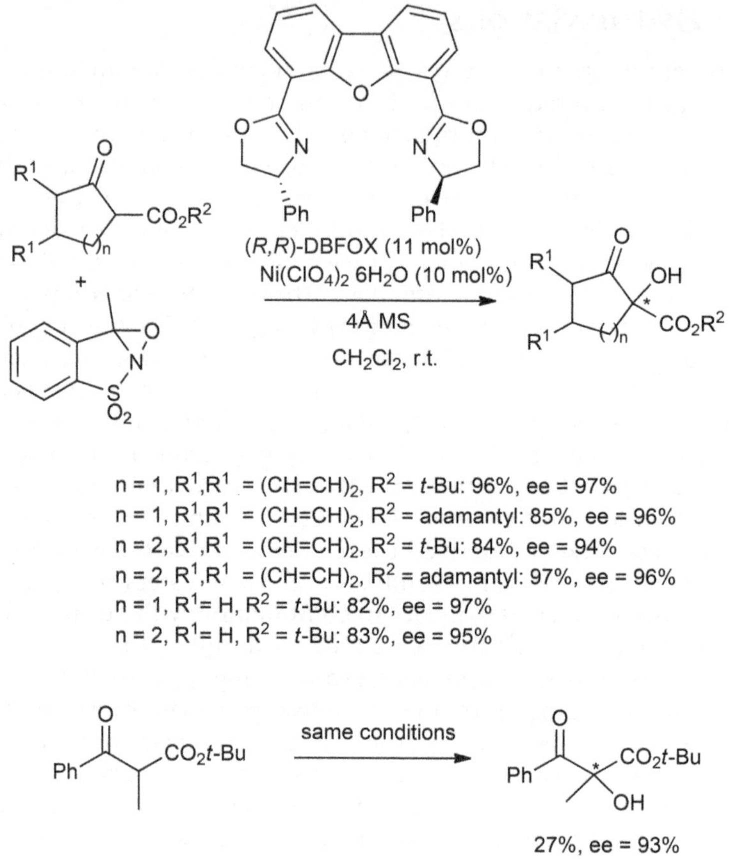

Scheme 6.14 α-Hydroxylation of β-keto esters with an *in situ* generated nickel catalyst derived from the DBFOX ligand.

6.5 α-Arylations and α-Alkylations

Optically active α-aryl carbonyl moieties are important structural features of many naturally occurring products, pharmaceutically attractive molecules, synthetically useful intermediates, and precursors to emissive polymers.[47] Particularly, direct carbon–carbon bond formation between an arene and the α-carbon adjacent to a carbonyl group remains a formidable challenge in organic synthesis.[4b] For a long time, successful metal-mediated coupling of enolates was achieved using stoichiometric quantities of nickel complexes. For example, Semmelhack *et al.* reported the nickel-mediated intramolecular arylation of an ester,[48] and Millard and Rathke described the nickel-mediated intermolecular arylation of lithium enolates.[49] The catalytic version of Reformatsky-type arylation of zinc enolates was reported by Fauvarque and Jutand with a limited substrate scope in 1979.[50] Significant development of this area was achieved by Migita *et al.*, who pioneered the palladium-catalysed coupling of aryl and vinyl halides with transmetallating zinc and tin

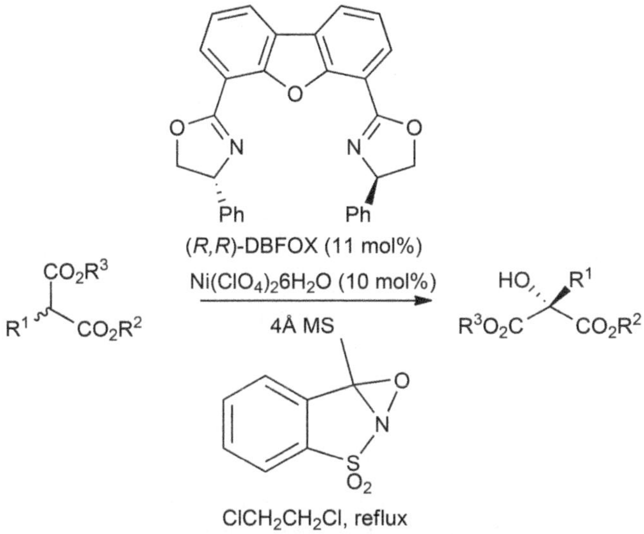

R^1 = Bn, R^2 = Me, R^3 = *t*-Bu: 82%, ee = 91%
R^1 = Ph, R^2 = Me, R^3 = *t*-Bu: 83%, ee = 93%
R^1 = R^2 = Me, R^3 = *t*-Bu: 84%, ee = 98%
R^1 = Et, R^2 = Me, R^3 = *t*-Bu: 74%, ee = 95%
R^1 = *n*-Bu, R^2 = Me, R^3 = *t*-Bu: 80%, ee = 94%
R^1 = Me, R^2 = Et, R^3 = *t*-Bu: 72%, ee = 94%
R^1 = Bn, R^2 = Me, R^3 = Et: 98%, ee = 12%
R^1 = Me, R^2 = Et, R^3 = Ph: 24%, ee = 81%
R^1 = Me, R^2 = Et, R^3 = *o*-FC_6H_4: 91%, ee = 88%
R^1 = Me, R^2 = Et, R^3 = 1-Naph: 93%, ee = 90%

Scheme 6.15 α-Hydroxylation of malonates with an *in situ* generated nickel catalyst derived from the DBFOX ligand.

enolates,[51] albeit applicable only to acetates or methyl ketones. Buchwald and Palucki elegantly demonstrated the foremost direct α-arylation of an enolate in the absence of a transmetallating agent.[52] A number of palladium complexes with bulky electron-rich ligands have been used successfully for the non-enantioselective intra- and intermolecular version of these processes. On the other hand, the first asymmetric arylation of ketone enolates was achieved by Buchwald *et al.* by using chiral palladium catalysts derived from BINAP or dialkylphosphino-binaphthyl ligands, which provided good yields and enantioselectivities (88% ee).[53] However, these reactions required high catalyst loadings of 20 mol% of palladium sources and 12–24 mol% of ligands. In 2006, Kwong and Chan described the first highly enantioselective nickel-catalysed α-arylation of ketone enolates with aryl halides in the presence of the atropisomeric bipyridylbisphosphine ligand P-Phos as supporting ligand (Scheme 6.16).[54] This nice process required only 2.4 mol% of the (*R*)-P-Phos ligand and 2 mol% of $Ni(cod)_2$ in the presence of NaO*t*-Bu as

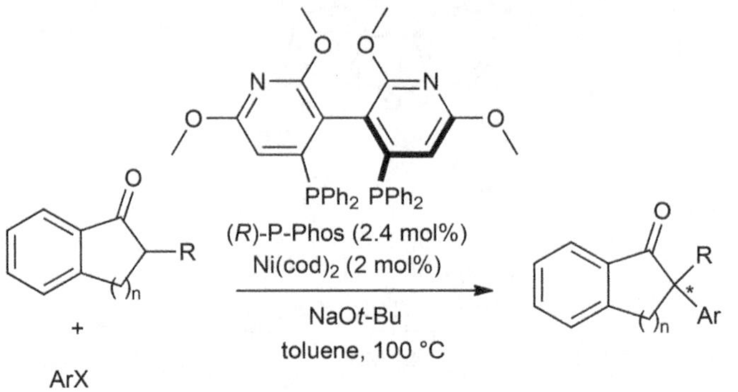

R = Me, ArX = PhBr, n = 2: 85%, ee = 92%
R = Me, ArX = *p*-TolBr, n = 2: 78%, ee = 93%
R = Me, ArX = *p*-(*t*-Bu)C_6H_4Br, n = 2: 92%, ee = 95%
R = Me, ArX = *p*-NCC_6H_4Br, n = 2: 94%, ee = 98%
R = Me, ArX = *m*-NCC_6H_4Br, n = 2: 79%, ee = 95%
R = Me, ArX = *p*-$CF_3C_6H_4$Br, n = 2: 66%, ee = 70%
R = Me, ArX = PhI, n = 2: 97%, ee = 92%
R = Me, ArX = PhCl, n = 2: 59%, ee = 91%
R = Me, ArX = PhBr, n = 1: 78%, ee = 88%
R = Me, ArX = *p*-TolBr, n = 1: 69%, ee = 87%
R = Me, ArX = *p*-FC_6H_4BrBr, n = 1: 80%, ee = 67%
R = Me, ArX = PhBr, n = 3: 70%, ee = 73%
R = Me, ArX = *p*-$CF_3C_6H_4$Br, n = 3: 74%, ee = 60%

Scheme 6.16 α-Arylation of cyclic ketones with haloarenes in the presence of an *in situ* generated nickel catalyst derived from a bisphosphine ligand.

a base. The reaction of a range of bicyclic ketones with various aryl halides afforded the corresponding coupling products in good yields (59–97%) along with moderate to excellent enantioselectivities of 60–98% ee. The best results were achieved in the case of tetralones reacting with *meta*- and *para*-substituted aryl bromides, with the best enantioselectivity of 98% ee reached with *para*-bromobenzonitrile. However, poor reactivity was observed with *ortho*-substituted aryl bromides (less than 10% yield). Indanones also underwent coupling, but in lower enantioselectivities (67–88% ee). The reaction of seven-membered cyclic ketones was also feasible, albeit with even lower enantioselectivities (60–73% ee). Remarkably, an unactivated aryl chloride (ArX = PhCl) could also be applied under the reaction conditions for the first time, with good enantioselectivity (91% ee). However, the arylation of acyclic 2-substituted propiophenones was unsuccessful. It must be noted that this novel methodology provided the best enantioselectivity accomplished so far for this transformation.

Later, Hartwig *et al.* investigated comparable reactions, albeit using aryl triflates as arylating agents and (*R*)-Difluorphos as the chiral ligand under related reaction conditions.[55] As shown in Scheme 6.17, very good enantioselectivities of up to 96% ee for 2-methylindanones, along with

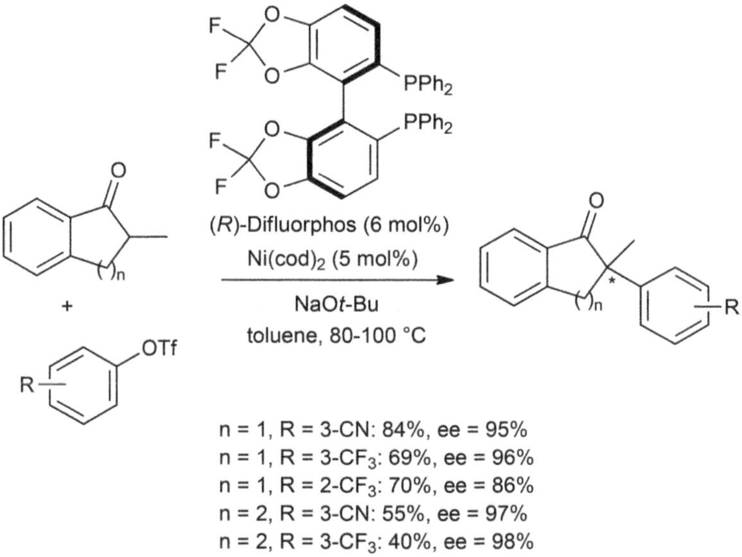

Scheme 6.17 α-Arylation of cyclic ketones with aryl triflates in the presence of an *in situ* generated nickel catalyst derived from the Difluorphos ligand.

even higher enantioselectivities of up to 98% ee for 2-methyltetralones, were achieved for reactions with electron-poor aryl triflates. In this study, the authors investigated the efficiency of the corresponding chiral palladium catalyst and demonstrated that moderate enantioselectivity (55% ee) was obtained in the reaction of an electron-poor aryl triflate, such as *para*-cyanophenyl trifluoromethanesulfonate, with 2-methylindanone. On the other hand, better enantioselectivities of 70–95% ee were achieved in the reactions of electron-neutral and electron-rich aryl triflates. Thus, an appropriate pairing of the catalyst (nickel or palladium) with the electronic properties of the aryl group of the triflate led to a set of α-arylation processes that encompassed reactions of electron-rich, electron-neutral, and electron-poor aryl triflates.

Having identified a dramatic effect of the electrophile on the enantioselectivity of nickel-catalysed α-arylations, these same authors investigated the reactions of cyclic ketones with chloroarenes.[56] It was demonstrated that α-arylations of both 2-alkyl-substituted indanones and tetralones with a range of chloroarenes catalysed by an *in situ* generated nickel catalyst from the (*R*)-BINAP ligand afforded the corresponding arylated products in moderate to high yields (53–89%) along with general excellent enantioselectivities of 92–99% ee, as shown in Scheme 6.18. Much better yields and enantioselectivities were reached with chloroarenes than with the corresponding bromoarenes as well as aryl triflates, demonstrating that the leaving group in the aryl electrophile significantly affects both the yield and the enantioselectivity of the reaction. The substrate scope of the process was found broad since a range of electron-rich, electron-neutral, and electron-deficient chloroarenes

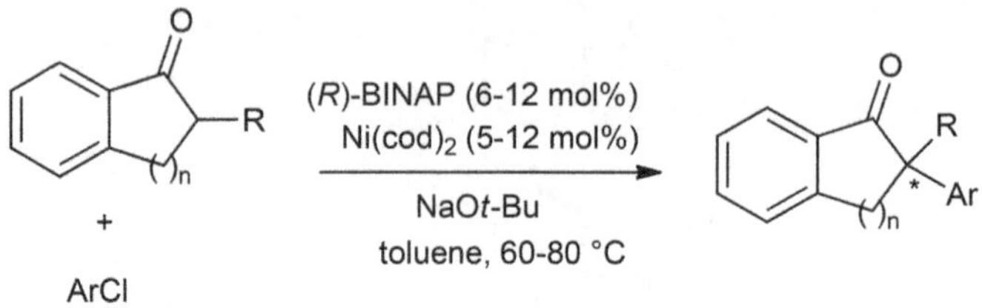

R = Me, n = 1, Ar = *m*-$CF_3C_6H_4$: 75%, ee = 96%
R = Me, n = 1, Ar = *p*-$CF_3C_6H_4$: 81%, ee = 95%
R = Me, n = 1, Ar = *p*-$(CO_2Me)C_6H_4$: 41%, ee = 99%
R = Me, n = 1, Ar = *p*-NCC_6H_4: 58%, ee = 92%
R = Et, n = 1, Ar = *p*-$CF_3C_6H_4$: 66%, ee = 92%
R = Et, n = 1, Ar = Ph: 77%, ee = 98%
R = Bn, n = 1, Ar = *p*-$CF_3C_6H_4$: 89%, ee = 99%
R = Bn, n = 1, Ar = *m*-$MeOC_6H_4$: 83%, ee = 98%
R = Me, n = 2, Ar = *m*-$CF_3C_6H_4$: 74%, ee = 96%
R = Me, n = 2, Ar = *p*-$MeOC_6H_4$: 72%, ee = 97%
R = Bn, n = 2, Ar = *p*-$CF_3C_6H_4$: 76%, ee = 94%
R = Bn, n = 2, Ar = Ph: 72%, ee = 98%
R = Bn, n = 2, Ar = *m*-$MeOC_6H_4$: 53%, ee = 99%

Scheme 6.18 α-Arylation of cyclic ketones with chloroarenes in the presence of an *in situ* generated nickel catalyst derived from the BINAP ligand.

provided comparable excellent results. Moreover, reactions of ketones containing various aliphatic substituents at the α-carbon of indanones and tetralones occurred with similarly high enantioselectivities.

Asymmetric α-heteroarylation is as important as asymmetric α-arylation because heteroaromatic units are ubiquitous in medicinal chemistry and can be reduced to saturated heterocycles. However, asymmetric α-heteroarylation is more challenging than asymmetric α-arylation because ligation of heteroarenes can lead to poisoning of the catalyst or displacement of the chiral ligand to form an achiral catalyst. Because BINAP is a bidentate ligand, it was investigated as the ligand in nickel-catalysed α-heteroarylation of cyclic ketones.[56] However, no reaction between 2-methylindan-1-one and 2-bromopyridine occurred by using this ligand. On the other hand, the authors obtained good results by using the (*R*)-Difluorphos ligand (12 mol%) in combination with $Ni(cod)_2$ (10 mol%). As shown in Scheme 6.19, a range of heteroaryl chlorides underwent α-heteroarylation with indanones and tetralones with moderate to high yields (54–93%) and with moderate to excellent enantioselectivities of 21–99% ee. Actually, generally high enantioselectivities of 90–99% ee were achieved; the few exceptions of lower enantioselectivities were observed for reactions of 2-chloro-6-methoxypyridine (35–41% ee). In contrast to the usual reactivity of haloarenes, bromoarenes reacted with lower yield and enantioselectivity than the corresponding chloroarenes, most likely because of the greater reactivity of the bromoarene through a less selective catalyst formed by decomposition of the nickel species.

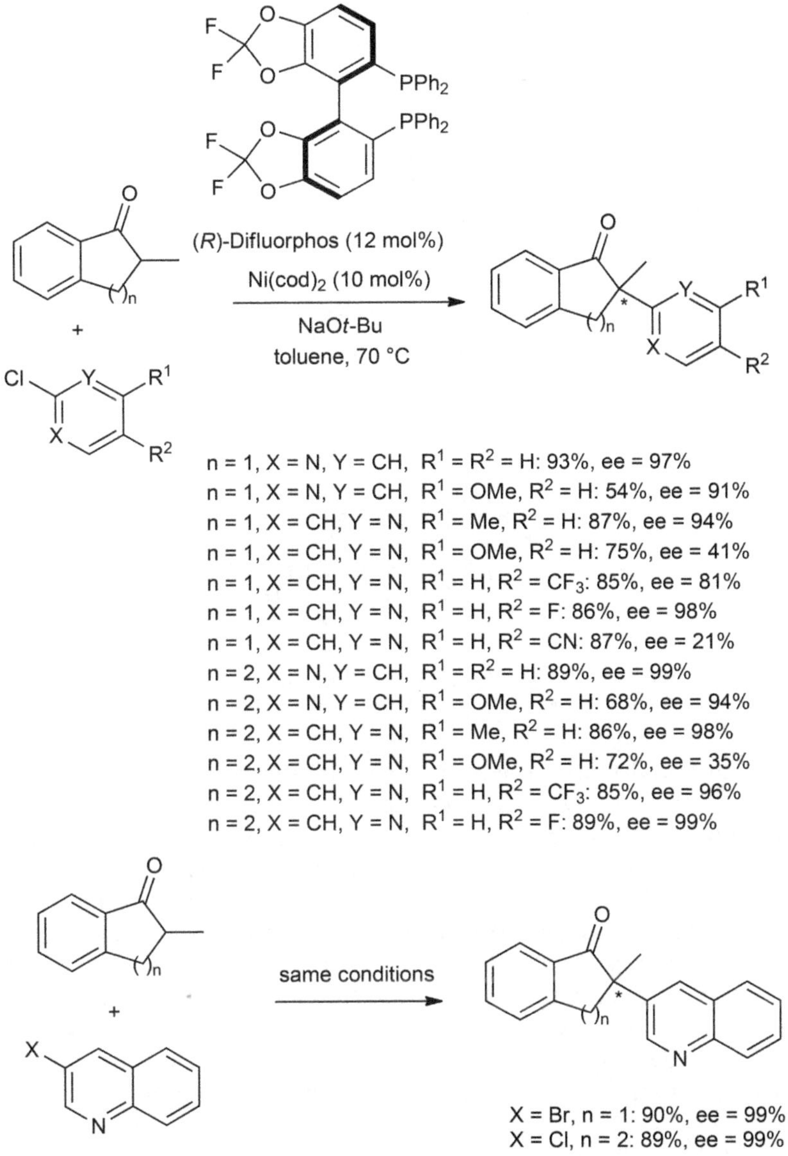

Scheme 6.19 α-Heteroarylations of cyclic ketones with haloarenes in the presence of an *in situ* generated nickel catalyst derived from the Difluorphos ligand.

The catalytic enantioselective direct alkylation reaction of enolates is a less developed field.[57] Early research from Evans' group demonstrated that preformed titanium enolates derived from chiral *N*-acyloxazolidinones reacted with orthoesters to provide the alkylated adducts with high levels of diastereocontrol.[58] In 2005, the same group reported the enantioselective nickel-catalysed

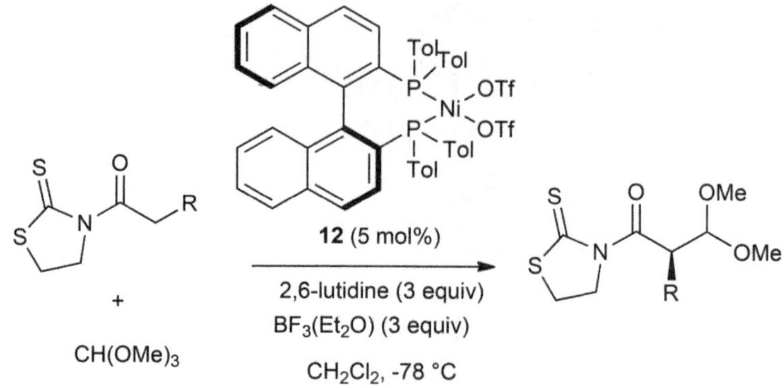

R = Me: 73%, ee = 97%
R = Et: 76%, ee = 95%
R = *i*-Bu: 62%, ee = 93%
R = allyl: 92%, ee = 94%
R = Bn: 91%, ee = 92%
R = Ph: 91%, ee = 98%
R = *p*-MeOC_6H_4: 91%, ee = 98%
R = *p*-ClC_6H_4: 63%, ee = 99%
R = OBn: 89%, ee = 90%

Scheme 6.20 α-Alkylation of *N*-acylthiazolidinethiones in the presence of a preformed nickel catalyst derived from the (*S*)-Tol-BINAP ligand.

alkylation of *N*-acylthiazolidinethiones with methyl orthoesters.[59] As shown in Scheme 6.20, the process catalysed by the chiral preformed nickel complex **12** derived from BINAP afforded the corresponding alkylated products in good yields (62–91%) and with uniformly high enantioselectivities of 90–99% ee. In general, saturated alkyl-substituted thiazolidinethiones provided the products in slightly lower yields than the corresponding allyl and benzyl derivatives. Substrates bearing aryl substituents also gave excellent results, as well as a substrate with a heteroatom substitution (R = OBn). Furthermore, the use of other orthoesters was investigated, and while ethyl orthoformate was competent under the developed conditions (68% yield, 98% ee), the use of trimethyl orthoacetate and trimethyl orthopropionate afforded no product.

6.6 Conclusions

The last decade has seen a number of highly enantioselective nickel-catalysed electrophilic halogenation reactions to be achieved. Most of them are fluorination reactions, involving several types of carbonyl substrates, such as cyclic as well as acyclic β-keto esters, *N*-acylthiazolidinones, acid chlorides, and α-chloro-β-keto esters. All these reactions provided near-perfect enantioselectivities by using various ligands, such as DBFOX, benzoylquinidine, (oxazolinyl)pyridine, or chiral diamines. It is important to note that the majority of these reactions generated quaternary stereogenic centres.

The bromination reactions remained challenging. On the other hand, remarkable enantioselective nickel-catalysed α-aminations of *N*-Boc-oxindoles with azodicarboxylates have been achieved with enantioselectivities of up to 99% ee by using a chiral Schiff base dinuclear catalyst. It must be noted that the use of other sources of electrophilic nitrogen, such as nitroso compounds and iodinanes, in these reactions has not been described. In the context of asymmetric α-hydroxylation reactions, early developments required multistep procedures, where the substrates were activated *via* enol or enolate intermediates. More recently, it has been demonstrated that the hydroxylation of β-keto esters could be achieved directly, by using oxaziridines. In this context, advances have been made with highly enantioselective nickel-catalysed α-hydroxylations of cyclic and acyclic β-keto esters as well as malonates, performed with the DBFOX ligand, providing excellent enantioselectivities of up to 97% and 98% ee, respectively. In the context of α-arylations, many excellent results have been achieved, such as nickel-catalysed arylations and heteroarylations of a wide variety of indanones and tetralones with chloroarenes or aryl triflates, which all provided enantioselectivities of up to 99% ee using Difluorphos or BINAP ligands. Finally, a nice α-alkylation of *N*-acylthiazolidinethiones catalysed by a BINAP–nickel catalyst was performed with up to 99% ee. It is important to note that many of these novel reactions (halogenations, aminations, hydroxylations, as well as arylations) generated challenging quaternary carbon stereogenic centres.

References

1. (a) K. L. Kirk, *J. Fluorine Chem.*, 2006, **127**, 1013–1029; (b) C. Isanbor and D. O'Hagan, *J. Fluorine Chem.*, 2006, **127**, 303–319.
2. (a) H. J. Bohm, D. Nanmer, S. Bendels, M. Kansy, B. Kuhn, K. Muller, U. Obst-Sander and M. Stahl, *ChemBioChem*, 2004, **5**, 637–643; (b) B. E. Smart, *J. Fluorine Chem.*, 2001, **109**, 3–11; (c) F. M. D. Ismail, *J. Fluorine Chem.*, 2002, **118**, 27–33.
3. (a) G. S. Lal, G. P. Pez and R. G. Syvret, *Chem. Rev.*, 1996, **96**, 1737–1756; (b) S. D. Taylor, C. C. Kotoris and G. Hum, *Tetrahedron*, 1999, **55**, 12431–12655.
4. (a) P. M. Pihko, *Angew. Chem., Int. Ed.*, 2006, **45**, 544–547; (b) C. C. Johansson and T. J. Colacot, *Angew. Chem., Int. Ed.*, 2010, **49**, 676–707; (c) K. Muniz, *Topics in Organometallic Chemistry*, Springer, Berlin, 2010, pp. 1–18; (d) Y. Takemoto and H. Miyabe, *Catalytic Asymmetric Synthesis*, Wiley, Weinheim, 3rd edn, 2010, pp. 227–267; (e) A. M. R. Smith and K. K. Hii, *Chem. Rev.*, 2011, **111**, 1637–1656; (f) S. Shirakawa and K. Maruoka, *Science of Synthesis: Water in Organic Synthesis*, Thieme, Stuttgart, 2012, pp. 601–615.
5. (a) V. A. Brunet and D. O'Hagan, *Angew. Chem., Int. Ed.*, 2008, **47**, 1179–1182; (b) J. A. Ma and D. Cahard, *Chem. Rev.*, 2004, **104**, 6119–6146.
6. L. Hintermann and A. Togni, *Angew. Chem., Int. Ed.*, 2000, **39**, 4359–4362.

7. L. Hintermann and A. Togni, *Helv. Chim. Acta*, 2000, **83**, 2425–2435.
8. (a) Y. Hamashima, K. Yagi, H. Takano, L. Tamas and M. Sodeoka, *J. Am. Chem. Soc.*, 2002, **124**, 14530–14531; (b) T. Suzuki, T. Goto, Y. Hamashima and M. Sodeoka, *J. Org. Chem.*, 2007, **72**, 246–250; (c) N. R. Lee, S. M. Kim and D. Y. Kim, *Bull. Korean Chem. Soc.*, 2009, **30**, 829–836.
9. M. Perseghini, M. Massaccesi, Y. Y. Liu and A. Togni, *Tetrahedron*, 2006, **62**, 7180–7190.
10. L. Bernardi and K. A. Jorgensen, *Chem. Commun.*, 2005, 1324–1326.
11. (a) M. Marigo, N. Kumaragurubaran and K. A. Jorgensen, *Chem.–Eur. J.*, 2004, **10**, 2133–2137; (b) M. Frings and C. Bolm, *Eur. J. Org. Chem.*, 2009, 4085–4090.
12. N. Shibata, T. Ishimura, T. Nagai, J. Kohno and T. Toru, *Synlett*, 2004, **10**, 1703–1706.
13. N. Shibata, J. Kohno, K. Takai, T. Ishimaru, S. Nakamura, T. Toru and S. Kanemasa, *Angew. Chem., Int. Ed.*, 2005, **44**, 4204–4207.
14. K. Shibatomi, Y. Tsuzuki, S.-i. Nakata, Y. Sumikawa and S. Iwasa, *Synlett*, 2007, **4**, 551–554.
15. K. Shibatomi, Y. Tsuzuki and S. Iwasa, *Chem. Lett.*, 2008, **37**, 1098–1099.
16. T. Suzuki, Y. Hamashima and M. Sodeoka, *Angew. Chem., Int. Ed.*, 2007, **46**, 5435–5439.
17. Y. Hamashima, T. Nagi, R. Shimizu, T. Tsuchimoto and M. Sodeoka, *Eur. J. Org. Chem.*, 2011, 3675–3678.
18. T. Ishimaru, N. Shibata, D. S. Reddy, T. Horikawa, S. Nakamura and T. Toru, *Beilstein J. Org. Chem.*, 2008, **4**, 16.
19. D. S. Reddy, N. Shibata, T. Horikawa, S. Suzuki, S. Nakamura, T. Toru and M. Shiro, *Chem.–Asian J.*, 2009, **4**, 1411–1415.
20. D. H. Paull, M. T. Scerba, E. Alden-Danorth, L. R. Widger and T. Lectka, *J. Am. Chem. Soc.*, 2008, **130**, 17260–17261.
21. R. Frantz, M. Hintermann, M. Perseghini, D. Broggini and A. Togni, *Org. Lett.*, 2003, **5**, 1709–1712.
22. M. J. Cho, Y. K. Kang, N. R. Lee and D. Y. Kim, *Bull. Korean Chem. Soc.*, 2007, **28**, 2191–2192.
23. S. H. Kang and D. Y. Kim, *Adv. Synth. Catal.*, 2010, **352**, 2783–2786.
24. (a) R. M. Williams, *Synthesis of Optically Active α-Amino Acids*, Pergamon, Oxford, 1989; (b) R. O. Duthaler, *Tetrahedron*, 1994, **50**, 1539–1650; (c) S. Hanessian, G. McNaughton-Smith, H.-G. Lombart and W. D. Lubell, *Tetrahedron*, 1997, **53**, 12789–12854; (d) M. Arend, *Angew. Chem., Int. Ed.*, 1999, **38**, 2873–2874; (e) S. Kotha, *Acc. Chem. Res.*, 2003, **36**, 342–351; (f) K. Maruoka and T. Ooi, *Chem. Rev.*, 2003, **103**, 3013–3028; (g) C. Najera and J. M. Sansano, *Chem. Rev.*, 2007, **107**, 4584–4671.
25. E. Erdik, *Tetrahedron*, 2004, **60**, 8747–8782.
26. D. A. Evans and S. G. Nelson, *J. Am. Chem. Soc.*, 1997, **119**, 6452–6453.
27. Z. Zhang, Z. Wang, S. Bai, K. Shen, D. Chen, X. Liu, L. Lin and X. Feng, *Chem.–Eur. J.*, 2010, **16**, 6632–6637.
28. (a) D. A. Evans and D. S. Johnson, *Org. Lett.*, 1999, **1**, 595–598; (b) M. Marigo, K. Juhl and K. A. Jørgensen, *Angew. Chem., Int. Ed.*, 2003, **42**,

1367–1369; (c) S. Ma, N. Jiao, Z. Zheng, Z. Ma, Z. Lu, L. Ye, Y. Deng and G. Chen, *Org. Lett.*, 2004, **6**, 2193–2196; (d) C. Foltz, B. Stecker, G. Marconi, S. Bellemin-Laponnaz, H. Wadepohl and L. H. Gade, *Chem. Commun.*, 2005, 5115–5117; (e) D. P. Huber, K. Stanek and A. Togni, *Tetrahedron: Asymmetry*, 2006, **17**, 658–664; (f) Y. K. Kang and D. Y. Kim, *Tetrahedron Lett.*, 2006, **47**, 4565–4568; (g) M. V. Nandakumar, S. Ghosh and C. Schneider, *Eur. J. Org. Chem.*, 2009, 6393–6398; (h) S. Ghosh, M. V. Nandakumar, H. Krautscheid and C. Schneider, *Tetrahedron Lett.*, 2010, **51**, 1860–1862.

29. L. Bernardi, W. Zhuang and K. A. Jørgensen, *J. Am. Chem. Soc.*, 2005, **127**, 5772–5778.
30. J. H. Lee, H. T. Bang and D. Y. Kim, *Synlett*, 2008, 1821–1824.
31. (a) Y. Yamashita, H. Hishitani and S. Kobayashi, *Can. J. Chem.*, 2000, **78**, 666–672; (b) Q. A. Chen, W. Zeng and Y. G. Zhou, *Tetrahedron Lett.*, 2009, **50**, 6866–6868.
32. (a) T. Mashiko, K. Hara, D. Tanaka, Y. Fujiwara, N. Kumagai and M. Shibasaki, *J. Am. Chem. Soc.*, 2007, **129**, 11342–11343; (b) T. Mashiko, N. Kumagai and M. Shibasaki, *J. Am. Chem. Soc.*, 2009, **131**, 14990–14999.
33. Y. Hasegawa, M. watanabe, I. D. Gridnev and T. Ikariya, *J. Am. Chem. Soc.*, 2008, **130**, 2158–2159.
34. J. Comelles, A. Pericas, M. Moreno-Manas, A. Vallribera, G. Drudis-Solé, A. Lledos, T. Parella, A. Roglans, S. Garcia-Granda and L. Roces-Fernandez, *J. Org. Chem.*, 2007, **72**, 2077–2087.
35. D. Y. Kim, *Bull. Korean Chem. Soc.*, 2008, **29**, 2036–2038.
36. J. Y. Mang, D. G. Kwon and D. Y. Kim, *J. Fluorine Chem.*, 2009, **130**, 259–262.
37. J. Y. Mang, D. G. Kwon and D. Y. Kim, *Bull. Korean Chem. Soc.*, 2009, **30**, 249–252.
38. S. Mouri, Z. Chen, H. Mitsunuma, M. Furutachi, S. Matsunaga and M. Shibasaki, *J. Am. Chem. Soc.*, 2010, **132**, 1255–1257.
39. S. Mouri, Z. Chen, S. Matsunaga and M. Shibasaki, *Heterocycles*, 2012, **84**, 879–892.
40. A. Paju, T. Pehk and M. Lopp, *Tetrahedron*, 2002, **58**, 7321–7326.
41. M. Lopp, A. Paju, T. Kanger and T. Pehk, *Tetrahedron Lett.*, 1997, **38**, 5051–5054.
42. P. Y. Toullec, C. Bonaccorsi, A. Mezzeti and A. Togni, *Proc. Natl. Acad. Sci. U. S. A.*, 2004, **101**, 5810–5814.
43. T. Ishimaru, N. Shibata, J. Nagai, S. Nakamura, T. Toru and S. Kanemasa, *J. Am. Chem. Soc.*, 2006, **128**, 16488–16489.
44. G. Olack and H. Morrison, *J. Org. Chem.*, 1991, **56**, 4969–4971.
45. (a) M. R. Acocella, O. G. Mancheno, M. Bella and K. A. Jørgensen, *J. Org. Chem.*, 2004, **69**, 8165–8167; (b) C. Bonaccorsi, M. Althaus, C. Becker, A. Togni and A. Mezzeti, *Pure Appl. Chem.*, 2006, **78**, 391–396.
46. D. S. Reddy, N. Shibata, J. Nagai, S. Nakamura and T. Toru, *Angew. Chem., Int. Ed.*, 2009, **48**, 803–806.
47. (a) K. Fuji, *Chem. Rev.*, 1993, **93**, 2037–2066; (b) C. Bolm, J. P. Hildebrand, K. Müniz and N. Hermanns, *Angew. Chem., Int. Ed.*, 2001, **40**, 3284–3308; (c) D. A. Culkin and J. F. Hartwig, *Acc. Chem. Res.*, 2003, **36**, 234–245.

48. M. F. Semmelhack, R. D. Stauffer and T. D. Rogerson, *Tetrahedron Lett.*, 1973, **14**, 4519–4522.
49. A. A. Millard and M. W. Rathke, *J. Am. Chem. Soc.*, 1977, **99**, 4833–4835.
50. J. F. Fauvarque and A. Jutand, *J. Organomet. Chem.*, 1979, **99**, 273–281.
51. M. Kogusi, I. Hagiwara, T. Sumiya and T. Migita, *Bull. Chem. Soc. Jpn.*, 1984, **57**, 242–246.
52. M. Palucki and S. L. Buchwald, *J. Am. Chem. Soc.*, 1997, **119**, 11108–11109.
53. (a) J. P. Ahman, M. V. Wolfe, M. Troutman, S. L. Palucki and S. L. Buchwald, *J. Am. Chem. Soc.*, 1988, **120**, 1918–1919; (b) T. Hamada, A. Chieffi, J. P. Ahman and S. L. Buchwald, *J. Am. Chem. Soc.*, 2002, **124**, 1261–1268.
54. G. Chen, F. Y. Kwong, H. O. Chan, W.-Y. Yu and A. S. C. Chan, *Chem. Commun.*, 2006, 1413–1415.
55. X. Liao, Z. Weng and J. F. Hartwig, *J. Am. Chem. Soc.*, 2008, **130**, 195–200.
56. S. Ge and J. F. Hartwig, *J. Am. Chem. Soc.*, 2011, **133**, 16330–16333.
57. (a) D. L. Hughes, in *Comprehensive Asymmetric Catalysis*, ed. E. N. Jacobsen, A. Pfaltz and H. Yamamoto, Springer-Verlag, Berlin, 1999, vol. 3, ch. 3, pp. 1273–1294; (b) D. L. Hughes, in *Comprehensive Asymmetric Catalysis, Supplement*, ed. E. N. Jacobsen, A. Pfaltz and H. Yamamoto, Springer-Verlag, Berlin, 2004, vol. 1, pp. 161–169; (c) A. G. Doyle and E. N. Jacobsen, *J. Am. Chem. Soc.*, 2005, **127**, 62–63.
58. D. A. Evans, F. Urpi, T. C. Somers, J. S. Clark and M. T. Bilodeau, *J. Am. Chem. Soc.*, 1990, **112**, 8215–8216.
59. D. A. Evans and R. J. Thomson, *J. Am. Chem. Soc.*, 2005, **127**, 10506–10507.

CHAPTER 7

Enantioselective Nickel-Catalysed Additions of Organometallic Reagents to Aldehydes

7.1 Introduction

The enantioselective addition of organometallic reagents to aldehydes in the presence of a chiral catalyst is one of the most established carbon–carbon bond-forming asymmetric processes, providing enantioenriched secondary alcohols which are highly valuable intermediates for preparing chiral pharmaceuticals and agricultural products.[1] Furthermore, these reactions often serve as test reactions for the investigation of novel catalysts. Among nucleophiles enantioselectively added to aldehydes, dialkylzincs, and especially diethylzinc, have been by far the most extensively studied, thus constituting a commonly used method for synthesising chiral secondary alcohols. On the other hand, the enantioselective nickel-catalysed additions of trialkyl- or triarylaluminum reagents to aldehydes first reported in 2005 have known several interesting advances only in recent years. In addition, the first highly efficient example of asymmetric nickel-catalysed arylation of aldehydes with a boron reagent was reported in 2009.

7.2 Additions of Organoaluminum Reagents

For alkylation reagents, trialkylaluminum compounds are more interesting than other organometallic reagents because they are economically obtained on an industrial scale from aluminum hydride and alkenes.[2] Despite this

RSC Catalysis Series No. 26
Enantioselective Nickel-Catalysed Transformations
By Hélène Pellissier

Published by the Royal Society of Chemistry, www.rsc.org

advantage, their use has been rare for a long time. In this respect, the few most successful catalysts for the enantioselective addition of trialkylaluminum to aldehydes have been titanium complexes bearing chiral diols or *N*-sulfonylated amino alcohols as ligands.[3] However, the high catalyst loadings needed and the slow turnover rate hampered the potential utility of these catalytic systems. In 2005, Woodward and co-workers reported the first example of asymmetric addition of trialkylaluminum reagents to aldehydes employing a nickel catalyst containing a phosphoramidite ligand.[4] As shown in Scheme 7.1, the addition of $AlMe_3$ to a range of aldehydes induced by a catalyst generated *in situ* from $Ni(acac)_2$ and phosphoramidite ligand **1** used at a low catalyst loading of 1–2 mol% afforded the corresponding chiral secondary alcohols in good yields (55–95%) and with moderate to high enantioselectivities of 61–95% ee. The best enantioselectivities were achieved for isopentanal and benzaldehydes bearing a substituent at the *para* position, while the lowest enantioselectivity (7% ee) was obtained for the addition of $AlEt_3$ to an aliphatic aldehyde with a cyclohexyl substituent. Both 1- and 2-naphthaldehydes did not participate in effective catalysis since they provided poor conversion and enantioselectivity (<10% ee). The results seemed to indicate that the structure of the chiral ligand required modification for each substrate. Almost no trace of β-elimination by-products was found.

Often, the reactivity of air-unstable $AlMe_3$ in catalytic synthesis is modified by the presence of oxo-containing by-products attained through

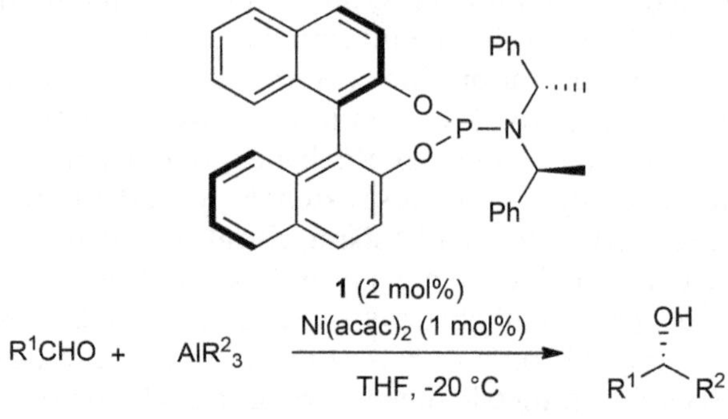

R^1 = Ph, R^2 = Me: 60%, ee = 85%
R^1 = Ph, R^2 = Et: 95%, ee = 83%
R^1 = Cy, R^2 = Et: 55%, ee = 7%
R^1 = *i*-Bu, R^2 = Me: 60%, ee = 95%
R^1 = *i*-Bu, R^2 = Et: 72%, ee = 48%
R^1 = *t*-BuCH$_2$, R^2 = Me: 95%, ee = 67%
R^1 = *n*-Hex, R^2 = Me: 91%, ee = 61%
R^1 = BnCH$_2$, R^2 = Me: 94%, ee = 77%
R^1 = *p*-(AcO)C$_6$H$_4$, R^2 = Me: 78%, ee = 90%
R^1 = *p*-(*t*-Bu)C$_6$H$_4$, R^2 = Me: 84%, ee = 92%
R^1 = (*E*)PhCH=CH, R^2 = Me: 73%, ee = 77%

Scheme 7.1 Alkylation of aldehydes with AlR^2_3 in the presence of an *in situ* generated nickel catalyst derived from a phosphoramidite ligand.

accidental exposure of stock solutions to traces of air or moisture. In looking for an alternative approach to preparing very pure $AlMe_3$ samples, Woodward *et al.* sought to use amine adducts ($R_3N{\cdot}AlR'_3$, with R′ = Me or Et) first synthesised by Brown and Davidson.[5] Screening a number of these compounds, the authors found that DABCO·$AlMe_3$ was stable in air. Then, this substrate was submitted to the enantioselective addition to aldehydes under comparable reaction conditions to those of Scheme 7.1, affording the corresponding chiral secondary alcohols in low to high enantioselectivities (2–95% ee), as shown in Scheme 7.2. Generally, aromatic aldehydes gave good results with the exception of electron-rich *para*- and *ortho*-methoxybenzaldehydes,

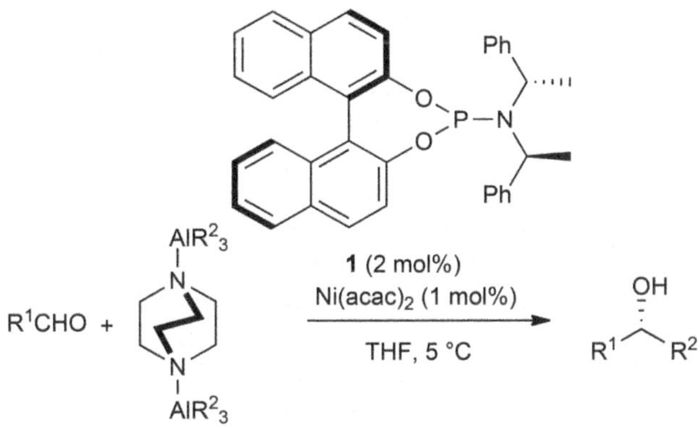

R^1 = Ph, R^2 = Me: 92%, ee = 91%
R^1 = Ph, R^2 = Et: 95%, ee = 86%
R^1 = *p*-BrC_6H_4, R^2 = Me: 98%, ee = 93%
R^1 = *p*-BrC_6H_4, R^2 = Et: 91%, ee = 89%
R^1 = *p*-(OAc)C_6H_4, R^2 = Me: 78%, ee = 90%
R^1 = *p*-(*t*-Bu)C_6H_4, R^2 = Me: 84%, ee = 92%
R^1 = *p*-ClC_6H_4, R^2 = Me: 79%, ee = 93%
R^1 = *p*-FC_6H_4, R^2 = Me: 84%, ee = 94%
R^1 = *m*-FC_6H_4, R^2 = Me: 77%, ee = 94%
R^1 = *o*-FC_6H_4, R^2 = Me: 81%, ee = 87%
R^1 = *p*-NCC_6H_4, R^2 = Me: 69%, ee = 94%
R^1 = *p*-$CF_3C_6H_4$, R^2 = Me: 58%, ee = 94%
R^1 = *p*-$CF_3C_6H_4$, R^2 = Et: 90%, ee = 95%
R^1 = *p*-Tol, R^2 = Me: 90%, ee = 81%
R^1 = *m*-Tol, R^2 = Me: 82%, ee = 93%
R^1 = *o*-Tol, R^2 = Me: 78%, ee = 90%
R^1 = *p*-$MeOC_6H_4$, R^2 = Me: 70%, ee = 2%
R^1 = *m*-$MeOC_6H_4$, R^2 = Me: 58%, ee = 89%
R^1 = *o*-$MeOC_6H_4$, R^2 = Me: 32%, ee = 2%
R^1 = Cy, R^2 = Me: 80%, ee = 94%
R^1 = Cy, R^2 = Et: 60%, ee = 91%
R^1 = (*E*)-PhCH=CH, R^2 = Me: 94%, ee = 66%

Scheme 7.2 Alkylation of aldehydes with $(R^2{}_3Al)_2{\cdot}$DABCO in the presence of an *in situ* generated nickel catalyst derived from a phosphoramidite ligand.

which both provided the products in only 2% ee. Changing the *para*-methoxy ether substituent to the less electron-releasing *para-O*-acetyl group or *para-tert*-butyl group resulted in recovery of the enantioselectivity (90–92% ee). In contrast to the poor result (7% ee) obtained in the addition of $AlMe_3$ to cyclohexanecarbaldehyde (Scheme 7.1), the use of DABCO·AlR_3 (R = Me or Et) allowed high enantioselectivities (91–94% ee) to be achieved. Steric limitations on the substrates which could be employed were demonstrated since neither 1- nor 2-naphthaldehydes afforded acceptable results.

To further extend the range of ligands and performance of this asymmetric nickel-catalysed addition of organoaluminum reagents to aldehydes, Woodward and co-workers designed a library of chiral monophosphate ligands derived from natural D-glucose, D-galactose, and D-fructose.[6] These ligands have the advantage of carbohydrate and phosphite ligands, such as availability at a low price from readily available alcohols and facile modular modifications. In addition, they are less sensitive to air than typical phosphines. With this library, the authors fully investigated the effects of systematically varying the configurations at C-3 and C-4 of the ligand backbone, different substituents/configurations in the biaryl phosphite moiety, the carbohydrate ring size, and the flexibility of the ligand backbone. From this study, D-glucofuranose ligand **2** was selected as optimal in the addition of $(AlMe_3)_2$·DABCO to benzaldehydes. As shown in Scheme 7.3, the corresponding alcohols were achieved in moderate yields (53–100%) and with moderate to good enantioselectivities of 41–94% ee. It was shown that the enantioselectivity was hardly affected by the presence of electron-withdrawing or electron-donating groups at the *para* position of the phenyl group. However, the best yield (100%) was achieved using benzaldehyde as substrate, while the substrate with a *para*-methoxy group gave the poorest (53%). The enantioselectivity was also significantly influenced by steric factors, since it was better when *para*-substituted arylaldehydes were used as substrates. The results of using $(AlEt_3)_2$·DABCO as alkylating agent indicated that the catalytic performance followed the same trend as for $(AlMe_3)_2$·DABCO.

D-Glucofuranose ligand **2** was also investigated in the addition of $(AlR_3)_2$·DABCO (R = Me or Et) to benzaldehydes.[6] As shown in Scheme 7.4, the corresponding alcohols were achieved in moderate yields (22–78%) and with moderate to good enantioselectivities of 35–91% ee. The results indicated that the catalytic performance followed the same trend as for the trialkylaluminum addition to aldehydes (see Scheme 7.3). However, the yields were lower than in the trimethylaluminum addition. Again, the catalytic precursor containing the phosphite ligand **2** provided the best enantioselectivity (91% ee). It must be noted that these two processes (Schemes 7.3 and 7.4) present the advantage of using a low catalyst loading of only 1 mol%.

With the aim of further expanding the range of ligands in these processes, the same authors investigated the performance of phosphite-oxazoline ligands, demonstrating that the enantioselectivity of $AlMe_3$ and $(AlMe_3)_2$·DABCO additions to benzaldehydes depended strongly on the steric properties of the oxazoline substituents and on the substrates.[7]

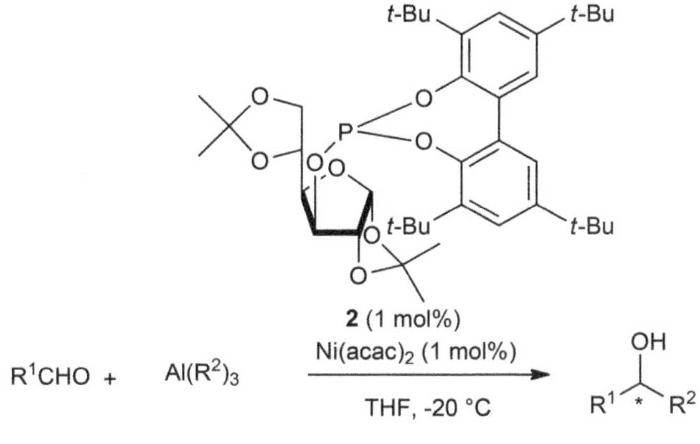

R^1 = Ph, R^2 = Me: 100%, ee = 90% (*S*)
R^1 = Ph, R^2 = Et: 96%, ee = 88% (*S*)
R^1 = *p*-ClC_6H_4, R^2 = Me: 82%, ee = 91% (*S*)
R^1 = *p*-ClC_6H_4, R^2 = Et: 83%, ee = 90% (*S*)
R^1 = *p*-$MeOC_6H_4$, R^2 = Me: 53%, ee = 94% (*S*)
R^1 = *p*-$CF_3C_6H_4$, R^2 = Me: 95%, ee = 93% (*S*)
R^1 = *p*-$CF_3C_6H_4$, R^2 = Et: 96%, ee = 94% (*S*)
R^1 = *p*-Tol, R^2 = Me: 84%, ee = 91% (*S*)
R^1 = *p*-Tol, R^2 = Et: 85%, ee = 88% (*S*)
R^1 = *p*-BrC_6H_4, R^2 = Me: 86%, ee = 92% (*S*)
R^1 = *m*-ClC_6H_4, R^2 = Me: 73%, ee = 74% (*S*)
R^1 = *o*-ClC_6H_4, R^2 = Me: 85%, ee = 41% (*R*)
R^1 = CH_2Bn, R^2 = Me: 89%, ee = 25% (*S*)
R^1 = (*E*)-PhCH=CH, R^2 = Me: 44%, ee = 25% (*R*)

Scheme 7.3 Alkylation of aldehydes with trialkylaluminum reagents in the presence of an *in situ* generated nickel catalyst from a D-glucofuranose-derived phosphite ligand.

Phosphite-oxazoline **3** was selected as the most efficient ligand in this study, albeit providing only moderate enantioselectivities of up to 59% ee, as shown in Scheme 7.5.

A new class of furanoside phosphite-phosphoramidite ligands was later investigated in the same reactions by Diéguez and Pamies.[8] The ligands studied were prepared from inexpensive D-xylose and D-glucose and had the advantage of carbohydrate and phosphite/phosphoramidate moieties. After systematic variation of the position of the phosphoramidite group at either the C-5 or C-3 positions, the configuration of C-3 and the substituents in the biaryl phosphite/phosphoramidite group moieties, enantioselectivities of up to 84% ee and good yields of up to 99% were obtained in the nickel-catalysed additions of AlR_3 and $(AlR_3)_2$·DABCO (R = Me or Et) to various benzaldehydes by using the D-glucose-derived phosphite-phosphoramidite ligand **4**, as shown in Scheme 7.6. The results showed that the catalytic performance

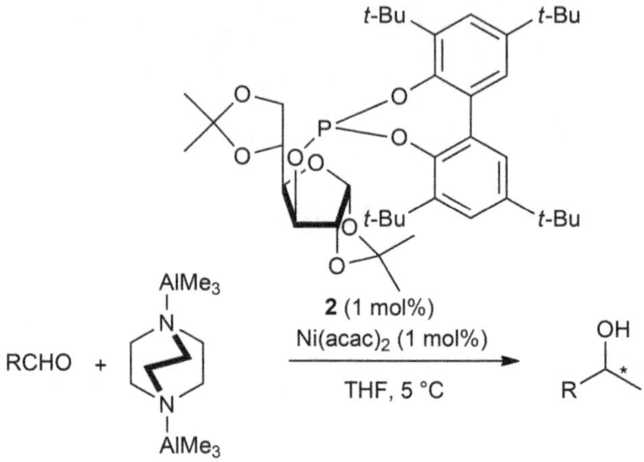

R = Ph: 78%, ee = 88% (*S*)
R = *p*-ClC$_6$H$_4$: 64%, ee = 91% (*S*)
R = *m*-ClC$_6$H$_4$: 58%, ee = 67% (*S*)
R = *o*-ClC$_6$H$_4$: 62%, ee = 35% (*R*)
R = *p*-MeOC$_6$H$_4$: 25%, ee = 78% (*S*)
R = *p*-CF$_3$C$_6$H$_4$: 53%, ee = 82% (*S*)
R = *p*-BrC$_6$H$_4$: 69%, ee = 90% (*S*)
R = *p*-Tol: 68%, ee = 88% (*S*)
R = (*E*)-PhCH=CH: 22%, ee = 30% (*R*)

Scheme 7.4 Methylation of aldehydes with $(AlMe_3)_2$·DABCO in the presence of an *in situ* generated nickel catalyst from a D-glucofuranose-derived phosphite ligand.

followed the same trend in the additions of both types of alkylating agents AlR_3 and $(AlR_3)_2$·DABCO; however, the yields were generally lower for the use of $(AlR_3)_2$·DABCO as substrates.

In 2009, another library of modular sugar-based phosphoramidite ligands was screened by these authors in comparable nickel-catalysed alkylations of aromatic aldehydes with $AlR^2{}_3$ and $(AlR^2{}_3)_2$·DABCO.[9] After systematic variation of the sugar backbone, the substituents at the phosphoramidite moieties and the flexibility of the ligand backbone, the monophosphoramidite ligand **5** derived from D-glucose was selected as optimal, providing moderate to almost quantitative yields at a low catalyst loading of 1 mol%, along with moderate enantioselectivities of up to 69% ee for both $AlR^2{}_3$ and $(AlR^2{}_3)_2$·DABCO additions (Scheme 7.7).

In 2011, the same authors reported slightly better enantioselectivities of up to 84% ee in the same reactions by using D-glucose-derived phosphite ligand **6** at 1 mol% of catalyst loading (Scheme 7.8).[10] In this study, the authors demonstrated that the introduction of a methyl substituent at C-3 of the sugar backbone in allofuranoside ligands was highly advantageous in terms of enantioselectivity.

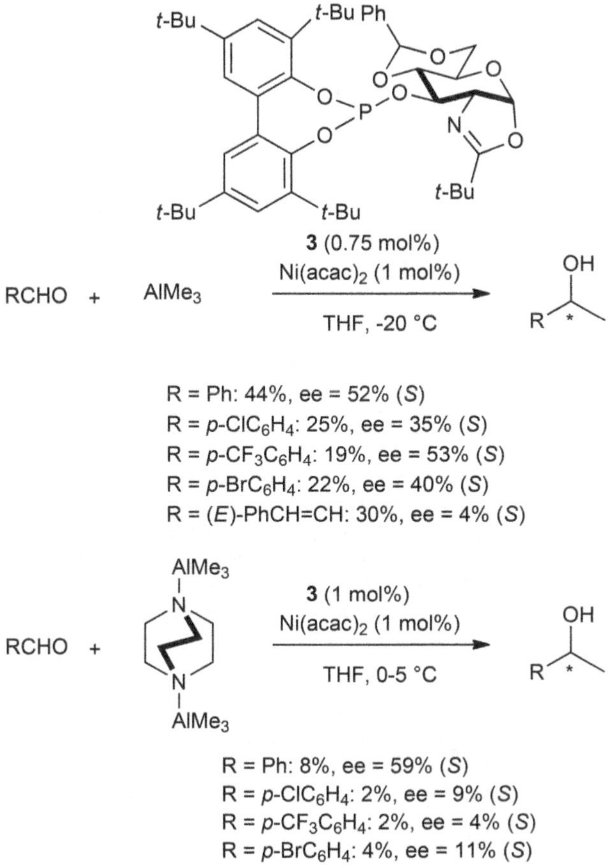

Scheme 7.5 Methylations of aldehydes with $AlMe_3$ and $(AlMe_3)_2$·DABCO in the presence of an *in situ* generated nickel catalyst from a D-glucosamine-derived phosphite-oxazoline ligand.

7.3 Additions of Organozinc Reagents

Among nucleophiles enantioselectively added to aldehydes, diethylzinc has been by far the most extensively studied. Since the first report by Oguni and Omi in 1984,[11] hundreds of chiral ligands of different families have been used for this type of reaction.[12] For example, a series of chiral α-amino acid amides were successfully applied as nickel ligands to these reactions by Burguete and co-workers, affording good to excellent enantioselectivities (85–97% ee).[13] The formation of the active complexes required basic media and was accompanied by deprotonation of the amide (NH) group and the generation of square-planar nickel species (colour change from blue to orange). The stoichiometry of the complexes was determined by the continuous variations method. For this purpose, chiral amino amides containing strong chromophoric groups were later selected as ligands by the same

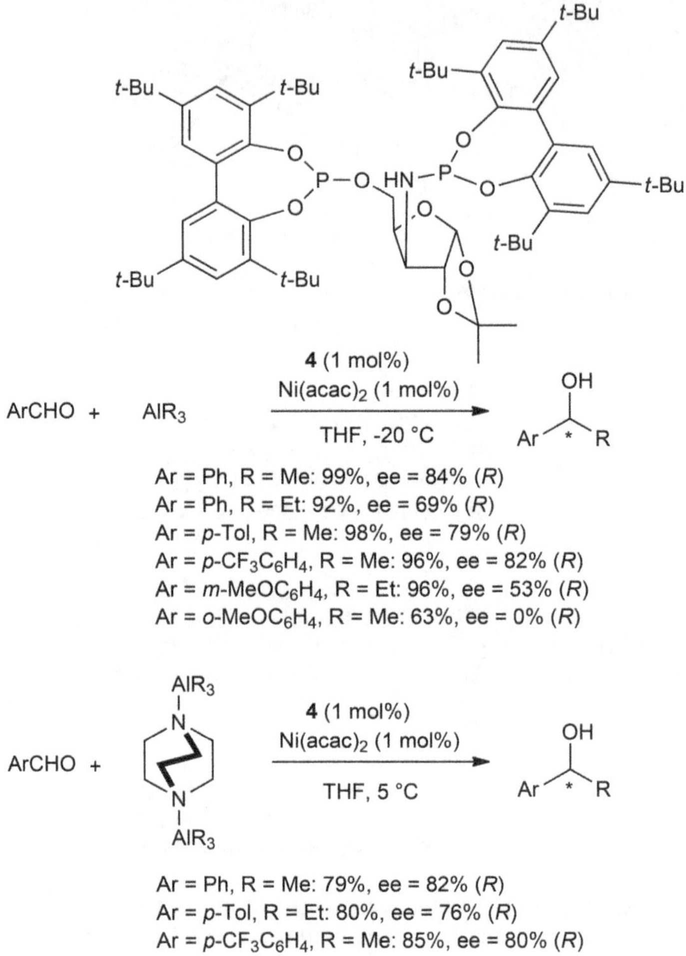

Scheme 7.6 Alkylations of aromatic aldehydes with AlR_3 and $(AlR_3)_2$·DABCO in the presence of an *in situ* generated nickel catalyst from a D-glucose-derived phosphite-phosphoramidite ligand.

authors.[14] The results from the absorbance at approximately 400 nm revealed the existence of the corresponding nickel complexes with two different stoichiometries, namely 1:1 and 1:2 (Ni/ligand). The data obtained on complexes from 1:1 metal/ligand ratios suggested the presence of monomeric and oligomeric species in equilibrium. However, the presence of the correct M/ligand ratio was confirmed through acidic treatment of the complex and separate UV determination of the ligand and the metal. For the 1:2 complexes, the square-planar structure was confirmed by NMR spectroscopy. All data indicated that the 1:2 complexes in the more stable *trans* configuration were of higher stability than the 1:1 complexes. The nickel complexes **7** and **8** were prepared under both metal/ligand stoichiometries (1:1 and 1:2), and

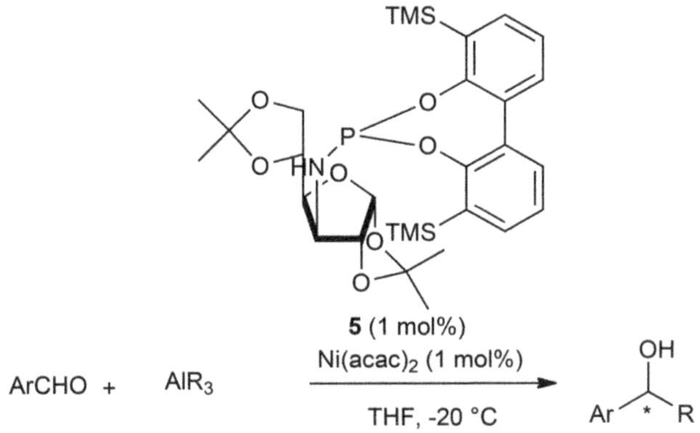

Ar = Ph, R = Me: 99%, ee = 69% (*R*)
Ar = Ph, R = Et: 93%, ee = 78% (*R*)
Ar = *p*-Tol, R = Me: 95%, ee = 62% (*R*)
Ar = *p*-Tol, R = Et: 96%, ee = 65% (*R*)
Ar = *p*-$MeOC_6H_4$, R = Me: 98%, ee = 64% (*R*)
Ar = *p*-$MeOC_6H_4$, R = Et: 95%, ee = 69% (*R*)
Ar = *p*-$CF_3C_6H_4$, R = Me: 26%, ee = 41% (*R*)
Ar = *p*-$CF_3C_6H_4$, R = Et: 29%, ee = 42% (*R*)
Ar = *m*-$MeOC_6H_4$, R = Me: 94%, ee = 37% (*R*)
Ar = *o*-$MeOC_6H_4$, R = Me: 17%, ee = 37% (*R*)

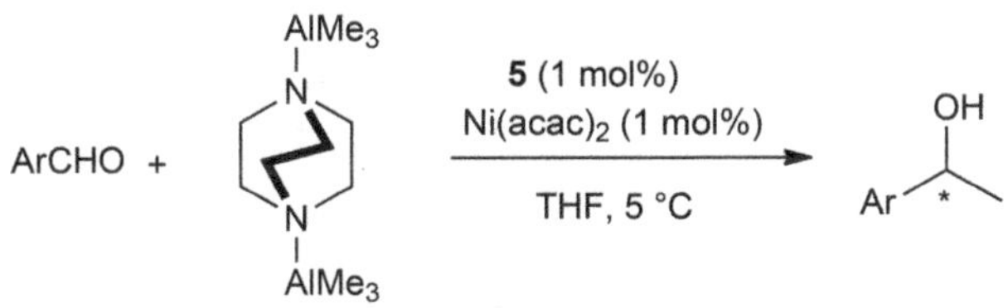

Ar = Ph: 82%, ee = 67% (*R*)
Ar = *p*-Tol: 80%, ee = 60% (*R*)
Ar = *p*-$CF_3C_6H_4$: 15%, ee = 43% (*S*)
Ar = *m*-$MeOC_6H_4$: 62%, ee = 33% (*R*)

Scheme 7.7 Alkylations of aromatic aldehydes with $AlR^2{}_3$ and $(AlR^2{}_3)_2$·DABCO in the presence of an *in situ* generated nickel catalyst from a D-glucose-derived phosphoramidite ligand.

further investigated as catalysts for the addition of diethylzinc to aromatic aldehydes (Scheme 7.9). Both complexes were found active at a low catalyst loading of 1 mol%. The most significant observation was that the 1:2 complexes afforded alcohols with the (*R*) configuration while the 1:1 complexes provided (*S*)-alcohols. Thus, a very effective chirality switching was achieved just by using the corresponding nickel complexes **7** or **8** with 1:1 or 1:2 stoichiometries as catalysts. This general dual stereocontrol was explained by the authors on the basis of calculating the relative energies of all reasonable

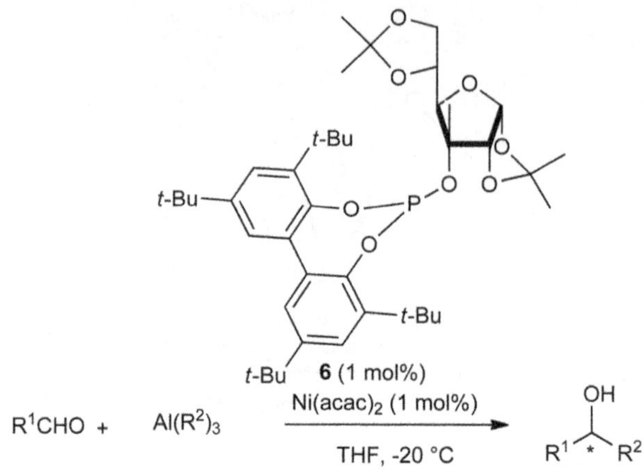

R^1 = Ph, R^2 = Me: 98%, ee = 75% (*S*)
R^1 = Ph, R^2 = Et: 99%, ee = 72% (*S*)
R^1 = *p*-$MeOC_6H_4$, R^2 = Me: 94%, ee = 66% (*S*)
R^1 = *p*-$MeOC_6H_4$, R^2 = Et: 90%, ee = 64% (*S*)
R^1 = *p*-Tol, R^2 = Me: 92%, ee = 77% (*S*)
R^1 = *p*-Tol, R^2 = Et: 91%, ee = 77% (*S*)
R^1 = *p*-FC_6H_4, R^2 = Me: 94%, ee = 74% (*S*)
R^1 = *p*-FC_6H_4, R^2 = Et: 94%, ee = 71% (*S*)
R^1 = 2-Naph, R^2 = Me: 91%, ee = 82% (*S*)
R^1 = 2-Naph, R^2 = Et: 94%, ee = 81% (*S*)
R^1 = *m*-$MeOC_6H_4$, R^2 = Me: 93%, ee = 78% (*S*)
R^1 = *m*-$MeOC_6H_4$, R^2 = Et: 95%, ee = 75% (*S*)
R^1 = *o*-$MeOC_6H_4$, R^2 = Me: 35%, ee = 45% (*R*)
R^1 = *o*-$MeOC_6H_4$, R^2 = Et: 33%, ee = 45% (*R*)

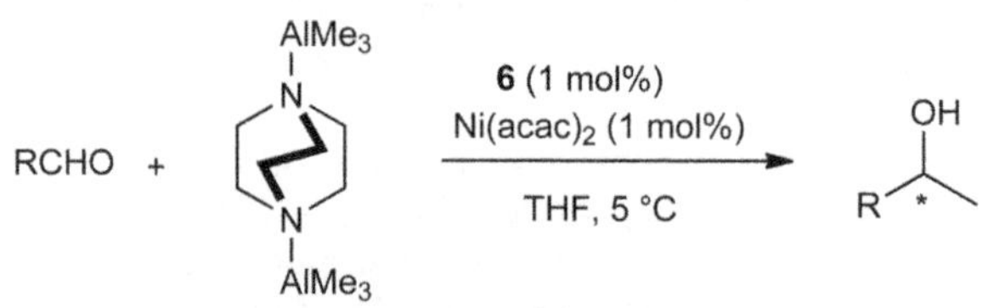

R = Ph: 95%, ee = 76% (*S*)
R = *p*-$MeOC_6H_4$: 73%, ee = 65% (*S*)
R = *p*-Tol: 91%, ee = 77% (*S*)
R = *p*-$CF_3C_6H_4$: 93%, ee = 76% (*S*)
R = *p*-FC_6H_4: 91%, ee = 73% (*S*)
R = 2-Naph: 88%, ee = 84% (*S*)
R = *m*-$MeOC_6H_4$: 82%, ee = 76% (*S*)
R = *o*-$MeOC_6H_4$: 11%, ee = 46% (*R*)

Scheme 7.8 Alkylations of aromatic aldehydes with AlR^2_3 and $(AlR^2_3)_2$·DABCO in the presence of an *in situ* generated nickel catalyst from a D-glucose-derived phosphite ligand.

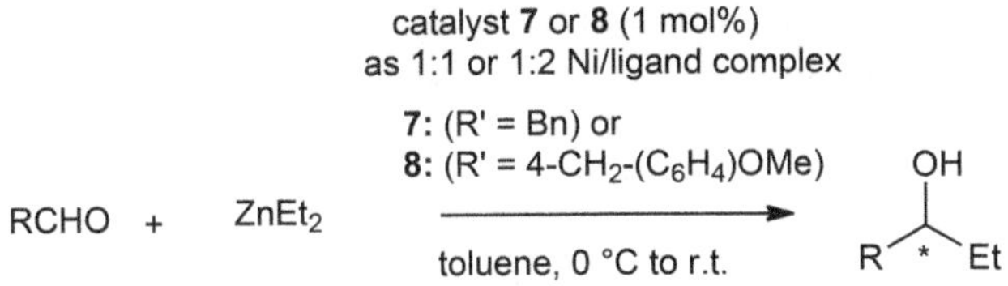

with catalyst **7** as 1:2 (Ni/ligand) complex:
R = Ph: 95%, ee = 87% (*R*)
R = *p*-ClC$_6$H$_4$: 96%, ee = 92% (*R*)
R = *p*-Tol: 95%, ee = 85% (*R*)
with catalyst **7** as 1:1 (Ni/ligand) complex:
R = Ph: 97%, ee = 87% (*S*)
R = *p*-ClC$_6$H$_4$: 92%, ee = 90% (*S*)
R = *p*-Tol: 95%, ee = 83% (*S*)
with ligand **8** as 1:2 (Ni/ligand) complex:
R = Ph: 89%, ee = 85% (*R*)
R = *p*-MeOC$_6$H$_4$: 96%, ee = 92% (*R*)
R = 1-Naph: 91%, ee = 84% (*R*)
with ligand **8** as 1:1 (Ni/ligand) complex:
R = Ph: 91%, ee = 68% (*S*)
R = *p*-MeOC$_6$H$_4$: 95%, ee = 85% (*S*)
R = 1-Naph: 89%, ee = 76% (*S*)

Structures of 1:1 and 1:2 (Ni/ligand) complexes:

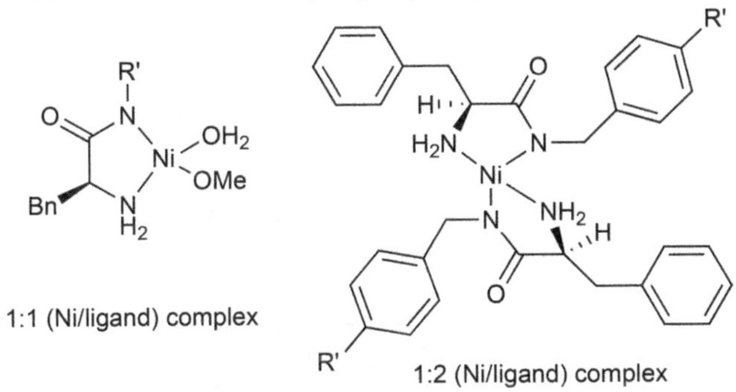

R' = Bn: **7**
R' = 4-CH$_2$-(C$_6$H$_4$)OMe): **8**

Scheme 7.9 Addition of $ZnEt_2$ to aldehydes in the presence of preformed nickel catalysts derived from simple α-amino amide ligands.

transition states. Thus, for 1 : 1 complexes, the coordination of benzaldehyde at one square-planar position, by substituting one weakly coordinated ligand, or at one octahedral position allowed a tricyclic transition state with an *anti-trans* disposition which could favour the formation of the (*S*)-enantiomer. For 1 : 2 complexes, coordination could occur at one of the octahedral positions. In this case, the presence of the R′ substituent could favour a *syn-trans* disposition of the tricyclic transition state, thereby affording the (*R*)-enantiomer. The results collected in Scheme 7.9 showed that each enantiomer of various aromatic alcohols could be prepared in high yields and with good to high enantioselectivities of up to 92% ee. The enantioselectivities were systematically higher by using the 1 : 2 complexes than the corresponding 1 : 1 complexes, whereas the yields were comparable.

Later, the same authors reported the synthesis of a series of other 1 : 2 (Ni/ligand) complexes derived from α-amino amides which were further investigated in the addition of diethylzinc to benzaldehyde.[15] Their modular structure allowed easy optimisation of the catalytic system. Different structural modifications were examined in this regard, resulting in the selection of the phenylalanine-based ligand **7** bearing a benzyl group in the amide moiety as the optimal ligand (Scheme 7.10). The corresponding 1 : 2 complex was applied to the addition of diethyl- and dimethylzinc reagents to a range of aldehydes, including aliphatic ones. The corresponding secondary alcohols were achieved in good to high enantioselectivities of up to 99% ee. The presence of electron-donating or electron-withdrawing substituents on the aromatic ring of benzaldehyde was compatible with the reaction. Aromatic

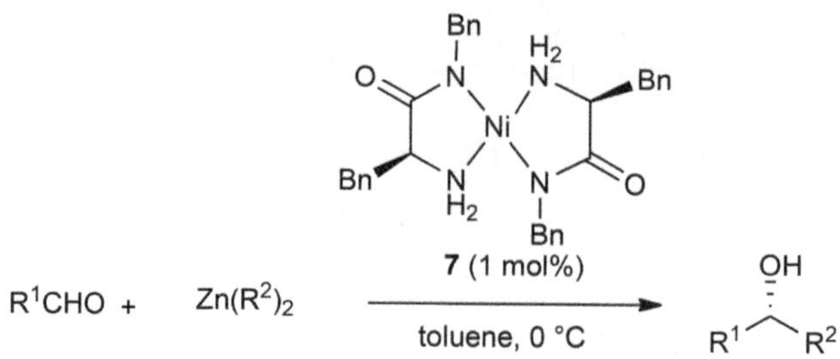

R^1 = Ph, R^2 = Et: 91%, ee = 92%
R^1 = 2-Naph, R^2 = Et: 91%, ee = 76%
R^1 = 1-Naph, R^2 = Et: 89%, ee = 80%
R^1 = *p*-$MeOC_6H_4$, R^2 = Et: 94%, ee = 99%
R^1 = *p*-ClC_6H_4, R^2 = Et: 93%, ee = 79%
R^1 = *p*-Tol, R^2 = Et: 90%, ee = 95%
R^1 = *m*-Tol, R^2 = Et: 96%, ee = 80%
R^1 = *o*-Tol, R^2 = Et: 89%, ee = 71%
R^1 = Cy, R^2 = Et: 72%, ee = 41%
R^1 = Cy, R^2 = Me: 89%, ee = 32%
R^1 = *n*-Hex, R^2 = Et: 88%, ee = 80%
R^1 = *n*-Hex, R^2 = Me: 80%, ee = 35%
R^1 = Ph, R^2 = Me: 95%, ee = 96%
R^1 = 2-Naph, R^2 = Me: 93%, ee = 81%
R^1 = 1-Naph, R^2 = Me: 89%, ee = 44%
R^1 = *p*-Tol, R^2 = Me: 90%, ee = 82%
R^1 = *p*-$MeOC_6H_4$, R^2 = Me: 95%, ee = 82%
R^1 = *p*-ClC_6H_4, R^2 = Me: 93%, ee = 76%

Scheme 7.10 Addition of dialkylzincs to aldehydes in the presence of a preformed nickel catalyst (1 : 2 Ni/ligand) derived from an α-amino amide ligand.

substrates including 1-naphthaldehyde, 2-naphthaldehyde, and 4-chlorobenzaldehyde reacted, with enantioselectivities ranging from 44 to 81% ee. An important steric substituent effect was observed at the *ortho* position. Furthermore, aliphatic aldehydes were also tolerated in enantioselectivities of 32–80% ee, albeit producing lower yields. According to calculations, the authors assumed that the coordination of the aldehyde to the complex did not take place through the metal but with a hydrogen atom of the amino groups of the ligand that became increasingly acidic upon formation of the nickel(II) complex.

7.4 Additions of Organoboron Reagents

The asymmetric arylation of aromatic aldehydes is one of the most important carbon–carbon bond-forming reactions,[16] because chiral diarylmethanols are important intermediates for the synthesis of biologically active compounds.[17] Metal-catalysed α-arylation of aldehydes is notoriously difficult because of its propensity to readily undergo aldol condensation under the basic conditions. Among the various arylmetal reagents used, arylboron reagents are more desirable due to the recent demand for safe and sustainable organic synthesis, because their reagents are less toxic and air-stable. In 1998, Miyaura's group found that rhodium(I) complexes catalysed the 1,2-addition to aldehydes with arylboronic acids with enantioselectivies of up to 41% ee;[18] later, attention was focused on arylation using a combination of the rhodium catalyst and arylboronic acids with enantioselectivities of up to 85% ee.[19] On the other hand, several groups have reported the use of less expensive palladium catalysts for the 1,2-addition of arylboronic acids to aromatic aldehydes.[20] Actually, the only highly efficient example of the catalytic asymmetric arylation of aldehydes was developed by Shibasaki by using chiral copper catalysts.[21] From the viewpoint of cost and practical convenience, the use of a more economical and naturally abundant metal catalyst such as nickel, rather than rhodium or palladium, is desirable. In 2005, Shirakawa reported the first successful example of nickel-catalysed arylation of aldehydes with arylboron reagents.[22] The development of an asymmetric version was achieved by Kondo and Aoyama in 2007.[23] As shown in Scheme 7.11, the authors employed (*R*,*R*)-Et-DUPHOS as a chiral ligand of $Ni(cod)_2$ in the addition of triarylboroxins to aromatic aldehydes to afford the corresponding chiral secondary alcohols in good yields of up to 93% and with moderate enantioselectivities of 65–78% ee. The process gave the best results when performed in the presence of NaO*t*-Bu as base in a 5 : 1 mixture of DME/H_2O as solvent at 100 °C. Investigating the substrate scope, the authors found that acceptable enantioselectivities of up to 78% ee were achieved for 1-naphthaldehyde and 2-substituted aromatic aldehydes, but low enantioselectivities were obtained for aromatic aldehydes without a 2-substituted group.

Later, the same authors investigated the enantioselective nickel-catalysed addition of aryltrifluoroborates to aromatic aldehydes in the presence of the same ligand [(*R*,*R*)-Et-DUPHOS].[24] Indeed, convenient potassium

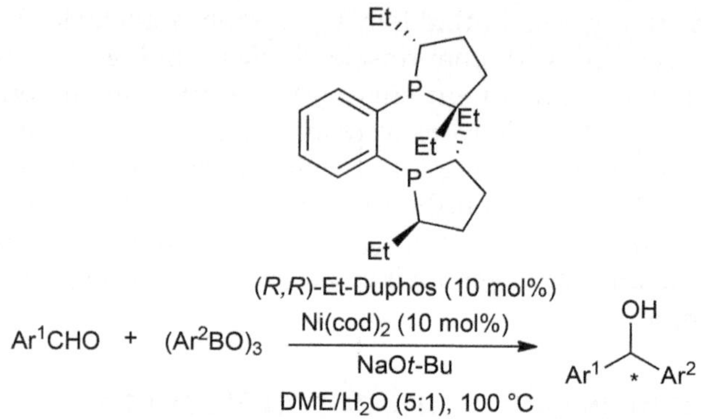

Ar^1 = 1-Naph, Ar^2 = Ph: 93%, ee = 68% (*R*)
Ar^1 = *o*-Tol, Ar^2 = Ph: 91%, ee = 78% (*R*)
Ar^1 = 2-Me-3-FC_6H_3, Ar^2 = Ph: 93%, ee = 75%
Ar^1 = *o*-PhC_6H_4, Ar^2 = Ph: 83%, ee = 72%
Ar^1 = 1-Naph, Ar^2 = *p*-FC_6H_4: 87%, ee = 65%

Scheme 7.11 Addition of triarylboroxins to aromatic aldehydes in the presence of an *in situ* generated nickel catalyst derived from the (*R*,*R*)-Et-DUPHOS ligand.

aryltrifluoroborates are easily handled and have a greater stability to water than triarylboroxins. The (*R*,*R*)-Et-DUPHOS ligand was again selected among a range of chiral bisphosphines, including BINAP, DIOP, Segphos, CHIRAPHOS, and various DUPHOS derivatives. Solvent screening showed that a 5 : 1 mixture of dioxane/H_2O was the best solvent at 80 °C. Under these reaction reactions, a series of chiral aromatic alcohols were synthesised in moderate to excellent yields (45–97%) along with low to good enantioselectivities of up to 81% ee, as shown in Scheme 7.12. Again, the best results were achieved with 1-naphthaldehyde and 2-substituted aromatic aldehydes, while substrates bearing a 4-methyl or a 4-methoxy substituent exhibited low enantioselectivities (17–41% ee). The utility of this novel methodology was demonstrated in the synthesis of an antihistaminic and cholinergic (*R*)-orphenadrine. This work represented the first example of asymmetric nickel-catalysed 1,2-addition of arylboron reagents to aromatic aldehydes.

A drawback of the above study was the requirement for *ortho*-alkyl or -aryl substituents on the benzene ring of the aromatic aldehydes in order to achieve acceptable enantioselectivities. With the aim of improving the enantioselectivity of these reactions, the same authors proposed to sterically tune the benzaldehyde substrates with an *ortho*-Me_2PhSi group, together with the use of potassium 1-aryl-4-methyl-2,6,7-trioxa-1-boranuidabicyclo[2.2.2]octanes (aryltriolborates) as aryl sources.[25] As shown in Scheme 7.13, the addition of various potassium aryltriolborates to (non)substituted *ortho*-silylated aromatic aldehydes afforded the corresponding chiral secondary alcohols in good to

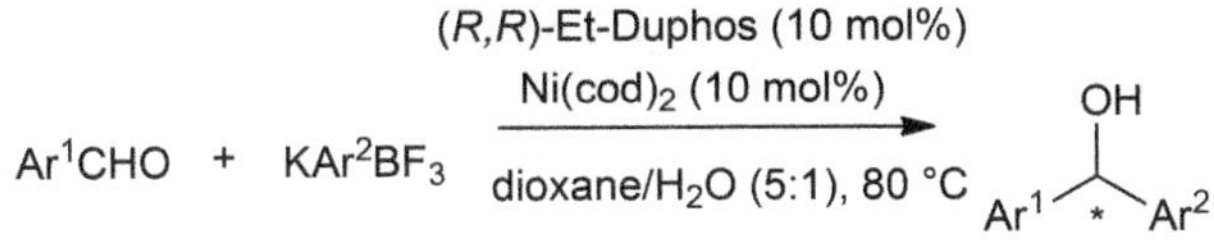

Ar^1 = 1-Naph, Ar^2 = Ph: 97%, ee = 72% (*R*)
Ar^1 = *o*-Tol, Ar^2 = *p*-(*t*-Bu)C_6H_4: 93%, ee = 81% (*R*)
Ar^1 = *o*-Tol, Ar^2 = *p*-FC_6H_4: 86%, ee = 75% (*R*)
Ar^1 = *o*-Tol, Ar^2 = *p*-$MeOC_6H_4$: 87%, ee = 66% (*R*)
Ar^1 = *o*-Tol, Ar^2 = Ph: 65%, ee = 76%
Ar^1 = *p*-FC_6H_4, Ar^2 = Ph: 88%, ee = 67% (*S*)
Ar^1 = *p*-Tol, Ar^2 = Ph: 90%, ee = 41% (*S*)
Ar^1 = 2-Me-3-FC_6H_3, Ar^2 = Ph: 89%, ee = 78%
Ar^1 = *o*-PhC_6H_4, Ar^2 = Ph: 65%, ee = 76%
Ar^1 = 1-Naph, Ar^2 = *o*-Tol: 84%, ee = 68%
Ar^1 = *p*-$MeOC_6H_4$, Ar^2 = Ph: 45%, ee = 17% (*S*)

Scheme 7.12 Addition of aryltrifluoroborates to aromatic aldehydes in the presence of an *in situ* generated nickel catalyst derived from the (*R*,*R*)-Et-DUPHOS ligand.

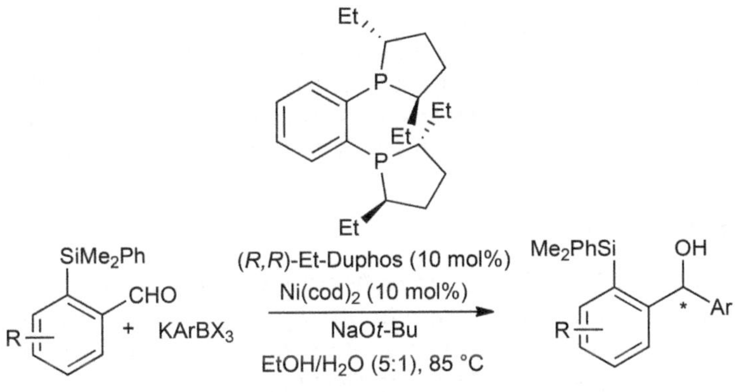

R = H, Ar = *p*-FC_6H_4: 90%, ee > 99% (*S*)
R = H, Ar = *p*-(*i*-Pr)C_6H_4: 73%, ee = 93%
R = H, Ar = *m*-Tol: 94%, ee = 93% (*S*)
R = H, Ar = 3,5-$Me_2C_6H_3$: 92%, ee > 99%
R = H, Ar = *m*-$MeOC_6H_4$: 90%, ee = 89% (*S*)
R = 4-Me, Ar = Ph: 70%, ee = 91% (*R*)
R = 5-F, Ar = *p*-Tol: 84%, ee = 91% (*R*)
R = 5-F, Ar = *p*-$MeOC_6H_4$: 72%, ee = 90%
R = 5-MeO, Ar = *p*-Tol: 98%, ee = 93%

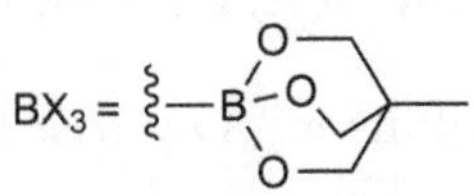

Scheme 7.13 Addition of potassium aryltriolborates to aromatic aldehydes in the presence of an *in situ* generated nickel catalyst derived from the (*R*,*R*)-Et-DUPHOS ligand.

excellent yields (70–98%) and with high enantioselectivities ranging from 89% to >99% ee. These good results were achieved by using the same catalyst system as above, based on the (*R*,*R*)-Et-DUPHOS ligand and Ni(cod)$_2$ in the presence of NaO*t*-Bu as base and a 5 : 1 mixture of EtOH/H_2O as solvent at 85 °C. It is important to note that this nice work represented the first highly efficient example of asymmetric nickel-catalysed arylation of aldehydes with a boron reagent.

7.5 Conclusions

Even if a lot of early work had been made in the area of metal-catalysed enantioselective additions of organometallic reagents to aldehydes, especially with dialkylzinc reagents as nucleophiles, the first successful examples of nickel-catalysed enantioselective additions of both organoaluminum and -boron reagents to aldehydes have been reported in the last decade. Indeed, high enantioselectivities of up to 95% ee were achieved in the addition of trialkyl- or triarylaluminum reagents to aldehydes by using phosphoramidite or sugar-based phosphite ligands. In addition, the first highly efficient example of asymmetric nickel-catalysed arylation of aldehydes with a boron reagent was reported in 2009. For example, enantioselectivities of up to >99% ee were reached in the addition of potassium aryltriolborates to aromatic aldehydes in the presence of an *in situ* generated nickel catalyst from the (*R*,*R*)-Et-DUPHOS ligand. In the area of the well-known dialkylzinc addition to aldehydes, an interesting example of very effective chirality switching was achieved just by using different stoichiometries of nickel complexes bearing α-amino amide ligands, providing very high enantioselectivities of up to 99% ee.

References

1. (a) K. Soai and T. Shibata, *Comprehensive Asymmetric Catalysis*, ed. E. N. Jacobsen, A. Pfaltz and H. Yamamoto, Springer-Verlag, New York, 1999, ch. 26.1; (b) L. Pu and H. B. Yu, *Chem. Rev.*, 2001, **101**, 757–824; (c) C. Garcia and V. S. Martin, *Curr. Org. Chem.*, 2006, **10**, 1849–1889; (d) M. R. Luderer, W. F. Bailey, M. R. Luderer, J. D. Fair, R. J. Dancer and M. B. Sommer, *Tetrahedron: Asymmetry*, 2009, **20**, 981–998; (e) C. M. Binder and B. Singaram, *Org. Prep. Proced. Int.*, 2011, **43**, 139–208; (f) A. Kolb and P. von Zezchwitz, *Topics in Organometallic Chemistry*, Springer, Heidelberg, 2013, vol. 41, pp. 245–276; (g) E. Vrancken, J.-M. Campagne and P. Mangeney, *Comprehensive Organic Synthesis*, 2nd edn, Elsevier, Amsterdam, 2014, vol. 1, pp. 74–123; (h) P. Knochel and G. A. Molander, *Comprehensive Organic Synthesis*, 2nd edn, Elsevier, Amsterdam, 2014, vol. 1, pp. 344–364.
2. F. A. Cotton and G. Wilkinson, *Advanced Inorganic Chemistry*, 5th edn, Wiley, New York, 1988, p. 224.
3. (a) A. S. C. Chan, F.-Y. Zhang and C.-W. Yip, *J. Am. Chem. Soc.*, 1997, **119**, 4080–4081; (b) B. L. Pagenkopf and E. M. Carreira, *Tetrahedron Lett.*, 1998, **39**, 9593–9596; (c) J.-F. Lu, J.-S. You and H.-M. Gau, *Tetrahedron:*

Asymmetry, 2000, **11**, 2531–2535; (d) J.-S. You, S.-J.-S. You and H.-M. Gau, *Chem. Commun.*, 2001, 1546–1547.
4. (a) K. Biswas, O. Prieto, P. J. Goldsmith and S. Woodward, *Angew. Chem., Int. Ed.*, 2005, **44**, 2232–2234; (b) K. Biswas, A. Chapron, T. Cooper, P. K. Fraser, A. Novak, O. Prieto and S. Woodward, *Pure Appl. Chem.*, 2006, **78**, 511–518; (c) V. E. Albrow, A. J. Blake, R. Fryatt, C. Wilson and S. Woodward, *Eur. J. Org. Chem.*, 2006, 2549–2557.
5. H. C. Brown and N. Davidson, *J. Am. Chem. Soc.*, 1942, **64**, 316–324.
6. Y. Mata, M. Diéguez, O. Pamies and S. Woodward, *J. Org. Chem.*, 2006, **71**, 8159–8165.
7. Y. Mata, M. Diéguez, O. Pamies and S. Woodward, *Inorg. Chim. Acta*, 2008, **361**, 1381–1384.
8. E. Raluy, M. Diéguez and O. Pamies, *Tetrahedron Lett.*, 2009, **50**, 4495–4497.
9. E. Raluy, M. Diéguez and O. Pamies, *Tetrahedron: Asymmetry*, 2009, **20**, 1575–1579.
10. S. Alegre, M. Diéguez and O. Pamies, *Tetrahedron: Asymmetry*, 2011, **22**, 834–839.
11. N. Oguni and T. Omi, *Tetrahedron Lett.*, 1984, **25**, 2823–2824.
12. (a) K. Soai and S. Niwa, *Chem. Rev.*, 1992, **92**, 833–856; (b) L. Pu, *Tetrahedron*, 2003, **59**, 9873–9886.
13. M. I. Burguete, M. Collado, J. Escorihuela, F. Galindo, E. Garcia-Verdugo, S. V. Luis and M. J. Vicent, *Tetrahedron Lett.*, 2003, **44**, 6891–6894.
14. M. I. Burguete, M. Collado, J. Escorihuela and S. V. Luis, *Angew. Chem., Int. Ed.*, 2007, **46**, 9002–9005.
15. J. Escorihuela, B. Altava, M. I. Burguete and S. V. Luis, *Tetrahedron*, 2013, **69**, 551–558.
16. (a) F. Schmidt, R. T. Stemmler, J. Rudolph and C. Bolm, *Chem. Soc. Rev.*, 2006, **35**, 454–470; (b) C. C. C. Johansson and T. J. Colacot, *Angew. Chem., Int. Ed.*, 2010, **49**, 676–707.
17. M. Seto, Y. Iizawa, M. Baba and M. Shiraishi, *Chem. Pharm. Bull.*, 2004, **52**, 818–829.
18. M. Sakai, M. Ueda and N. Miyaura, *Angew. Chem., Int. Ed.*, 1998, **37**, 3279–3281.
19. (a) C. Moreau, C. Hague, A. S. Weller and C. G. Frost, *Tetrahedron Lett.*, 2001, **42**, 6957–6960; (b) K. Suzuki, S. Iishi, K. Kondo and T. Aoyama, *Synlett*, 2006, 1360–1364.
20. (a) T. Yamamoto, T. Ohta and Y. Ito, *Org. Lett.*, 2005, **7**, 4153–4155; (b) K. Suzuki, T. Arao, S. Ishii, Y. Maeda, K. Kondo and T. Aoyama, *Tetrahedron Lett.*, 2006, **47**, 5789–5792; (c) P. He, Y. Lu, C.-G. Dong and Q.-S. Hu, *Org. Lett.*, 2007, **9**, 343–346; (d) A. Novodomska, M. Dudicova, F. R. Leroux and F. Colobert, *Tetrahedron: Asymmetry*, 2007, **18**, 1628–1634; (e) R. Shirai, R. Shimazawa and M. Kuriyama, *J. Org. Chem.*, 2008, **73**, 1597–1600.
21. (a) D. Tomita, R. Wada, M. Kanai and M. Shibasaki, *J. Am. Chem. Soc.*, 2005, **127**, 4138–4139; (b) D. Tomita, M. Kanai and M. Shibasaki, *Chem.–Asian J.*, 2006, **1–2**, 161–166.

22. G. Takahashi, E. Shirakawa, T. Tsuchimoto and Y. Kawakami, *Chem. Commun.*, 2005, 1459–1461.
23. T. Arao, K. Kondo and T. Aoyama, *Tetrahedron*, 2007, **63**, 5261–5264.
24. K. Yamamoto, K. Tsurumi, F. Sakurai, K. Kondo and T. Aoyama, *Synthesis*, 2008, **22**, 3585–3590.
25. F. Sakurai, K. Kondo and T. Aoyama, *Tetrahedron Lett.*, 2009, **50**, 6001–6003.

CHAPTER 8

Enantioselective Nickel-Catalysed Aldol-Type and Mannich-Type Reactions

8.1 Introduction

The use of chiral metal catalysts in aldol- and Mannich-type reactions has become a major area of study. The intrinsic efficiency of catalytic methods is the major driving force behind research in this area. The need for more cost-effective and 'greener' synthetic methods, especially for industrial applications, has been widely recognised. By the very fact of the lower costs of nickel catalysts in comparison with other transition metals, in addition to their high abundance, these catalysts have been recently widely investigated in these reactions. Catalytic methods based on nickel for the aldol and Mannich reactions potentially offer very mild reaction conditions, with the attendant tolerance for a range of functional groups that this implies. Further, the ability to 'tune' a catalytic metal centre by judicious ligand design is an attractive feature. Despite early success in developing catalysts for the Mukaiyama aldol addition, only recently has the direct catalytic aldol and Mannich reactions received serious attention. Indeed, this situation is changing as catalysts capable of promoting powerful direct asymmetric aldol and Mannich reactions are developed, especially dinuclear Schiff base nickel complexes from Shibasaki's group and *N,N'*-dioxide nickel catalysts from Feng's group.

RSC Catalysis Series No. 26
Enantioselective Nickel-Catalysed Transformations
By Hélène Pellissier

Published by the Royal Society of Chemistry, www.rsc.org

8.2 Aldol-Type Reactions

The direct catalytic asymmetric aldol reaction is a powerful and atom-economical method for synthesising chiral β-hydroxy carbonyl compounds.[1] To date, many chiral metal and organocatalysts have been developed for reactions of various donors with aldehydes. The use of formaldehyde as a useful C_1 unit in direct catalytic asymmetric aldol reactions, however, has been relatively limited,[2] possibly due to its high reactivity. Highly enantioselective chiral catalysts for indirect aldol reactions of formaldehyde with preformed silyl enolates have been developed,[3] but for direct aldol reactions there remains room for improvement in catalyst loading, catalyst reactivity, formaldehyde amount, and substrate scope. In this context, Shibasaki *et al.* have reported the use of the powerful homodinuclear nickel Schiff base catalyst **1** in a direct asymmetric aldol reaction of β-keto esters with formaldehyde to give the corresponding chiral hydroxymethylated products (Scheme 8.1).[4] This catalyst was proven to be much more efficient than the corresponding cobalt, manganese, zinc, or palladium complexes under the same reaction conditions, providing enantioselectivities of 1% ee for the palladium complex, 9% ee for the zinc complex, 22% ee for the cobalt complex, and 30% ee for the manganese complex, along with lower yields ranging from 61 to 82% (*vs.* 78–94% with

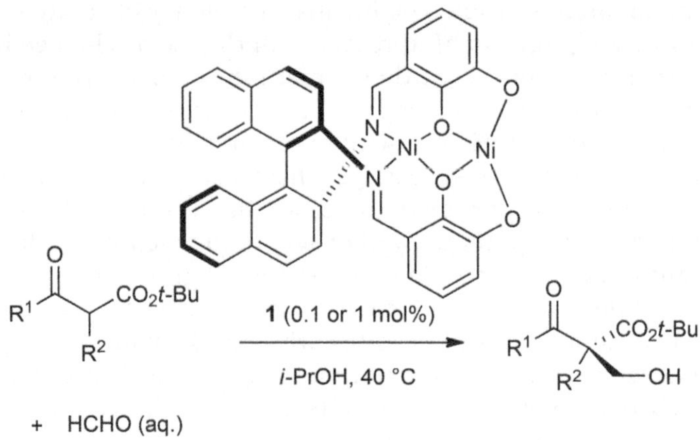

with 1 mol% of catalyst:
$R^1,R^2 = (CH_2)_4$: 91%, ee = 85%
with 0.1 mol% of catalyst:
$R^1,R^2 = (CH_2)_3$: 94%, ee = 93%
$R^1,R^2 = (CH_2)_5$: 81%, ee = 66%
$R^1 = R^2$ = Me: 79%, ee = 81%
R^1 = Me, R^2 = Et: 79%, ee = 89%
R^1 = Me, R^2 = Bn: 84%, ee = 90%
R^1 = Ph, R^2 = Me: 22%, ee = 94%

Scheme 8.1 Aldol reaction of formaldehyde with β-keto esters in the presence of a preformed dinuclear nickel catalyst derived from a Schiff base ligand.

the nickel catalyst). The reaction was generally performed with a low catalyst loading of 0.1 mol% to avoid the undesirable retroaldol reaction and to obtain the product under kinetic control. Under these conditions, a range of products were obtained in low to high yields (22–94%) and with good to high enantioselectivities of up to 94% ee, as shown in Scheme 8.1. The investigation of the substrate scope showed that substrates with a six-membered ring gave lower yields (91% *vs.* 94%) and enantioselectivities (85% ee *vs.* 93% ee) than the corresponding five-membered substrate. With the seven-membered substrate, the reaction proceeded smoothly, but the enantioselectivity was modest (66% ee). The catalyst system was also applicable to various acyclic β-keto esters, providing enantioselectivities of 81–94% ee. Substrates bearing bulky substituents were found much less reactive than the others, giving products in only 22–43% yields.

Chiral β-hydroxy-α-amino acids are important structural motifs for the preparation of various bioactive compounds. Many biologically active natural products, such as vancomycin, ristocetin, and biphenomycin A, contain β-hydroxy-α-amino acids within their structural frameworks.[5] A number of highly efficient catalytic enantioselective variants for the construction of these molecules have been established, although the majority of methods require the use of preformed enolate equivalents.[6] The direct catalytic enantioselective aldol reaction between a glycine equivalent and carbonyl compounds, involving the creation of a C–C bond and two stereogenic centres in a single operation, is one of the most attractive and atom-efficient methods to obtain such chiral building blocks, and intense effort has been devoted to this area. Early in 2005, Willis *et al.* described the first highly enantioselective aldol reaction of aldehydes with an α-isothiocyanato imide by using a chiral magnesium–PYBOX complex.[7] Later, Shibasaki *et al.* achieved a breakthrough with magnesium–Schiff base complexes in the direct asymmetric aldol reaction of α-isothiocyanato esters with ketones.[8] In 2010, Feng *et al.* developed a powerful chiral catalyst system based on an easily available *N,N′*-dioxide–nickel complex for the direct asymmetric aldol reaction of an α-isothiocyanato imide with aldehydes.[9] As shown in Scheme 8.2, high yields combined with both excellent diastereo- and enantioselectivities of up to >98% de and >99% ee, respectively, were achieved. The authors have investigated other metals in these reactions, such as magnesium, copper, iron, and cobalt, finding in comparison with the corresponding nickel complex much lower enantioselectivities (0–13% ee *vs.* 57% ee for the nickel complex). L-Proline-derived ligand **2** was selected among variously substituted five- and six-membered *N,N′*-dioxide ligands. The investigation of the substrate scope showed that a wide range of aldehydes were compatible with the reaction. Therefore, benzaldehydes with different electron-donating as well as electron-withdrawing substituents on the aromatic ring were suitable, along with α,β-unsaturated, heteroaromatic, and aliphatic aldehydes. On the other hand, ketones did not lead to the corresponding aldol products.

On the other hand, Maheswaran *et al.* have investigated the direct aldol reaction between methyl vinyl ketone and various aromatic aldehydes in the

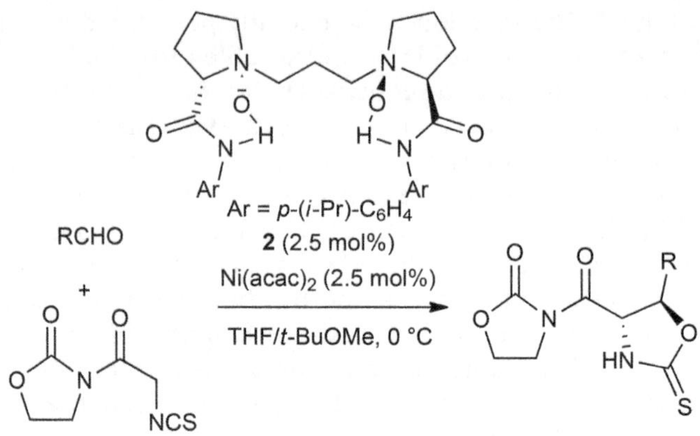

R = Ph: 92%, *trans/cis* = 97:3, ee = 96%
R = *p*-FC_6H_4: 95%, *trans/cis* = 97:3, ee = 96%
R = *p*-BrC_6H_4: 98%, *trans/cis* = 97:3, ee = 95%
R = *p*-ClC_6H_4: 98%, *trans/cis* = 97:3, ee = 95%
R = *p*-Tol: 90%, *trans/cis* = 97:3, ee = 97%
R = *p*-PhC_6H_4: 96%, *trans/cis* > 99:1, ee = 93%
R = *p*-$CF_3C_6H_4$: 90%, *trans/cis* = 95:5, ee = 91%
R = *m*-BrC_6H_4: 97%, *trans/cis* = 93:7, ee > 99%
R = *m*-ClC_6H_4: 96%, *trans/cis* = 93:7, ee > 99%
R = *m*-$MeOC_6H_4$: 87%, *trans/cis* > 95:5, ee = 92%
R = *m*-$PhOC_6H_4$: 95%, *trans/cis* > 95:5, ee = 93%
R = *p*-$MeOC_6H_4$: 88%, *trans/cis* = 91:9, ee = 96%
R = 2-Naph: 90%, *trans/cis* = 94:6, ee = 95%
R = 1-Naph: 86%, *trans/cis* = 93:7, ee = 93%
R = (*E*)-PhCH=CH: 82%, *trans/cis* = 85:15, ee = 90%
R = 3-furyl: 86%, *trans/cis* = 92:8, ee = 93%
R = 3-thienyl: 89%, *trans/cis* = 95:5, ee = 95%
R = *n*-Hex: 77%, *trans/cis* = 92:8, ee = 82%
R = Cy: 68%, *trans/cis* = 85:15, ee = 81%
R = *i*-Bu: 72%, *trans/cis* = 90:10, ee = 80%

Scheme 8.2 Aldol reaction of aldehydes with an α-isothiocyanato imide in the presence of an *in situ* generated nickel catalyst derived from an *N,N'*-dioxide ligand.

presence of nickel and copper (−)-sparteine complexes.[10] They found that an enantioreversal occurred under TEA-promoted reaction conditions. A quantitative analysis of the X-ray structures of the copper and nickel (−)-sparteine catalysts has been carried out using group theoretical analysis *via* symmetry deformation coordinates, and the results from the study unambiguously showed that the (−)-sparteine ligand configuration in the copper complex mimicked its antipodal structure (+)-sparteine in the nickel (−)-sparteine complex or *vice versa*. As shown in Scheme 8.3, moderate enantioselectivities were reached in both cases of complexes, but generally those obtained from nickel were slightly higher than those achieved by using the copper complex.

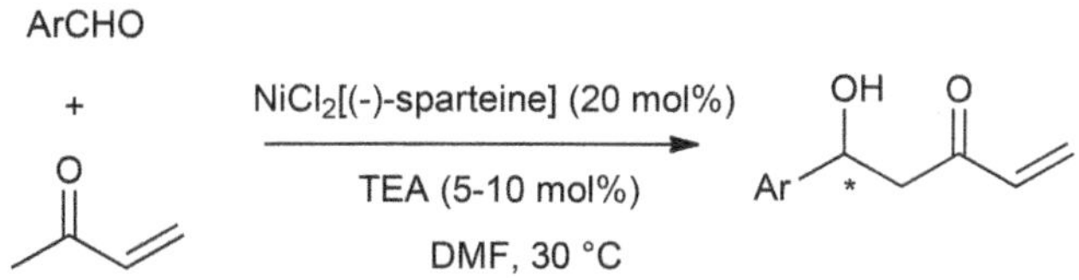

with nickel catalyst:
Ar = *p*-$NO_2C_6H_4$: 80%, ee = 76% (*S*)
Ar = *m*-$NO_2C_6H_4$: 75%, ee = 35% (*S*)
Ar = *p*-NCC_6H_4: 65%, ee = 49% (*S*)
Ar = *o*-$NO_2C_6H_4$: 81%, ee = 65% (*R*)
Ar = 4-Cl-3-$NO_2C_6H_3$: 71%, ee = 54% (*S*)
with $CuCl_2$[(-)-sparteine] (20 mol%):
Ar = *p*-$NO_2C_6H_4$: 61%, ee = 79% (*R*)
Ar = *m*-$NO_2C_6H_4$: 55%, ee = 31% (*R*)
Ar = *p*-NCC_6H_4: 48%, ee = 24% (*R*)
Ar = *o*-$NO_2C_6H_4$: 58%, ee = 37% (*S*)
Ar = 4-Cl-3-$NO_2C_6H_3$: 54%, ee = 44% (*R*)

Scheme 8.3 Aldol reaction of aromatic aldehydes with methyl vinyl ketone in the presence of a preformed nickel catalyst derived from the sparteine ligand.

Moreover, better yields were achieved when using nickel rather than copper complexes.

The enantioselective aldol reaction of enolsilanes with aldehydes and ketones (Mukaiyama aldol reaction) is one of the most important synthetic tools for C–C bond formations.[1j,11] Since the pioneering work reported by Mukaiyama *et al.* in 1990,[12] a number of chiral catalysts have been developed for this reaction, including both metal complexes[13] and organocatalysts.[14] Thus far, the aldol reaction of 1,2-dicarbonyl compounds, which constitutes an approach to 2-hydroxy-1,4-dicarbonyl compounds, has been well investigated,[15] but most of the work has focused on the reaction of pyruvates. In 2011, Feng *et al.* reported the asymmetric aldol reaction between glyoxal derivatives and enolsilanes catalysed by a chiral nickel catalyst generated *in situ* from $Ni(BF_4)_2{\cdot}6H_2O$ and chiral *N*,*N*′-oxide ligand **3**.[16] As shown in Scheme 8.4, uniformly very good enantioselectivities of 90–95% ee were achieved in combination with good yields for the corresponding aldol products after subsequent hydrolysis with aqueous HCl, demonstrating that both the electronic and steric substituents in the substrates had no obvious effect on the enantioselectivity.

Enamides and enecarbamates are considered as versatile synthetic building blocks in organic synthesis, but they have been rarely used as nucleophiles, presumably due to their lower reactivity compared with enamines and enols. In 2004, Kobayashi *et al.* demonstrated the utility of enecarbamates as nucleophiles for C–C bond formation in asymmetric copper-catalysed reactions of aldehydes and aldimines.[17] Later in 2005, these authors reported the enantioselective nickel-catalysed reactions of a simple diketone, such

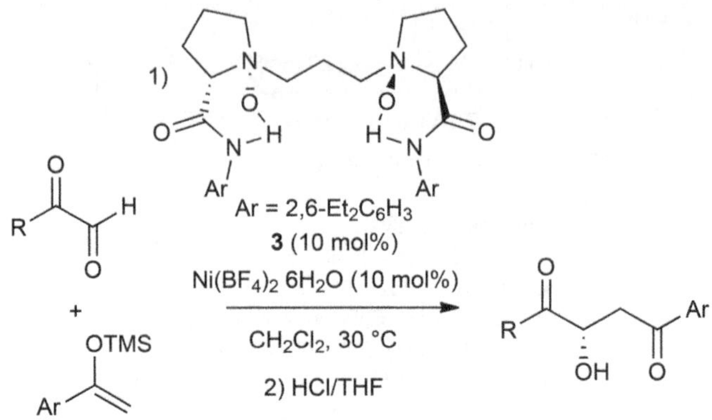

R = Ar = Ph: 94%, ee = 92%
R = *o*-Tol, Ar = Ph: 88%, ee = 92%
R = *m*-Tol, Ar = Ph: 95%, ee = 93%
R = *p*-Tol, Ar = Ph: 65%, ee = 90%
R = *m*-MeOC_6H_4, Ar = Ph: 74%, ee = 93%
R = *m*-ClC_6H_4, Ar = Ph: 74%, ee = 93%
R = *p*-ClC_6H_4, Ar = Ph: 70%, ee = 91%
R = *p*-FC_6H_4, Ar = Ph: 85%, ee = 92%
R = *p*-BrC_6H_4, Ar = Ph: 77%, ee = 95%
R = 2-furyl, Ar = Ph: 63%, ee = 91%
R = OEt, Ar = Ph: 93%, ee = 94%
R = Ph, Ar = *p*-ClC_6H_4: 67%, ee = 92%

Scheme 8.4 Mukaiyama aldol reaction of glyoxal derivatives with enolsilanes in the presence of an *in situ* generated nickel catalyst derived from an *N*,*N*′-dioxide ligand.

as butane-2,3-dione, with enecarbamates performed in the presence of chiral C_2-symmetric diamine **4**.[18] As shown in Scheme 8.5, the corresponding tertiary alcohols were achieved in good yields and with moderate to good enantioselectivities (65–84% ee). In this study, the authors compared the efficiency of the nickel complex with that of the corresponding scandium, cobalt, copper, and zinc complexes, finding that the use of the zinc catalyst allowed comparable enantioselectivity with that achieved with the corresponding nickel complex (75% ee for zinc *vs.* 76% ee for nickel), albeit with much lower yield (31% *vs.* 67%). The copper catalyst provided a slightly lower enantioselectivity of 71% ee than nickel (76%), combined with a better yield of 72% yield (*vs.* 67% for nickel). On the other hand, the corresponding cobalt complex showed much lower reactivity (5% yield) and only 45% ee, while no reaction occurred with the scandium complex.

In 2010, Feng *et al.* reported remarkable results in the enantioselective nickel-catalysed aldol-type reactions of glyoxal derivatives with enecarbamates and enamides by using chiral *N*,*N*′-dioxide ligand **5** (Scheme 8.6).[19] Various *N*,*N*′-oxides were investigated as ligands in this work, and intensive studies on the amide moiety revealed that ligand **5** with bromine at the *ortho*-position of aniline allowed the best results to be achieved. A remarkably wide

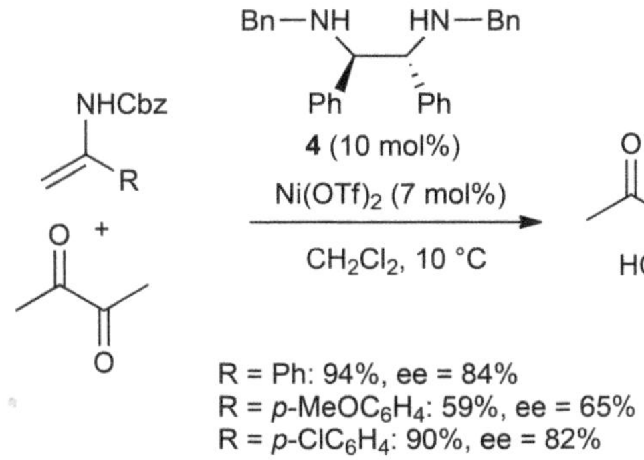

Scheme 8.5 Enecarbamate addition to butane-2,3-dione in the presence of an *in situ* generated nickel catalyst derived from a C_2-symmetric diamine.

range of aromatic, aliphatic, and heterocyclic glyoxal derivatives, including glyoxylate, were investigated with phenyl enecarbamate, all providing high reactivities and excellent enantioselectivities of 98 to >99% ee. Moreover, similar excellent results were achieved in the reaction of various aromatic enamides with phenylglyoxal (Scheme 8.6). Generally, an electron-donating substituent, such as OMe or Me, on the aromatic ring slightly enhanced the reactivity, while electron-withdrawing substituents, such as F or Cl, at the *para* position diminished the reactivity. It must be noted that this novel synthesis of chiral 2-hydroxy-1,4-dicarbonyl compounds is remarkable by the extremely high enantioselectivity achieved, broad substrate scope, and mild reaction conditions.

In 2014, the same authors applied a closely related catalyst system **6** to the first highly enantioselective aldol-type reaction of α-keto esters with 5-methyleneoxazolines, providing a novel remarkable route to chiral 2,5-disubstituted oxazole derivatives bearing a quaternary stereogenic centre (Scheme 8.7).[20] The investigation of the substrate scope of the process showed that a wide range of α-keto esters were compatible, since methyl, ethyl, isopropyl, and *tert*-butyl esters gave identical excellent results. Generally, the reactions were remarkably tolerant of functional groups in terms of enantioselectivity, regardless of the electronic properties and steric hindrance of the substituents on the α-aryl group of the α-keto esters. The catalyst system based on *N*,*N*′-oxide ligand **6** was also applicable to heteroaryl α-keto esters, which delivered the corresponding products in excellent outcomes. Moreover, aliphatic α-keto esters also reacted in excellent enantioselectivity (>99% ee), even though the yield was decreased (40%). Three different aromatic 5-methyleneoxazolines were proven to give comparable excellent results. The scope of the process could also be extended to glyoxal derivatives, which reacted with phenyl 5-methyleneoxazoline to give the corresponding products in comparable excellent yields and enantioselectivities of up to 99% and >99% ee, respectively. In view of the importance of oxazoles as structural motifs in a wide variety of natural products, pharmaceuticals, and synthetic

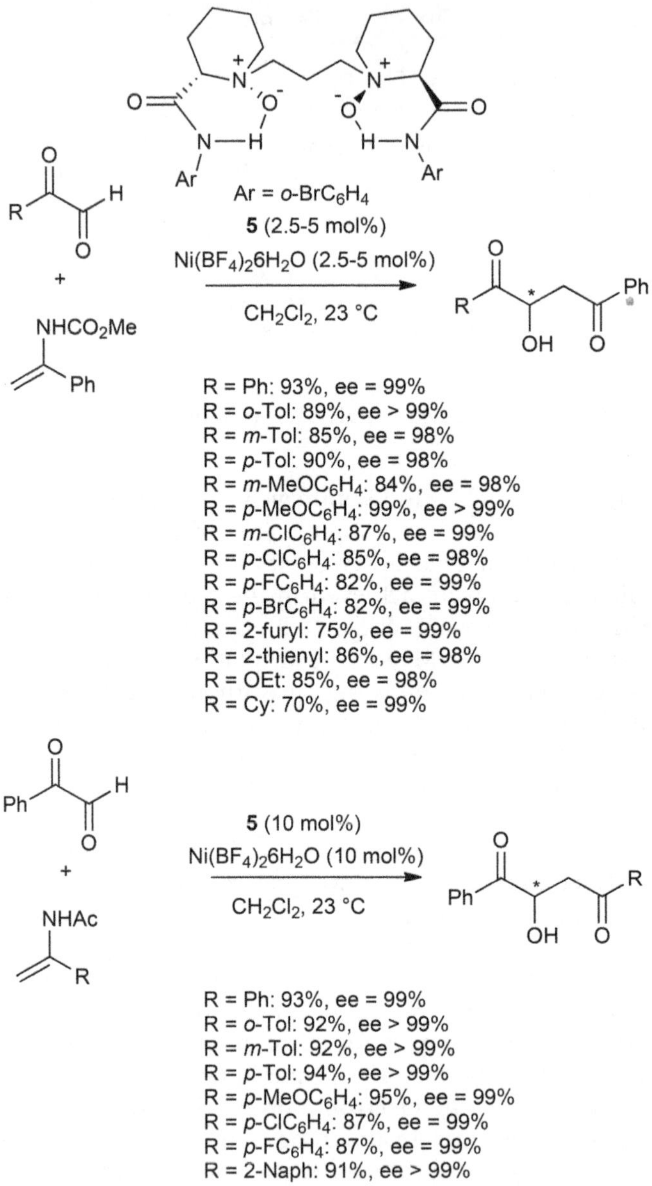

Scheme 8.6 Aldol-type reactions of glyoxal derivatives with enecarbamates and enamides in the presence of an *in situ* generated nickel catalyst derived from an *N,N′*-dioxide ligand.

intermediates,[21] this remarkable process offers a novel route for the preparation of a broad variety of enantiopure 2,5-disubstituted oxazole derivatives under mild reaction conditions.

The catalytic asymmetric nitroaldol (Henry) reaction is a useful carbon–carbon bond-forming reaction.[22] It affords enantiomerically enriched β-hydroxy

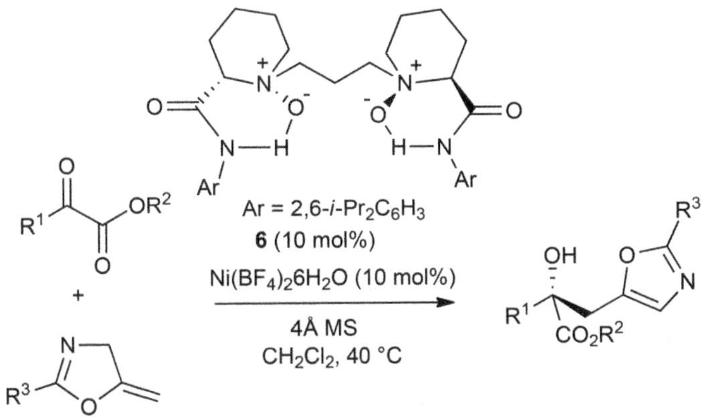

$R^1 = R^3 = Ph$, $R^2 = Me$: 97%, ee > 99%
$R^1 = R^3 = Ph$, $R^2 = Et$: 99%, ee > 99%
$R^1 = R^3 = Ph$, $R^2 =$ *i*-Pr: 98%, ee > 99%
$R^1 = R^3 = Ph$, $R^2 =$ *t*-Bu: 98%, ee > 99%
$R^1 =$ *m*-Tol, $R^2 = Me$, $R^3 = Ph$: 95%, ee > 99%
$R^1 =$ *p*-Tol, $R^2 = Me$, $R^3 = Ph$: 77%, ee > 99%
$R^1 =$ *m*-$MeOC_6H_4$, $R^2 = Me$, $R^3 = Ph$: 97%, ee > 99%
$R^1 =$ *p*-$MeOC_6H_4$, $R^2 = Me$, $R^3 = Ph$: 63%, ee > 99%
$R^1 =$ *m*-FC_6H_4, $R^2 = Me$, $R^3 = Ph$: 99%, ee > 99%
$R^1 =$ *p*-FC_6H_4, $R^2 = Me$, $R^3 = Ph$: 95%, ee > 99%
$R^1 =$ *p*$CF_3C_6H_4$, $R^2 = Me$, $R^3 = Ph$: 97%, ee = 99%
$R^1 =$ *m*-(*t*-Bu)C_6H_4, $R^2 = Me$, $R^3 = Ph$: 96%, ee > 99%
$R^1 =$ *p*-PhC_6H_4, $R^2 = Me$, $R^3 = Ph$: 95%, ee > 99%
$R^1 =$ 2-Naph, $R^2 = Me$, $R^3 = Ph$: 98%, ee > 99%
$R^1 =$ 2-thienyl, $R^2 = Me$, $R^3 = Ph$: 98%, ee = 99%
$R^1 =$ 2-furyl, $R^2 = Me$, $R^3 = Ph$: 96%, ee > 99%
$R^1 = Cy$, $R^2 = Me$, $R^3 = Ph$: 40%, ee > 99%
$R^1 = R^2 = Me$, $R^3 = Ph$: 84%, ee > 99%
$R^1 = Ph$, $R^2 = Me$, $R^3 =$ *p*-Tol: 98%, ee > 99%
$R^1 = Ph$, $R^2 = Me$, $R^3 =$ *m*-$MeOC_6H_4$: 91%, ee > 99%

Scheme 8.7 Aldol-type reaction of α-keto esters with 5-methyleneoxazolines in the presence of an *in situ* generated nickel catalyst derived from an *N*,*N*′-dioxide ligand.

nitroalkanes, which are key intermediates and building blocks. Owing to its synthetic utility, increasing efforts have been directed towards developing an efficient catalytic asymmetric version of this reaction. Since the pioneering application of Shibasaki's heterometallic catalyst system in the Henry reaction,[23] various types of catalyst systems have been studied,[24] including the highly efficient Shibasaki's multimetallic complex,[25] Trost's dinuclear zinc complex,[26] and Savoia's dinuclear copper complex.[27] Very recently, Zhou *et al.* investigated the nitroaldol reaction of aldehydes with nitromethane in the presence of an *in situ* generated nickel catalyst from a polyfunctionalised α-amino acid-derived

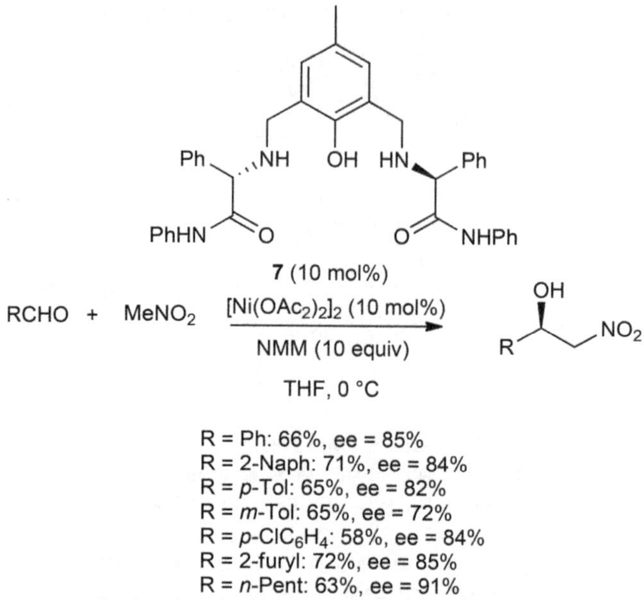

Scheme 8.8 Nitroaldol reaction of aldehydes with nitromethane in the presence of an *in situ* generated nickel catalyst from an α-amino acid-derived ligand.

ligand 7 (Scheme 8.8).[28] This ligand bearing two phenyl groups was selected as optimal among a series of variously substituted ligands of the same type. In this study, the authors also investigated several metal salts. The zinc and copper complexes did not work well in the reaction, but high activity was obtained for the corresponding manganese complex, albeit in combination with a low chiral induction (3% ee). The best results were achieved by using the *in situ* generated nickel complex of ligand 7 at 10 mol% of catalyst loading in the presence of *N*-methylmorpholine (NMM) as an additive. Under these optimised reaction conditions, the investigation of the substrate scope showed that arylaldehydes with a *meta* or *para* substituent or an electron-withdrawing substituent, as well as heteroarylaldehydes, afforded the corresponding products in moderate yields and with good enantioselectivities of up to 85% ee. It is worth noting that an aliphatic aldehyde was also suitable for this catalyst system, providing a higher enantioselectivity (91% ee) than benzaldehyde.

8.3 Mannich-Type Reactions

The classic direct Mannich reaction, discovered in 1912,[29] is an aminoalkylation of carbonyl compounds involving ammonia (or a primary or secondary amine derivative), a non-enolisable aldehyde (usually formaldehyde) or a ketone, and an enolisable carbonyl compound, leading to β-amino carbonyl derivatives.[30] Catalytic asymmetric Mannich-type reactions of aldehydes, ketones, esters, and other donors for the synthesis of β-amino carbonyl compounds have been investigated intensively over the past decade.[1i,31] Traditionally, asymmetric Mannich reactions are catalysed by chiral transition-metal complexes.[32] In the last few years, several groups have developed efficient enantioselective Mannich-type reactions by using chiral catalysts of various metals, such as scandium,[33] silver,[34] tin,[35] zirconium,[36] copper,[37] or nickel. For example, Shibasaki *et al.* reported in 2008 a novel preformed dinuclear nickel catalyst from Schiff base ligand **1** to be applied in enantioselective Mannich-type reactions of α-substituted nitroacetates with *N*-Boc-imines, which afforded the corresponding α,α,α,α-tetrasubstituted *anti*-α,β-diamino acid surrogates in very high yields and with enantioselectivities of 91–99% ee, along with good to high diastereoselectivities of 72–94% de (Scheme 8.9).[38] Studying the substrate scope of the process, the authors showed that several α-alkyl-substituted nitroacetates were compatible, as well as non-isomerisable aryl and heteroaryl imines which gave 91–99% ee at 0 °C; for isomerisable aliphatic imines the reactions were performed at a lower temperature (−20 or −40 °C) to prevent the undesired isomerisation

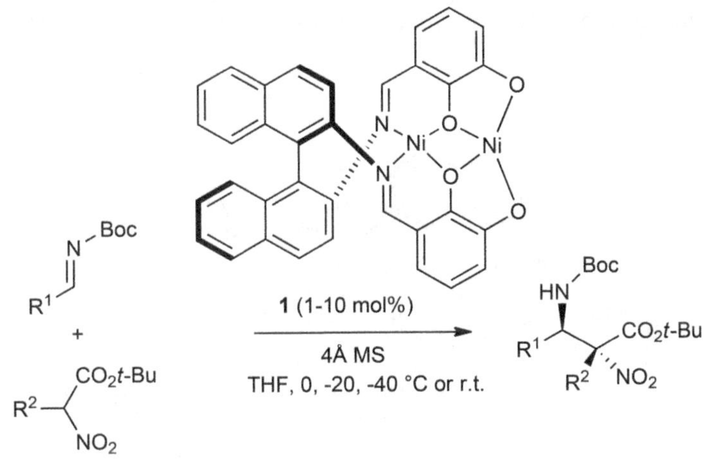

R^1 = Ph, R^2 = Me: 95%, dr = 91:9, ee = 98%
R^1 = *p*-$MeOC_6H_4$, R^2 = Me: 92%, dr = 87:13, ee = 98%
R^1 = *p*-Tol, R^2 = Me: 90%, dr = 89:11, ee = 97%
R^1 = *p*-ClC_6H_4, R^2 = Me: 87%, dr = 86:14, ee = 97%
R^1 = *p*-FC_6H_4, R^2 = Me: 91%, dr = 90:10, ee = 91%
R^1 = 3-thienyl, R^2 = Me: 96%, dr = 91:9, ee = 99%
R^1 = Ph, R^2 = Et: 91%, dr = 94:6, ee = 99%
R^1 = Ph, R^2 = *n*-Pr: 92%, dr = 92:8, ee > 99%
R^1 = Ph, R^2 = Bn: 94%, dr = 88:12, ee = 94%
R^1 = $BnCH_2$, R^2 = Me: 73%, dr > 97:3, ee = 95%
R^1 = *n*-Bu, R^2 = Me: 67%, dr > 97:3, ee = 93%
R^1 = *i*-Bu, R^2 = Me: 85%, dr > 97:3, ee = 91%

Scheme 8.9 Mannich-type reaction of nitroacetates with *N*-Boc-imines in the presence of a preformed dinuclear nickel catalyst derived from a Schiff base ligand.

of the imines to enamides, affording the corresponding products in enantioselectivities of 91–95% ee. In this work, the authors also investigated other metal combinations using the same ligand, such as one nickel atom and one samarium atom (59% ee), two copper atoms (9% ee), and two palladium atoms (0% ee), which all gave much less satisfactory results.

Few asymmetric catalytic Mannich-type reactions have been applied to the synthesis of β-aminophosphonic acid derivatives. In 2005, Jørgensen *et al.* reported the first catalytic enantio- and diastereoselective direct Mannich-type reaction of β-ketophosphonates, giving β-aminophosphonates in 43–84% ee.[39] The imine in this study was, however, limited to an *N*-Ts-imino ester. In 2008, Shibasaki *et al.* applied their powerful dinuclear nickel catalyst from Schiff base ligand **1** (see Scheme 8.9) to the enantioselective direct Mannich-type reaction of a β-ketophosphonate.[40] As shown in Scheme 8.10, this substrate reacted smoothly with various aryl and heteroaryl *N*-Boc-imines, providing the corresponding Mannich products in moderate to good yields (43–90%), moderate to high diastereoselectivities (34–90% de), and good to excellent enantioselectivities (84–99% ee). In the case of aryl imines bearing substituents as substrates, the reactivity was much lower than with the imine

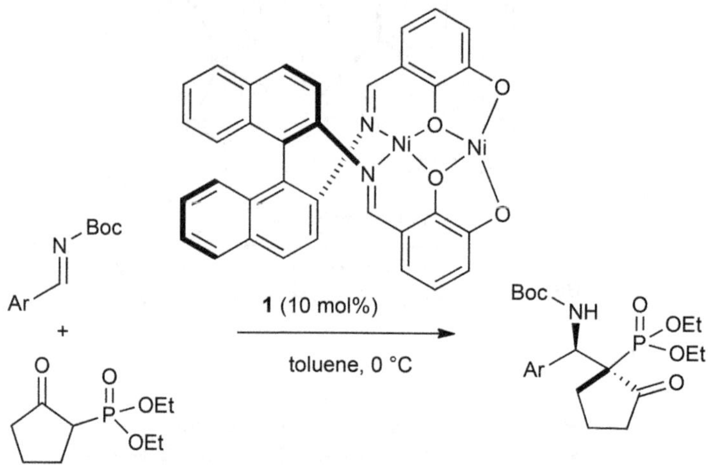

Ar = Ph: 85%, dr = 94:6, ee = 98%
Ar = 3-thienyl: 90%, dr = 94:6, ee = 98%
Ar = 3-furyl: 82%, dr = 93:7, ee = 99%
Ar = 2-furyl: 75%, dr = 86:14, ee = 98%
Ar = *p*-Tol: 63%, dr = 91:9, ee = 95%
Ar = *p*-$MeOC_6H_4$: 86%, dr = 88:16, ee = 87%
Ar = *p*-FC_6H_4: 64%, dr = 86:14, ee = 85%
Ar = *p*-ClC_6H_4: 43%, dr = 92:8, ee = 84%
Ar = *m*-$MeOC_6H_4$: 73%, dr = 67:33, ee = 92%
Ar = 2-Naph: 71%, dr = 95:5, ee = 84%

Scheme 8.10 Mannich-type reaction of a β-ketophosphonate with *N*-Boc-imines in the presence of a preformed dinuclear nickel catalyst derived from a Schiff base ligand.

derived from benzaldehyde. One drawback of this novel process was its limitations on β-ketophosphonate donors. Indeed, when using other cyclic β-ketophosphonates, such as six-membered ones, only modest reactivity and enantioselectivity were observed (47–51% yield, 47–55% ee).

On the other hand, Mannich-type reactions of homoenolates or their synthetic equivalents for the production of γ-amino acids have been less studied than those involving enolates. In this context, the same authors applied a closely related preformed dinuclear nickel catalyst derived from Schiff base ligand **8** to induce the asymmetric direct Mannich-type reaction of α-keto anilides with *o*-Ns-protected imines (Scheme 8.11).[41] The corresponding

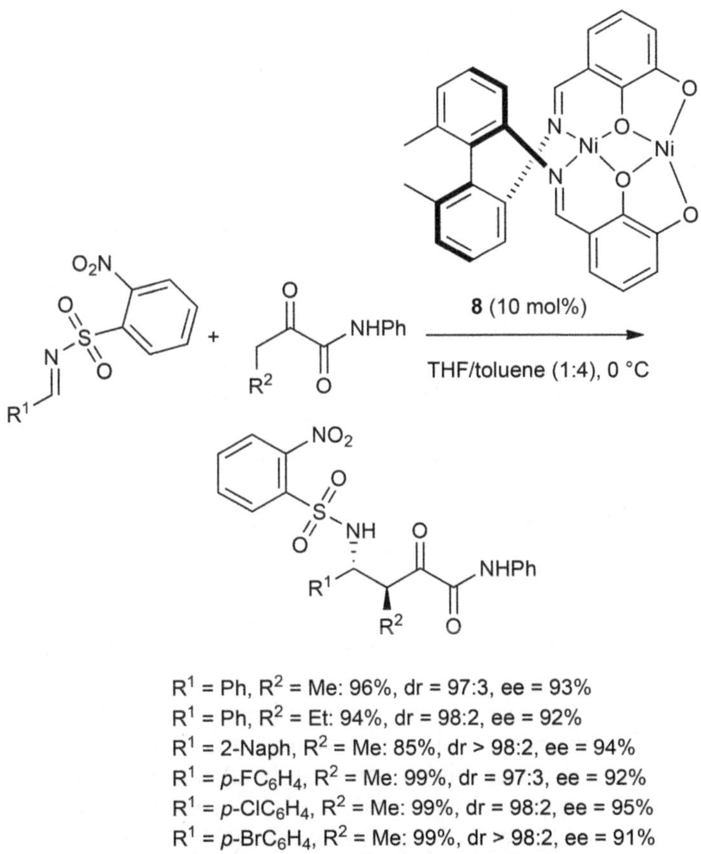

R^1 = Ph, R^2 = Me: 96%, dr = 97:3, ee = 93%
R^1 = Ph, R^2 = Et: 94%, dr = 98:2, ee = 92%
R^1 = 2-Naph, R^2 = Me: 85%, dr > 98:2, ee = 94%
R^1 = *p*-FC_6H_4, R^2 = Me: 99%, dr = 97:3, ee = 92%
R^1 = *p*-ClC_6H_4, R^2 = Me: 99%, dr = 98:2, ee = 95%
R^1 = *p*-BrC_6H_4, R^2 = Me: 99%, dr > 98:2, ee = 91%
R^1 = *p*-$MeOC_6H_4$, R^2 = Me: 86%, dr > 98:2, ee = 93%
R^1 = *p*-Tol, R^2 = Me: 76%, dr > 98:2, ee = 91%
R^1 = 3-thienyl, R^2 = Me: 90%, dr = 94:6, ee = 93%
R^1 = Cy, R^2 = Me: 87%, dr = 91:9, ee = 91%
R^1 = Ph, R^2 = Me: 96%, dr = 97:3, ee = 93%

Scheme 8.11 Mannich-type reaction of α-keto anilides with *o*-Ns-protected imines in the presence of a preformed dinuclear nickel catalyst derived from a Schiff base ligand.

Mannich products were achieved in very high yields (up to 99%), *anti*/*syn* ratios often greater than 98 : 2, and with high enantioselectivities of 91–95% ee. A range of non-isomerisable aryl imines with either an electron-donating or an electron-withdrawing substituent on the aromatic ring gave the Mannich products in 76–99% yield with at least 97 : 3 and in some cases greater than 98 : 2 *anti* selectivity, along with 91–95% ee. An heteroaryl imine also led to the corresponding product in high yield (90%) and enantioselectivity (93% ee), albeit with slightly decreased *anti* selectivity (*anti*/*syn* = 94 : 6). Moreover, an isomerisable aliphatic imine provided the corresponding Mannich product in 87% yield, 91% ee, and an *anti*/*syn* ratio of 91 : 9. The synthetic utility of this novel methodology was demonstrated in the subsequent stereoselective reduction of the Mannich products, which afforded the corresponding α-hydroxy-β-alkyl-δ-amino amides bearing three contiguous stereocentres with an *anti*/*anti* relative configuration. The latter constituted good precursors for fully substituted azetidine-2-amides, which are useful nonnatural amino acid derivatives.

Catalytic asymmetric vinylogous reactions of γ-butenolides and related compounds have been intensively studied, giving versatile functionalised chiral γ-butenolide skeletons.[42] In contrast, the use of their aza analogues, α,β-unsaturated γ-butyrolactams, as donors in catalytic asymmetric reactions is rare, despite their synthetic utility. To address this issue, Shibasaki *et al.* have developed direct catalytic asymmetric vinylogous Mannich-type reactions of α,β-unsaturated γ-butyrolactams with *N*-Boc-imines (Scheme 8.12).[43] The reactions were catalysed by dinuclear nickel catalyst **1** in the presence of $CaSO_4$ as an additive, and occurred selectively at the γ-position, providing the corresponding vinylogous Mannich adducts in good yields, with moderate to high diastereoselectivities of 66–94% de, along with remarkable general enantioselectivity of 99% ee in all cases of substrates studied. The catalytic system was applicable to various non-isomerisable aryl and heteroaryl imines. High enantioselectivities were achieved for aryl imines with either an electron-withdrawing or an electron-donating substituent at the *ortho*, *meta*, or *para* position. With heteroaryl imines, the reactivity was somewhat decreased (61–83% yields). Unfortunately, isomerisable aliphatic imines resulted in low yields (<20%), due to competitive isomerisation to enamides over nucleophilic activation of the α,β-unsaturated γ-butyrolactams.

Complex molecules containing an indole skeleton, which are widely found in nature, are promising candidates for drug development. In particular, a number of biologically active compounds include a 3-substituted 3-amino-2-oxindole as a core structure. For constructing a chiral quaternary aminocarbon centre at the C-3 position of oxindoles, catalytic asymmetric nucleophilic addition to isatin-derived ketimines is the most direct and rational approach.[44] In this context, Arai *et al.* recently reported enantioselective nitro-Mannich reactions of isatin-derived *N*-Boc-ketimines with nitroalkanes (Scheme 8.13).[45] These reactions were mediated by an *in situ* generated nickel catalyst from $NiCl_2$ and bis(imidazolidine)-pyridine ligand **9**, in the presence of DIPEA as additive. Before selecting this catalyst system, the authors

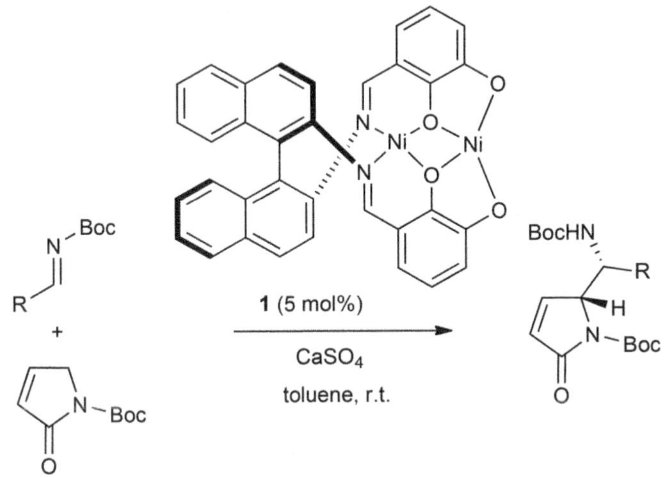

R = Ph: 95%, dr > 97:3, ee = 99%
R = 2-Naph: 76%, dr = 97:3, ee = 99%
R = 1-Naph: 93%, dr = 96:4, ee = 99%
R = *o*-ClC_6H_4: 95%, dr > 97:3, ee = 99%
R = *m*-Tol: 87%, dr > 97:3, ee = 99%
R = *p*-ClC_6H_4: 87%, dr > 97:3, ee = 99%
R = *p*-$MeOC_6H_4$: 85%, dr = 96:4, ee = 99%
R = 2-furyl: 61%, dr = 83:17, ee = 99%
R = 3-thienyl: 83%, dr = 95:5, ee = 99%

Scheme 8.12 Mannich-type reaction of α,β-unsaturated γ-butyrolactams with *N*-Boc-imines in the presence of a preformed dinuclear nickel catalyst derived from a Schiff base ligand.

investigated the corresponding complexes of different metals, such as copper, cobalt, iron, and zinc. For example, the complex derived from $Cu(OTf)_2$ and ligand **9** provided the product (R^1 = Me, X = Y = Z = R^2 = H) in quantitative yield and with 55% ee, while the complex from $CoCl_2$ gave the same product in 78% yield and 87% ee (*vs.* 98% yield and 90% ee with $NiCl_2$). On the other hand, the use of the complex from $ZnCl_2$ provided 75% yield and 35% ee, whereas $FeCl_3$ led to the racemic product. Studying the substrate scope of the reaction, it was found that the isatin-derived *N*-Boc-ketimines with alkyl substituents at R^1 (*e.g.*, R^1 = Me, Bn, allyl) were smoothly converted by reaction with nitromethane (R^2 = H) into the corresponding products while maintaining high enantioselectivity of up to 95% ee, although the acetyl substituent significantly reduced the reactivity of the substrate (19% yield) as well as the enantioselectivity (57% ee). *N*-Methylisatins containing not only electron-deficient but also electron-donating substituents on the benzene ring reacted successfully to give the nitro-Mannich products, with enantioselectivities ranging from 78 to 95% ee. Nitroethane was also tolerated, giving the product in a diastereomeric ratio of 80 : 20, with both diastereomers obtained in a highly enantioselective manner (85 and 90% ee).

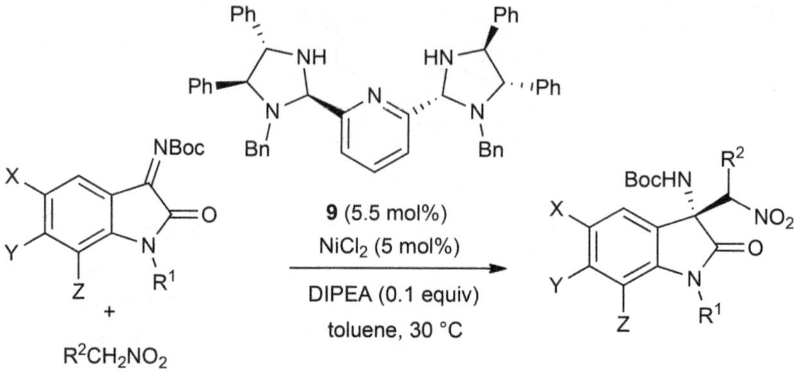

R^1 = Me, R^2 = X = Y = Z = H : 99%, ee = 94%
R^1 = Bn, R^2 = X = Y = Z = H : 71%, ee = 84%
R^1 = allyl, R^2 = X = Y = Z = H : 90%, ee = 84%
R^1 = Ac, R^2 = X = Y = Z = H : 19%, ee = 57%
R^1 = Me, R^2 = Y = Z = H, X = F : 79%, ee = 82%
R^1 = Me, R^2 = Y = Z = H, X = Cl : 86%, ee = 88%
R^1 = Me, R^2 = X = Z = H, Y = Cl : 80%, ee = 86%
R^1 = Me, R^2 = X = Z = H, Y = Br : 77%, ee = 81%
R^1 = Me, R^2 = X = Y = H, Z = Cl : 72%, ee = 78%
R^1 = Me, R^2 = Y = Z = H, X = Me : 99%, ee = 95%
R^1 = Me, R^2 = Y = Z = H, X = OMe : 74%, ee = 92%
R^1 = R^2 = Me, X = Y = Z = H: 97%, dr = 80:20, ee = 90% and 85%

Scheme 8.13 Nitro-Mannich reaction of isatin-derived *N*-Boc-ketimines with nitroalkanes in the presence of an *in situ* generated nickel catalyst derived from a bis(imidazolidine)-pyridine ligand.

8.4 Conclusions

The last decade has seen remarkable results achieved in enantioselective nickel-catalysed aldol-type reactions as well as Mannich-type reactions, especially from the results independently reported by the groups of Shibasaki and Feng. Among the most important advances are direct asymmetric aldol reactions of an α-isothiocyanato imide with aldehydes performed with an *N,N′*-dioxide–nickel catalyst, which provided the corresponding aldol products in high yields and both excellent diastereo- and enantioselectivities of up to >98% de and >99% ee, respectively. Another nice result was achieved with the asymmetric aldol reaction between glyoxal derivatives and enolsilanes catalysed by a chiral nickel complex generated *in situ* from another chiral *N,N′*-oxide ligand. Very good enantioselectivities of 90–95% ee were reported in combination with good yields for the corresponding aldol products. Always in the context of aldol-type reactions, enantioselective nickel-catalysed aldol-type reactions of glyoxal derivatives with enecarbamates as well as enamides catalysed by a chiral *N,N′*-dioxide–nickel complex provided high reactivities and excellent enantioselectivities of 98 to >99% ee for a wide range of substrates. Finally, a novel remarkable route to chiral 2,5-disubstituted oxazole derivatives bearing a quaternary stereogenic centre was based on the first highly enantioselective aldol-type reaction of α-keto esters with 5-methyleneoxazolines. This powerful process gave comparable excellent yields and enantioselectivities of up to 99% and >99% ee, respectively, for a broad substrate scope.

In the context of enantioselective nickel-catalysed Mannich-type reactions, several excellent advances have also been reported, such as the use of a dinuclear Schiff base nickel catalyst applied in enantioselective Mannich-type reactions of α-substituted nitroacetates with *N*-Boc-imines, which afforded the corresponding α,α,α,α-tetrasubstituted *anti*-α,β-diamino acid surrogates in very high yields and with enantioselectivities of 91–99% ee, along with good to high diastereoselectivities of up to 94% de. In addition, the asymmetric direct Mannich-type reaction between α-keto anilides and *o*-Ns-protected imines was achieved in very high yields (up to 99%), *anti*/*syn* ratios often greater than 98 : 2, and with high enantioselectivities of 91–95% ee by using the same chiral dinuclear Schiff base–nickel catalyst. Finally, direct catalytic asymmetric vinylogous Mannich-type reactions of α,β-unsaturated γ-butyrolactams with *N*-Boc-imines were induced by another dinuclear chiral Schiff base nickel complex, leading to the corresponding vinylogous Mannich adducts in good yields, high diastereoselectivities of up to 94% de, along with remarkable general enantioselectivity of 99% ee in all cases of substrates studied.

In conclusion, an important amount of work has been done in the last 10 years to develop asymmetric control in the aldol- and Mannich-type reactions catalysed by chiral nickel complexes. This catalysis is intrinsically elegant and economical, but it appears that, at least for the time being, it is still limited to simpler substrates in most cases. In addition, further studies will have to focus on better understanding asymmetric control in these reactions.

References

1. (a) T. Bach, *Angew. Chem., Int. Ed.*, 1994, **33**, 417–450; (b) S. G. Nelson, *Tetrahedron: Asymmetry*, 1998, **9**, 357–389; (c) J. S. Johnson and D. A. Evans, *Acc. Chem. Res.*, 2000, **33**, 325–335; (d) T. D. Machajewski and C.-H. Wong, *Angew. Chem., Int. Ed.*, 2000, **39**, 1352–1375; (e) B. Alcaide and P. Almendros, *Eur. J. Org. Chem.*, 2002, 1595–1601; (f) C. Palomo, M. Oiarbide and J. M. Garcia, *Chem. Soc. Rev.*, 2004, **33**, 65–75; (g) W. Notz, F. Tanaka and C. F. Barbas, *Acc. Chem. Res.*, 2004, **37**, 580–591; (h) *Modern Aldol Reactions*, ed. R. Mahrwald, Wiley-VCH, Weinheim, 2004, vol. 1 and 2; (i) M. Shibasaki, N. Yoshikawa and S. Matsunaga, *Comprehensive Asymmetric Catalysis, Supplement*, Springer, Heidelberg, 2004, vol. 1, pp. 135–141; (j) E. M. Carreira, A. Fettes and C. Marti, *Org. React.*, 2006, **67**, 1–216; (k) S. Mukherjee, J. W. Yang, S. Hoffmann and B. List, *Chem. Rev.*, 2007, **107**, 5471–5569; (l) L. M. Geary and P. G. Hultin, *Tetrahedron: Asymmetry*,

2009, **20**, 131–173; (m) B. M. Trost and C. S. Brindle, *Chem. Soc. Rev.*, 2010, **39**, 1600–1632; (n) L. C. Dias, E. C. de Lucca Jr., M. A. B. Ferreira and E. C. Polo, *J. Braz. Chem. Soc.*, 2012, **23**, 2137–2158.

2. With a Pd-catalyst: I. Fukuchi, Y. Hamashima and M. Sodeoka, *Adv. Synth. Catal.*, 2007, **349**, 509–512, with a Rh-catalyst; R. Kuwano, H. Miyazaki and Y. Ito, *J. Organomet. Chem.*, 2000, **603**, 18–29.
3. (a) S. Ishikawa, T. Hamada, K. Manabe and S. Kobayashi, *J. Am. Chem. Soc.*, 2004, **126**, 12236–12237; (b) S. Kobayashi, T. Ogino, H. Shimizu, S. Ishikawa, T. Hamada and K. Manabe, *Org. Lett.*, 2005, **7**, 4729–4731; (c) M. Kokubo, C. Ogawa and S. Kobayashi, *Angew. Chem., Int. Ed.*, 2008, **47**, 6909–6911.
4. S. Mouri, Z. Chen, S. Matsunaga and M. Shibasaki, *Chem. Commun.*, 2009, 5138–5140.
5. *Glycopeptide Antibiotics*, ed. R. Nagarajan, Marcel Dekker, New York, 1994.
6. C. Najera and J. M. Sansano, *Chem. Rev.*, 2007, **107**, 4584–4671.
7. M. C. Willis, G. A. Cutting, V. J.-D. Piccio, M. J. Durbin and M. P. John, *Angew. Chem., Int. Ed.*, 2005, **44**, 1543–1545.
8. T. Yoshino, H. Morimoto, G. Lu, S. Matsunaga and M. Shibasaki, *J. Am. Chem. Soc.*, 2009, **131**, 17082–17083.
9. X. Chen, Y. Zhu, Z. Qiao, M. Xie, L. Lin, X. Liu and X. Feng, *Chem.-Eur. J.*, 2010, **16**, 10124–10129.
10. H. Maheswaran, P. J. A. Joseph, K. L. Prasanth, S. Priyadarshini, P. Satyanarayana, P. R. Likhar and M. L. Kantam, *Tetrahedron: Asymmetry*, 2010, **21**, 2158–2166.
11. (a) H. Gröger, E. M. Vogl and M. Shibasaki, *Chem.-Eur. J.*, 1998, **4**, 1137–1141; (b) C. Palomo, M. Oiarbide and J. M. Garcia, *Chem.-Eur. J.*, 2002, **8**, 36–44; (c) B. Schetter and R. Mahrwald, *Angew. Chem., Int. Ed.*, 2006, **45**, 7506–7525; (d) S. Adachi and T. Harada, *Eur. J. Org. Chem.*, 2009, 3661–3671.
12. S. Kobayashi, Y. Fujishita and T. Mukaiyama, *Chem. Lett.*, 1990, 1455–1458.
13. (a) K. Furuta, T. Maruyama and H. Yamamoto, *J. Am. Chem. Soc.*, 1991, **113**, 1041–1042; (b) E. M. Carreira, R. A. Singer and W. Lee, *J. Am. Chem. Soc.*, 1994, **116**, 8837–8838; (c) K. Mikami and S. Matsukawa, *J. Am. Chem. Soc.*, 1994, **116**, 4077–4078; (d) G. E. Keck and D. Krischnamurthy, *J. Am. Chem. Soc.*, 1995, **117**, 2363–2364; (e) D. A. Evans, M. C. Kozlowski, J. A. Murry, C. S. Burgey, K. R. Campos, B. T. Connell and R. J. Staples, *J. Am. Chem. Soc.*, 1999, **121**, 669–685; (f) K. Oisaki, D. Zhao, M. Kanai and M. Shibasaki, *J. Am. Chem. Soc.*, 2006, **128**, 7164–7165.
14. (a) S. E. Denmark and Y. Fan, *J. Am. Chem. Soc.*, 2002, **124**, 4233–4235; (b) V. B. Gondi, K. Hagihara and V. H. Rawal, *Angew. Chem., Int. Ed.*, 2009, **48**, 776–779.
15. (a) D. A. Evans, D. W. C. MacMillan and K. R. Campos, *J. Am. Chem. Soc.*, 1997, **119**, 10859–10860; (b) V. B. Gondi, K. Hagihara and V. H. Rawal, *Chem. Commun.*, 2010, **46**, 904–906.
16. J. Zhao, K. Zheng, S. J. Yang, L. Lin, X. Liu and X. Feng, *Synlett*, 2011, **7**, 903–906.

17. (a) R. Matsubara, Y. Nakamura and S. Kobayashi, *Angew. Chem., Int. Ed.*, 2004, **43**, 1679–1681; (b) R. Matsubara, P. Vital, Y. Nakamura, H. Kiyohara and S. Kobayashi, *Tetrahedron*, 2004, **60**, 9769–9784.
18. J. S. Fossey, R. Matsubara, P. Vital and S. Kobayashi, *Org. Biomol. Chem.*, 2005, **3**, 2910–2913.
19. K. Zheng, X. Liu, J. Zhao, Y. Yang, L. Lin and X. Feng, *Chem. Commun.*, 2010, **46**, 3771–3773.
20. W. Luo, J. Zhao, C. Yin, X. Liu, L. Lin and X. Feng, *Chem. Commun.*, 2014, **50**, 7524–7526.
21. (a) P. Wipf, *Chem. Rev.*, 1995, **95**, 2115–2134; (b) Z. Jin, *Nat. Prod. Rep.*, 2013, **30**, 869–915.
22. (a) L. Henry, *Hebd. Séances Acad. Sci.*, 1895, **120**, 1265–1268; (b) C. Palomo, M. Oiarbide and A. Laso, *Eur. J. Org. Chem.*, 2007, 2561–2574.
23. H. Sasai, T. Suzuki, S. Arai, T. Arai and M. Shibasaki, *J. Am. Chem. Soc.*, 1992, **114**, 4418–4420.
24. N. Ananthi and S. Velmathi, *Indian J. Chem.*, 2013, 87–108.
25. (a) M. Shibasaki, H. Arai and T. Arai, *Angew. Chem., Int. Ed.*, 1997, **36**, 1236–1256; (b) S. Matsunaga and M. Shibasaki, *Synthesis*, 2013, **45**, 421–437.
26. (a) B. M. Trost and V. S. C. Yeh, *Angew. Chem., Int. Ed.*, 2002, **41**, 861–863; (b) B. M. Trost and C. Müller, *J. Am. Chem. Soc.*, 2006, **128**, 2778–2779.
27. A. Gualandi, L. Cerisoli, H. Stoeckli-Evans and D. Savoia, *J. Org. Chem.*, 2011, **76**, 3399–3408.
28. Y. Liu, P. Deng, X. Li, Y. Xiong and H. Zhou, *Synlett*, 2014, **25**, 1735–1738.
29. C. Mannich and W. Krosche, *Arch. Pharm.*, 1912, **250**, 647–667.
30. (a) M. Arend, B. Westermann and N. Risch, *Angew. Chem., Int. Ed.*, 1998, **37**, 1044–1070; (b) A. Cordova, *Acc. Chem. Res.*, 2004, **37**, 102–112; (c) M. Ueno and S. Kobayashi, *Enantioselective Synthesis of β-Amino Acids*, Wiley, Hoboken, 2nd edn, 2005, pp. 139–157; (d) J. M. M. Verkade, L. J. C. Van Hemert, P. J. L. M. Quaedflieg and F. P. J. T. Rutjes, *Chem. Soc. Rev.*, 2008, **37**, 29–41; (e) A. Cordova and R. Rios, *Amino Group Chemistry*, Wiley, Weinheim, 2008, pp. 185–205; (f) R. Gomez Arrayas and J. C. Carretero, *Chem. Soc. Rev.*, 2009, **38**, 1940–1948; (g) X.-H. Cai and B. X. Guo, *Eur. J. Org. Chem.*, 2012, **3**, 258–266; (h) B. Karimi, D. Enders and E. Jafari, *Synthesis*, 2013, **45**, 2769–2812.
31. G. K. Friestad and A. K. Mathies, *Tetrahedron*, 2007, **63**, 2541–2569.
32. (a) A. B. Charette, A. A. Boezio, A. Cote, E. Moreau, J. Pytkowicz, J. N. Desrosiers and C. Legault, *Pure Appl. Chem.*, 2005, **77**, 1259–1267; (b) M. Shibasaki and S. Matsunaga, *J. Organomet. Chem.*, 2006, **691**, 2089–2100; (c) J. Wang, T. Shi, H. H. Deng, H. L. Jiang and H. Liu, *J. Org. Chem.*, 2008, **73**, 8563–8570; (d) Y. K. Kang and D. Y. Kim, *Tetrahedron Lett.*, 2011, **52**, 2356–2358.
33. Q. Zhang, Y. Hui, X. Zhou, L. Lin, X. Liu and X. Feng, *Adv. Synth. Catal.*, 2010, **352**, 976–980.
34. H. Mandai, K. Mandai, M. L. Snapper and H. Hoveyda, *J. Am. Chem. Soc.*, 2008, **130**, 17961–17969.

35. A. Izumiseki, K. Yoshida and A. Yanagisawa, *Org. Lett.*, 2009, **11**, 5310–5313.
36. S. Kobayashi, J. Kobayashi, R. Yazaki and M. Ueno, *Chem.-Asian J.*, 2007, **2**, 135–144.
37. Y. Du, L.-W. Xu, Y. Shimizu, K. Oisaki, M. Kanai and M. Shibasaki, *J. Am. Chem. Soc.*, 2008, **130**, 16146–16147.
38. Z. Chen, H. Morimoto, S. Matsunaga and M. Shibasaki, *J. Am. Chem. Soc.*, 2008, **130**, 2170–2171.
39. A. Kjaersgaaer and K. A. Jørgensen, *Org. Biomol. Chem.*, 2005, **3**, 804–808.
40. Z. Chen, K. Yakura, S. Matsunaga and M. Shibasaki, *Org. Lett.*, 2008, **10**, 3239–3242.
41. Y. Xu, G. Lu, S. Matsunaga and M. Shibasaki, *Angew. Chem., Int. Ed.*, 2009, **48**, 3353–3356.
42. (a) G. Casiraghi, F. Zanardi, L. Battistini and G. Rassu, *Synlett*, 2009, 1525–1542; (b) G. Casiraghi, F. Zanardi, G. Appendino and G. Rassu, *Chem. Rev.*, 2000, **100**, 1929–1972.
43. N. E. Shepend, H. Tanabe, Y. Xu, S. Matsunaga and M. Shibasaki, *J. Am. Chem. Soc.*, 2010, **132**, 3666–3667.
44. P. Chauhan and S. S. Chimni, *Tetrahedron: Asymmetry*, 2013, **24**, 343–356.
45. T. Arai, E. Matsumura and H. Masu, *Org. Lett.*, 2014, **16**, 2768–2771.

CHAPTER 9

Enantioselective Nickel-Catalysed Hydrogenation Reactions

9.1 Introduction

The catalytic enantioselective reduction of prochiral unsaturated organic molecules, such as ketones,[1] using molecular hydrogen is widely recognised as one of the most efficient methods for installing chirality into target compounds, providing an environmentally benign synthetic process to prepare pharmaceuticals, perfumes, and agrochemicals.[2] Ruthenium has become the dominant choice among the central transition metals since the first-generation BINAP–Ru catalyst for asymmetric hydrogenation of ketones made its debut in 1987.[3] Other precious metals, such as osmium, rhodium, iridium, and palladium,[4] have also been successfully employed, and the recent trend is to replace these metals with inexpensive base metals, such as iron, cobalt, nickel, or copper. There is no single catalyst for asymmetric hydrogenation of ketones that can be used for any types of substrate, and a trial-and-error method is still the major approach even now. The screening efficiency has, however, been dramatically enhanced by the use of robotics technology. This digest organises the developments in asymmetric hydrogenation of ketones on the basis of the following six categories of chiral ligands: (i) bisphosphine-based PP-ligands, (ii) bisphosphine/diamine-based P_2/N_2-ligands, (iii) tridentate or tetradentate phosphine/amine-based P_mN_n-ligands, (iv) diamine-based N,N-ligands, (v) tetradentate amine-based N_4-ligands, and (vi) tetradentate thioether/amine-based S_2N_2-ligands. The first report of phosphine-containing transition metal

RSC Catalysis Series No. 26
Enantioselective Nickel-Catalysed Transformations
By Hélène Pellissier

Published by the Royal Society of Chemistry, www.rsc.org

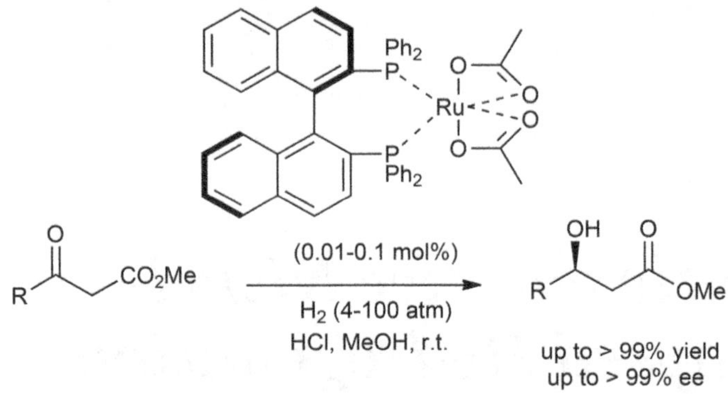

Scheme 9.1 Pioneering report on BINAP–Ru methods for highly enantioselective hydrogenation of β-keto esters.

catalysts dates back to 1970, when Schrock and Osborn discovered the catalytic activity of monocationic $RhH_2(PPhMe_2)_2(solvent)_2$.[5] This work was later extended to the asymmetric hydrogenation of prochiral ketones using chiral phosphines, eventually leading to the discovery of BINAP–Ru chemistry (Scheme 9.1).[3,6] Ever since, abundant and inexpensive base transition metals, such as copper and nickel, have attracted much attention in the area of catalytic hydrogenation.

9.2 Hydrogenations of Ketones

The first example of the homogeneous nickel-catalysed asymmetric hydrogenation of ketones was reported by Hamada and co-workers in 2008.[7] Using a 1:4 mixture of CF_3CH_2OH/acetic acid as solvent, aromatic α-amino-β-keto esters could be reduced to the corresponding *anti*-β-hydroxy-α-amino esters with high diastereoselectivities of up to 98% de and excellent enantioselectivities of up to 95% ee through dynamic kinetic resolution. The process was catalysed by an *in situ* generated nickel catalyst from $Ni(OAc)_2$ and a chiral ferrocenylphosphine **1** employed at 5 mol% of catalyst loading (Scheme 9.2). The reaction was optimal when performed in a mixture of trifluoroethanol (TFE) and acetic acid as solvent in the presence of sodium acetate, while most other solvents were ineffective for this hydrogenation. A series of substrates with different substituents were tolerated, but the best diastereo- and enantioselectivities were obtained in the case of aromatic substrates. In contrast to aromatic substrates, the reaction of aliphatic substrates was slow and the enantioselectivities were moderate (54–81% ee). It is important to note that this work represented the first use of homogeneous chiral nickel–phosphine complexes in asymmetric hydrogenation, but also the first example of nickel-catalysed dynamic kinetic resolution.

Later, these authors extended this methodology to the hydrogenation of substituted aromatic α-amino ketone hydrochlorides by using the same

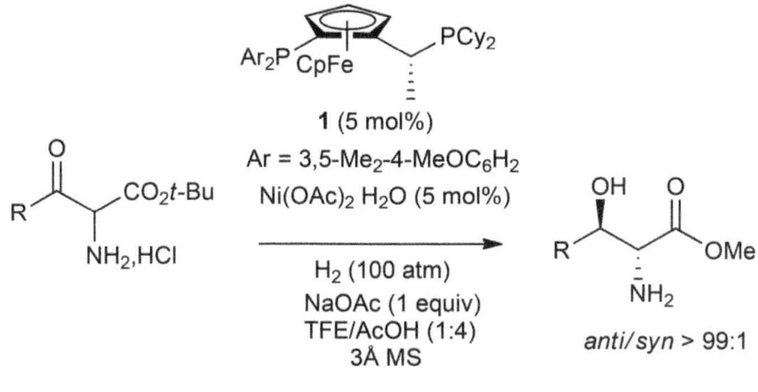

R = Ph: 90%, ee = 81%
R = *m*-Tol: 90%, ee = 92%
R = *m*-FC_6H_4: 88%, ee = 89%
R = *m*-ClC_6H_4: 91%, ee = 92%
R = *m*-BrC_6H_4: 91%, ee = 92%
R = *p*-Tol: 82%, ee = 93%
R = *p*-(*t*-Bu)C_6H_4: 90%, ee = 92%
R = *p*-$BnOC_6H_4$: 94%, ee = 91%
R = *p*-$NO_2C_6H_4$: 80%, ee = 91%
R = *p*-$CO_2MeC_6H_4$: 82%, ee = 88%
R = 2-Naph: 92%, ee = 90%
R = 2-thienyl: 79%, ee = 95%
R = Cy: 8%, ee = 81%
R = *t*-Bu: 21%, ee = 54%

Scheme 9.2 Hydrogenation of α-amino-β-keto ester hydrochlorides in the presence of an *in situ* generated nickel catalyst derived from a ferrocenylphosphine ligand.

homogeneous *in situ* generated chiral nickel catalyst.[8] In this case, the reaction required a 10 mol% of catalyst loading, and the presence of NaBArF at 10 mol% was found essential for the hydrogenation in toluene as solvent. Surprisingly, when the reaction was carried out in the previously used mixture of TFE and acetic acid as solvent, it did not take place. As shown in Scheme 9.3, the reaction afforded a range of chiral *anti*-β-amino alcohols with excellent diastereoselectivities of >98% de in almost all cases of substrates studied, in combination with good yields and enantioselectivities of 61–96% ee. In the case of substrates with an electron-donating or bulky group, the enantioselectivities increased, but the reactions were sluggish. It is important to note that this work constituted the first example of asymmetric hydrogenation through dynamic kinetic resolution for substituted aromatic α-amino ketones with a primary amino group.

Among the most spectacular recent developments in catalytic asymmetric synthesis, asymmetric transfer hydrogenation is an attractive method for the preparation of optically active alcohols, because of its operational simplicity and wide substrate scope. Therefore, the design and synthesis of new chiral ligands are always a challenge for researchers. In recent decades, numerous

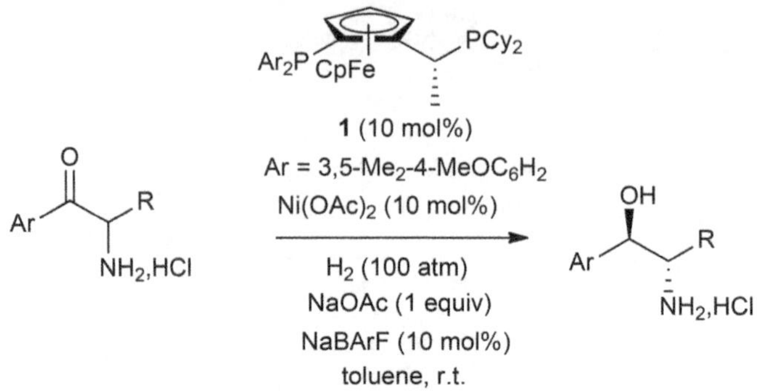

Ar = Ph, R = Et: 86%, *anti/syn* > 99:1, ee = 76%
Ar = Ph, R = Me: 94%, *anti/syn* > 99:1, ee = 85%
Ar = Ph, R = *n*-Pr: 91%, *anti/syn* > 99:1, ee = 82%
Ar = *p*-Tol, R = Me: 92%, *anti/syn* > 99:1, ee = 86%
Ar = *p*-MeOC$_6$H$_4$, R = Me: 92%, *anti/syn* > 99:1, ee = 85%
Ar = *p*-CF$_3$C$_6$H$_4$, R = Me: 82%, *anti/syn* = 80:20, ee = 63%
Ar = *m*-Tol, R = Me: 90%, *anti/syn* > 99:1, ee = 85%
Ar = *o*-Tol, R = Me: 76%, *anti/syn* > 99:1, ee = 61%
Ar = *p*-Tol, R = *i*-Pr: 21%, *anti/syn* > 99:1, ee = 96%
Ar = *p*-Tol, R = Bn: 92%, *anti/syn* > 99:1, ee = 84%

Scheme 9.3 Hydrogenation of substituted aromatic α-amino ketone hydrochlorides in the presence of an *in situ* generated nickel catalyst derived from a ferrocenylphosphine ligand.

different PP-, NN-, NO-, NPN-, and PNNP-type ligands have been applied to this reaction.[9] These mixed-type ligands have attracted more and more attention because of their better ability to stabilise the metal centre and special chiral coordination environments surrounding the metal centre. In previous studies of asymmetric transfer hydrogenation of ketones, however, most research was focused on the expensive noble platinum-metal based complexes, such as those of ruthenium,[10] rhodium,[11] and iridium.[12] Compared with these precious metals, the first-row transition metals, which are more abundant and benign, are now attracting more and more research interest because they are cost-effective and "green" catalysts. In 2004, Gao *et al.* reported the first example of an iron catalyst used in the asymmetric transfer hydrogenation of ketones.[13] Later, Morris and co-workers developed other iron catalysts for this reaction.[14] Although there have been some successful examples of iron complexes as chiral catalysts for the asymmetric transfer hydrogenation of ketones, other first-row transition metals, such as nickel or cobalt complexes, are still rarely reported. A recent example was described by Dong and Gao, using novel PNO-type chiral ligands.[15] Among several ligands of this type investigated, the catalyst system generated *in situ* from Ni(PPh$_3$)Cl$_2$ and the PNO ligand **2** was found the most efficient to induce the transfer hydrogenation of various ketones, providing the corresponding alcohols in good to quantitative yields and moderate to good enantioselectivities of

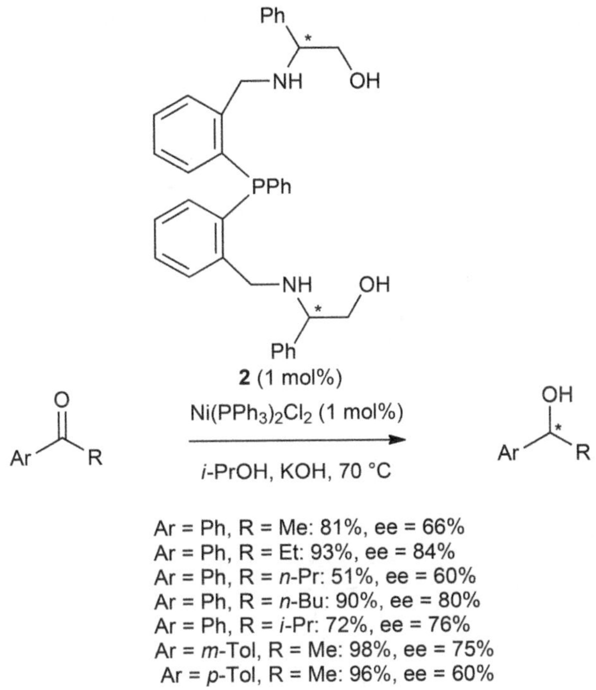

Scheme 9.4 Transfer hydrogenation of ketones in the presence of an *in situ* generated nickel catalyst derived from a PNO-type ligand.

10–84% ee, as shown in Scheme 9.4. Generally, the enantioselectivity was improved with increase of the bulkiness of the alkyl substituents of the ketones, but *n*-butyrophenone was not a good substrate for this catalytic system since the conversion was only 51% with enantioselectivity of 60% ee in the case of this substrate. Moreover, it was found that the introduction of a group on the aromatic ring of the ketones made the catalytic system less effective to reduce the ketones (ee = 10–60%).

In another context, heterogeneous chiral catalysts have several advantages, such as easy preparation, easy separation from the reaction mixture and low energy for separation, and easy reuse. Although enantio-differentiating homogeneous catalysts have recently enjoyed a great success for attaining high enantioselectivity, it is sometimes preferable to switch to heterogeneous ones. In this context, the group of Osawa,[1b] among others,[16] has widely investigated the use of tartaric acid-modified nickel catalysts in the enantioselective hydrogenation of ketones. This heterogeneous catalyst was invented by Izumi *et al.* in 1963.[17] It is a solid catalyst on the surface of which a modifier (tartaric acid) and co-modifier (typically NaBr) are adsorbed on a base nickel catalyst. The base nickel catalyst is defined as that with hydrogenation activity, and is converted into an enantio-differentiating catalyst *via* modification. The asymmetric hydrogenation of ketones over a tartaric acid-modified nickel catalyst is performed through the interaction between the substrate, the nickel surface, and a tartaric acid adsorbed on the nickel surface. The tartaric acid-modified nickel catalyst has been widely applied to hydrogenate β-keto esters and alkan-2-ones, with high enantioselectivities.[18] For example, an enantioselectivity of 91% ee was reported by Osawa *et al.* for the hydrogenation of methyl acetoacetate by using a tartaric acid-modified nickel catalyst.[19] This type of methodology has been applied by several groups to a wide range of β-keto esters and (functionalised) alkanones,[20] providing enantioselectivities of up to 98% ee and 85% ee, respectively (Scheme 9.5).[1b,21] In the same area, Qu and co-workers recently reported the preparation of a reduced graphene oxide-supported tartaric acid-modified nickel catalyst, allowing enantioselectivities of >98.5% ee in the asymmetric hydrogenation of methyl acetoacetate.[22]

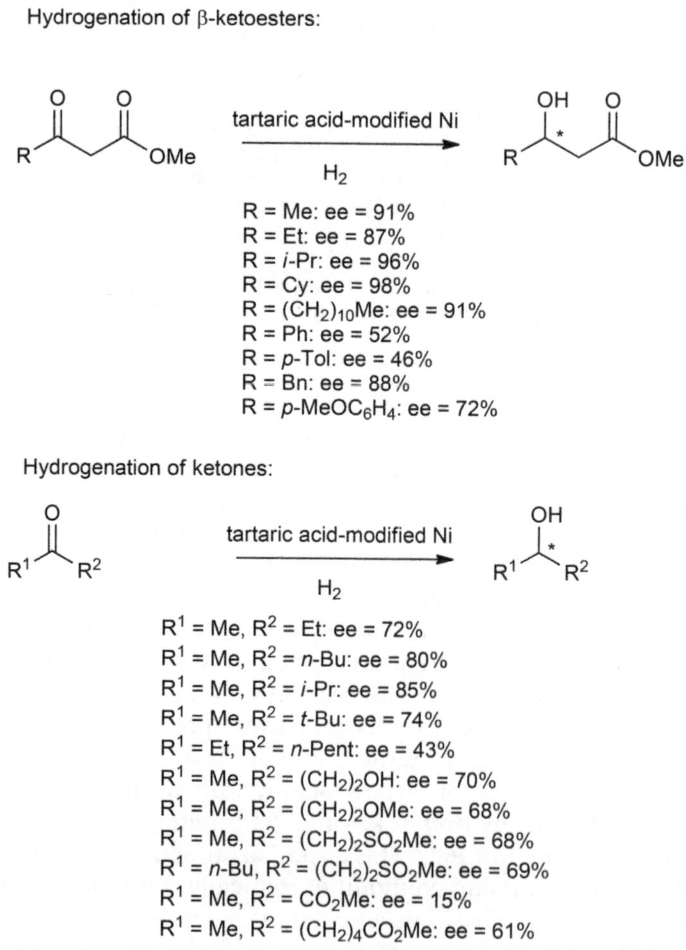

Scheme 9.5 Hydrogenations of β-keto esters and (functionalised) ketones in the presence of a heterogeneous tartaric acid-modified nickel catalyst.

9.3 Hydrogenations of Alkenes

While the enantioselective hydrogenation of C=O bonds using heterogeneous catalysts has been widely studied, a few studies on the asymmetric hydrogenation of C=C bonds[23] over a modified heterogeneous nickel catalyst have been reported. Whereas moderate to high enantioselectivities for the C=C hydrogenation have been attained over cinchona-modified palladium (72–94% ee),[24] an enantioselectivity of 17% ee was reported in the hydrogenation of sodium 2-phenylcinnamate over a tartaric acid pre-modified Raney nickel catalyst (Scheme 9.6).[25]

On the other hand, Zhou *et al.* recently disclosed a simple nickel/Binapine catalytic system for asymmetric transfer hydrogenation reactions, leading to α- and β-amino acids and employing formic acid as the hydrogen equivalent.[26] Thus, the safety hazard associated with the storage and handling of high-pressure hydrogen gas was avoided. In addition, nickel is 100- to 1000-fold cheaper than ruthenium, rhodium, or iridium, when the price of their chlorides is compared.[27] When the *in situ* generated nickel catalyst derived from $Ni(OAc)_2$ and (*S*)-Binapine was applied to induce the transfer hydrogenation of a range of β-acetamidoacrylates, it afforded the corresponding β-acylamido esters in remarkable yields and enantioselectivities of up to 99% and 99% ee, respectively. As shown in Scheme 9.7, the scope of the alkenes that could be hydrogenated under these reaction conditions was broad since various alkyl and aryl groups could be present at the β-position of the alkenes. Alkenes with both electron-rich and electron-poor aryl groups provided the corresponding products in very high enantioselectivities. Importantly, some heteroaryl rings were also tolerated. Both amides and imides seemed to function as directing groups and induced excellent enantioselectivities. The (*S*)-Binapine ligand was selected among other less-donating phosphine ligands, such as BINAP, Segphos, DIPAMP, and PHOX, which were found completely inactive. Moreover, the authors have also investigated many other metal salts of (*S*)-Binapine, such as those of iron, cobalt, and copper, which were all found inactive in the model reaction. This catalytic system was also successfully applied to the synthesis of α-acetamido esters, albeit in lower enantioselectivities of up to 86% ee. It is important to note that this remarkable work constituted the first highly enantioselective (transfer) hydrogenation of alkenes using nickel catalysts.

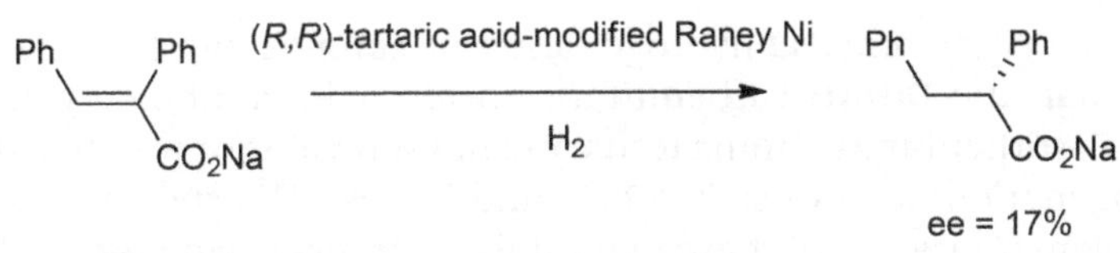

Scheme 9.6 Hydrogenation of an alkene in the presence of a heterogeneous tartaric acid-modified Raney nickel catalyst.

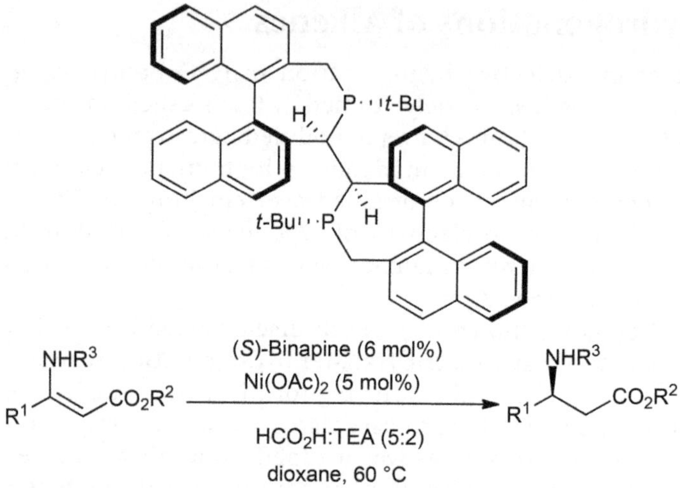

$R^1 = R^2 = Me$, $R^3 = Ac$: 94%, ee = 98%
$R^1 = Me$, $R^2 = Et$, $R^3 = Ac$: 97%, ee = 96%
$R^1 = Me$, $R^2 = i\text{-}Pr$, $R^3 = Ac$: 97%, ee = 95%
$R^1 = Ph$, $R^2 = Me$, $R^3 = Ac$: 96%, ee = 98%
$R^1 = p\text{-}MeOC_6H_4$, $R^2 = Me$, $R^3 = Ac$: 99%, ee = 98%
$R^1 = p\text{-}MeOC_6H_4$, $R^2 = Me$, $R^3 = Ac$: 95%, ee = 98%
$R^1 = p\text{-}Tol$, $R^2 = Me$, $R^3 = Ac$: 99%, ee = 98%
$R^1 = 2\text{-}Naph$, $R^2 = Et$, $R^3 = Ac$: 99%, ee = 98%
$R^1 = 1\text{-}Naph$, $R^2 = Et$, $R^3 = Ac$: 99%, ee = 99%
$R^1 = p\text{-}FOC_6H_4$, $R^2 = Me$, $R^3 = Ac$: 95%, ee = 99%
$R^1 = p\text{-}CF_3C_6H_4$, $R^2 = Me$, $R^3 = Ac$: 97%, ee = 99%
$R^1 = 3\text{-}pyridyl$, $R^2 = Me$, $R^3 = Ac$: 99%, ee = 98%
$R^1 = 2\text{-}thienyl$, $R^2 = Me$, $R^3 = Ac$: 96%, ee = 99%
$R^1 = 2\text{-}furyl$, $R^2 = Me$, $R^3 = Ac$: 96%, ee = 98%
$R^1 = Et$, $R^2 = Me$, $R^3 = Ac$: 96%, ee = 96%
$R^1 = R^2 = Me$, $R^3 = Bz$: 98%, ee = 94%
$R^1 = Ph$, $R^2 = Et$, $R^3 = Phth$: 78%, ee = 96%

Scheme 9.7 Transfer hydrogenation of alkenes in the presence of an *in situ* generated nickel catalyst derived from Binapine.

9.4 Conclusions

Homogeneous asymmetric hydrogenation of ketones has a 40 year history. In particular, the BINAP–ruthenium/Brønsted acid combined catalyst and the BINAP–ruthenium/diamine ternary catalyst revolutionised the asymmetric hydrogenation of ketones in 1987[3] and in 1995,[28] respectively. Although the complementary use of these two catalysts covers a wide range of ketonic substrates, there is still no universal catalyst for the asymmetric hydrogenation of ketones. Because the production of chiral secondary alcohols through this methodology is so important in asymmetric synthesis, the efficiency of asymmetric hydrogenation of ketones should be further pursued. At present, the vast majority of catalysts are based on precious metals, including ruthenium, osmium, rhodium, iridium, and palladium. Replacement of these expensive and toxic elements with more abundant base metals such as iron, cobalt, nickel, or copper should also be thoroughly investigated from the viewpoints of cost and safety. Among the very good results involving nickel catalysts reported in the last few years is the first use of homogeneous chiral nickel–phosphine complexes in asymmetric hydrogenation of α-amino-β-keto ester hydrochlorides, achieved in remarkable diastereoselectivities of up to 98% de and excellent enantioselectivities of up to 95% ee.

In the area of asymmetric hydrogenation of alkenes, there is also a renewed interest in developing cheap, abundant, and less toxic metals. In this context, a remarkable result was recently reported with the first highly enantioselective (transfer) hydrogenation of β-acetamidoacrylates using nickel catalysts, providing β-acylamido esters in excellent yields and enantioselectivities of up to 99% and 99% ee, respectively.

References

1. (a) T. Osawa, T. Harada and O. Takayasu, *Top. Catal.*, 2000, **13**, 155–168; (b) T. Osawa, T. Harada and O. Takayasu, *Curr. Org. Chem.*, 2006, **10**, 1513–1531; (c) E. Bouwman, *Handbook of Homogeneous Hydrogenation*, ed. J. G. de Vries and C. J. Elsevier, Wiley, Weinheim, 2007, p. 93; (d) A. M. Palmer and A. Zanotti-Gerosa, *Curr. Opin. Drug Discovery Dev.*, 2010, **13**, 698–716; (e) H.-U. Blaser, B. Pugin and F. Spindler, *Top. Organomet. Chem.*, 2012, **42**, 65–102; (f) M. Yoshimura, S. Tanaka and M. Kitamura, *Tetrahedron Lett.*, 2014, **55**, 3635–3640; (g) J.-H. Xie, D.-H. Bao and Q.-L. Zhou, *Synthesis*, 2015, **47**, 460–471.
2. (a) M. Kitamura and R. Noyori, in *Organic Synthesis*, ed. S. Murahashi, Wiley, Weinheim, 2004, pp. 3–52; (b) G. Shang, W. Li and X. Zhang, Asymmetric Hydrogenation, in *Catalytic Asymmetric Synthesis*, ed. I. Ojima, Wiley, New York, 3rd edn, 2010, pp. 343–436; (c) N. Arai and T. Ohkuma, *Chem. Rec.*, 2012, **12**, 284–289.
3. R. Noyori, T. Ohkuma, M. Kitamura, H. Takaya, N. Sayo, H. Kumobayashi and S. Akutagawa, *J. Am. Chem. Soc.*, 1987, **109**, 5856–5858.
4. Q.-A. Chen, Z.-S. Ye, Y. Duan and Y.-G. Zhou, *Chem. Soc. Rev.*, 2013, **42**, 497–511.
5. R. R. Schrock and J. A. Osborn, *Chem. Commun.*, 1970, 567–568.
6. M. Kitamura and H. Nakatsuka, *Chem. Commun.*, 2011, 842–846.
7. Y. Hamada, Y. Koseki, T. Fujii, T. Maeda, T. Hibino and K. Makino, *Chem. Commun.*, 2008, 6206–6208.
8. T. Hibino, K. Makino, T. Sugiyama and Y. Hamada, *ChemCatChem*, 2009, **1**, 237–240.
9. H.-L. Kwong, H.-L. Yeung, C. T. Yeung, W.-S. Lee, C.-S. Lee and W.-L. Wong, *Coord. Chem. Rev.*, 2007, **251**, 2188–2222.

10. A. M. Hayes, D. J. Morris, G. J. Clarkson and M. Wills, *J. Am. Chem. Soc.*, 2005, **127**, 7318–7319.
11. S. H. Kwak, S. A. Lee and K. I. Lee, *Tetrahedron: Asymmetry*, 2010, **21**, 800–804.
12. R. Malacea, R. Poli and E. Manoury, *Coord. Chem. Rev.*, 2010, **254**, 729–752.
13. J. S. Chen, L. L. Chen, Y. Xing, G. Chen, W.-Y. Shen, Z.-R. Dong, Y.-Y. Li and J.-X. Gao, *Helv. Chim. Sin.*, 2004, **62**, 1745–1748.
14. P. O. Lagaditis, A. J. Lough and R. H. Morris, *J. Am. Chem. Soc.*, 2011, **133**, 9662–9665.
15. Z. R. Dong, Y. Y. Li, S. L. Yu, G. S. Sun and J. X. Gao, *Chin. Chem. Lett.*, 2012, **23**, 533–536.
16. (a) H. Chen, R. Li, H. Wang, L. Yin, F. Wang and J. Ma, *Chem. Lett.*, 2006, **35**, 910–911; (b) H. Chen, R. Li, H. Wang, J. Liu, F. Wang and J. Ma, *J. Mol. Catal. A*, 2007, **269**, 125–132.
17. Y. Izumi, M. Imaida, H. Fukawa and S. Akabori, *Bull. Chem. Soc. Jpn.*, 1963, **36**, 155–160.
18. (a) T. Osawa, Y. Amaya, T. Harada and O. Takayasu, *J. Mol. Catal. A*, 2004, **211**, 93–96; (b) D. Jo, J. S. Lee and K. H. Lee, *J. Mol. Catal. A*, 2004, **222**, 199–205; (c) T. Osawa, A. Iwai, C. Honda, S. Mita, T. Harada and O. Takayasu, *React. Kinet. Catal. Lett.*, 2005, **84**, 279–286; (d) T. Osawa, Y. Hagino, T. Harada and O. Takayasu, *Catal. Lett.*, 2006, **112**, 57–61; (e) T. Osawa, K. Yoshino, K. Takimoto, O. Takayasu and T. Harada, *Catal. Lett.*, 2006, **112**, 167–171; (f) T. Osawa, T. Kizawa, I.-Y. S. Lee, S. Ikeda, T. Kitamura, Y. Inoue and V. Borovkov, *Catal. Commun.*, 2011, **15**, 15–17.
19. (a) T. Osawa, M. Kitano, T. Harada and O. Takayasu, *Catal. Lett.*, 2009, **128**, 413–417; (b) T. Osawa, I.-Y. S. Lee, S. Ikeda, T. Kitamura, Y. Inoue and V. Borovkov, *Appl. Catal., A*, 2012, **445–446**, 269–273; (c) T. Osawa, T. Kizawa, F. Takano, S. Ikeda, T. Kitamura, Y. Inoue and V. Borovkov, *ChemCatChem*, 2014, **6**, 170–178.
20. (a) S. Murakami, T. Harada and A. Tai, *Bull. Chem. Soc. Jpn.*, 1980, **53**, 1356–1360; (b) Y. Hikaki, K. Ito, T. Harada and A. Tai, *Chem. Lett.*, 1981, 131–132; (c) A. Tai, T. Harada, Y. Hiraki and S. Murakami, *Bull. Chem. Soc. Jpn.*, 1983, **56**, 1414–1419; (d) T. Osawa and T. Harada, *Bull. Chem. Soc. Jpn.*, 1984, **57**, 1518–1521; (e) T. Osawa, E. Mieno, T. Harada and O. Takayasu, *J. Mol. Catal. A*, 2003, **200**, 315–321; (f) T. Sugimura, J. Watanabe, S. Nakagawa and T. Okuyama, *J. Mol. Catal. A*, 2006, **248**, 233–236.
21. (a) S. Nakagawa, T. Sugimura and A. Tai, *Chem. Lett.*, 1997, 859–860; (b) T. Osawa, Y. Nakagawa, T. Harada and O. Takayasu, *Catal. Lett.*, 2006, **112**, 163–166; (c) T. Sugimura, T. Matsuda and T. Osawa, *Tetrahedron: Asymmetry*, 2009, **20**, 1877–1880.
22. C. Ding, W. Wei, H. Sun, J. Ding, J. Ren and X. Qu, *Carbon*, 2014, **79**, 615–622.
23. X. Cui and K. Burgess, *Chem. Rev.*, 2005, **105**, 3272–3296.
24. (a) Y. Nitta and K. Kobiro, *Chem. Lett.*, 1996, 897–898; (b) W. R. Huck, T. Mallat and A. Baiker, *Catal. Lett.*, 2002, **80**, 87–92.

25. M. Bartok, G. Wittmann, G. Göndös and G. V. Smith, *J. Org. Chem.*, 1987, **52**, 1139–1141.
26. P. Yang, H. Xu and J. Zhou, *Angew. Chem., Int. Ed.*, 2014, **53**, 12210–12213.
27. F. Alonso, P. Riente and M. Yus, *Acc. Chem. Res.*, 2011, **44**, 379–391.
28. T. Ohkuma, H. Ooka, S. Hashiguchi, T. Ikariya and R. Noyori, *J. Am. Chem. Soc.*, 1995, **117**, 2675–2676.

CHAPTER 10

Enantioselective Nickel-Catalysed Miscellaneous Reactions

10.1 Introduction

This chapter collects miscellaneous enantioselective nickel-catalysed reactions that could not be included in other chapters, such as cyclisation reactions, amination reactions, ring-opening reactions, Friedel–Crafts reactions, allylation reactions of aldehydes, and other reactions. It well demonstrates the remarkably diverse chemical reactivity exhibited by chiral nickel catalysts.

10.2 Cyclisation Reactions

Chiral carbo- and heterocycles represent an important structural motif in biologically active substances. Therefore, the development of efficient stereoselective synthetic methods to generate this class of compound is highly desirable. The transition metal-catalysed cycloisomerisation of 1,6-dienes and enynes offers an atom-economic route[1] to five- and six-membered carbo- and heterocycles.[2] Since the first example reported by Trost *et al.* in 1989,[3] a number of efficient catalytic systems has been developed for enyne cyclisation and a high level of chiral induction can be achieved in many cases.[4] The alternative synthetic strategy starting from dienes constitutes a powerful route to five-membered carbo- and heterocyclic compounds bearing a new stereogenic centre. A number of achiral transition metal catalyst systems, for example based on palladium,[5] nickel,[6] rhodium,[7] ruthenium,[8] platinum,[9] and titanium,[10] are known to allow high levels of chemo- and regioselectivity towards these five-membered ring products. In

RSC Catalysis Series No. 26
Enantioselective Nickel-Catalysed Transformations
By Hélène Pellissier

Published by the Royal Society of Chemistry, www.rsc.org

contrast, only a few examples of enantioselective catalysts have been reported so far.[11] A particular challenge is the design of catalytic systems that show high chemo-, regio-, and enantioselectivity for the formation of the thermodynamically least-favoured products, providing an exocyclic methylene group for further transformation. In 2005, Leitner *et al.* reported a nickel-based catalyst system for the asymmetric cycloisomerisation of diethyl diallylmalonate, showing high activities, very high regioselectivities towards diethyl 4-methyl-3-methylenecyclopentane-1,1-dicarboxylate, and enantioselectivities of up to 80% ee, as shown in Scheme 10.1.[12] This result was achieved by using Wilke's azaphospholene ligand **1**,[13] which was selected as the most efficient among various ligands, including monodentate phosphoramidites. Indeed, the cycloisomerisation of diethyl diallylmalonate catalysed by 0.5 mol% of a nickel chiral complex generated *in situ* from Wilke's azaphospholene ligand **1** and [Ni(allyl)(cod)][BAr_4] [with Ar = 3,5-$(CF_3)_2C_6H_3$] in dichloromethane at 0 °C afforded the corresponding methyl-substituted *exo*-methylenecyclopentane derivative with a regioselectivity of 97% and an enantioselectivity of 80% ee, opening a promising synthetic strategy for the formation of chiral five-membered carbocycles.

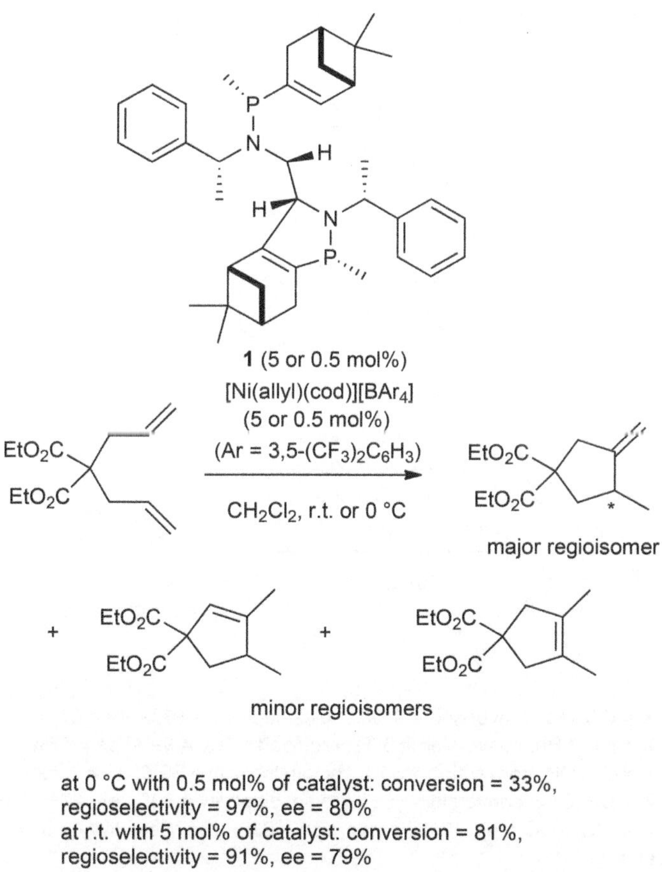

at 0 °C with 0.5 mol% of catalyst: conversion = 33%, regioselectivity = 97%, ee = 80%
at r.t. with 5 mol% of catalyst: conversion = 81%, regioselectivity = 91%, ee = 79%

Scheme 10.1 Cycloisomerisation of diethyl diallylmalonate.

In 2008, the same authors extended the scope of the above methodology to other symmetrical 1,6-dienes.[14] Unprecedented high enantioselectivities of up to 91% ee were achieved for the formation of a range of five-membered cyclic products, as summarised in Scheme 10.2. The reactions were performed at room temperature by using between 0.5 mol% and 5 mol% of the same catalyst system as in Scheme 10.1. Dimethyl and di-*tert*-butyl diallylmalonates provided the corresponding cycloadducts in good enantioselectivities of 67 and 88% ee, respectively. The conversion in the case of the sterically more demanding di-*tert*-butyl substrate was lower (55%) than for the dimethyl substrate (91%), while the regioselectivities were comparable (89% and 97%, respectively). The scope was also extended to substrates bearing other functional groups on the carbon backbone. For example, the cycloisomerisation of 4,4-bis(hydroxymethyl)hepta-1,6-diene provided the corresponding product in 39% conversion, >99% regioselectivity, and enantioselectivity of 91% ee. The activity was regained by protecting the hydroxy functionalities by acetylation since a full conversion was reached in the case of this substrate, but affording the corresponding cycloadduct in a lower enantioselectivity of 71% ee. An excellent result was also achieved using acetone as the protecting agent for the hydroxy functionalities. Indeed, in this

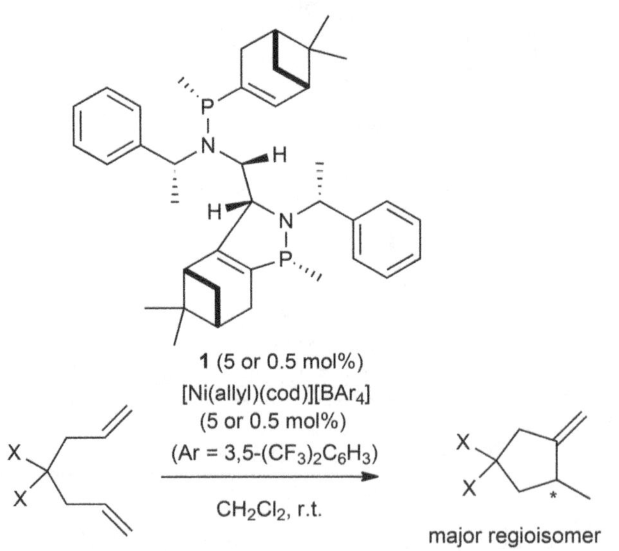

X = CO_2Me: conversion = 91%, regioselectivity = 97%, ee = 67%
X = CO_2*t*-Bu: conversion = 55%, regioselectivity = 89%, ee = 88%
X = CH_2OH: conversion = 39%, regioselectivity > 99%, ee = 91%
X = CH_2OAc: conversion > 99%, regioselectivity = 83%, ee = 71%
X,X = $CH_2OC(Me)_2OCH_2$: conversion > 99%, regioselectivity = 87%, ee = 86%

Scheme 10.2 Cycloisomerisation of other symmetrical 1,6-dienes.

case the product was produced in complete conversion, with a regioselectivity of 87% along with an enantioselectivity of 86% ee. On the other hand, no reaction occurred with diallylmalononitrile as substrate.

The prevalence of five-membered carbocycles in natural products and other bioactive compounds has provided a major impetus for the development of efficient methods for their construction. Over the years, the Nazarov reaction has been increasingly refined to meet this need.[15] The Nazarov cyclisation in its original form involves the cyclisation of divinyl ketones to cyclopentenones under the influence of very strong acids. The recognition that it belongs to a general class of cationic electrocyclic reactions and that even mild Lewis acids can promote the cyclisation has contributed significantly to the development of the reaction. Somewhat surprisingly, it was not until the end of 2003 that catalytic asymmetric versions of the Nazarov cyclisation began to surface in the literature. The reason for this may lie in the complex mechanism of the reaction, involving steps such as loss of proton and reprotonation, which are fraught with regio- and stereoselectivity problems. Furthermore, the final cyclopentenone is potentially subject to racemisation if the cyclisation proceeds slowly. In addition, the catalyst turnover may be a concern, as evidenced by the fact that most reported Nazarov cyclisations require acidic solvents or stoichiometric amounts of a Lewis acid. In order to address these issues, Trauner *et al.* developed, in 2003, the first example of an asymmetric Nazarov cyclisation of an alkoxydienone catalysed by 20 mol% of a chiral scandium triflate–PYBOX (pyridine-bisoxazoline) complex, affording an enantiomerically enriched tricycle in moderate enantioselectivity (61% ee).[16] Concomitantly, Aggarwal *et al.* disclosed the asymmetric cyclisation of divinyl ketones bearing α-ester or α-amide groups using 50–100 mol% of copper–bisoxazoline Lewis acid complexes, with moderate to good enantioselectivities (up to 88% ee).[17] Later, a sterically demanding indane–PYBOX–scandium complex was applied by Trauner *et al.* to an enantioselective Nazarov cyclisation, providing moderate enantioselectivities of up to 79% ee.[18] More recently, Togni and Walz reported the use of a dicationic nickel(II) complex containing a chiral tridentate phosphine, Pigiphos, in enantioselective Nazarov cyclisations of various dialkenyl ketones.[19] This catalyst was generated *in situ* from $[Ni(H_2O)_6][ClO_4]_2$ and the Pigiphos ligand **2** in THF at room temperature (Scheme 10.3). The reaction of dialkenyl ketones bearing a trimethoxyphenyl group as well as a 4-methoxyphenyl group afforded the corresponding chiral cyclopentenones in good yields and moderate to good enantioselectivities (45–88% ee). No significant effect appeared to derive from the nature of R^1, both in steric and electronic terms. However, a significant drop in both reactivity and enantioselectivity resulted in going from ethyl and propyl to benzyl and 1-naphthyl esters, respectively. In the latter case (R^2 = Naph), no reaction occurred. Thus, a size match between the ester group and the aromatic substituent R^3 was essential in order to maintain reactivity and to obtain high enantioselectivity. It is important to note that these Nazarov substrates have been previously investigated only with stoichiometric amounts of a chiral Lewis acid.[17]

In 2008, Jacobsen and Watson reported nickel-catalysed asymmetric intramolecular alkene arylcyanations, providing chiral indanes with quaternary carbon

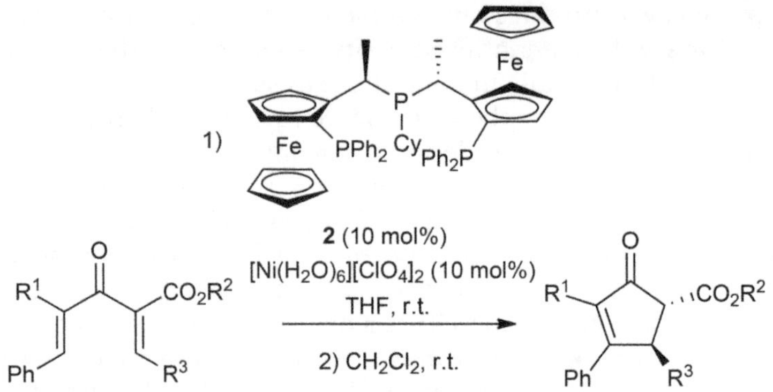

R^1 = Me, R^2 = Et, R^3 = 2,4,6-$(MeO)_3C_6H_2$: 84%, ee = 86%
R^1 = Ph, R^2 = Et, R^3 = 2,4,6-$(MeO)_3C_6H_2$: 85%, ee = 87%
R^1 = Me, R^2 = Et, R^3 = *p*-$MeOC_6H_4$: 32%, ee = 71%
R^1 = Ph, R^2 = Et, R^3 = *p*-$MeOC_6H_4$: 96%, ee = 83%
R^1 = Me, R^2 = Pr, R^3 = 2,4,6-$(MeO)_3C_6H_2$: 80%, ee = 82%
R^1 = Ph, R^2 = Pr, R^3 = 2,4,6-$(MeO)_3C_6H_2$: 82%, ee = 88%
R^1 = Me, R^2 = Bn, R^3 = 2,4,6-$(MeO)_3C_6H_2$: 58%, ee = 45%

Scheme 10.3 Nazarov cyclisation of dialkenyl ketones.

stereogenic centres from readily available benzonitrile precursors.[20] Among a range of chiral ligands screened, including (*S*)-MOP, (*S*,*S*,*R*,*R*)-TANGPHOS, (*R*)-i-Pr-PHOX, (*R*,*R*)-Me-BPE, and others, (*S*,*S*,*R*,*R*)-TANGPHOS was selected as the most efficient ligand when employed at 5 mol% of catalyst loading in combination with $NiCl_2$(DME) as the nickel source. Investigation of the scope of the reaction revealed that this methodology provided access to a range of substituted indane structures in good to high enantioselectivities of up to 96% ee, as shown in Scheme 10.4. Substrates bearing varying substitution on the benzonitrile (R^1 = F, OMe) and on the alkene (R^2 = Me, *n*-Pr, i-Bu, Ph, aryl) all underwent cyclisation with consistently high enantioselectivities (92–96% ee). However, attainment of useful product yields from substrates bearing sterically demanding or electron-deficient alkene substituents necessitated elevated catalyst loadings (10 mol%). The reaction also needed the use of 10 mol% of Zn and BPh_3 as co-catalysts to be successful. The reaction of an analogous allylic ether (X = O) under similar reaction conditions failed to provide the corresponding cyclisation product, probably because of complete catalyst inhibition.

The development of metal-catalysed asymmetric halolactonisation reactions has been rather limited so far. One of the few examples was reported by Gao and co-workers, who examined the use of a cobalt–salen complex as a Lewis acid catalyst for asymmetric iodolactonisation.[21] In this context, Arai *et al.* have developed an asymmetric iodolactonisation of alkenyl carboxylic acids catalysed by a chiral nickel catalyst generated *in situ*.[22] As shown in Scheme 10.5, the combination of $Ni(OAc)_2$ with chiral PyBidine allowed a range of chiral iodolactones to be achieved in moderate to good yields and with enantioselectivities of up to 89% ee, starting from the corresponding

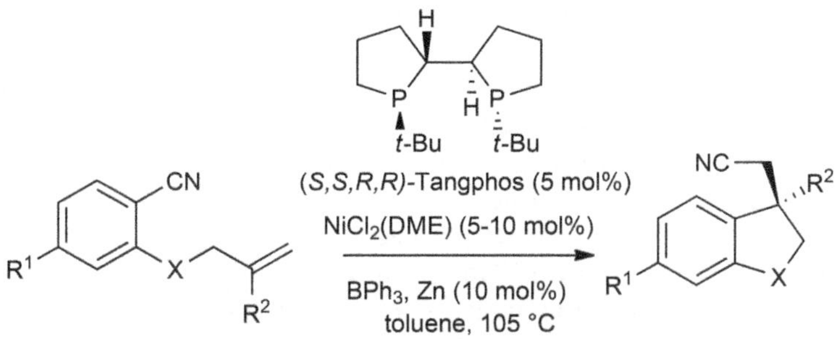

R^1 = H, R^2 = Me, X = CH_2: 85%, ee = 93%
R^1 = F, R^2 = Me, X = CH_2: 84%, ee = 92%
R^1 = OMe, R^2 = Me, X = CH_2: 75%, ee = 93%
R^1 = H, R^2 = *n*-Pr, X = CH_2: 69%, ee = 94%
R^1 = H, R^2 = *i*-Bu, X = CH_2: 75%, ee = 95%
R^1 = H, R^2 = Ph, X = CH_2: 72%, ee = 95%
R^1 = H, R^2 = *p*-$CF_3C_6H_4$, X = CH_2: 49%, ee = 92%
R^1 = H, R^2 = *p*-$MeOC_6H_4$, X = CH_2: 65%, ee = 96%
R^1 = H, R^2 = Me, X = OCH_2: 47%, ee = 77%
R^1 = H, R^2 = *p*-$CF_3C_6H_4$, X = O: 0%

Scheme 10.4 Intramolecular arylcyanation of unactivated alkenes.

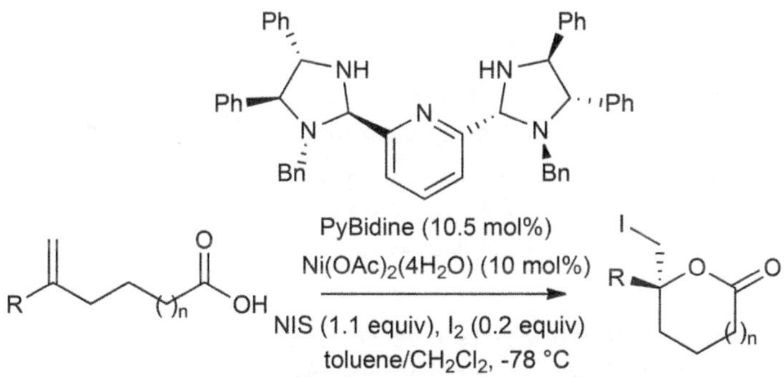

R = Ph, n= 1: 85%, ee = 78%
R = Ph, n = 0: 76%, ee = 53%
R = *p*-ClC_6H_4, n = 1: 81%, ee = 64%
R = *p*-FC_6H_4, n = 1: 67%, ee = 75%
R = *p*-Tol, n = 1: 95%, ee = 89%
R = *p*-$MeOC_6H_4$, n = 1: 82%, ee = 87%
R = *m*-Tol, n = 1: 78%, ee = 72%
R = 2-Naph, n = 1: 45%, ee = 75%
R = *n*-Oct, n = 1: 82%, ee = 48%
R = Cy, n = 1: 79%, ee = 57%

Scheme 10.5 Iodolactonisation of alkenyl carboxylic acids.

alkenyl carboxylic acids. Each of the 5-arylhex-5-enoic acid substrates ($n = 1$) was efficiently converted into the corresponding chiral gluconolactones in reasonable enantiomeric excesses, although the reaction generating γ-butyrolactones resulted in a somewhat lower enantioselectivity ($n = 0$, 53% ee). The presence of electron-donating substituents on the benzene ring was found to increase the enantioselectivity (89% ee, with R = *p*-Tol). It was found that 5-alkyl-substituted hexenoic acids were also tolerated, albeit providing moderate enantioselectivities (48–57% ee).

In another area, Cramer and Donets have developed a novel class of chiral diaminophosphine oxide ligands, enabling asymmetric hydrocarbamoylations of homoallylic formamides by a bimetallic activation mode.[23] As shown in Scheme 10.6, a combination of $AlMe_3$ with a chiral nickel catalyst, generated *in situ* from $Ni(cod)_2$ and chiral diaminophosphine oxide ligand **3**, allowed the C–H activation of a series of formamides, providing a novel access to chiral γ-lactams in high yields and with enantioselectivities of up to 95% ee. The selectivity of the cyclisation was found largely independent of the substitution of the nitrogen atom, allowing alkyl, aryl, or benzyl groups

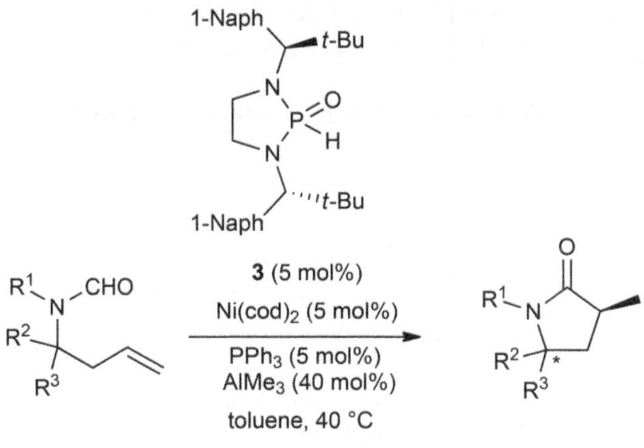

R^1 = Bn, $R^2 = R^3$ = H: 89%, ee = 88%
R^1 = CH_2Bn, $R^2 = R^3$ = H: 79%, ee = 94%
R^1 = Ph, $R^2 = R^3$ = H: 91%, ee = 93%
R^1 = CH_2CO_2Et, $R^2 = R^3$ = H: 75%, ee = 92%
R^1 = Bn, $R^2 = R^3$ = Me: 95%, ee = 79%
R^1 = Me, R^2,R^3 = $(CH_2)_5$: 82%, ee = 84%
R^1 = PMB, R^2 = H, R^3 = allyl: 81%, dr = 5:1, ee = 95%
R^1,R^2 = $(CH_2)_3$, R^3 = allyl: 98%, dr = 4:1, ee = 89%
R^1,R^2 = $(CH_2)_4$, R^3 = allyl: 83%, dr > 20:1, ee = 88%
R^1,R^2 = $(CH_2)_2O(CH_2)$, R^3 = allyl: 90%, dr = 9:1, ee = 92%
R^1,R^2 = $(CH_2)_2N(Ns)(CH_2)_2$, R^3 = allyl: 94%, dr = 5:1, ee = 92%
R^1,R^2 = $(CH_2)_2O(CH_2)_2$, R^3 = allyl: 81%, dr > 20:1, ee = 92%

Scheme 10.6 Hydrocarbamoylation of alkenes.

as well as esters. Bis-allylated formamides were also tolerated, giving the corresponding lactams in high yields and enantioselectivities. The diastereoselectivities ranged from 4:1 to >20:1, evidencing a good enantiotopic allyl group selection.

10.3 Amination Reactions

In 2004, Berkowitz and Maiti reported the first example of asymmetric nickel-catalysed allylic amination.[24] The work consisted of the synthesis of a chiral protected vinylglycinol through intramolecular allylic substitution catalysed by a chiral nickel catalyst generated *in situ* from Ni(cod)$_2$ and (*R*)-MeO-BIPHEP as ligand (Scheme 10.7). The product was obtained in good yield (88%) and with an enantioselectivity of 75% ee. The utility of this novel methodology was demonstrated by the conversion of the product into natural L-vinylglycine and the important anti-epileptic drug (*S*)-vigabatrin.

Indole-derived pyrroloindolines widely exist in a large number of natural products and pharmaceutically important compounds. However, a small collection of molecules are linked through N1–C3 bonds. These indole alkaloids show unique structures and interesting bioactivity profiles. Methods to build this type of C–N bond directly in an asymmetric and catalytic fashion are rare. Recently, Wang *et al.* disclosed an enantioselective nickel-catalysed amination reaction of 3-bromooxindoles with indolines for construction of the N1–C3 linkage quaternary stereogenic centres (Scheme 10.8).[25] This process was induced by an *in situ* generated nickel catalyst derived from bisoxazoline **4** and Ni(OAc)$_2$ in the presence of DABCO in MTBE at room temperature. The study of the substrate scope showed that a broad range of 3-bromooxindoles and indolines could readily participate in the reaction. Indolines with electron-withdrawing

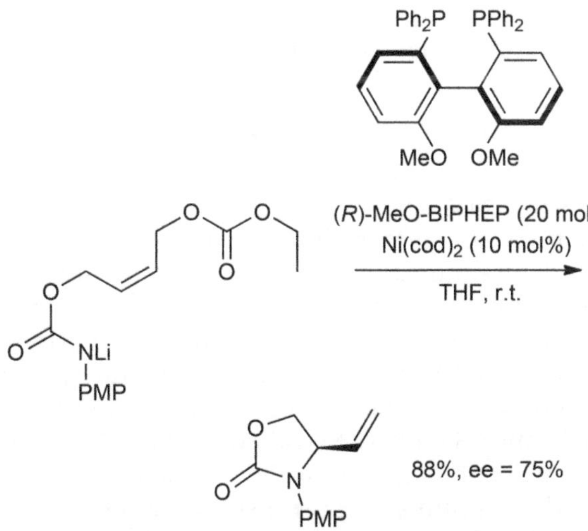

Scheme 10.7 First nickel-catalysed allylic amination.

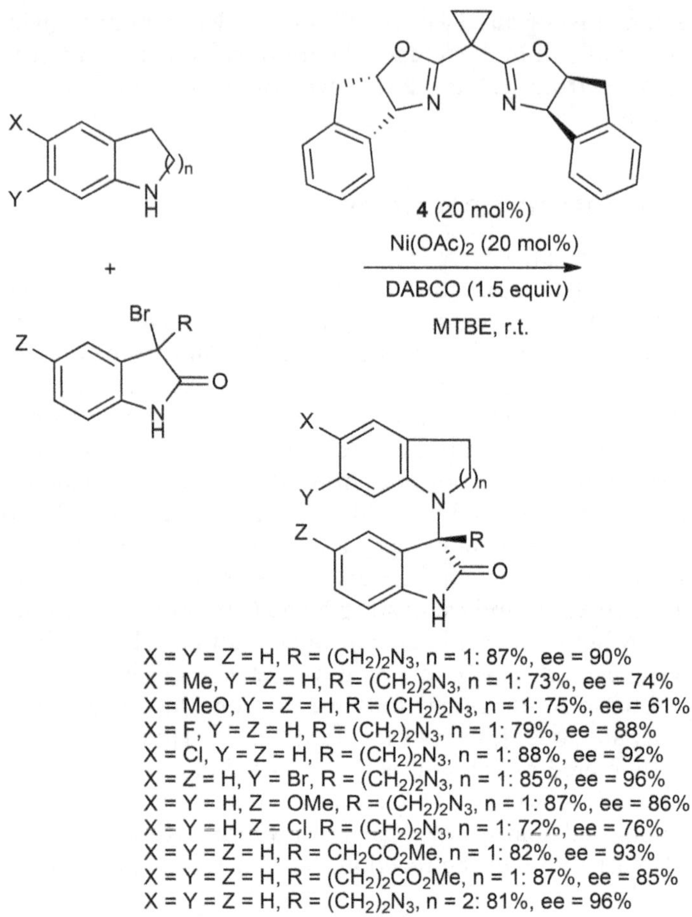

Scheme 10.8 Amination of 3-bromooxindoles with indolines.

groups gave adducts with high enantioselectivities (88–96%) and high yields (71–88%), but indolines with electron-donating groups experienced lower enantioselectivities (61–74% ee). Furthermore, the scope of the 3-bromooxindoles was surveyed, showing that the bromooxindole with a methoxy group at C5 produced the indoline amination product with good results (87% yield, 86% ee). In contrast, only moderate enantioselectivity (76% ee) and yield (72%) were observed with a substrate containing an electron-withdrawing C5–Cl on the bromooxindole core. Remarkably, in addition to indolines, a tetrahydroquinoline was also found to be suitable for the reaction, leading to the respective adduct in 81% yield and 96% ee. The synthetic utility of this novel highly efficient methodology was demonstrated by the production of a key intermediate in the synthesis of the antitumor agent (+)-psychotrimine.

Since palladium on carbon or on alkaline earth supports is generally not effective for the dynamic kinetic resolution of aliphatic amines, and

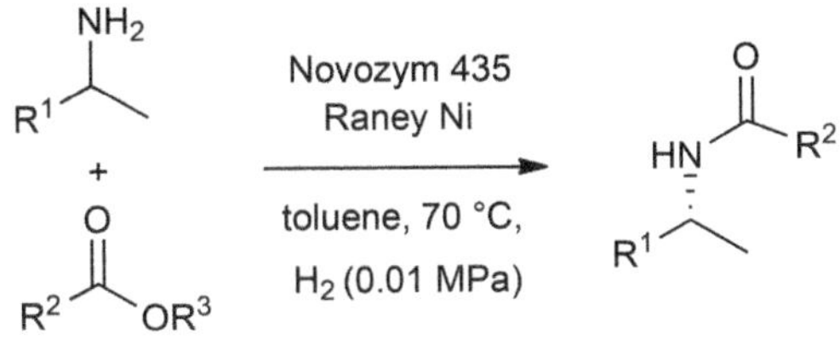

R^1 = *n*-Bu, R^2 = CH_2OMe, R^3 = Et: 99% ee = 98%
R^1 = *n*-Pent, R^2 = CH_2OMe, R^3 = Et: 74% ee = 98%
R^1 = *n*-Hex, R^2 = CH_2OMe, R^3 = Et: 75% ee = 82%
R^1 = Cy, R^2 = CH_2OMe, R^3 = Et: 70% ee = 96%
R^1 = CH_2Bn, R^2 = CH_2OMe, R^3 = Et: 79% ee = 98%
R^1 = Ph, R^2 = *n*-Non, R^3 = Me: 62% ee = 94%
R^1 = *p*-Tol, R^2 = CH_2OMe, R^3 = Et: 72% ee = 80%

Scheme 10.9 Synthesis of amides through the dynamic kinetic resolution of amines.

in the search for less expensive heterogeneous racemisation catalysts, Jacobs *et al.* have shown that heterogeneous Raney nickel could be applied to the racemisation of aliphatic amines in addition to the more usually employed benzylic amines.[26] As an extension, when combined with Novozym 435, Raney nickel allowed the dynamic kinetic resolution of a range of amines to be efficiently achieved, as shown in Scheme 10.9. For aliphatic amines, racemisation and enzymatic resolution could be combined in one pot, resulting in an efficient dynamic kinetic resolution process. When ethyl methoxyacetate was used as the acyl donor, the reaction allowed the corresponding amides to be obtained in good yields and with excellent enantioselectivities. For benzylic amines which reacted less fast with the enzyme, it could be demonstrated that the slow enzymatic conversion of the amine in the presence of the nickel catalyst was the main effect impeding efficient one-pot dynamic kinetic resolution. Consequently, a two-pot process was proposed in which the liquid was alternately shuttled between two vessels containing the solid racemisation catalyst and the biocatalyst. After four such cycles, the corresponding amides were isolated in good yields and high enantioselectivities (Scheme 10.9).

10.4 Ring-Opening Reactions

Recently, several ring-opening reactions have been successfully catalysed by chiral nickel catalysts. As an example, Shibasaki *et al.* have developed catalytic enantioselective desymmetrisation of *meso*-glutaric anhydrides using the stable and commercially available dinuclear Schiff base nickel catalyst **5**.[27] The ring-opening reaction of various anhydrides by a range of alcohols provided the corresponding chiral hemiesters in both high yields and with

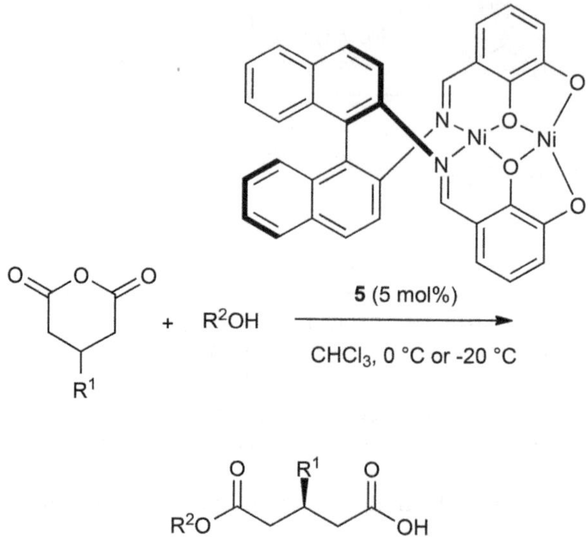

$R^1 = R^2 = Me$: 91%, ee = 94%
$R^1 = Et$, $R^2 = Me$: 99%, ee = 90%
$R^1 = Ph$, $R^2 = Me$: 93%, ee = 80%
$R^1 = OMe$, $R^2 = Me$: 83%, ee = 83%
$R^1 = p\text{-}FC_6H_4$, $R^2 = Me$: 92%, ee = 88%
R^1 = 1-furyl, $R^2 = Me$: 85%, ee = 89%
$R^1 = Me$, $R^2 = Et$: 88%, ee = 90%
$R^1 = Me$, $R^2 = n\text{-}Pr$: 90%, ee = 92%
$R^1 = Me$, $R^2 = Bn$: 80%, ee = 91%
$R^1 = Me$, R^2 = allyl: 88%, ee = 90%

Scheme 10.10 Ring-opening reaction of *meso*-glutaric anhydrides with alcohols.

enantioselectivities of up to 99% and 94% ee, respectively. The results collected in Scheme 10.10 demonstrated the generality of this novel methodology. Furthermore, the authors have shown that using the opposite enantiomer of the catalyst allowed the formation of the opposite enantiomer of the products in the same yields and enantioselectivities, thereby gaining access to both hemiester enantiomers.

The ring-opening reaction of activated donor–acceptor cyclopropanes with nucleophiles provides versatile access to various functionalised carbon skeletons. Of the strategies developed, Lewis acids have been shown to promote such reactions under mild reaction conditions for most nucleophiles. Amine-initiated nucleophilic ring-opening represents a very useful transformation, affording γ-substituted γ-amino acid derivatives. In 2012, the first asymmetric version of the ring-openings of donor–acceptor cyclopropanes with amines was reported by Tang *et al.*[28] This process was based

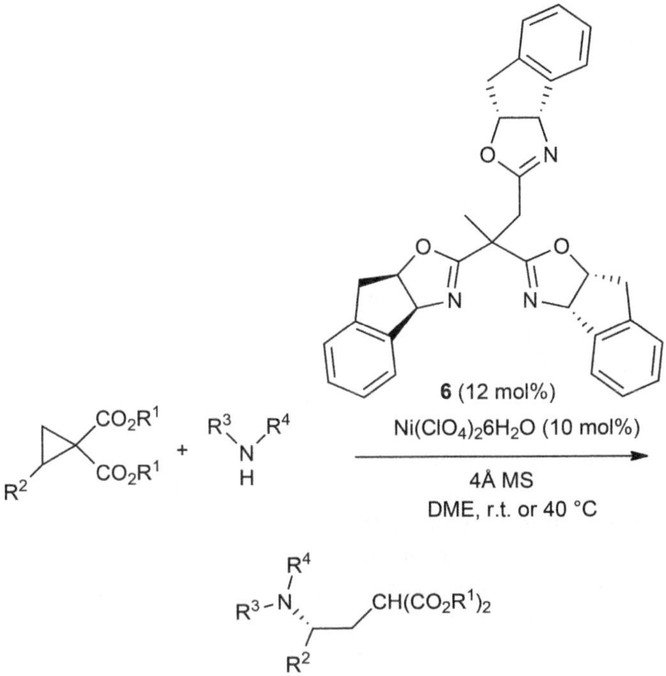

R^1 = CH_2*t*-Bu, R^2 = *p*-$NO_2C_6H_4$, R^3 = H, R^4 = Bn: 75%, ee = 98%
R^1 = CH_2*t*-Bu, R^2 = *o*-$MeOC_6H_4$, R^3 = H, R^4 = Bn: 59%, ee = 96%
R^1 = CH_2*t*-Bu, R^2 = *p*-$MeOC_6H_4$, R^3 = H, R^4 = Bn: 92%, ee = 93%
R^1 = CH_2*t*-Bu, R^2 = 2-thienyl, R^3 = H, R^4 = Bn: 95%, ee = 94%
R^1 = CH_2*t*-Bu, R^2 = Ph, R^3 = H, R^4 = Bn: 90%, ee = 90%
R^1 = CH_2*t*-Bu, R^2 = *p*-$MeOC_6H_4$, R^3 = R^4 = Bn: 98%, ee = 94%
R^1 = CH_2*t*-Bu, R^2 = furyl, R^3 = R^4 = Bn: 99%, ee = 90%
R^1 = CH_2*t*-Bu, R^2 = *p*-$MeOC_6H_4$, R^3 = Bn, R^4 = $(CH_2)_2$OTBS: 94%, ee = 91%
R^1 = CH_2*t*-Bu, R^2 = *p*-$MeOC_6H_4$, R^3 = Bn, R^4 = CH_2CO_2Et: 82%, ee = 88%
R^1 = CH_2*t*-Bu, R^2 = *p*-$MeOC_6H_4$, R^3 = Bn, R^4 = allyl: 85%, ee = 91%
R^1 = CH_2*t*-Bu, R^2 = *p*-$MeOC_6H_4$, R^3 = Ph, R^4 = allyl: 95%, ee = 91%
R^1 = CH_2*t*-Bu, R^2 = R^3 = *p*-$MeOC_6H_4$, R^4 = allyl: 88%, ee = 90%

Scheme 10.11 Ring-opening reaction of donor–acceptor cyclopropanes with amines.

on the use of a chiral nickel catalyst generated *in situ* from $Ni(ClO_4)_2{\cdot}6H_2O$ and chiral trisoxazoline **6**. As shown in Scheme 10.11, a series of 2-substituted cyclopropane-1,1-dicarboxylates were found to react smoothly with primary benzylamine, affording the corresponding products in high enantioselectivities of up to 98% ee as well as in high yields. The substituents on the benzene ring of the 2-arylcyclopropane diesters had a slight effect on the enantioselectivity, but all of them gave enantioselectivities >90% ee.

However, the position of the substituent on the aryl group influenced the yield. For example, *p*-$MeOC_6H_4$- and *o*-$MeOC_6H_4$-substituted cyclopropanes both gave excellent enantioselectivities (96% ee and 93% ee, respectively), but the former showed much better reactivity (59% yield *vs.* 92% yield). The reaction could be extended to a 2-thienylcyclopropane, which also provided both excellent yield and enantioselectivity (95% yield and 94% ee). The process was tolerable with a range of secondary aliphatic amines, which gave uniform good results with enantioselectivities ranging from 88 to 94% ee (Scheme 10.11). It is important to note that this novel methodology opened a novel promising route to chiral γ-substituted γ-amino acid derivatives.

Transition metal-catalysed asymmetric ring-opening reactions of oxa- and azabicyclic alkenes have emerged as important methods for the construction of carbon–carbon and carbon–heteroatom bonds.[29] These transformations are especially valuable because multiple stereocentres can be established in a single step, and the resulting hydronaphthalene scaffolds exist in a wide range of natural products and bioactive molecules. Numerous metal catalysts have been explored in the asymmetric ring-opening reactions of heterobicyclic alkenes with hydride reagents, as well as heteroatom and carbon nucleophiles. However, it is only recently that the first nickel-catalysed enantioselective ring-opening of oxabicyclic alkenes by arylboronic acids has been reported by Yang and Long.[30] As shown in Scheme 10.12, this novel, versatile, and highly efficient nickel-catalysed asymmetric ring-opening of oxabenzonorbornadienes with arylboronic acids afforded *cis*-2-aryl-1,2-dihydronaphthalen-1-ols in high yields of up to 99% and with good to excellent enantioselectivities of up to 99% ee. Compared to the previous rhodium,[31] palladium,[32] and platinum catalysts,[33] this catalyst, generated *in situ* from $Ni(cod)_2$ and (*S*,*S*)-Me-DUPHOS, featured a cheaper precursor, lower catalyst loading, and higher efficiency and enantioselectivity, as well as better functional group tolerance. The study of the substrate scope showed that a wide range of mono- and disubstituted arylboronic acids were compatible with the reaction. It appeared that the electronic property of the substituents on the phenyl ring of the arylboronic acids had little effect on the enantioselectivity. Except for *ortho*-chlorophenylboronic acid, which gave the corresponding product in lower enantioselectivity (66% ee), other monosubstituted phenylboronic acids with electron-donating, electron-withdrawing, or neutral substituents reacted smoothly with various oxabenzonorbornadienes with high enantioselectivities. However, the positional property of the substituents had a significant impact on the reactivity. In general, *para*- and *meta*-substituted arylboronic acids gave better results than their *ortho*-substituted counterparts. Moreover, disubstituted arylboronic acids also showed remarkable reactivity and good enantioselectivity. Concerning the scope of the oxabenzonorbornadienes, substrates with different substituents on the phenyl ring all reacted with various arylboronic acids smoothly to generate the corresponding products in good to excellent enantioselectivities of up to 99% ee.

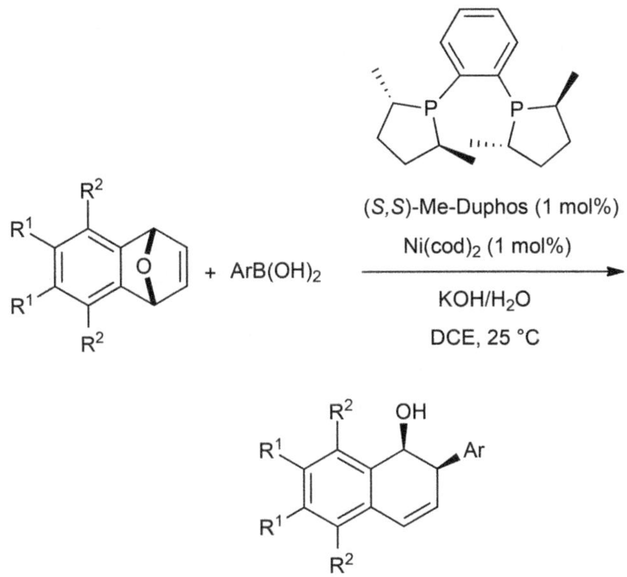

$R^1 = R^2$ = H, Ar = Ph: 98%, ee = 89%
$R^1 = R^2$ = H, Ar = *p*-ClC_6H_4: 56%, ee = 90%
$R^1 = R^2$ = H, Ar = *m*-ClC_6H_4: 99%, ee = 86%
$R^1 = R^2$ = H, Ar = *o*-ClC_6H_4: 40%, ee = 66%
$R^1 = R^2$ = H, Ar = *p*-$CF_3C_6H_4$: 80%, ee = 90%
$R^1 = R^2$ = H, Ar = *p*-Tol: 91%, ee = 89%
$R^1 = R^2$ = H, Ar = *m*-Tol: 63%, ee = 90%
$R^1 = R^2$ = H, Ar = *o*-Tol: 44%, ee = 90%
$R^1 = R^2$ = H, Ar = 2,4-$Me_2F_3C_6H_3$: 98%, ee = 72%
$R^1 = R^2$ = H, Ar = 3,4-$OCH_2OC_6H_3$: 99%, ee = 92%
R^1 = Br, R^2 = H, Ar = Ph: 93%, ee = 90%
R^1 = Br, R^2 = H, Ar = *m*-ClC_6H_4: 94%, ee = 88%
R^1 = Br, R^2 = H, Ar = *p*-Tol: 90%, ee = 98%
R^1 = Br, R^2 = H, Ar = *m*-Tol: 97%, ee = 99%
R^1 = H, R^2 = OMe, Ar = 2,4-$Me_2F_3C_6H_3$: 98%, ee = 65%
R^1 = H, R^2 = OMe, Ar = 3,4-$OCH_2OC_6H_3$: 97%, ee = 97%
R^1,R^2 = $(CH{=}CH)_2$, Ar = Ph: 92%, ee = 85%

Scheme 10.12 Ring-opening reaction of oxabenzonorbornadienes with arylboronic acids.

10.5 Friedel–Crafts Reactions

Recently, the Michael-type asymmetric Friedel–Crafts reaction of electron-deficient alkenes has been established as an important route to chiral benzylic stereocentres.[34] However, applications of this methodology for the synthesis of all-carbon quaternary stereocentres are conspicuously limited. β-Monosubstituted nitroalkenes have turned out to be active substrates in asymmetric Friedel–Crafts alkylations of indoles, among other heterocyclic compounds. Utilisation of the corresponding β,β-disubstituted nitroalkenes as substrates was reported for the first time in 2013 by Jia *et al.*[35] This reaction occurred between indoles and β-CF_3-β-substituted nitroalkenes upon catalysis with an *in situ* generated chiral nickel catalyst derived from $Ni(ClO_4)_2{\cdot}6H_2O$ and chiral bisoxazoline 7 (Scheme 10.13). This ligand was selected among several bisoxazoline ligands variously (un)substituted at the C-4 positions of the oxazoline rings. The reaction afforded the corresponding indole-bearing chiral compound with trifluoromethylated all-carbon quaternary stereocentres in good yields and with excellent enantioselectivities of up to 97% ee. The effect of the Lewis acid was screened in this study, demonstrating that the corresponding chiral zinc complex was a less efficient catalyst, providing an enantioselectivity of 84% ee and yield of 80% (*vs.* 97% ee and 87%

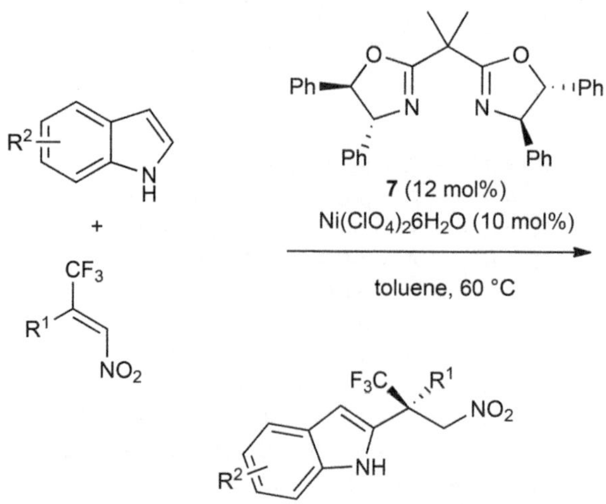

R^1 = Ph, R^2 = H: 95%, ee = 96%
R^1 = *p*-Tol, R^2 = H: 91%, ee = 96%
R^1 = *p*-$MeOC_6H_4$, R^2 = H: 96%, ee = 96%
R^1 = *p*-ClC_6H_4, R^2 = H: 72%, ee = 92%
R^1 = *m*-$CF_3C_6H_4$, R^2 = H: 86%, ee = 95%
R^1 = 3-thienyl, R^2 = H: 96%, ee = 96%
R^1 = Bn, R^2 = H: 78%, ee = 33%
R^1 = Ph, R^2 = 4-MeO: 92%, ee = 97%
R^1 = Ph, R^2 = 5-MeO: 92%, ee = 96%
R^1 = Ph, R^2 = 5-Me: 89%, ee = 97%
R^1 = Ph, R^2 = 5-Br: 42%, ee = 96%
R^1 = Ph, R^2 = 6-Cl: 86%, ee = 90%
R^1 = Ph, R^2 = 7-Me: 95%, ee = 96%
R^1 = Ph, R^2 = 2-Me: 85%, ee = 15%

Scheme 10.13 Friedel–Crafts reaction of indoles with β-CF_3-β-substituted nitroalkenes.

yield with nickel). On the other hand, the reaction performed with the chiral copper complex of ligand **7** did not proceed. Examination of the substrate scope showed that generally high enantioselectivities were reached for the reactions of all variously substituted nitroalkenes with the exception of the substrate bearing a benzyl group, which gave low enantioselectivity (33% ee). High enantioselectivities were also achieved with variously substituted indoles except in the case of 2-methylindole as substrate, which provided only 15% ee.

Soon after, Feng *et al.* reported enantioselective nickel-catalysed Friedel–Crafts reactions of indoles with β,γ-unsaturated α-keto esters, which afforded the corresponding 2-substituted chiral indoles in generally high yields and with excellent enantioselectivities of up to 96% and 90–99% ee, respectively.[36] This remarkable process was catalysed by a combination of $Ni(OTf)_2$ and chiral *N,N'*-oxide **8**, which was selected as optimal ligand among variously substituted *N,N'*-oxides. The results summarised in Scheme 10.14 well highlight the generality of the reaction, with comparable yields and enantioselectivities for all types of both β,γ-unsaturated α-keto esters and indoles. Only the reactions of 3-methylindoles with *N*-protective groups, such as *N*-benzyl, *N*-Boc, *N*-Ts, and *N*-Ac groups, and 3-benzyl-substituted indoles failed.

Inspired by their previous work dealing with Friedel–Crafts reactions of indoles with β-CF_3-β-substituted nitroalkenes (see Scheme 10.13), Jia *et al.* recently applied almost the same reaction conditions to the enantioselective Friedel–Crafts reaction of indoles with α-substituted β-nitroacrylates.[37] In this case, the reaction needed a considerably much lower catalyst loading (1.2 *vs.* 12 mol%), giving the corresponding chiral indolic β-nitro esters bearing all-carbon quaternary stereocentres in good to high yields and with enantioselectivities of up to 97% ee, as shown in Scheme 10.15. In this study, the authors demonstrated that chiral zinc complexes also promoted the reaction, albeit resulting in lower enantioselectivities (75% ee *vs.* 81% ee for the corresponding nickel catalyst). The study of the substrate scope showed that generally excellent enantioselectivities were achieved for all α-aryl-β-nitroacrylates bearing either electron-withdrawing or electron-donating groups on the phenyl ring, but no reaction occurred for *ortho*-methyl-substituted substrates. Concerning the indole substrates, the lowest enantioselectivities (54–76% ee) were observed in the case of 1-methyl-, 2-methyl-, and 1-allyl-substituted indoles. The authors demonstrated that the reaction still worked with comparable enantioselectivities in the presence of only 0.1 mol% of catalyst loading, albeit with lower yields. These remarkable results opened a novel reliable access to potential biologically active β-tryptophan derivatives.

10.6 Allylation Reactions of Aldehydes

Asymmetric carbonyl allylation represents one of the most important types of reaction in current organic synthesis.[38] In this context, Morken and Zhang described enantioselective nickel-catalysed allylation of dienals in 2009.[39]

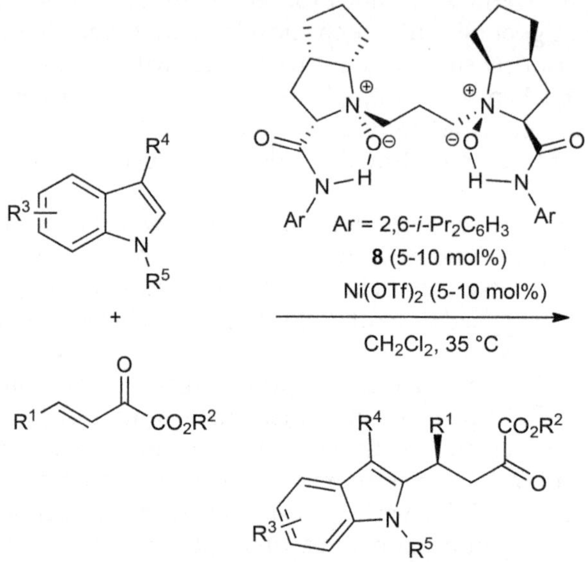

R^1 = Ph, R^2 = R^3 = R^4 = R^5 = Me: 96%, ee = 96%
R^1 = Ph, R^2 = Et, R^3 = R^4 = R^5 = Me: 90%, ee = 95%
R^1 = Ph, R^2 = *i*-Pr, R^3 = R^4 = R^5 = Me: 88%, ee = 95%
R^1 = Ph, R^2 = Bn, R^3 = R^4 = R^5 = Me: 94%, ee = 97%
R^1 = Ph, R^2 = allyl, R^3 = R^4 = R^5 = Me: 85%, ee = 96%
R^1 = Ph, R^2 = *t*-Bu, R^3 = R^4 = R^5 = Me: 95%, ee = 99%
R^1 = *o*-FC_6H_4, R^2 = Et, R^3 = R^4 = R^5 = Me: 90%, ee = 96%
R^1 = *m*-ClC_6H_4, R^2 = Et, R^3 = R^4 = R^5 = Me: 90%, ee = 95%
R^1 = *p*-BrC_6H_4, R^2 = Et, R^3 = R^4 = R^5 = Me: 84%, ee = 95%
R^1 = *p*-Tol, R^2 = Et, R^3 = R^4 = R^5 = Me: 92%, ee = 96%
R^1 = 2-Naph, R^2 = Et, R^3 = R^4 = R^5 = Me: 74%, ee = 96%
R^1 = 2-thienyl, R^2 = Et, R^3 = R^4 = R^5 = Me: 88%, ee = 96%
R^1 = 3,4-$Cl_2C_6H_3$, R^2 = Et, R^3 = R^4 = R^5 = Me: 83%, ee = 93%
R^1 = (*E*)-PhCH=CH, R^2 = Et, R^3 = R^4 = R^5 = Me: 64%, ee = 95%
R^1 = Cy, R^2 = Et, R^3 = R^4 = R^5 = Me: 82%, ee = 90%
R^1 = Ph, R^2 = R^4 = R^5 = Me, R^3 = 4-Me: 90%, ee = 93%
R^1 = Ph, R^2 = R^4 = R^5 = Me, R^3 = 5-Me: 91%, ee = 93%
R^1 = Ph, R^2 = R^4 = R^5 = Me, R^3 = 5-Cl: 87%, ee = 97%
R^1 = Ph, R^2 = R^4 = R^5 = Me, R^3 = 5-OMe: 91%, ee = 88%
R^1 = Ph, R^2 = R^4 = R^5 = Me, R^3 = 6-F: 92%, ee = 96%
R^1 = Ph, R^2 = R^4 = R^5 = Me, R^3 = 7-Me: 93%, ee = 95%
R^1 = Ph, R^2 = R^5 = Me, R^3 = H, R^4 = Et: 86%, ee = 95%
R^1 = Ph, R^2 = R^4 = Me, R^3 = H, R^5 = allyl: 50%, ee = 95%

Scheme 10.14 Friedel–Crafts reaction of indoles with β,γ-unsaturated α-keto esters.

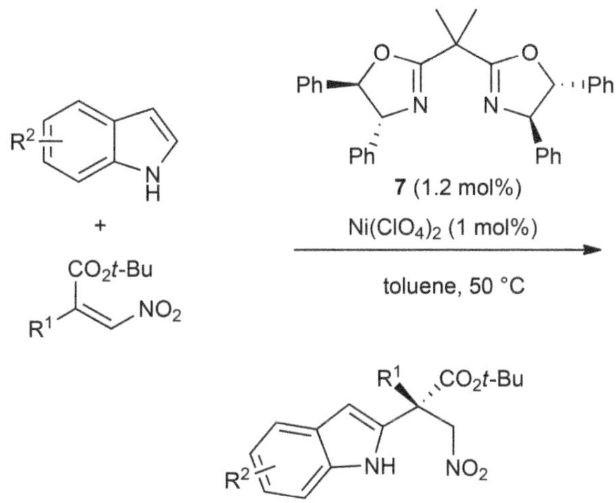

R^1 = Ph, R^2 = H: 98%, ee = 96%
R^1 = *m*-Tol, R^2 = H: 98%, ee = 95%
R^1 = *m*-$MeOC_6H_4$, R^2 = H: 98%, ee = 97%
R^1 = *p*-FC_6H_4, R^2 = H: 97%, ee = 97%
R^1 = *p*-$CF_3C_6H_4$, R^2 = H: 98%, ee = 93%
R^1 = 2-thienyl, R^2 = H: 97%, ee = 96%
R^1 = 2-furyl, R^2 = H: 97%, ee = 95%
R^1 = Bn, R^2 = H: 87%, ee = 88%
R^1 = Ph, R^2 = 4-MeO: 61%, ee = 77%
R^1 = Ph, R^2 = 5-MeO: 97%, ee = 97%
R^1 = Ph, R^2 = 5-Br: 75%, ee = 95%
R^1 = Ph, R^2 = 6-Cl: 95%, ee = 93%
R^1 = Ph, R^2 = 7-Me: 98%, ee = 96%
R^1 = Ph, R^2 = 2-Me: 43%, ee = 54%
R^1 = Ph, R^2 = 1-allyl: 57%, ee = 76%

Scheme 10.15 Friedel–Crafts reaction of indoles with α-substituted β-nitroacrylates.

The reactions occurred between dienals and allylB(pin) in the presence of a combination of $Ni(cod)_2$ and chiral phosphonite **9** at 10 mol% of catalyst loading (Scheme 10.16). It afforded the corresponding 1,2-addition products in good yields and with enantioselectivities of up to 94% ee, along with generally high diastereoselectivities of up to >90% de. The stereoselectivity was dependent upon the diene substituents even when these groups were five atoms away from the newly formed stereocentre. Notably, the reaction delivered the (*E*,*Z*)-diastereomer as the predominant product. The (*E*,*E*)-alkene diastereomer, when observed, was racemic and assumed to arise from a noncatalysed reaction that occurred during room-temperature workup. To explain these results, the authors have proposed that the reaction evolved through unsaturated π-allyl complexes.

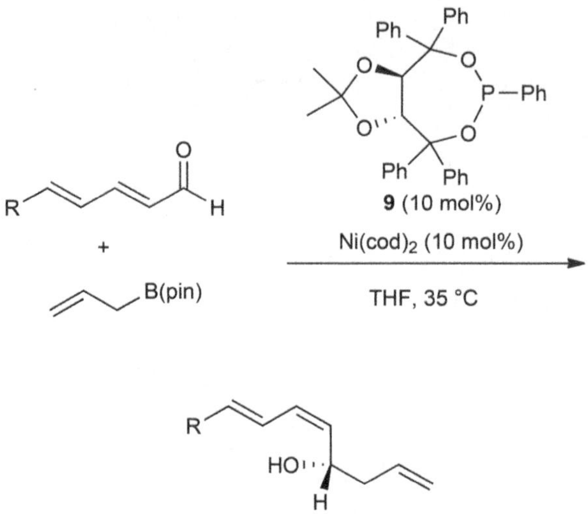

R = Me: 84%, (*E*,*Z*):(*E*:*E*) > 20:1, ee = 88%
R = *n*-Pent: 84%, (*E*,*Z*):(*E*:*E*) > 20:1, ee = 87%
R = Ph: 68%, (*E*,*Z*):(*E*:*E*) > 20:1, ee = 91%
R = CH_2OBn: 86%, (*E*,*Z*):(*E*:*E*) = 15:1, ee = 73%
R = CH_2OTBS: 81%, (*E*,*Z*):(*E*:*E*) = 7:1, ee = 85%
R = Cy: 92%, (*E*,*Z*):(*E*:*E*) > 20:1, ee = 93%
R = furyl: 73%, (*E*,*Z*):(*E*:*E*) = 15:1, ee = 94%
R = $(CH_2)_2OTBS$: 83%, (*E*,*Z*):(*E*:*E*) = 16:1, ee = 90%

Scheme 10.16 Allylation of dienals with allylB(pin).

On the other hand, an enantioselective nickel-catalysed reductive allylation of aldehydes with 2-aryl allylic carbonates was recently developed by Qian and co-workers.[40] The process employed zinc as the terminal reductant and afforded the corresponding homoallylic alcohols in moderate to good yields and enantioselectivities of up to 96% and 91% ee, respectively (Scheme 10.17). Among several bisoxazoline ligands tested, tridentate PYBOX **10** was selected as the most efficient ligand. The electronic nature of the allylic partners seemed to be important in the control of the enantioselectivities, and the aromatic aldehydes that did not bear electron-withdrawing groups generally produced high enantioselectivities.

10.7 Other Reactions

In 2007, Frauenrath and Flock described the double bond isomerisation of 5-methyl-4*H*-1,3-dioxin performed upon catalysis with [NiI_2(*R*,*R*)-Me-DUPHOS], which afforded the corresponding chiral dioxins in excellent enantioselectivities of up to 95% ee, along with good yields (Scheme 10.18).[41] These products were further aziridinated and then ring-opened to lead to chiral 4-methyl-1,3-oxazolidine-4-carbaldehydes with 73% de.

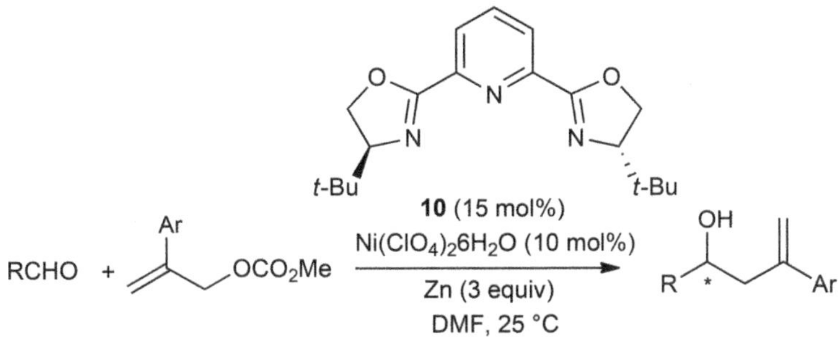

R = *o*-MeOC$_6$H$_4$, Ar = Ph: 96%, ee = 88%
R = *p*-(NMe$_2$)C$_6$H$_4$, Ar = Ph: 95%, ee = 85%
R = *p*-Tol, Ar = Ph: 95%, ee = 88%
R = *o*-Tol, Ar = Ph: 85%, ee = 86%
R = furyl, Ar = Ph: 95%, ee = 80%
R = Ar = Ph: 94%, ee = 87%
R = 1-Naph, Ar = Ph: 95%, ee = 73%
R = *o*-ClC$_6$H$_4$, Ar = Ph: 94%, ee = 77%
R = (*E*)-MeCH=CH, Ar = Ph: 91%, ee = 85%
R = Ph, Ar = *p*-CF$_3$C$_6$H$_4$: 90%, ee = 82%
R = Ph, Ar = *o*-Tol: 53%, ee = 41%
R = (*E*)-PhCH=CMe, Ar = Ph: 87%, ee = 91%

Scheme 10.17 Reductive allylation of aldehydes with allylic carbonates.

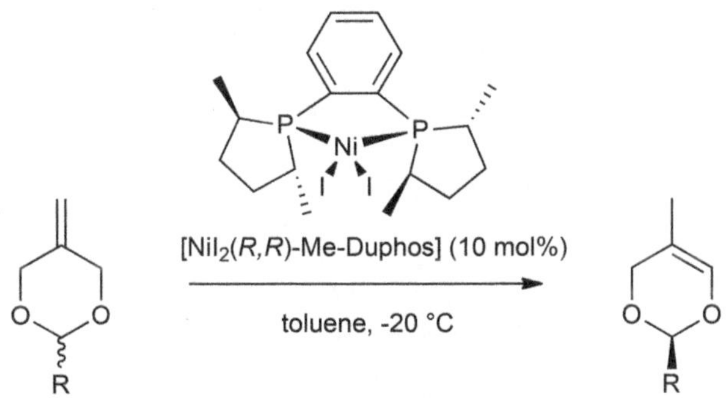

R = *i*-Pr: 79%, ee = 93%
R = *t*-Bu: 83%, ee = 95%

Scheme 10.18 Isomerisation of alkenes.

In another context, a nickel catalyst generated *in situ* from chiral *N*,*N*′-dioxide **11** was applied by Feng *et al.* to induce the asymmetric carbonyl–ene reaction of glyoxal derivatives with alkenes to provide the corresponding γ,δ-unsaturated α-hydroxy carbonyl compounds.[42] As shown in Scheme 10.19, these products were obtained in high yields and with excellent enantioselectivities of up to >99% ee. It was noteworthy that this catalyst system exhibited a remarkably broad substrate scope. For example, neither the electronic properties nor the

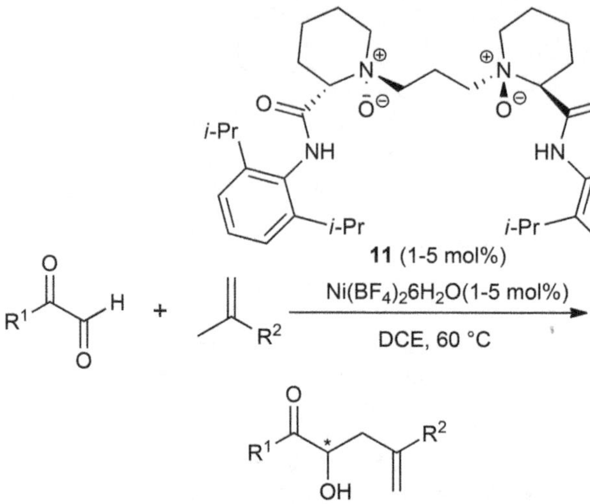

$R^1 = R^2$ = Ph: 83%, ee = 99%
R^1 = *o*-Tol, R^2 = Ph: 78%, ee = 99%
R^1 = *m*-Tol, R^2 = Ph: 82%, ee = 99%
R^1 = *p*-Tol, R^2 = Ph: 80%, ee > 99%
R^1 = *m*-$MeOC_6H_4$, R^2 = Ph: 78%, ee > 99%
R^1 = *p*-$MeOC_6H_4$, R^2 = Ph: 85%, ee > 99%
R^1 = *m*-ClC_6H_4, R^2 = Ph: 70%, ee = 99%
R^1 = *p*-ClC_6H_4, R^2 = Ph: 75%, ee > 99%
R^1 = *p*-FC_6H_4, R^2 = Ph: 73%, ee = 99%
R^1 = *p*-BrC_6H_4, R^2 = Ph: 70%, ee = 99%
R^1 = *p*-$NO_2C_6H_4$, R^2 = Ph: 78%, ee > 99%
R^1 = 2-Naph, R^2 = Ph: 77%, ee = 99%
R^1 = 2-furyl, R^2 = Ph: 80%, ee > 99%
R^1 = 2-thienyl, R^2 = Ph: 83%, ee = 98%
R^1 = Cy, R^2 = Ph: 80%, ee = 97%
R^1 = Me, R^2 = Ph: 75%, ee = 99%
R^1 = Ph, R^2 = *p*-FC_6H_4: 82%, ee > 99%
R^1 = Ph, R^2 = *o*-Tol: 73%, ee > 99%
R^1 = Ph, R^2 = *t*-$BuCH_2$: 84%, ee = 98%
R^1 = Ph, R^2 = *i*-Pr: 86%, ee > 99%
R^1 = OEt, R^2 = Ph: 99%, ee = 99%
R^1 = OEt, R^2 = *p*-FC_6H_4: 94%, ee = 97%
R^1 = OEt, R^2 = *o*-Tol: 77%, ee = 99%
R^1 = OEt, R^2 = *t*-$BuCH_2$: 87%, ee = 98%

Scheme 10.19 The carbonyl–ene reaction.

steric hindrance of the substituents on the aromatic ring of the glyoxal derivatives had any obvious effect on the enantioselectivity. Moreover, excellent enantioselectivities were achieved for the first time in the asymmetric carbonyl–ene reaction of heteroaromatic glyoxals and aliphatic glyoxals. In addition, alkenes bearing aliphatic as well as aromatic substituents provided comparable enantioselectivities. The scope of this methodology could also be extended to ethyl glyoxylate, which also provided excellent results.

The Reformatsky reaction was first reported in 1887 and is still widely used in organic synthesis.[43] It involves the zinc-induced formation of β-hydroxyalkanoates from α-halo carbonyl compounds and aldehydes or ketones. The mild reaction conditions, the excellent functional group tolerance, and the use of inexpensive non-toxic metals have made it an important alternative to the base-catalysed aldol reaction. However, the usual heterogeneous reaction conditions have made the development of catalytic stereoselective variants quite difficult. Recently, significant improvements have been achieved in the enantioselective Reformatsky reaction with aldehydes,[44] but only a few examples of ketones have been reported due to the low reactivity and decreased steric discrimination, even though the resulting chiral tertiary alcohols are important structural units present in many biologically active compounds and synthetic intermediates. In this context, Lu *et al.* have developed an enantioselective nickel-catalysed Reformatsky reaction of an α-bromo ester with ketones.[45] They used chiral indolinylmethanol ligands applied for the first time in such reactions (Scheme 10.20). Ligand **12** was selected as the most efficient among a range of chiral indole-derived ligands investigated. Performed in the presence of zinc powder, with CF_3CO_2H as an additive and $NiBr_2$ as the source of nickel,

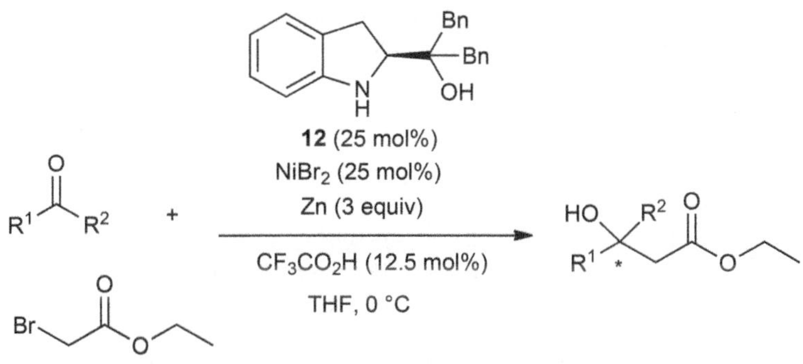

R^1 = Ph, R^2 = Me: 61%, ee = 81%
R^1 = *p*-Tol, R^2 = Me: 60%, ee = 87%
R^1 = *p*-$MeOC_6H_4$, R^2 = Me: 56%, ee = 81%
R^1 = *o*-ClC_6H_4, R^2 = Me: 73%, ee = 87%
R^1 = *m*-ClC_6H_4, R^2 = Me: 67%, ee = 78%
R^1 = *p*-ClC_6H_4, R^2 = Me: 68%, ee = 83%
R^1 = *p*-BrC_6H_4, R^2 = Me: 65%, ee = 86%
R^1 = *p*-FC_6H_4, R^2 = Me: 70%, ee = 82%
R^1 = Ph, R^2 = Et: 65%, ee = 70%
R^1 = 2-thienyl, R^2 = Me: 64%, ee = 62%
R^1 = $BnCH_2$, R^2 = Me: 75%, ee = 28%

Scheme 10.20 Reformatsky reaction of an α-bromo ester with ketones.

the reaction led to synthetically useful chiral β-hydroxy esters from aromatic and aliphatic ketones in moderate to good yields and with enantioselectivities of up to 87% ee. It was shown that changing the nickel salt to $Ni(acac)_2$ allowed enantioselectivity of 96% ee to be achieved, albeit in low yield.

On the other hand, Hayashi *et al.* have reported an example in which a five-membered ring was cleaved atropoenantioselectively.[46] Hence, the asymmetric nickel-catalysed cross coupling of dibenzothiophenes with Grignard reagents in the presence of chiral phosphines **13** or **14** delivered the corresponding biaryl-2-thiols (Scheme 10.21). In some cases, both the chemical and optical yields were excellent, but the success of this intriguing reaction varied both with the size of the Grignard reagent employed and the nature of the original *ortho* substituents. It is believed that the nickel catalyst first inserts into the C–S bond, with the stereochemically determining step being the transmetallation of the Grignard reagent or the following reaction.

In 2012, Murakami *et al.* reported the asymmetric intramolecular alkene insertion reaction of 3-(2-styryl)cyclobutanones catalysed by an *in situ* generated chiral nickel complex bearing a BINOL-derived phosphoramidite ligand **15** (Scheme 10.22).[47] The reaction provided a unique and straightforward access to

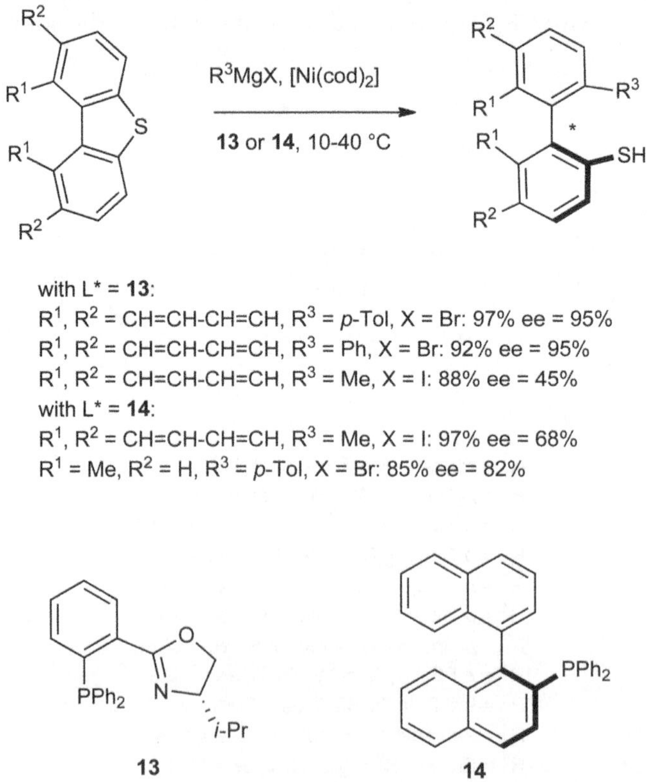

Scheme 10.21 Ring-cleaving biaryl synthesis through *C*-alkylation or *C*-arylation.

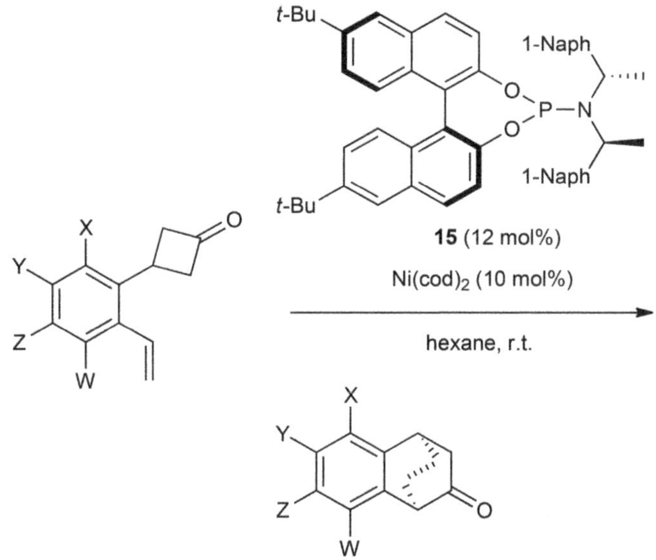

X = Y = H, Z,W = CH=CH-C(OMe)=CH: 84%, ee = 92%
X = H, Y = Me, Z,W = $(CH=CH)_2$: 77%, ee = 84%
X = Y = Z = W = H: 94%, ee = 84%
X = Z = W = H, Y = OMe: 93%, ee = 83%
X = Z = W = H, Y = F: 90%, ee = 80%
X = Y = W = H, Z = F: 90%, ee = 88%
X = Y = Z = H, W = Me: 95%, ee = 93%
X = F, Y = Z = H, W = Me: 97%, ee = 83%

Scheme 10.22 Intramolecular alkene insertion reaction of 3-(2-styryl) cyclobutanones.

chiral benzobicyclo[2.2.2]octenones in high yields and with enantioselectivities of up to 93% ee. These products constituted useful intermediates for the synthesis of biologically active molecules such as calcium channel blockers.

Optically active secondary alcohols are highly valuable intermediates in organic synthesis, especially when their significance as chiral building blocks for numerous natural products, pharmaceuticals, and other biologically active molecules is taken into account.[48] The non-noble-metal-catalysed enantioselective hydrosilylation of prochiral ketones represents a rewarding transformation towards chiral alcohols, owing to the economic benefits and the operational simplicity of such methods.[49] Thus, in recent decades, a variety of effective transition metal catalysts based on titanium, zinc, tin, copper, iron, and cobalt have been accordingly developed and applied in the asymmetric hydrosilylation of ketones, with good to excellent enantioselectivities. In 2012, Wu *et al.* reported the first example of this type of reaction enantioselectively catalysed by a chiral nickel catalyst.[50] The process employed $PhSiH_3$ as the hydride donor, 4 Å MS as additives, and (*S*)-Xyl-P-Phos as a chiral ligand of NiF_2

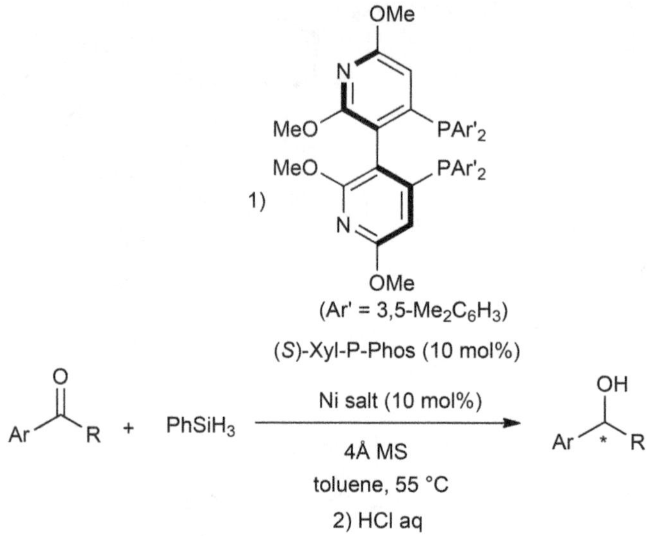

with Ni salt = $Ni(OAc)_2 4H_2O$:
Ar = *p*-$CF_3C_6H_4$, R = Me: 91%, ee = 80% (*S*)
Ar = *p*-$CO_2MeC_6H_4$, R = Me: 71%, ee = 73% (*S*)
Ar = 3,5-$(CF_3)_2C_6H_3$, R = Me: 97%, ee = 87% (*S*)
with Ni salt = NiF_2:
Ar = *p*-NCC_6H_4, R = Me: 87%, ee = 82% (*S*)
Ar = *m*-NCC_6H_4, R = Me: 42%, ee = 61%
Ar = *m*-$CF_3C_6H_4$, R = Me: 34%, ee = 71% (*S*)
Ar = 3,5-$(CF_3)_2C_6H_3$, R = Et: 78%, ee = 86%
Ar = 2,4,5-$F_3C_6H_2$, R = Me: 68%, ee = 67%

Scheme 10.23 Hydrosilylation of ketones.

or $Ni(OAc)_2{\cdot}4H_2O$ (Scheme 10.23). This ligand was selected after screening of a range of other chiral bisphosphines. As for cobalt-catalysed reactions, in nickel-catalysed hydrosilylations of alkyl aryl ketones the nature of the substituents on the phenyl ring of the ketones had a dramatic effect on the reaction activities, and the substrates bearing an electron-withdrawing group reacted favourably to give higher yields and enantioselectivities of up to 87% ee. Moreover, the outcomes of the reaction depended on the positioning of the substituents on the aryl ring of the ketones. Substrates with a *para*-substituted electron-deficient aryl group reacted favourably to afford the desired alcohols with higher enantioselectivities and better yields when compared with the substrates with *meta* substitution. Noticeably, the reaction presented the advantage of being carried out in an air atmosphere without special precautions, which highlights the practical potential of this novel protocol.

The Claisen rearrangement and its variants have enjoyed unparalleled success because of the utility of the products in the synthesis of complex organic structures.[51] The development of a general array of catalytic asymmetric rearrangements represents a highly desirable goal. Asymmetric versions of the classic Claisen rearrangement of allyl vinyl ethers have been successfully developed using various types of catalysts, including transition metals. Comparatively, the enantioselective catalytic version of the propargyl vinyl ether rearrangement continues to be relatively rare, although it provides a useful route to synthetically valuable functionalised allenes. The first example was recently reported by Kozlowski *et al.* and employed a chiral palladium(II) complex of BINAP as the catalyst.[52] Later in 2014, Feng *et al.* disclosed a general asymmetric propargyl vinyl ether rearrangement to give a series of allenyl-substituted cyclic β-keto esters by using an *in situ* generated nickel catalyst derived from chiral *N,N'*-dioxide **16**.[53] As shown in Scheme 10.24, a range of propargyl

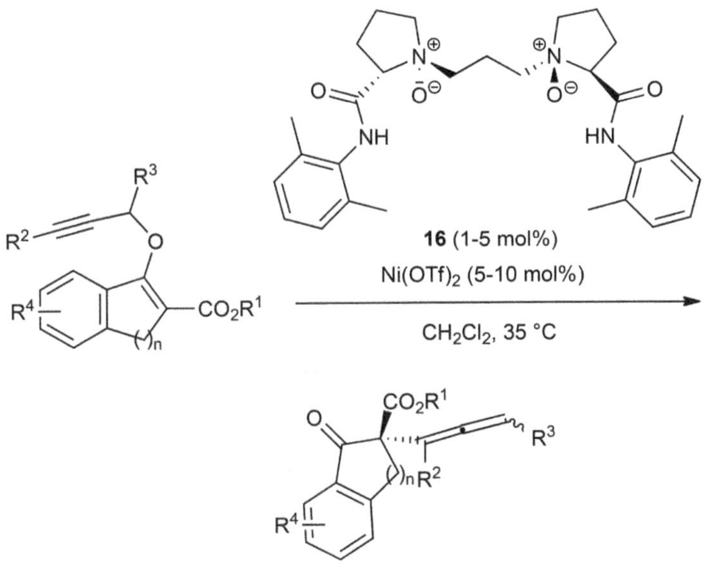

R^1 = Et, R^2 = Ph, R^3 = R^4 = H, n = 1: 88%, ee = 98%
R^1 = Et, R^2 = Ph, R^3 = H, R^4 = 5-F, n = 1: 93%, ee = 98%
R^1 = Et, R^2 = Ph, R^3 = H, R^4 = 5-Br, n = 1: 92%, ee = 98%
R^1 = Et, R^2 = Ph, R^3 = H, R^4 = 6-Me, n = 1: 90%, ee = 99%
R^1 = Me, R^2 = Ph, R^3 = R^4 = H, n = 1: 91%, ee = 98%
R^1 = *t*-Bu, R^2 = Ph, R^3 = R^4 = H, n = 1: 84%, ee = 97%
R^1 = Et, R^2 = 4-OMe, R^3 = R^4 = H:, n = 1 87%, ee = 99%
R^1 = Et, R^2 = 4-Me, R^3 = R^4 = H, n = 1: 93%, ee = 98%
R^1 = *t*-Bu, R^2 = *p*-ClC$_6$H$_4$, R^3 = R^4 = H, n = 1: 85%, ee = 98%
R^1 = Et, R^2 = R^3 = R^4 = H, n = 1: 93%, ee = 90%
R^1 = Et, R^2 = Me, R^3 = R^4 = H, n = 1: 91%, ee = 98%
R^1 = R^2 = Et, R^3 = R^4 = H, n = 1: 90%, ee = 98%
R^1 = Et, R^2 = Ph, R^3 = R^4 = H, n = 2: 90%, ee = 94%
R^1 = Et, R^2 = Ph, R^3 = R^4 = H, n = 3: 95%, ee = 98%
R^1 = Et, R^2 = Ph, R^3 = Me, R^4 = H, n = 1: 95%, dr = 45:55, ee = 98%/75%

Scheme 10.24 Propargyl vinyl ether Claisen rearrangement.

vinyl ethers could be rearranged into the corresponding chiral allenic derivatives in excellent yields and with enantioselectivities of up to 99% ee. The substrate scope of the reaction was surveyed with various substituents at the propargyl group and β-keto ester unit. Both electron-donating and electron-withdrawing substituents on the 1*H*-indene backbone of the substrate were tolerated. Varying the ester groups of the keto esters delivered similar levels of enantioselectivity. Aryl substituents (R^2) at the terminal position of the alkynyl group had no obvious influence on the yields or enantioselectivities.

Applying closely related reaction conditions to allyl vinyl ethers allowed the corresponding Claisen rearrangement to be achieved in high yields and enantioselectivities, as shown in Scheme 10.25. Remarkably, a variety of allyl rearrangement products bearing continuous tertiary-quaternary stereocentres was obtained in 90–99% yields, high diastereoselectivities of 88–98% de, and high enantioselectivities of 91–97% ee. For these substrates the catalyst

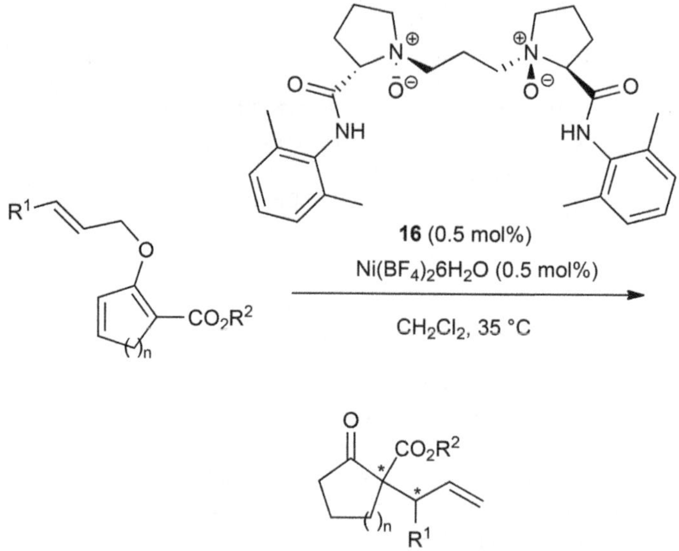

R^1 = Ph, R^2 = Et, n = 1: 90%, dr = 97:3, ee = 94%
R^1 = Ph, R^2 = Me, n = 1: 90%, dr = 99:1, ee = 97%
R^1 = Ph, R^2 = *t*-Bu, n = 1: 97%, dr = 93:7, ee = 96%
R^1 = *p*-$NO_2C_6H_4$, R^2 = Et, n = 1: 93%, dr = 98:2 ee = 96%
R^1 = *p*-Tol, R^2 = Et, n = 1: 90%, dr = 94:6 ee = 95%
R^1 = *m*-$CF_3C_6H_4$, R^2 = Et, n = 1: 90%, dr = 99:1 ee = 91%
R^1 = 1-furyl, R^2 = Et, n = 1: 95%, dr = 98:2 ee = 97%
R^1 = 1-thienyl, R^2 = Et, n = 1: 95%, dr = 99:1 ee = 92%
R^1 = 2-Naph, R^2 = Et, n = 1: 95%, dr = 99:1 ee = 96%
R^1 = (*E*)-PhCH=CH, R^2 = Et, n = 1: 95%, dr = 94:6 ee = 97%
R^1 = *p*-$NO_2C_6H_4$, R^2 = Et, n = 2: 90%, dr = 97:3 ee = 95%

Scheme 10.25 Allyl vinyl ether Claisen rearrangement.

loading was lowered to 0.5 mol%. Neither the ester group of the β-keto esters nor the substituents at the allyl group had an adverse effect on the yield and the stereoselectivity of this nice process.

Recently, Gade *et al.* reported a nickel-catalysed enantioselective hydrodehalogenation of geminal dihalides with moderate enantioselectivities of up to 74% ee (Scheme 10.26).[54] The process was performed with a chiral preformed nickel catalyst and $LiBEt_3H$ as reductant. In a first step, the chiral nickel complex **17** abstracted a halogen atom from the germinal dihalide, generating a nickel(II) chloride complex and a configurationally labile radical species. The latter could be stabilised by reversible coordination to the generated Ni–Cl complex. In a subsequent step, the liberated radical was then trapped by the nickel(II) hydrido complex, present in a large excess under the catalytic conditions, which in turn induced the enantioselectivity during the hydrogen atom transfer onto the radical intermediate.

Finally, highly congested vicinal all-carbon quaternary stereocentres were produced by Wang *et al.* on the basis of an asymmetric alkylation reaction of 3-bromooxindoles with 3-substituted indoles catalysed by a chiral nickel complex generated *in situ* from $Ni(OAc)_2$ and chiral diamine ligand **18**.[55] As shown in Scheme 10.27, the process employed K_3PO_4 as a base and afforded the corresponding coupling products in both high yields and enantioselectivities of

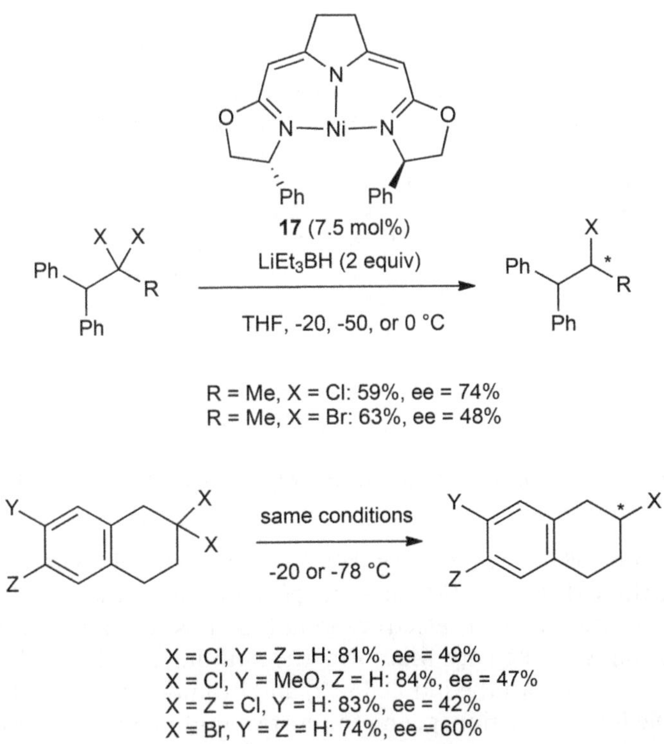

Scheme 10.26 Hydrodehalogenations of geminal dihalides.

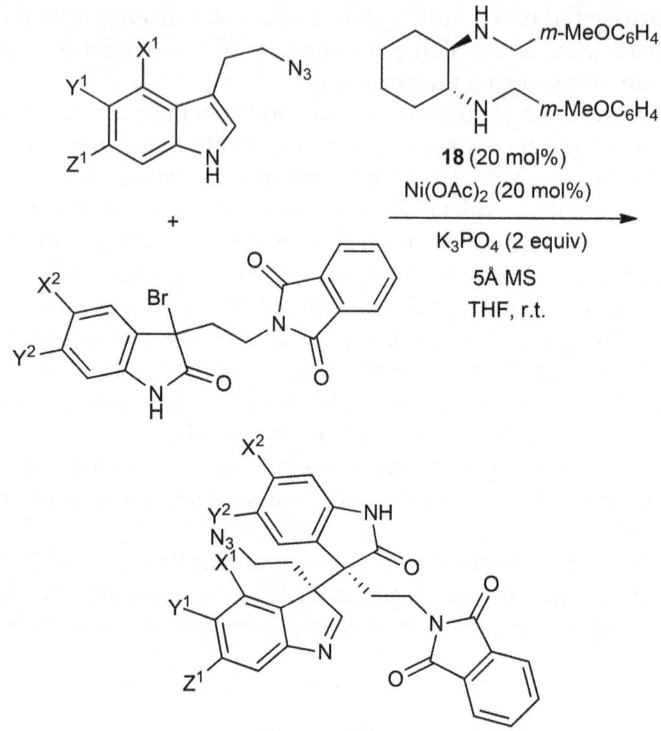

$X^1 = Z^1 = Y^1 = X^2 = Y^2 =$ H: 94%, de = 82%, ee = 92%
$X^1 =$ Me, $Y^1 = Z^1 = X^2 = Y^2 =$ H: 92%, de > 90%, ee = 86%
$X^1 =$ F, $Y^1 = Z^1 = X^2 = Y^2 =$ H: 86%, de = 84%, ee = 96%
$X^1 =$ Cl, $Y^1 = Z^1 = X^2 = Y^2 =$ H: 81%, de = 86%, ee = 99%
$X^1 = Z^1 = X^2 = Y^2 =$ H, $Y^1 =$ Me: 93%, de = 80%, ee = 85%
$X^1 = Z^1 = X^2 = Y^2 =$ H, $Y^1 =$ OMe: 94%, de > 90%, ee = 80%
$X^1 = Z^1 = X^2 = Y^2 =$ H, $Y^1 =$ Cl: 83%, de > 90%, ee = 97%
$X^1 = Z^1 = X^2 = Y^2 =$ H, $Y^1 =$ Me: 93%, de = 80%, ee = 85%
$X^1 = Z^1 = X^2 = Y^2 =$ H, $Z^1 =$ F: 73%, de > 90%, ee = 95%
$X^1 = Z^1 = X^2 = Y^2 =$ H, $Z^1 =$ Br: 86%, de = 84%, ee = 98%
$X^1 = Y^1 = Z^1 = X^2 =$ H, $Y^2 =$ OMe: 91%, de = 88%, ee = 94%
$X^1 = Z^1 = Y^2 =$ H, $Y^1 =$ Cl, $X^2 =$ Br: 51%, de = 84%, ee = 90%

Scheme 10.27 Alkylation of 3-bromooxindoles with 3-substituted indoles.

up to 94% and 99% ee, respectively. When electron-withdrawing groups were present at the C-4, C-5, or C-6 position of the indoles, the reaction worked well, whereas indoles with electron-donating substituents showed slightly decreased enantioselectivity. Moreover, the authors found that the presence of substituents on both the indole (5-Cl) and bromooxindole (6-Br) moieties was suitable for the reaction, giving the corresponding product with high levels of stereochemical control (84% de, 90% ee), which could be used for a further asymmetric total synthesis of the natural product (+)-perophoramidine.

10.8 Conclusions

Efforts to develop new asymmetric transformations have focused mainly on the use of a few metals, such as titanium, copper, nickel, ruthenium, rhodium, palladium, iridium, and more recently gold. However, by the very fact of the lower costs of nickel catalysts in comparison with other transition metals, enantioselective nickel-mediated transformations have received a continuous ever-growing attention during the last decade that has led to exciting and fruitful research.[56,57] This interest might also be related to the fact that nickel complexes are of high abundance, and exhibit a remarkably diverse chemical reactivity which is well demonstrated in this chapter. Among the most efficient enantioselective nickel-catalysed miscellaneous reactions developed in the last decade are hydrocarbamoylations of homoallylic formamides to give γ-lactams in up to 95% ee using a chiral diaminophosphine oxide ligand; aminations of 3-bromooxindoles with indolines to afford indole-derived pyrrolidines in up to 96% ee upon catalysis with a chiral bisoxazoline ligand; the first asymmetric ring-opening of cyclopropanes with amines performed with a chiral trisoxazoline ligand which gave enantioselectivities of up to 98% ee; the first asymmetric ring-opening of oxabicyclic alkenes with arylboronic acids which provided *cis*-2-aryl-1,2-dihydronaphthalen-1-ols in up to 99% ee in the presence of (*S*,*S*)-Me-DUPHOS; the first asymmetric Friedel-Crafts reaction of indoles with β,β-disubstituted nitroalkenes which yielded indole-bearing chiral compounds with trifluoromethylated all-carbon quaternary stereocentres in up to 97% ee using a chiral bisoxazoline ligand; Friedel–Crafts reactions of indoles with β,γ-unsaturated α-keto esters leading to 2-substituted chiral indoles in up to 97% ee with a chiral *N*,*N*′-oxide ligand; carbonyl–ene reactions of glyoxal derivatives with alkenes providing γ,δ-unsaturated α-hydroxyl carbonyl compounds in up to >99% ee with a chiral *N*,*N*′-oxide ligand; intramolecular alkene insertions of 3-(2-styryl)cyclobutanones into benzobicyclo[2.2.2]octenones in up to 92% ee with a BINOL-derived phosphoramidite ligand; and propargyl vinyl ether and allyl vinyl ether Claisen rearrangements achieved in up to 99% ee with a chiral *N*,*N*′-oxide ligand.

References

1. B. M. Trost, *Acc. Chem. Res.*, 2002, **35**, 695–705.
2. (a) B. M. Trost, *Acc. Chem. Res.*, 1990, **23**, 34–42; (b) B. M. Trost and M. J. Krische, *Synlett*, 1998, 1–16; (c) R. A. Widenhoefer, *Acc. Chem. Res.*, 2002, **35**, 905–913; (d) C. Aubert, O. Buisine and M. Malacria, *Chem. Rev.*, 2002, **102**, 813–841; (e) G. C. Lloyd-Jones, *Org. Biomol. Chem.*, 2003, **1**, 215–236; (f) B. Schmidt, *Angew. Chem., Int. Ed.*, 2003, **42**, 4996–4999; (g) S. Ma, S. Yu and Z. Gu, *Angew. Chem., Int. Ed.*, 2006, **45**, 200–203; (h) L. Zhang, J. Sun and S. S. Kozmin, *Adv. Synth. Catal.*, 2006, **348**, 2271–2296.
3. B. M. Trost, D. C. Lee and F. Rise, *Tetrahedron Lett.*, 1989, **30**, 651–654.

4. I. J. S. Fairlamb, *Angew. Chem., Int. Ed.*, 2004, **43**, 1048–1052.
5. (a) P. Kisanga, L. A. Goj and R. A. Widenhoefer, *J. Org. Chem.*, 2001, **66**, 635–637; (b) K. L. Bray, I. J. S. Fairlamb, J.-P. Kaiser, G. C. Lloyd-Jones and P. A. Slatford, *Top. Catal.*, 2002, **19**, 49–59; (c) A. Heumann, L. Giordano and A. Tenaglia, *Tetrahedron Lett.*, 2003, **44**, 1515–1518.
6. (a) A. Behr, U. Freudenberg and W. Keim, *J. Mol. Catal.*, 1986, **35**, 9–17; (b) B. Radetich and T. V. RajanBabu, *J. Am. Chem. Soc.*, 1998, **120**, 8007–8008; (c) D. Necas, M. Tursky, I. Tislerova and M. Kotora, *New J. Chem.*, 2006, **30**, 671–674.
7. R. Crigg, J. F. Malone, T. R. B. Mitchell, A. Ramasubbu and R. M. Scott, *J. Chem. Soc., Perkin Trans. 1*, 1984, 1745–1754.
8. (a) Y. Yamamoto, Y. Nakagai, N. Ohkoshi and K. Itoh, *J. Am. Chem. Soc.*, 2001, **123**, 6372–6380; (b) I. Öhkoshi, E. Cetinkaya, M. Cicek, D. Sémeril, C. Bruneau and P. H. Dixneuf, *Eur. J. Inorg. Chem.*, 2004, 418–422; (c) Y. Terada, M. Arisawa and A. Nishida, *Angew. Chem., Int. Ed.*, 2004, **43**, 4063–4067; (d) I. J. S. Fairlamb, G. P. McGlacken and F. Weissberger, *Chem. Commun.*, 2006, 988–990.
9. W. D. Kerber, J. H. Koh and M. R. Gagné, *Org. Lett.*, 2004, **6**, 3013–3015.
10. S. Okamoto and T. Livinghouse, *J. Am. Chem. Soc.*, 2000, **122**, 1223–1224.
11. (a) B. Bogdanovic, *Adv. Organomet. Chem.*, 1979, **17**, 105–140; (b) A. Heumann and M. Moukhliss, *Synlett*, 1998, 1211–1222.
12. (a) C. Böing, G. Francio and W. Leitner, *Chem. Commun.*, 2005, 1456–1458; (b) C. Böing, G. Francio and W. Leitner, *Adv. Synth. Catal.*, 2005, **347**, 1537–1541; (c) C. J. Diez-Holz, C. Böing, G. Francio, M. Hölscher and W. Leitner, *Eur. J. Org. Chem.*, 2007, 2995–3002.
13. G. Wilke, *Angew. Chem., Int. Ed. Engl.*, 1988, **27**, 185–206.
14. C. Böing, J. Hahne, G. Francio and W. Leitner, *Adv. Synth. Catal.*, 2008, **350**, 1073–1080.
15. (a) K. L. Habermas and S. E. Denmark, *Org. React.*, 1994, **45**(11971), 1–158; (b) S. E. Denmark, *Comprehensive Organic Synthesis*, ed. L. A. Paquette, Pergamon Press, Oxford, 1991, vol. 5, pp. 751–784; (c) C. Santelli-Rouvier and M. Santelli, *Synthesis*, 1983, 429–442; (d) H. Pellissier, *Tetrahedron*, 2005, **61**, 6479–6517.
16. G. Liang, S. N. Gradl and D. Trauner, *Org. Lett.*, 2003, **5**, 4931–4934.
17. V. K. Aggarwal and A. J. Belfield, *Org. Lett.*, 2003, **5**, 5075–5078.
18. G. Liang and D. Trauner, *J. Am. Chem. Soc.*, 2004, **126**, 9544–9545.
19. I. Walz and A. Togni, *Chem. Commun.*, 2008, 4315–4317.
20. M. P. Watson and E. N. Jacobsen, *J. Am. Chem. Soc.*, 2008, **130**, 12594–12595.
21. Z. Ning, R. Jin, J. Ding and L. Gao, *Synlett*, 2009, 2291–2294.
22. T. Arai, S. Kajikawa and E. Matsumura, *Synlett*, 2013, **24**, 2045–2048.
23. P. A. Donets and N. Cramer, *J. Am. Chem. Soc.*, 2013, **135**, 11772–11775.
24. D. B. Berkowitz and G. Maiti, *Org. Lett.*, 2004, **6**, 2661–2664.
25. H. Zhang, H. Kang, L. Hong, W. Dong, G. Li, X. Zheng and R. Wang, *Org. Lett.*, 2014, **16**, 2394–2397.
26. A. N. Parvulescu, P. A. Jacobs and D. E. De Vos, *Adv. Synth. Catal.*, 2008, **350**, 113–121.

27. P. Gopinath, T. Watanabe and M. Shibasaki, *Org. Lett.*, 2012, **14**, 1358–1361.
28. Y.-Y. Zhou, L.-J. Wang, J. Li, X.-L. Sun and Y. Tang, *J. Am. Chem. Soc.*, 2012, **134**, 9066–9069.
29. K. Fagnou and M. Lautens, *Chem. Rev.*, 2003, **103**, 169–196.
30. Z. Zeng, D. Yang, Y. Long, X. Pan, G. Huang, X. Zuo and W. Zhou, *J. Org. Chem.*, 2014, **79**, 5249–5257.
31. M. Lautens, C. Dockendorff, K. Fagnou and A. Malicki, *Org. Lett.*, 2002, **4**, 1311–1314.
32. T.-K. Zhang, D.-L. Mo, L.-X. Dai and X.-L. Hou, *Org. Lett.*, 2008, **10**, 3689–3692.
33. X.-J. Pan, G.-B. Huang, Y.-H. Long, X.-J. Zuo, X. Xu, F. L. Gu and D.-Q. Yang, *J. Org. Chem.*, 2014, **79**, 187–196.
34. (a) T. B. Poulsen and K. A. Jørgensen, *Chem. Rev.*, 2008, **108**, 2903–2915; (b) M. Bandini and A. Eichholzer, *Angew. Chem., Int. Ed.*, 2009, **48**, 9608–9644; (c) V. Terrasson, R. M. Figueiredo and J. M. Campagne, *Eur. J. Org. Chem.*, 2010, 2635–2655.
35. J.-R. Gao, H. Wu, B. Xiang, W.-B. Yu, L. Han and Y.-X. Jia, *J. Am. Chem. Soc.*, 2013, **135**, 2983–2986.
36. Y. Zhang, X. Liu, X. Zhao, J. Zhang, L. Zhou, L. Lin and X. Feng, *Chem. Commun.*, 2013, **49**, 11311–11313.
37. J.-Q. Weng, Q.-M. Deng, L. Wu, K. Xu, H. Wu, R.-R. Liu, J.-R. Gao and Y.-X. Jia, *Org. Lett.*, 2014, **16**, 776–779.
38. (a) M. Yus, J. C. Gonzalez-Gomez and F. Foubelo, *Chem. Rev.*, 2011, **111**, 7774–7854; (b) I. Marek and G. Sklute, *Chem. Commun.*, 2007, 1683–1691; (c) S. E. Denmark and J. Fu, *Chem. Rev.*, 2003, **103**, 2763–2793; (d) W. J. Kennedy and D. G. Hall, *Angew. Chem., Int. Ed.*, 2003, **42**, 4732–4739.
39. P. Zhang and J. P. Morken, *J. Am. Chem. Soc.*, 2009, **131**, 12550–12551.
40. Z. Tan, X. Wan, Z. Zang, Q. Qian, W. Deng and H. Gong, *Chem. Commun.*, 2014, **50**, 3827–3830.
41. S. Flock and H. Frauenrath, *ARKIVOC*, 2007, **x**, 245–259.
42. K. Zheng, J. Shi, X. Liu and X. Feng, *J. Am. Chem. Soc.*, 2008, **130**, 15770–15771.
43. (a) A. Fürstner, *Synthesis*, 1989, 571–590; (b) R. Ocampo and W. R. Dolbier, *Tetrahedron*, 2004, **60**, 9325–9374.
44. C. M. R. Ribeiro and F. M. C. de Farias, *Mini-Rev. Org. Chem.*, 2006, **3**, 1–10.
45. N. Lin, M.-M. Chen, R.-S. Luo, Y.-Q. Deng and G. Lu, *Tetrahedron: Asymmetry*, 2010, **21**, 2816–2824.
46. (a) Y.-H. Cho, A. Kina, T. Shimada and T. Hayashi, *J. Org. Chem.*, 2004, **69**, 3811–3823; (b) T. Shimada, Y.-H. Cho and T. Hayashi, *J. Am. Chem. Soc.*, 2002, **124**, 13396–13397.
47. L. Liu, N. Ishida and M. Murakami, *Angew. Chem., Int. Ed.*, 2012, **51**, 2485–2488.
48. V. Farina, J. T. Reeves, C. H. Senanayake and J. J. Song, *Chem. Rev.*, 2006, **106**, 2734–2793.
49. C. G. Arena, *Mini-Rev. Org. Chem.*, 2009, **6**, 159–167.

50. F.-F. Wu, J.-N. Zhou, Q. Fang, Y.-H. Hu, S. Li, X.-C. Zhang, A. S. C. Chan and J. Wu, *Chem.–Asian J.*, 2012, **7**, 2527–2530.
51. (a) J. Rehbein and M. Hiersemann, *Synthesis*, 2013, **45**, 1121–1159; (b) E. A. Ilardi, C. E. Stivala and A. Zakarian, *Chem. Soc. Rev.*, 2009, **38**, 3133–3148; (c) A. M. Martin Castro, *Chem. Rev.*, 2004, **104**, 2939–3002.
52. T. Cao, J. Deitch, E. C. Linton and M. C. Kozlowski, *Angew. Chem., Int. Ed.*, 2012, **51**, 2448–2451.
53. Y. Liu, H. Hu, H. Zheng, Y. Xia, X. Liu, L. Lin and X. Feng, *Angew. Chem., Int. Ed.*, 2014, **53**, 11579–11582.
54. C. Rettenmeier, H. Wadepohl and L. H. Gade, *Chem.–Eur. J.*, 2014, **20**, 9657–9665.
55. H. Zhang, L. Hong, H. Kang and R. Wang, *J. Am. Chem. Soc.*, 2013, **135**, 14098–14101.
56. For general reviews on organonickel chemistry, see: (a) G. Wilke, *Angew. Chem., Int. Ed. Engl.*, 1988, **27**, 185–206; (b) R. Shintani and T. Hayashi, in *Modern Organonickel Chemistry*, ed. Y. Tamaru, Wiley-VCH, Weinheim, 2005, pp. 240–272; (c) S. Z. Tasker, E. A. Standley and T. F. Jamison, *Nature*, 2014, **509**, 299–309; (d) For a special issue on recent advances in organonickel chemistry, see: T. F. Jamison, *Tetrahedron*, 2006, **62**, 7503.
57. For specific reviews on nickel-catalysed reductive coupling reactions, see: (a) J. Montgomery, *Angew. Chem., Int. Ed.*, 2004, **43**, 3890–3908; (b) R. M. Moslin, K. Miller-Moslin and T. F. Jamison, *Chem. Commun.*, 2007, 4441–4449; (c) B. M. Rosen, K. W. Quasdorf, D. A. Wilson, N. Zhang, A.-M. Resmerita, N. K. Garg and V. Percec, *Chem. Rev.*, 2011, **111**, 1346–1416; (d) B. L. H. Taylor and E. R. Jarvo, *Synlett*, 2011, **19**, 2761–2765.

CHAPTER 11

General Conclusions

The enantioselective production of compounds is a central theme in current research. The broad utility of synthetic chiral molecules as single-enantiomer pharmaceuticals, in electronic and optical devices, as components in polymers with novel properties, and as probes of biological function has made asymmetric synthesis a prominent area of investigation. Nearly all natural products are chiral and their physiological or pharmacological properties depend upon their recognition by chiral receptors, which will interact only with molecules of the proper absolute configuration. Indeed, the use of chiral drugs in enantiopure form is now a standard requirement for virtually every new chemical entity and the development of new synthetic methods to obtain enantiopure compounds has become a key goal for pharmaceutical companies.

The search for new and efficient methods for the synthesis of optically pure compounds constitutes an active area of research in organic synthesis. Of the methods available for preparing chiral compounds, catalytic asymmetric synthesis has attracted most attention. In particular, asymmetric transition-metal catalysis has emerged as a powerful tool to perform reactions in a highly enantioselective fashion over the past few decades. Efforts to develop new asymmetric transformations have focused mainly on the use of a few metals, such as titanium, nickel, copper, ruthenium, rhodium, palladium, iridium, and more recently gold. However, by the very fact of the lower cost of nickel catalysts in comparison with other transition metals, and their abundance, enantioselective nickel-catalysed transformations have received continuous ever-growing attention during recent decades that has led to exciting and fruitful research. This interest might also be related to the fact

RSC Catalysis Series No. 26
Enantioselective Nickel-Catalysed Transformations
By Hélène Pellissier

Published by the Royal Society of Chemistry, www.rsc.org

that nickel complexes exhibit a remarkably diverse chemical reactivity, and constitute one of the most useful Lewis acids in asymmetric catalysis. This book illustrates how much enantioselective nickel catalysis has contributed to the development of various types of enantioselective economical transformations. It updates the major progress in the field of enantioselective reactions catalysed by chiral nickel catalysts, well illustrating the power of these cheap catalysts to promote all types of organic reactions, from basic ones, such as cycloadditions, conjugate additions, cross-couplings, hydrovinylations, hydrophosphinations, hydrocyanations, hydroalkynylations, α-functionalisations, α-arylations and α-alkylations of carbonyl compounds, additions of organometallic reagents to aldehydes, aldol- and Mannich-type reactions, hydrogenations, and miscellaneous reactions, to completely novel methodologies including domino reactions, for example.

The first chapter of the book was dedicated to the advances in enantioselective nickel-catalysed cycloaddition reactions. Among the metals used to catalyse cycloadditions, nickel has been found to be competent to promote enantioselectively the formation of carbo- and heterocycles of various ring sizes. In particular, 1,3-dipolar cycloadditions have become one of the most powerful tools for the construction of enantiomerically pure five-membered heterocycles. In the last decade, effective catalysis by the use of a wide variety of chiral Lewis acid catalysts, including nickel complexes, has been reported for nitrone cycloaddition reactions using both electron-deficient and electron-rich alkene dipolarophiles. Early in 1998, Kanemasa reported highly diastereo- and enantioselective 1,3-dipolar cycloadditions of nitrones with 3-crotonoyloxazolidin-2-one catalysed by the (*R*,*R*)-DBFOX ligand.[1] Comparable excellent results were later described by Iwasa by using chiral PYBOX ligands.[2] Ever since, and especially in the last decade, a range of chiral nickel complexes, predominantly based on various nitrogen-containing chelating ligands, have been successfully applied as highly efficient catalysts in various enantioselective cycloadditions, including many 1,3-dipolar cycloadditions, several Diels–Alder cycloadditions, and other cycloadditions. Notably, the asymmetric 1,3-dipolar cycloaddition reaction has known the most developments in the last 10 years, with high levels of stereocontrol which is extremely important for constructing heterocyclic compounds from the viewpoint of the synthesis of biologically active compounds. The best results in enantioselective nickel-catalysed 1,3-dipolar cycloadditions have been reached by using ligands such as bisoxazolines, *N*,*N*′-dioxides, aminophenols, and BINIM derivatives. In these studies, nickel has often been compared with the corresponding complexes of other metals, and it was found that it generally provided better results than metals such as magnesium, zinc, cobalt, copper, manganese, scandium, ruthenium, silver, titanium, or palladium. More generally, the privileged chiral ligands of cobalt are bis(tris)oxazoline and Salen derivatives, along with a few bisphosphines (privileged ligands are those that effect a wide variety of transformations under exceptional enantiocontrol and with high productivity). Bisoxazoline ligands are often ligated to copper and magnesium, while BINOL-derived ligands are the privileged ligands of titanium and ytterbium. On the other hand, BINAP derivatives are the privileged ligands of rhodium, palladium, and gold, whereas PHOX ligands gave the best results for silver complexes. Catalytic enantioselective Diels–Alder reactions can be achieved by various Lewis acid transition metals. Among them, chiral nickel complexes have been used as efficient catalysts, based on various nitrogen-containing chelating ligands. Pioneering results in this area were reported by Kanemasa in 1997, who introduced the DBFOX ligand as a novel tridentate ligand providing perfect enantioselectivity in the Diels–Alder reactions of cyclopentadiene with 3-alkenoyloxazolidin-2-ones.[3]

Inspired by these pioneering studies, a number of bisoxazoline ligands have provided excellent results in the last few years, along with *N,N'*-oxide ligands. In these studies, the authors often compared the efficiency of the nickel catalysts with the corresponding complexes of other metals, finding that nickel was generally more efficient than iron, cobalt, zinc, magnesium, copper, silver, or ytterbium. More generally, the privileged chiral cobalt ligands for Diels–Alder reactions are Salen derivatives, but some bisoxazolines have also provided good results in this field. Bisoxazoline ligands have also given good results in the last few years in asymmetric magnesium-, zinc-, copper-, and even palladium-catalysed Diels–Alder cycloadditions. At the same time, BINOL was the privileged ligand for titanium, aluminum, indium, and boron. PHOX ligands have been successfully used in combination with metals such as copper and palladium. The latter has also encountered success with BINAP. In addition, high enantioselectivities were provided by Salen–chromium complexes. In the last decade, excellent results have also been described in enantioselective nickel-catalysed [2 + 2 + 2] cycloadditions by using oxazoline ligands, which were found to be superior to the corresponding cobalt, ruthenium, or rhodium complexes. Furthermore, bisoxazolines were proven to be excellent nickel ligands in asymmetric [3 + 3] cycloadditions, giving higher enantioselectivities than analogous ytterbium catalysts.

All the novel procedures for cycloaddition achieved by using chiral nickel complexes have greatly improved the structural scope and synthetic utility of cycloadditions, providing enantioselective access to various functionalised important (poly)(hetero)cyclic compounds with high enantioselectivities. Further progress in this area would include the discovery of more reactive catalyst systems, allowing the use of lower catalyst loadings, and the cycloadditions of even more challenging substrates, such as non-activated alkenes or highly substituted dipolarophiles, as well as the development of applications in the synthesis of natural products and bioactive compounds.

The second chapter of the book covered the advances in enantioselective nickel(II)-catalysed conjugate additions. An important breakthrough in the history of the Michael reaction was the achievement of the Michael addition of β-dicarbonyl compounds to a broad variety of electrophiles catalysed by nickel acetylacetonate, described by Nelson in 1979–1980.[4] Since this pioneering contribution, nickel has become one of the preferred metals in catalytic Michael reactions. Later, in 1988, Soai developed the first asymmetric conjugate additions performed under chiral nickel(II) catalysis.[5] Ever since, a wide variety of metals, including boron, magnesium, aluminum, calcium, cobalt, nickel, copper, zinc, ruthenium, rhodium, palladium, silver, iridium, and lanthanoids, have been successfully applied to the catalysis of a variety of asymmetric conjugate additions. In particular, during the last 10 years, an important number of novel highly efficient asymmetric conjugate additions of various nucleophiles to a wide variety of acceptor-activated alkenes have been developed on the basis of asymmetric nickel(II) catalysis. For example, asymmetric nickel-catalysed conjugate additions of various 1,3-dicarbonyl compounds to nitroalkenes, including complex and functionalised ones such as 3-nitro-2*H*-chromenes, nitroenynes, and nitrodienynes, were reported, with one beautiful example employing a recyclable mesoporous catalyst. Other nucleophiles, such as γ-butyrolactams, α-keto esters, α-keto anilides, 3-substituted oxindoles, azaarylacetates, highly functionalised acetamides, and acetylazaarenes, *etc.*, also gave excellent results in additions to nitroalkenes. Furthermore, organozinc reagents, β-keto esters, 2-siloxyfurans, malononitriles, nitromethane, nitroacetates, and cyclic amines, among other nucleophiles, have been successfully added to various α,β-unsaturated carbonyl compounds and derivatives.

Highly enantioselective intramolecular oxa-Michael additions to activated enones have also been described. Even more importantly, a range of powerful nickel-catalysed asymmetric domino reactions initiated by Michael additions, including multicomponent ones, have been successfully developed in the last 10 years. The best ligands for asymmetric nickel-catalysed conjugate additions are bisoxazolines, phosphoramidites, *N*,*N*′-dioxides, and Schiff bases, along with some bisphosphines. On the other hand, bisoxazolines have also been widely employed as highly efficient ligands of copper, rhodium, calcium, zinc, and cobalt. Moreover, the latter also provided excellent enantioselectivities when combined with various Salen ligands, along with tetrakisoxazolines, PYBOX, and bisphosphine ligands. On the other hand, rhodium, copper, and palladium complexes of BINAP, zinc and lanthanoid complexes of BINOL, rhodium and palladium complexes of bisphosphacycle ligands, copper complexes of the Josiphos ligand, and Salen complexes of aluminum recently afforded excellent results in a range of enantioselective conjugate additions. The future direction in the field of enantioselective nickel(II)-catalysed conjugate additions is to continue expanding their scope through the employment of novel chiral nickel catalysts, and apply these powerful strategies, including fascinating domino processes based on Michael reactions, to the synthesis of biologically interesting molecules, including natural products.

The third chapter of the book was devoted to the advances in enantioselective nickel-catalysed cross-coupling reactions. Transition metal-catalysed cross-coupling reactions represent a powerful approach for the construction of carbon–carbon bonds; consequently, these reactions have been widely studied during the last few decades. Their development has reached a level of sophistication that allows for a wide range of coupling partners to be combined efficiently. The emergence of cross-coupling as a popular method in synthesis arises from both the diversity of organometallic reagents used in these reactions and the broad range of functional groups which can be incorporated into these reagents. This paradigm for carbon–carbon bond construction has allowed chemists to assemble complex molecular frameworks of diversified interests, encompassing the total syntheses of natural products, medicinal chemistry, and industrial process development, as well as chemical biology, materials, and nanotechnology.

Among the transition metals employed in cross-coupling reactions, such as iron, cobalt, copper, zinc, silver, vanadium, chromium, and zirconium, among others, the majority of the investigations have focused on nickel- and palladium-catalysed cross-couplings of halides with various organometallic reagents. Advances in the use of these metals can be attributed to their versatility and high functional-group tolerance, as well as the readiness and selectivity with which electrophiles react (ease of oxidative addition and absence of β-hydride elimination pathways). Early in the development of homocoupling, the joining of identical chemical fragments, and cross-couplings, early transition metals such as nickel were identified as useful catalysts. However, more attention was firstly invested in the development of later transition metal catalysis, particularly palladium-catalysed Heck, Hiyama, Kumada, Negishi, Suzuki–Miyaura, Sonogashira, and Stille coupling techniques, due to some advantages in terms of the diversity and tenability of preparable catalysts, their oxidative and aqueous stability, and relatively facile isolation and structural analysis of their complexes, which aided mechanistic and methodologic developments. These typically palladium-catalysed coupling reactions used a diversity of organometallic transmetallating reagents or unsaturated functionalities for coordination/insertion chemistry, but nearly uniformly required aryl, vinyl, allyl, or sometimes alkyl halides as electrophiles.

On the other hand, while palladium-catalysed methods are limited to primary alkyl halide substrates, cheaper nickel complexes have been found to be uniquely suited to the catalysis of the cross-coupling of secondary alkyl halides. Bi- and tridentate nitrogen ligands, many of which are commercially available, have been found by Fu's group to be the key to the success of enantioselective nickel-catalysed cross-coupling reactions for a wide number of secondary alkyl halides and other electrophiles with a variety of organometallic reagents, including zinc, boron, silicon, magnesium, tin, zirconium, and indium compounds, among others. Nickel is not just an alternative to palladium; from the perspective of economics, it is clearly more desirable than the later elements in the d^{10} group. Indeed, nickel is a commodity metal with a cost of roughly $1.20 per mole, whereas palladium is a precious metal, with a significantly higher price of $1500 per mole.[6] Thus, unless a process is viable with very low levels of palladium that can be used and recycled, or very high levels of nickel are required, a nickel-catalysed approach would be preferred on a cost basis. In addition to often advantageous economics, palladium catalysis is not readily applied to coupling reactions involving phenol-derived electrophiles. Early transition metal nickel is more nucleophilic on account of its smaller size and can harness phenol-derived as well as other less reactive electrophiles, typically using less exotic ligands.

Moreover, nickel is particularly effective for reactions involving C–O derived electrophiles. Indeed, nickel is an outstanding reagent for cross-coupling from the standpoints of economics and versatility, and its use has particularly allowed a dramatic rise in the development of cross-coupling reactions of secondary electrophiles, making C–C bond formation through these methodologies a strategy of choice in the design for the total syntheses of biologically active and natural products. The development of these methodologies is significantly expanding the scope of transition-metal-catalysed reactions.

The privileged chiral ligands for nickel-catalysed cross-coupling reactions for the past decade have been mostly bisoxazolines (DBFOX, PYBOX) and phosphoramidites, along with *N*,*N*′-dioxides. Their use has allowed a broad variety of secondary electrophiles, including aryl bromides, allylic chlorides, alkyl iodides, propargylic bromides, bromoindanes, propargylic carbonates, benzylic alcohols, α-bromo ketones, α-bromo nitriles, α-zincated *N*-Boc-pyrrolidines, α,α-dihalo ketones, α-bromosulfonamides and α,α-dibromo sulfones, as well as methoxypyridinium salts, to considerably expand the scope of asymmetric Negishi reactions in particular. In addition, they allowed the first enantioselective versions of the Kumada, Hiyama, and Suzuki reactions to be recently achieved with excellent enantioselectivities. In addition to organozinc, -silyl, -magnesium, and -borane reagents employed respectively in nickel-catalysed asymmetric Negishi, Hiyama, Kumada, and Suzuki cross-coupling reactions with various electrophiles, organozirconium and trialkynylindium reagents have been recently successfully involved for the first time in this type of enantioselective reaction.

In the last decade, important advances have also been made in the area of enantioselective nickel-catalysed reductive coupling reactions. For example, excellent regio- and enantioselectivities were achieved in reductive couplings of alkynes and aldehydes. Moreover, the first asymmetric reductive coupling of dienes with aldehydes was successfully described, as well as the first reductive coupling of acid chlorides with secondary benzyl chlorides, and the first highly enantio- and regioselective reductive coupling of vinyl bromides with benzyl chlorides. Other interesting couplings have generated excellent enantioselectivities, such as that of dinaphthothiophene with Grignard reagents generating chiral biaryls, and that of xanthene with β-keto esters which was multicatalysed by a combination of nickel and iron.

Although nickel has been somewhat overlooked for a long time in favour of the more popularly studied and therefore well-understood palladium, nickel is now back in the limelight, and hopefully its full potential will be unlocked in the future. Whereas palladium-catalysed methods are often limited to primary alkyl halide substrates, cheaper nickel complexes have been found to be uniquely suited to catalysis of the cross-coupling of secondary alkyl halides. Indeed, the door has been opened to the development of general methods for the asymmetric coupling of any type of secondary electrophile. Such methodologies will certainly have a large impact on the synthesis of complex molecules and natural products. While progress has been considerable over the past decade in this field, better mechanistic and stereochemical understanding is needed in the future. Furthermore, the development of more active catalysts is still necessary to enable lower catalyst loadings than those currently used.

The fourth chapter of the book covered advances in enantioselective nickel-catalysed domino, multicomponent, and tandem reactions. Since the first definition of domino reactions by Tietze in 1993,[7] an explosive number of these fascinating one-pot reactions of two or more bond-forming reactions, evolving under identical conditions in which the subsequent transformation takes place at the functionalities obtained in the former transformation, have been developed, allowing the easy building of complex chiral molecular architectures from simple materials to be achieved in a single step. Even more interesting, the possibility to join two or more reactions in one asymmetric domino process catalysed by chiral metal catalysts has rapidly become a challenging goal for chemists, due to economic advantages such as avoiding costly protecting groups and time-consuming purification procedures after each step. The explosive development of enantioselective metal-catalysed domino (including multicomponent) reactions is a consequence of the considerable impact of the advent of asymmetric transition-metal catalysis. The wide variety of these novel highly efficient domino processes well reflects that of the metals employed to induce them. Indeed, an increasing number of different metals, such as magnesium, scandium, titanium, ruthenium, cobalt, rhodium, iridium, nickel, palladium, platinum, copper, silver, gold, zinc, aluminum, and tin, have been found to be effective catalysts. Among them, rhodium and gold catalysts in particular have allowed a range of novel highly efficient enantioselective domino reactions to be achieved.

Furthermore, during the last 10 years, an impressive number of novel powerful asymmetric domino and multicomponent processes have been developed on the basis of asymmetric nickel catalysis. For example, a number of enantioselective Michael-initiated domino reactions have been described, involving nitroalkenes as well as various α,β-unsaturated carbonyl compounds as acceptor-activated alkenes, which have provided a wide variety of chiral functionalised (poly)cyclic products in uniformly excellent enantioselectivities. Moreover, other types of enantioselective novel two-component domino reactions have been successfully catalysed by chiral nickel complexes, such as the first domino aldol-type/cyclisation reactions between aldehydes and isothiocyanatooxindoles, the first domino carbonyl-ylide formation/reverse-electron-demand 1,3-dipolar cycloaddition reactions between α,α′-dicarbonyldiazo compounds and cyclohexyl vinyl ether, domino denitrogenative annulation reactions of 2*H*-1,2,3,4-benzothiatriazine 1,1-dioxides with allenes, and domino cyclisation/cross-coupling reactions of alkylboron reagents bearing a pendant alkene with unactivated alkyl bromides, all providing excellent enantioselectivities.

In the context of enantioselective nickel-catalysed multicomponent reactions, many excellent results have also been achieved. For example, three-component reactions between 1,3-dienes, carbonyl compounds such as aldehydes or carbon dioxide, and various reducing agents such as organozinc reagents, silanes, or silaboranes have provided a variety of cyclic as well as acyclic chiral products in very high enantioselectivities. Furthermore, three-component reactions between allenes, aldehydes, and silanes have allowed chiral allylic alcohols to be easily produced with excellent enantioselectivities. These chiral products, along with chiral allylic amines, were also generated with high enantioselectivities on the basis of three-component reactions between alkynes, aldehydes or imines, and reducing agents such as boranes, silanes, or dialkylzincs.

Other types of multicomponent reactions have also been successfully developed, such as the first practical three-component imino-Reformatsky reaction, and a pseudo-three-component reaction between allenes and isocyanates, providing excellent levels of enantioselectivity. Finally, excellent results were described for several novel enantioselective tandem sequences. For example, very high enantioselectivities were reached in tandem Michael/intramolecular cyclisation sequences, as well as in a remarkable multicatalytic Michael sequence occurring between enones, alkynes, and DIBAL, which stereoselectively afforded a range of chiral β-alkenyl ketones bearing an all-carbon-substituted quaternary stereogenic centre in excellent enantioselectivities.

The privileged chiral ligands in nickel-catalysed domino reactions include phosphoramidites, bisoxazolines, Schiff bases, diamines, amino alcohols, bisoxazolidines, *N,N'*-dioxides, phosphines, FOXAP, and N-heterocyclic carbene ligands, while Salen ligands are typically employed in enantioselective cobalt-catalysed domino reactions, and BINOL-derived ligands in asymmetric titanium-catalysed domino reactions. Undoubtedly, the future direction in the field of nickel-catalysed asymmetric domino reactions is to continue expanding the scope of these elegant reactions through the combination of different types of reactions, the employment of novel chiral nickel catalysts, and to apply these powerful strategies to the syntheses of biologically interesting molecules, including natural products, and that of novel chiral ligands and functional materials.

The fifth chapter of the book described the advances in enantioselective nickel-catalysed hydrovinylation, hydrophosphination, hydrocyanation, and hydroalkynylation reactions. Outstanding results have been reported for the asymmetric hydrovinylation reaction of many vinylarenes and little room for further optimisation is left. The stereochemistry of the newly generated stereogenic centre in these reactions can be controlled by a wide range of chiral ligands using various metals. Mostly catalyst systems based on nickel and palladium are applied in these reactions, while the use of catalysts based on cobalt, platinum, iridium, ruthenium, and iron is less widespread. Chiral ligands of different types have been used. Among them are bidentate binaphthyl, phosphoramidite, and aminophosphine phosphinites, as well as monodentate *N*-hydroxy carbene ligands or phosphine ligands. In the most efficient results, nickel or palladium complexes with chiral phosphoramidite or binaphthyl-derived ligands have been used. It must be noted that in the last decade, phosphoramidites have been by far the privileged ligands in asymmetric nickel-catalysed hydrovinylations. In this field, several powerful protocols for the enantioselective nickel-catalysed hydrovinylation reaction of various alkenes have been recently described in which nearly quantitative yields of the desired products can be obtained using low catalytic amounts of nickel complexes along with high levels of chemo-, regio-, and enantioselectivities, often >95% ee. Nickel catalysts have been by far the most used catalysts in this type of reaction. All these novel procedures have used highly versatile phosphoramidite ligands derived from Feringa's ligand,[8] with NaBArF as catalyst activator, providing remarkable enantioselectivities in all cases. For example, excellent enantioselectivities were achieved in the asymmetric nickel-catalysed hydrovinylation of a range of vinylarenes to give the corresponding 3-arylbutenes. Moreover, when this reaction was followed by an oxidative degradation, the sequence offered a novel route to important anti-inflammatory chiral 2-arylpropionic acids, such as naproxen, ibuprofen, fenoprofen, and flurbiprofen. Comparable excellent enantioselectivities were also achieved in the case of other substrates, such as (α-alkylvinyl)arenes, as well as cyclic 1,3-dienes. It is important to note that the hydrovinylation of (α-alkylvinyl)arenes provided a new efficient access to the construction of chiral all-carbon quaternary centres. Furthermore, functionalised alkenes, such as silyl-protected allylic alcohols, could also be submitted to asymmetric hydrovinylation, with high enantioselectivities. Even higher enantioselectivities were recently reached for the nickel-catalysed hydrovinylation of strained alkenes such as norbornenes and heterobicyclo[2.2.1]heptenes and cyclobutenes.

In the area of nickel-catalysed asymmetric hydrophosphination of alkenes, the last decade has seen the first highly enantioselective reaction developed using the C_1-symmetric trisphosphine Pigiphos, which provided very good enantioselectivities. In addition, a highly efficient enantioselective nickel-catalysed hydrocyanation of arylalkenes was performed with very good enantioselectivities by employing a TADDOL-derived phosphine/phosphite ligand. Finally, the first nickel-catalysed hydroalkynylation of 1-arylbuta-1,3-dienes was achieved by using a phosphoramidite ligand. All the formed chiral products from hydrovinylation, hydrophosphination, hydrocyanation, and hydroalkynylation reactions of alkenes constitute useful building blocks for the total synthesis of natural products and biologically active compounds, since they can be readily transformed into a variety of other functionalised compounds.

The sixth chapter of the book was devoted to advances in enantioselective nickel-catalysed α-functionalisation, and to α-alkylation/arylation reactions of carbonyl compounds. A prochiral carbonyl compound can be activated toward electrophilic substitution *via* the formation of an enol or enolate intermediate, generating a tertiary or quaternary centre at the α-carbon. The use of a non-carbon electrophile leads to heterofunctionalised products, while that of carbon electrophiles affords α-arylated/alkylated carbonyl compounds, and the generation of a new stereogenic centre in these reactions makes them amenable to the development of asymmetric methodologies. In particular, α-heterofunctionalisation of a carbonyl compound is a highly direct and strategically simple method for the synthesis of a large number of interesting molecules and synthetic building blocks, such as amino acids, α-hydroxy acids, and α-fluorinated products. Asymmetric metal-catalysed electrophilic substitution at the α-carbon of a carbonyl compound by heteroatoms, particularly F, O, and N, can be achieved with excellent yields and enantioselectivities. Related chlorination, bromination, and sulfenylation reactions have also been studied to a lesser extent, but the selectivities of these reactions are not yet at a synthetically useful level (ee values of >90% are rare). The majority of catalysts for these reactions are Lewis acids, which are particularly useful for the heterofunctionalisation of certain β-functionalised carbonyl moieties, typically β-keto esters, malonates, and Boc-protected oxindoles, which are able to form a chelate with the reactive metal centre. Generally, nickel, copper, and zinc complexes of bisoxazoline and bisoxazolidine ligands, bisphosphine palladium complexes (BINAP), and titanium TADDOL complexes have shown a broad range of applicability across different substrates. Although other types of metal complexes have been found to display very good reactivities and selectivities, these tend to be restricted to a narrower range of substrates.

In the last decade, a number of highly enantioselective nickel-catalysed electrophilic halogenation reactions have been developed. Most of them are fluorination reactions of several types of carbonyl substrate, including cyclic as well as acyclic β-keto esters, *N*-acetylthiazolidinones, acid chlorides, and α-chloro-β-keto esters, which all provided near-perfect enantioselectivities. The privileged chiral ligands for nickel-catalysed halogenations used in the past decade are, by far, bisoxazolines (*e.g.*, DBFOX), but excellent results have also been reported with oxazolinylpyridine, benzoylquinidine, and diamine chiral ligands. Notably, the majority of these reactions generated quaternary stereogenic centres, but the asymmetric nickel-catalysed bromination of carbonyl compounds remained challenging.

In another context, excellent enantioselective nickel-catalysed α-aminations of *N*-Boc-oxindoles with azodicarboxylates have been achieved by using chiral Schiff base nickel catalysts. BINAP ligands have also encountered success in asymmetric nickel-catalysed electrophilic α-aminations and also in combination with other metals such as palladium. On the other hand, the use of other sources of electrophilic nitrogen, such as nitroso compounds and iodinanes, in reactions catalysed by nickel has so far not been described.

In the field of asymmetric α-hydroxylation reactions, early developments required multistep procedures, where the substrates were activated *via* enol or enolate intermediates. More recently, it has been demonstrated that the hydroxylation of β-keto esters could be achieved directly, by using oxaziridines. In this area, advances have been made with highly enantioselective nickel-catalysed α-hydroxylations of cyclic and acyclic β-keto esters, as well as malonates performed with the DBFOX ligand, which constitutes the privileged ligand for nickel in the past decade, providing excellent enantioselectivities.

In the field of α-arylations, many excellent results have been achieved, such as nickel-catalysed arylations and heteroarylations of a wide variety of indanones and tetralones with chloroarenes or aryl triflates, which provided excellent enantioselectivities using Difluorphos or BINAP ligands. The utility of these reactions is obvious since optically active α-aryl carbonyl moieties constitute important structural features of many natural occurring products, pharmaceutically attractive molecules, synthetically useful intermediates, and precursors to emissive polymers. While the privileged chiral ligands for asymmetric nickel-catalysed α-arylations in the past decade have been bisphosphines, such as BINAP and Difluorphos, other good results have been described by using PHOX-type ligands in combination with palladium. In another context, an interesting advance in the field of asymmetric nickel-catalysed α-alkylation reactions was recently reported with a nice α-alkylation of *N*-acylthiazolidinethiones catalysed by a BINAP-derived nickel catalyst, performed with remarkable enantioselectivities. It is important to note that many of these novel reactions (halogenations, aminations, hydroxylations, as well as arylations) generated challenging quaternary carbon stereogenic centres.

The seventh chapter of the book was dedicated to the advances in enantioselective nickel-catalysed additions of organometallic reagents to aldehydes. The enantioselective addition of organometallic reagents to aldehydes in the presence of a chiral catalyst is one of the most established carbon–carbon bond-forming asymmetric processes, providing enantioenriched secondary alcohols which are highly valuable intermediates for preparing chiral pharmaceuticals and agricultural products. Furthermore, these reactions often serve as test reactions for the investigation of novel catalysts. Among nucleophiles enantioselectively added to aldehydes, dialkylzincs, and especially diethylzinc, have been by far the most extensively studied, thus constituting a commonly used method for synthesising chiral secondary alcohols. In this area, an interesting example of very effective chirality switching was achieved just by using different stoichiometries of nickel complexes bearing α-amino amide ligands, providing very high enantioselectivities.

Even if a lot of early work was made in the area of metal-catalysed enantioselective additions of organometallic reagents to aldehydes, especially with dialkylzinc reagents as nucleophiles, the first successful examples of nickel-catalysed enantioselective additions of organoaluminum reagents to aldehydes have been reported only in the last decade. Indeed, high enantioselectivities were achieved in the addition of trialkyl(aryl)aluminum reagents to aldehydes by using phosphoramidite or sugar-based phosphite ligands. In addition, the first highly efficient example of asymmetric nickel-catalysed arylation of aldehydes with a boron reagent was recently reported. For example, perfect enantioselectivities were reached in the addition of potassium aryltriolborates to aromatic aldehydes in the presence of an *in situ* generated nickel catalyst from a DUPHOS ligand. The privileged chiral ligands for nickel-catalysed additions of aluminum reagents to aldehydes in the past decade have been phosphoramidites, while bisphosphines were used for nickel-catalysed additions of organoboranes. On the other hand, other good results in asymmetric metal-catalysed additions of organometallic reagents have been described with titanium complexes of BINOL, disulfonamides, and TADDOL derivatives, copper complexes of biphosphacycles, and chromium Salen complexes, among others.

The eighth chapter of the book covered advances in enantioselective nickel-catalysed aldol-type and Mannich-type reactions. The use of chiral metal catalysts in these reactions has become a major area of study. Despite early success in developing catalysts for the Mukaiyama aldol addition, only recently has the direct catalytic aldol and Mannich reactions received serious attention. Indeed, this situation is changing as catalysts capable of promoting powerful direct asymmetric aldol and Mannich reactions are developed, especially dinuclear Schiff base nickel complexes from Shibasaki's group and *N*,*N*′-dioxide nickel catalysts from Feng's group. These powerful catalytic aldol and Mannich reactions based on nickel present the advantages of offering very mild reaction conditions, and the attendant tolerance for a range of functional groups that this implies.

The direct catalytic asymmetric aldol reaction is a powerful and atom-economical method for synthesising chiral β-hydroxy carbonyl compounds. To date, many chiral metal and organocatalysts have been developed for reactions of various donors with aldehydes. On the other hand, catalytic asymmetric Mannich-type reactions of aldehydes, ketones, esters, and other donors for the synthesis of β-amino carbonyl compounds have been investigated intensively over the past decade. Traditionally, these reactions are catalysed by chiral transition-metal complexes. In recent years, several groups have developed efficient enantioselective Mannich-type reactions by using chiral catalysts of various metals, such as scandium, silver, tin, zirconium, copper, and nickel. The latter has achieved remarkable results in the past decade in the enantioselective nickel-catalysed aldol-type reactions, as well as in Mannich-type reactions, especially from the important results independently reported by the groups of Shibasaki and Feng. Among the most important advances are direct asymmetric aldol reactions of an α-isothiocyanato imide with aldehydes performed with an *N*,*N*′-dioxide–nickel catalyst, which provided the corresponding aldol products in high yields and with both excellent diastereo- and enantioselectivities. Another nice result was the asymmetric aldol reaction between glyoxal derivatives and enolsilanes catalysed by a chiral nickel complex generated *in situ* from another chiral *N*,*N*′-oxide ligand. Always in the context of aldol-type reactions, enantioselective nickel-catalysed aldol-type reactions of glyoxal derivatives with enecarbamates as well as enamides catalysed by a chiral *N*,*N*′-dioxide–nickel complex provided high reactivities and excellent enantioselectivities for a wide range of substrates. Finally, a novel remarkable route to chiral 2,5-disubstituted oxazole derivatives bearing a quaternary stereogenic centre was based on the first highly enantioselective aldol-type reaction of α-keto esters with 5-methyleneoxazolines. This powerful process gave comparable excellent yields and enantioselectivities for a broad substrate scope.

In the context of enantioselective nickel-catalysed Mannich-type reactions, several excellent advances have also been reported, such as the use of a Shibasaki's dinuclear Schiff base nickel catalyst applied in enantioselective Mannich-type reactions of α-substituted nitroacetates with *N*-Boc-imines, which afforded the corresponding α,α,α,α-tetrasubstituted *anti*-α,β-diamino acid surrogates in very high yields and enantioselectivities, along with good to high diastereoselectivities. In addition, the asymmetric direct Mannich-type reaction between α-keto anilides and *o*-Ns-protected imines was achieved in very high yields and enantioselectivities by using the same Schiff base nickel catalyst. Finally, direct catalytic asymmetric vinylogous Mannich-type reactions of α,β-unsaturated γ-butyrolactams with *N*-Boc-imines were induced by another dinuclear chiral Schiff base nickel complex, leading to the corresponding vinylogous Mannich adducts in good yields and high diastereoselectivities, along with remarkable general enantioselectivities in all cases of substrates studied.

To summarise, the privileged chiral ligands for asymmetric nickel-catalysed aldol reactions in the past decade have been *N*,*N*′-dioxides. On the other hand, bisoxazolines have been successfully employed as ligands in these reactions in combination with metals such as tin and copper. Titanium, zinc, zirconium, and lanthanoids gave the best results with BINOL, while BINAP is the privileged ligand for palladium, platinum, silver, and copper. In the field of asymmetric Mannich-type reactions, the privileged ligands for nickel catalysis are Schiff bases (in dinuclear complexes), whereas BINOL is better for titanium, zinc, and zirconium, and BINAP for palladium. Nickel catalysis is intrinsically elegant and economic; however, it appears that it is still limited to simpler substrates in most cases. Consequently, future studies will have to focus not only on improvement of the substrate scope but also on better understanding of asymmetric control in these reactions. Furthermore, especially for direct aldol reactions, there remains room for improvement in catalyst loading and catalyst reactivity.

The ninth chapter of the book described advances in enantioselective nickel-catalysed hydrogenation reactions. The catalytic enantioselective reduction of prochiral unsaturated organic molecules, such as ketones or alkenes, using molecular hydrogen is widely recognised as one of the most efficient methods for installing chirality into target compounds, providing an environmentally benign synthetic process to prepare pharmaceuticals, perfumes, and agrochemicals. Ruthenium has become the dominant choice among the central transition metals since the first generation of BINAP–Ru catalyst systems for asymmetric hydrogenation of ketones was reported in 1987.[9] It must be recognised that there is still no universal catalyst for the asymmetric hydrogenation of ketones that can be used for any types of substrate, and a trial-and-error method is still the major approach even now. Most of the time, asymmetric hydrogenations of ketones have been undertaken in the presence of six categories of chiral ligands, such as bisphosphine-based PP-ligands, bisphosphine/diamine-based P_2/N_2-ligands, tridentate or tetradentate phosphine/amine-based P_mN_n-ligands, diamine-based N,N-ligands, tetradentate amine-based N_4-ligands, and tetradentate thioether/amine-based S_2N_2-ligands.

The vast majority of catalysts used in hydrogenations are based on precious metals, including ruthenium, osmium, rhodium, iridium, and palladium. Replacement of these expensive and toxic elements with more abundant base metals such as iron, cobalt, nickel, or copper should be thoroughly investigated from the viewpoints of cost and safety. Among very good results involving nickel catalysts reported in the last 10 years, the first use of homogeneous chiral nickel–phosphine complexes in asymmetric hydrogenation of α-amino-β-keto ester hydrochlorides was achieved in excellent diastereo- and enantioselectivities. In the area of asymmetric hydrogenation of alkenes, there is also a renewed interest in developing cheap, abundant, and less toxic metals. In this context, a remarkable result was recently reported with the first highly enantioselective (transfer) hydrogenation of β-acetamidoacrylates using nickel catalysts, providing β-acylamido esters in excellent yields and enantioselectivities.

The privileged ligands for nickel-catalysed hydrogenations in the past decade have been bisphosphines, while analogous reactions performed with cobalt gave the best results with bisiminopyridine ligands, those with rhodium provided the best enantioselectivities by using BINAP, biphosphacycles, or Josiphos ligands, those catalysed by ruthenium have used BINAP, biphosphacycles, or PHOX derivatives as the privileged ligands, and those induced by iridium provided good results with PHOX ligands. Because the production of chiral secondary alcohols is so important in asymmetric synthesis, the efficiency of the asymmetric hydrogenation of ketones should be further pursued in particular.

The tenth chapter covered advances in enantioselective nickel-catalysed miscellaneous reactions. Efforts to develop new asymmetric transformations focused mainly on the use of a few metals, such as titanium, copper, nickel, ruthenium, rhodium, palladium, iridium, and, more recently, gold. However, by the very fact of the lower costs of nickel catalysts in comparison with other transition metals, a variety of other enantioselective nickel-mediated transformations have received continuous ever-growing attention during the last decade. This interest might also be related to the fact that nickel complexes exhibit a remarkably diverse chemical reactivity, which is well demonstrated in this chapter. Among the most efficient enantioselective nickel-catalysed miscellaneous reactions recently developed are hydrocarbamoylations of homoallylic formamides to give γ-lactams in very high enantioselectivities using a chiral diaminophosphine oxide ligand; aminations of 3-bromooxindoles with indolines to afford indole-derived pyrrolidines in excellent levels of enantioselectivity upon catalysis with a chiral bisoxazoline ligand; the first asymmetric ring-opening of cyclopropanes with amines performed with a chiral trisoxazoline ligand, which gave excellent enantioselectivities; the first asymmetric ring-opening of oxabicyclic alkenes with arylboronic acids, which provided almost enantiopure *cis*-2-aryl-1,2-dihydronaphthalen-1-ols in the presence of (*S*,*S*)-Me-DUPHOS; the first asymmetric Friedel–Crafts reaction of indoles with β,β-disubstituted nitroalkenes, which yielded indole-bearing chiral compounds with trifluoromethylated all-carbon quaternary stereocentres in excellent enantioselectivities using a chiral bisoxazoline ligand; Friedel–Crafts reactions of indoles with β,γ-unsaturated α-keto esters, leading to almost enantiopure 2-substituted chiral indoles with a chiral *N*,*N*′-oxide ligand; carbonyl–ene reactions of glyoxal derivatives with alkenes, providing enantiopure γ,δ-unsaturated α-hydroxyl carbonyl compounds with a chiral *N*,*N*′-oxide ligand; intramolecular alkene insertions of 3-(2-styryl)cyclobutanones into benzobicyclo[2.2.2]octenones in high enantioselectivities with a BINOL-derived phosphoramidite ligand; and propargyl vinyl ether and allyl vinyl ether Claisen rearrangements achieved in almost complete enantioselectivity with a chiral *N*,*N*′-oxide ligand.

While the privileged ligands for nickel-catalysed asymmetric Friedel-Crafts reactions have been bisoxazolines and *N*,*N*′-dioxides for the past decade, those for copper-catalysed analogous reactions were bisoxazolines. The latter were also the privileged ligands for copper-catalysed enantioselective Claisen rearrangements, although nickel gave the best enantioselectivities in this type of reaction when *N*,*N*′-dioxide ligands were used. These ligands were also privileged for nickel-catalysed carbonyl–ene reactions, while the same type of reactions catalysed by titanium provided their better results when combined with the BINOL ligand.

The ever-growing need for environmentally friendly catalytic processes has prompted organic chemists to focus on more abundant first-row transition metals such as nickel to develop new catalytic systems to perform various reactions, such as C–C bond formation, C–heteroatom bond formations, or C–H functionalisation. As demonstrated in this book, during the last decade a steadily growing number of novel asymmetric nickel-catalysed reactions have been developed on the basis of the outstanding ability of low-cost nickel catalysts to adopt new reaction pathways to achieve cyclic as well as acyclic chiral products in very high enantioselectivities under relatively mild conditions. A bright future is undeniable for more sustainable novel and enantioselective nickel-catalysed transformations. Nickel is being more and more found suitable and often a more effective alternative to precious metal catalysts for a range of enantioselective reactions. Although nickel has been somewhat overlooked for a long time in favour of more popularly studied and therefore well-understood metals, nickel is now back in the limelight. While progress has been considerable over the past decade, however, there clearly is a dearth of mechanistic understanding. Through the exploration of mechanistic studies and detailed understanding in addition to the continued exploration of nickel catalysis, it is expected that the full potential of this metal will be unlocked in the future.

References

1. S. Kanemasa, Y. Oderaotoshi, J. Tanaka and E. Wada, *J. Am. Chem. Soc.*, 1998, **120**, 12355–12356.
2. (a) S. Iwasa, S. Tsushima, T. Shimada and H. Nishiyama, *Tetrahedron Lett.*, 2001, **42**, 6715–6717; (b) S. Iwasa, S. Tsushima, T. Shimada and H. Nishiyama, *Tetrahedron*, 2002, **58**, 227–232.
3. S. Kanemasa, Y. Oderaotoshi, H. Yamamoto, J. Tanaka, E. Wada and D. P. Curran, *J. Org. Chem.*, 1997, **62**, 6454–6455.
4. (a) J. H. Nelson, P. N. Howells, G. L. Landen, G. C. DeLullo and R. A. Henry, *Fundam. Res. Homogeneous Catal.*, 1979, **3**, 921–939; (b) J. H. Nelson, P. N. Howells, G. C. DeLullo, G. L. Landen and R. A. Henry, *J. Org. Chem.*, 1980, **45**, 1246–1249.
5. (a) K. Soai, S. Yokoyama, T. Hayasaka and K. Ebihara, *J. Org. Chem.*, 1988, **53**, 4148–4149; (b) K. Soai, T. Hayasaka, S. Ogajin and S. Yokoyama, *Chem. Lett.*, 1988, 1571–1572.
6. Examples of prices for July 2010.
7. (a) L. F. Tietze and U. Beifuss, *Angew. Chem., Int. Ed.*, 1993, **32**, 131–163; (b) L. F. Tietze, *Chem. Rev.*, 1996, **96**, 115–136; (c) L. F. Tietze, G. Brasche and K. Gericke, *Domino Reactions in Organic Synthesis*, Wiley-VCH, Weinheim, 2006; (d) L. F. Tietze, *Domino Reactions-Concepts for Efficient Organic Synthesis*, Wiley-VCH, Weinheim, 2014.
8. B. L. Feringa, *Acc. Chem. Res.*, 2000, **33**, 346–353.
9. R. Noyori, T. Ohkuma, M. Kitamura, H. Takaya, N. Sayo, H. Kumobayashi and S. Akutagawa, *J. Am. Chem. Soc.*, 1987, **109**, 5856–5858.

Subject Index

References to figures are given in *italic* type.